LIBRARY
USE
ONLY

KV-575-382

GCSE
geography
in focus

GCSE geography

in focus

John Widdowson

John Smith

Roger Knill

JOHN MURRAY

The authors would like to thank everyone who contributed to GCSE *Geography in Focus*, and especially those who agreed to be interviewed or photographed and who appear in the book. Thanks also to Bob Jones for help on the ICT tasks.

Words in SMALL CAPITALS are defined in the Glossary on pages 376–82.

First published in 2001 by
John Murray (Publishers) Ltd
50 Albemarle Street
London W1S 4BD

Layouts by Amanda Hawkes
Artwork by Oxford Designers and Illustrators
Typeset in 10.5/12.5 pt Goudy by Wearset, Boldon, Tyne & Wear
Printed and bound in Spain by Bookprint, S.L., Barcelona

A catalogue entry for this book is available from the British Library

Student's Book ISBN 0 7195 7558 3

Acknowledgements

The Publishers would like to thank the following for permission to reproduce copyright material:

Photo credits:

Cover: *cr*, *br* Tony Stone, *tl*, *bc* PA Photos, *cl*, *c* Telegraph Colour Library; **p.1** *t* Corbis UK Ltd; *b* Image Bank; **p.2** and **p.3** NHPA; **p.5** *b* Bruce Coleman Collection; **p.6** *tr*, *clb*, *crb* Art Directors & Trip, *cla*, *cra*, *b* Bruce Coleman Collection; **p.8** *t* Still Pictures, *b* John Widdowson; **p.9** *t* Still Pictures, *b* John Widdowson; **p.10** *b* Geo Science Features Picture Library; **p.13** Science Photo Library; **p.14** *t* London Aerial Photo Library, *c* Topham Picturepoint; **p.16** *bl* John Widdowson, *br* John Townson/Creation; **p.18** *t* NHPA, *ca* Collections, *c* and *bl* John Widdowson; **p.20** Panos Pictures; **p.22** *t* Will Critchley/Oxfam, *b* Panos Pictures; **p.23** *t* Art Directors & Trip, *b* NHPA; **p.24** *t* NHPA, *b* Will Critchley/ Oxfam; **p.26** and **p.27** *t* Will Critchley/Oxfam; **p.29** *t* and *b* Bruce Coleman Collection; **p.30** *l* NHPA, *r* Panos Pictures; **p.31** *l* Still Pictures, *r* Art Directors & Trip; **p.32** NHPA; **p.33** Art Directors & Trip; **p.35** *c* Corbis UK Ltd, *b* Bruce Coleman Collection; **p.36** *tl* and *bl* John Widdowson, *tr* and *br* Corbis UK Ltd; **p.38** *tl* London Aerial Photo Library, *tr* Panos Pictures, *bl* and *br* Bruce Coleman Collection; **p.39** *t* PA News Photos, *b* and **p.40** Popperfoto; **p.41** Science Photo Library; **p.48** courtesy "Evening Press", York, published March 15, 1999; **p.49** *t* and *r*, courtesy "Evening Press", York, published March 15, 1999, *bl* Collections; **p.51** Peter Smith Photography, Malton, North Yorks; **p.61** Collections; **p.62** *t* NOAA/National Climatic Data Center, *b* Popperfoto; **p.64** *t*, *b* and **p.70** *t*, *b* Popperfoto; **p.74** Topham Picturepoint; **p.75** Rex Features; **p.79** *l* Rex Features, *r* Collections; **p.80** Science Photo Library; **p.87** Jon Arnold www.jonarnold.com; **p.98** Gettyone Stone; **p.99** *t* Leslie Garland Picture Library, *b* Collections; **p.106** *t* Popperfoto, *b* Collections; **p.108** Science Photo Library; **p.109** *tl* and *cl* Leslie Garland Picture Library, *bl* Topham Picturepoint, *br* and *tr* Collections; **p.113** *l* Topham Picturepoint, *r* Corbis UK Ltd; **p.124** *t* John Townson/ Creation, *b* Popperfoto; **p.126** John Townson/Creation; **p.128** *t* Bruce Coleman Collection, *b* Collections; **p.134–135** Richard Austin, Lyme Regis, Dorset; **p.138** *l* and *r* Roger Knill; **p.142** *b* and **p.143** Collections; **p.145** Richard Austin, Lyme Regis, Dorset; **p.148** *t* Science Photo Library, *bl* Panos Pictures, *br* Collections; **p.149** *t* Panos Pictures, *cr* Popperfoto, *bc* Leslie Garland Picture Library; **p.151** *t* Art Directors & Trip, *b* Panos Pictures; **p.157** and **p.158** Panos Pictures; **p.163** Topham Picturepoint; **p.164** *l* Panos Pictures, *r* Eye Ubiquitous; **p.170** *t* Corbis UK Ltd, *b* Collections; **p.172** *t* Collections, *b* Science Photo Library; **p.175** Corbis UK Ltd; **p.176** *t* Panos Pictures, *b* Image Bank; **p.177** Jon Arnold www.jonarnold.com; **p.178** *t* Corbis UK Ltd, *b* Gettyone Stone; **p.179** *t* and *b* Popperfoto; **p.180** Corbis UK Ltd; **p.182** Frank Spooner Pictures; **p.183** Associated Press; **p.184** *t* Still Pictures, *c* and *b* Panos Pictures; **p.185** *t* Popperfoto, *b* Panos Pictures; **p.186** Popperfoto; **p.189** *t* Collections, *b* John Townson/ Creation; **p.190** Gettyone Stone; **p.191** Bruce Coleman Collection; **p.192** Art Directors & Trip; **p.194** *tl* Art Directors & Trip, **p.194** *tr* and *b* Peter Kellett, University of Newcastle upon Tyne; **p.195** *l* and *r* Peter Kellett, University of Newcastle upon Tyne; **p.198** Cumbria Picture Library; **p.201** Panos Pictures; **p.203** *t* Panos Pictures, *bl* and *br* Peter Kellett, University of Newcastle upon Tyne; **p.204** Panos Pictures; **p.205** *l* Panos Pictures, *r* Peter Kellett, University of Newcastle upon Tyne; **p.207** *t* Panos Pictures, *bl* Peter Kellett, University of Newcastle upon Tyne, *br* John Widdowson; **p.217** *t* Corbis UK Ltd, *b* Art Directors & Trip; **p.219** *t* Collections, *b* John Townson/Creation; **p.220** and **p.221** John Townson/ Creation; **p.224** Image State; **p.225** *all* John Widdowson; **p.227** *t* Collections, *b* Skyscan; **p.228** *t* and *b* John Widdowson; **p.229** *t* and *b* John Widdowson; **p.230** *t* Collections, *b* John Widdowson; **p.232** Panos Pictures; **p.234** *t* and *b* Art Directors & Trip; **p.236** *t* Corbis Stock Market, *b* John Townson/Creation; **p.238** *t* Ecoscene; **p.239** Corbis UK Ltd; **p.240** Gettyone Stone; **p.241** *t* Collections, *b* Panos Pictures; **p.242** and **p.243** Collections; **p.244** Topham Picturepoint; **p.245** Network; **p.246** *t* Art Directors & Trip, *b* Collections; **p.248** *l* Birmingham City Council, www.birmingham.gov.uk; **p.251** Collections; **p.254** *all* RSPB Images; **p.264** *t*, *ca* and *cb* AGCO Ltd, Coventry (Massey Ferguson), *b* Panos Pictures; **p.265** *bl* AGCO Ltd, Coventry (Massey Ferguson); **p.266** *t* and *b* Panos Pictures; **p.268** *t* Panos Pictures, *b* PA News Photos; **p.271** *l* Hulton Getty, *r* Martyn Pitt www.mrpphotography.co.uk; **p.272** Panos Pictures; **p.273** Art Directors & Trip; **p.277** *all* Panos Pictures; **p.274** Earth Centre, Doncaster; **p.282** Collections; **p.284–285** and **p.287** Panos Pictures; **p.291** and **p.292** Collections; **p.294** *t* Ronald Grant Archive, *b* PA News Photos; **p.295** Edifice; **p.298** Corbis UK Ltd; **p.301** PA News Photos; **p.303** Corbis UK Ltd; **p.306** *tl* and *bc* Collections, *tc* Image State, *tr* Panos Pictures, *bl* Art Directors & Trip, *br* Corbis UK Ltd; **p.311** Bruce Coleman Collection; **p.319** *t* and *b* Collections; **p.315** *all* Collections; **p.317** *t* and *ca* Collections, *cb* Leeds City Council, *b* Corbis UK Ltd; **p.323** *t* Panos Pictures, *b* Impact Photos; **p.324** Corbis UK Ltd; **p.325** Art Directors & Trip; **p.327** *l* Leslie Garland Picture Library, *r* John Widdowson; **p.329** Swindon Borough Council; **p.331** *t* and *c* Image State, *b* Corbis UK Ltd; **p.332** *t* Panos Pictures, *cl* Corbis UK Ltd, *b* Corbis Stock Market; **p.332–323** *c* Panos Pictures; **p.333** *t* Science Photo Library, *b* Panos Pictures; **p.336** *t* Jon Arnold www.jonarnold.com; *b* Panos Pictures; **p.340** John Widdowson; **p.343** Corbis UK Ltd; **p.345** Panos Pictures; **p.346** Gettyone Stone; **p.348** Popperfoto; **p.349** *all* John Widdowson; **p.350** *t*, *b* and **p.351** John Widdowson; **p.352** Network Photographers; **p.353** John Townson/Creation; **p.354** and **p.355** John Widdowson; **p.356** *bl* John Widdowson, *c* Art Directors & Trip, *r* Corbis UK Ltd; **p.357** Art Directors & Trip; **p.358** Associated Press; **p.360** John Widdowson; **p.361** *t* Popperfoto, *b* John Widdowson; **p.362** *t*, *b* and **p.363** John Widdowson; **p.368** Image State/ AGE; **p.370** John Widdowson; **p.371** and **p.373** Corbis UK Ltd; **p.374** *all* John Widdowson; **p.375** *all* John Widdowson.

(*t* = top, *b* = bottom, *r* = right, *l* = left, *a* = above, *c* = *centre*)

Written sources:

p.13 © *Guardian*, 12 June 1999; **p.28** 'The tourist' adapted from © *Guardian*; 'The scientist' adapted from *The Geographical Review*; **p.41** © *Guardian*; **p.65** Source 3 adapted from a map courtesy of EQE International; **p.71** © *Guardian*, 30 January 1999; **p.93** © *Guardian*, 14 November 1999; **p.113** © *Guardian*, 26 April 2000; **p.119** Extract from The Greater Common Good, an essay in *The Cost of Living* by Arundhati Roy, April 1999, HarperCollins Publishers Ltd; **p.223** Newcastle City Council; **p.234** National Youth Homeless Charity, Centrepoint; **p.244** *Birmingham Post*, 1964; **p.250** Birmingham City Council; **p.255** *Catch-22* by Joseph Heller (© Joseph Heller 1955) reproduced by permission of A. M. Heath & Co. Ltd; **p.257** With permission of MAFF; **p.324** From *Faith in the Poor* by Bob Holman, Lion Publishers, 1999; **p.325** From *Gregory Cool* by Caroline Binch, published by Frances Lincoln Ltd, © Frances Lincoln Limited 1994. Text and illustrations © Caroline Binch 1994. Reproduced by permission of Frances Lincoln Ltd, 4 Torriano Mews, Torriano Avenue, London NW5 2RZ; **p.326** © *Observer*, 7 July 1999; **p.330** Source 3 adapted from a map in *A Geography of the European Union* by Garratt Nagle and Kris Spencer, by permission of Oxford University Press, 1996; **p.341** GNP/HDI table from United Nations Development Programme, Human Development Report 1999, Oxford University Press, New York; **pp.342–43** Dan Smith, *The State of the World Atlas*, 6th edition (Penguin 1999) © Myriad Editions Ltd; **p.344** Table from United Nations Development Programme, Human Development Report 1999, Oxford University Press, New York; **pp.349, 351, 367** Cartography by Philip's Copyright +, 1999, George Philip Ltd; **p.369** Extract taken from the *Rough Guide to Jamaica* published November 2000 by Rough Guides Ltd.

Maps on the following pages (**pp.50, 126, 130, 138, 142, 226, 250, 296, 321**) are reproduced from Ordnance Survey mapping with the permission of the Controller of Her Majesty's Stationery Office, © Crown copyright, License No. 100017789.

Crown copyright material appears on the following pages (**pp.17, 232, 238, 270, 274, 279, 308, 328, 329**) and is reproduced with the permission of the Controller of Her Majesty's Stationery Office.

Every effort has been made to contact copyright holders, and the publishers apologise for any omissions which they will be pleased to rectify at the earliest opportunity.

Contents

1 Ecosystems and Fragile Environments

Why do we need to protect the environment?

2 Natural Hazards

How do natural disasters affect our lives?

3 Physical Environment

Should we try to stop environmental change, or simply go with the flow?

4 Population

Why is population changing?

5 Settlement

How can we manage our cities?

6 Economy and the enviroment

How do our economic activities interact with the environment?

7 Quality of life and development

How can we get rid of poverty?

Use your case study

★ THINK!

This book is full of case studies, but your exam paper will never ask you to 'Write down everything you know about X or Y case study'. It will ask you general questions based on themes and issues. Your task will then be to illustrate your answers from the case studies you have worked on. So an important skill is to decide which category each case study fits into. The catch is that many case studies could fit into more than one category. Take Carlos' story (on pages 200—201), for instance: Carlos moved from his rural village to the sprawling shanty towns on the edge of Mexico City. He earns a living in the informal sector as a trucker's mate and unskilled bricklayer. Already this one case study could fit into **urbanisation**, **migration**, **development**, **work**, and there will be more ...

People and environment

Humans as planetary rulers

— doing what they want and keeping control

Humans as pawns

— buffeted by the elements and at their mercy

Humans as polluters

— a disease on the face of the Earth

Humans as partners

— intimately linked to every other part of the global system

Which cartoon fits best with your view of the relationship between people and their environment? Write a paragraph about each cartoon saying whether you agree with it, and explaining why. You will return to this at the end of the book and say whether you have changed your views.

Ecosystems and Fragile Environments

Why do we need to protect the environment?

1.1 Why do we need to understand systems?

DO YOU KNOW?

- An ECOSYSTEM is a community of plants and animals together with the environment in which they live. Ecosystems can be at any scale, from a village pond to an entire forest.
- Ecosystems are being destroyed by people. Globally we have lost half of the world's rainforest since 1950. In the UK we have lost 95 per cent of our meadows over the same time.

AIMS

- **To understand SYSTEMS and how they work.**
- **To investigate systems at a range of scales.**
- **To recognise that it is vital to understand natural systems if we are to use them sustainably.**

Spaceship Earth

Over the past 50 years the Earth has come to seem like a much smaller place. It was only in the 1950s that the first satellite went into space and sent back pictures of the Earth. It put our planet into perspective. We could confirm that the Earth itself is no more than a large satellite orbiting the Sun. But, unlike an artificial satellite that orbits the Earth, our planet now has 6 billion people living on it!

Since those first pictures arrived from space we have learned a lot about the Earth. We now know that our actions have an impact on people elsewhere, and can even affect the global environment. When we drop a piece of litter someone has to pick it up, when we use up paper more trees are chopped down, and when we drive our cars or switch on the electricity we are adding to GLOBAL WARMING. The term 'Spaceship Earth' has been coined to describe our planet. It reminds us that, like people on a spacecraft, we need to think about the consequences of our actions.

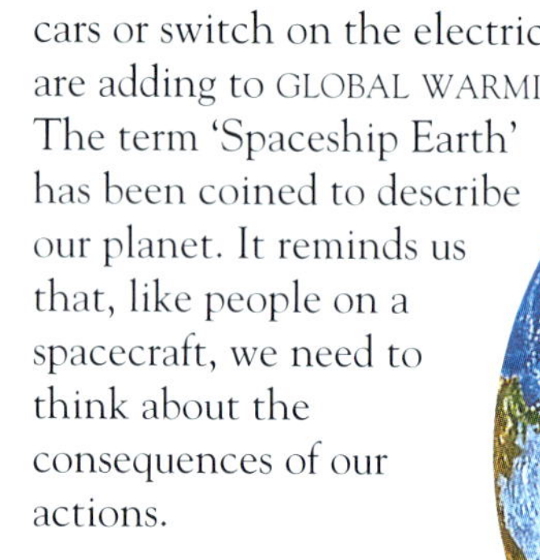

1 The Earth seen from a satellite in space

2 Nightmare in space! But could this be Spaceship Earth too?

What is an ecosystem?

3 An ecosystem in the UK: a hill farm

We can think of the Earth as a giant natural system. All systems require INPUTS to make them work, and they all produce OUTPUTS. The Earth gets energy from the Sun and then radiates heat back into space. PROCESSES, like the HYDROLOGICAL CYCLE, transfer this energy around the system and link all its components. As a result, a change in one place affects other places on the Earth.

On a smaller scale, Figure 3 shows a hill farm in the UK. It is an example of an ECOSYSTEM. All the components of this system – both living and non-living – are linked. The arrows between each part of the photo represent the processes that connect the atmosphere, plants, animals, soil and rocks within the ecosystem.

★ THINK!

Systems

This chapter is all about systems. Systems are a useful way to understand many geographical ideas. For example, in this book you will study farming systems, river systems and industry systems. Are there any other examples you can think of? What are the inputs, outputs and processes in these systems?

Tasks

1 Look at Cartoon 2.

a) Describe three problems that people on the spacecraft are creating.

b) How could these be related to problems on Earth?

c) What do you understand by the term 'Spaceship Earth'?

SAVE AS . . .

2 Look at Figure 3.

a) Write six sentences to describe the ways in which each pair of components in the ecosystem is linked – animals and plants, plants and atmosphere, and so on. For example, sheep manure the ground helping to keep the soil fertile.

b) What role do people have in this system? Would you describe it as a natural system? Give reasons.

Focus task

ICT

Why protect the environment?

3 People, too, are part of the natural system. They play an important role in changing ecosystems and, sometimes, in destroying them. In this unit you will look at the way people depend on their natural environment, and at the impact they can have on it. As you work through the unit, collect ideas and information to support the view that we need to protect our environment. Try visiting the websites of organisations such as Greenpeace and Friends of the Earth.

You will use this information to produce your own ICT presentation or magazine article to sum up your work on this topic.

What systems can you find in a city?

Half of the world's population now live in cities. In the UK, 90 per cent of us live in urban areas. Cities have an impact on the global environment far beyond the actual urban area. You will study cities in detail in Unit 5.

We can think of a city itself as a system with inputs, outputs and processes. Look at the urban area shown in Figure 4. It takes its resources from, and discharges its wastes over, a large area. As cities grow, and living standards continue to rise, the amount of land required to meet urban needs may be greater than the land that is available. In this case the cities become UNSUSTAINABLE. The challenge to urban planners of the future will be to design cities that are more SUSTAINABLE. This could involve changes in the way that the urban system works, such as fewer car journeys, more recycling of resources, and local production of energy and food.

★ **FACTFILE**

The surface area that is needed to feed a city, provide it with other resources and re-absorb its waste is known as the ECOLOGICAL FOOTPRINT of a city. The ecological footprint of London is estimated to be 125 times its own surface area – not much smaller than the whole of Britain.

5 London's ecological footprint compared with the size of Britain

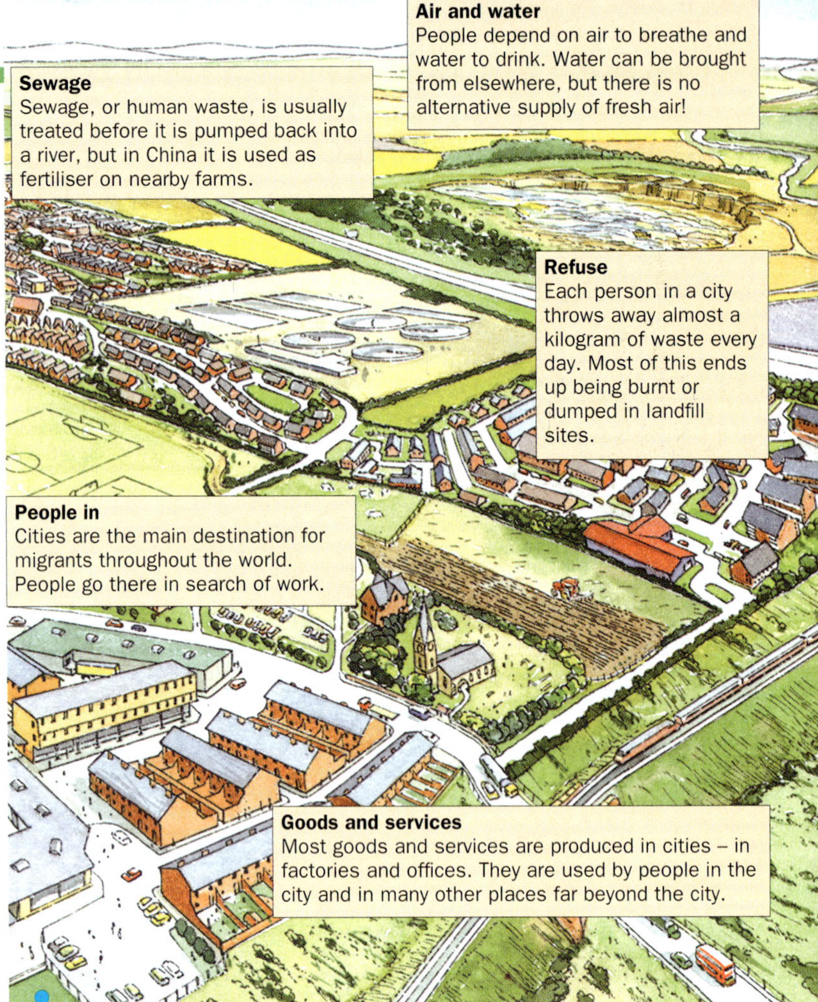

Air and water
People depend on air to breathe and water to drink. Water can be brought from elsewhere, but there is no alternative supply of fresh air!
Sewage
Sewage, or human waste, is usually treated before it is pumped back into a river, but in China it is used as fertiliser on nearby farms.
Refuse
Each person in a city throws away almost a kilogram of waste every day. Most of this ends up being burnt or dumped in landfill sites.
People in
Cities are the main destination for migrants throughout the world. People go there in search of work.
Goods and services
Most goods and services are produced in cities – in factories and offices. They are used by people in the city and in many other places far beyond the city.

4 Resources and wastes of an urban area

Tasks

1 Look at Figure 4. Read all of the text in the boxes.
   a) Sort the boxes into two groups – inputs and outputs.
   b) Think of up to ten processes (i.e. things that people do in cities, such as working in an office, shopping, travelling, and so on).
   c) Draw your own large copy of a diagram like this to show the inputs, outputs and processes in an urban system.

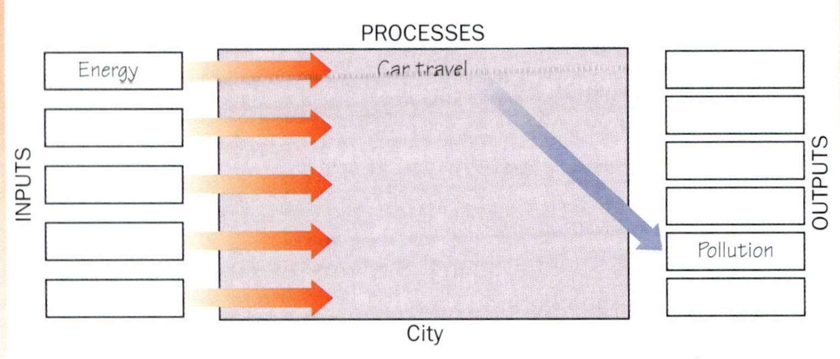

   d) Suggest how the urban system could be made more sustainable. For example, think about how resources could be recycled to reduce waste. You could draw another flow diagram to show your ideas.

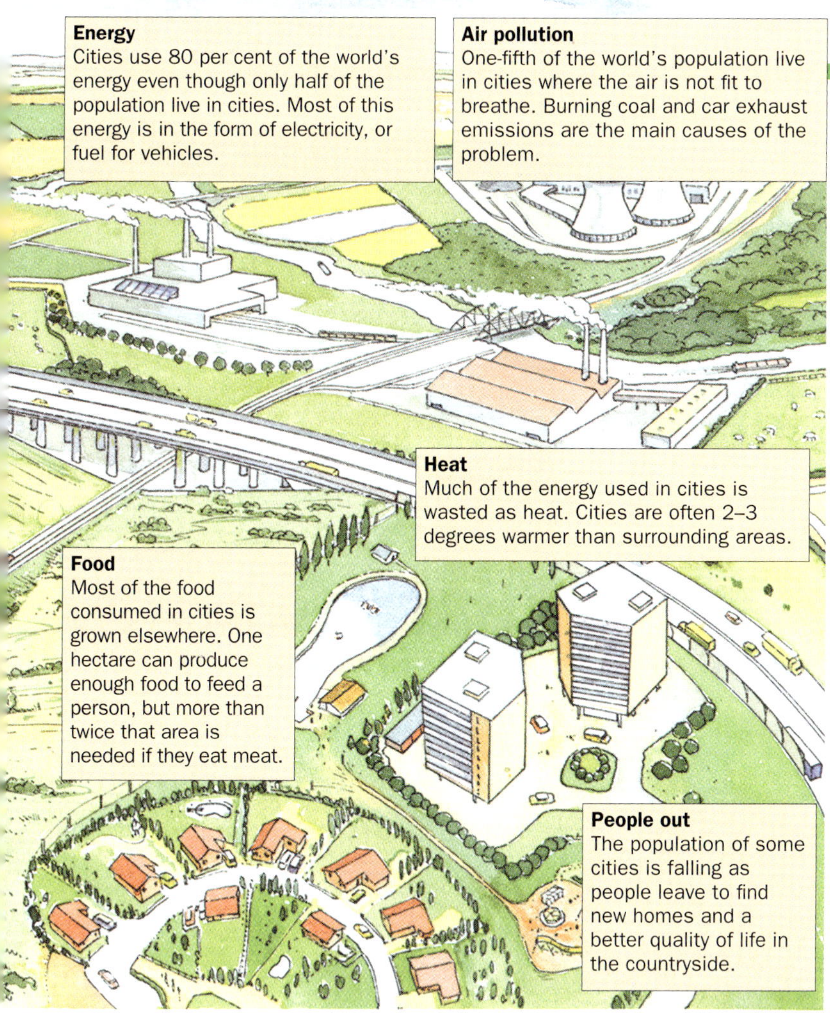

Energy
Cities use 80 per cent of the world's energy even though only half of the population live in cities. Most of this energy is in the form of electricity, or fuel for vehicles.
Air pollution
One-fifth of the world's population live in cities where the air is not fit to breathe. Burning coal and car exhaust emissions are the main causes of the problem.
Heat
Much of the energy used in cities is wasted as heat. Cities are often 2–3 degrees warmer than surrounding areas.
Food
Most of the food consumed in cities is grown elsewhere. One hectare can produce enough food to feed a person, but more than twice that area is needed if they eat meat.
People out
The population of some cities is falling as people leave to find new homes and a better quality of life in the countryside.

The last place that you might expect to find ecosystems would be in a city. But if you live in an urban area you might be surprised if you look carefully! The most obvious place to look is in your own back garden or the nearest park or waste ground. Look further in your own city, or at the urban area in Figure 4, and you will find other examples of ecosystems.

Cities are home to a variety of plants and animals that have managed to adapt to the concrete jungle. In recent years, the number of some species living in cities appears to have grown. Plants colonise motorway verges, while mammals, like the fox, learn to scavenge on human refuse rather than hunt in the wild. Even rare birds adapt to cities; for example, the peregrine falcon can live at the top of tall city buildings which are similar to its normal cliff habitat.

People have done much to encourage the appearance of wildlife in cities. Pollution controls, introduced in the UK in the 1960s, have helped to clean the atmosphere and the rivers in our cities. Salmon, not seen in many rivers for generations, have started to make a comeback. The demolition of older buildings and the creation of more green space in cities have also helped.

2 Look at Figure 4 again.

a) How many types of ecosystem can you identify?

b) Choose one ecosystem that you could investigate in your nearest urban area. It could even be within your own school grounds. Work with a partner. Look for an example of each of the following in your ecosystem. Something . . .

- old
- new
- alive
- that has never lived
- moving
- fixed
- edible
- that eats other things
- decaying
- growing
- shrinking
- common
- rare
- invisible
- hidden
- that is always there
- that is sometimes there

Record your findings.

c) Back in your classroom, work out links between all the components in your ecosystem. Draw a diagram to show how it works.

d) Consider what impact people have on this ecosystem. Does it need to be protected from damage by human activity? How would you do this?

6 An urban fox – an increasingly common sight in British cities

Ecosystems around the world

The UK was once covered by forest. It is hard for us to imagine this, when most of the landscape is dominated by farming and urban development. Today, just 10 per cent of the land area in the UK is forest. It is easy to forget that natural ecosystems still cover much of the Earth. Forests alone cover 30 per cent of the Earth's land surface.

Natural ecosystems that cover a large area of the Earth are known as BIOMES. The biome in any part of the world is determined mainly by the CLIMATE. TEMPERATE DECIDUOUS FOREST is the natural ecosystem in the UK and is the biome covering much of northern Europe. It grows in a TEMPERATE climate – one that is neither extremely hot nor very cold. Deciduous trees grow during the summer and lose their leaves in winter when it is too cold for trees to grow.

In other parts of the world there are different types of forest. The distribution of four of the world's most common forest biomes is shown on Map 11. Each one is adapted to the climate in that part of the world.

7 Coniferous (boreal) forest

8 Temperate deciduous forest in summer and in winter

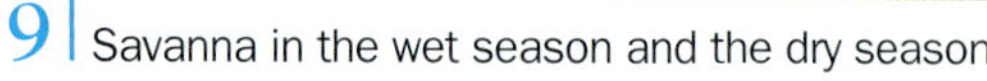

9 Savanna in the wet season and the dry season

10 Tropical rainforest

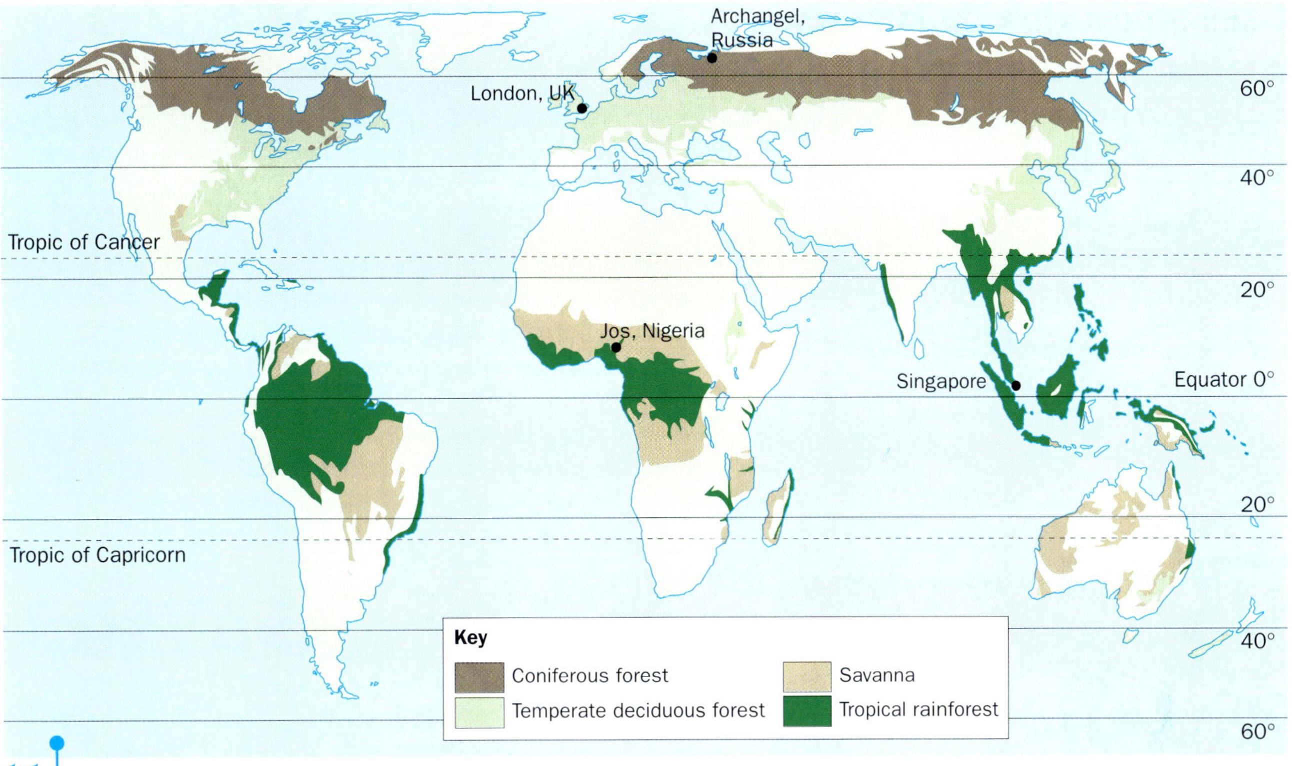

11 Worldwide distribution of four biomes

Tasks

1 Look at Photos 7–10 and Map 11. Complete a table with three columns to compare the four biomes. You could do this in a group of four, and divide the tasks.

a) In the first column, write a sentence to describe the ecosystem that you can see in each photo. Choose some of these words to use in your description: dense, sparse, broadleaved, needle-leaved, deciduous, evergreen.

b) In the second column, describe the distribution of each biome on the world map. Mention the range of latitude in each continent where it is found, for example: *coniferous forest is found between 55° and 65° North in Asia, Europe and North America.*

c) Find a world climate map in your atlas. Match each biome with the correct climate graph in Figure 12. In the third column, describe the climate for each biome. Mention average temperature, total rainfall, monthly range of temperature and rainfall and seasonal variations.

2 In which of the four ecosystems shown in Photos 7–10 would you be most likely to find each of the things in the box below? Work with a partner to discuss the ideas and give reasons for your answers.

- covering the trees
- Very few animals
- Lions hunting an antelope
- Creepers growing up trees
- Ground covered in dead leaves
- Fungus growing on a rotting tree trunk
- Bluebells growing in spring
- Plants that can survive drought

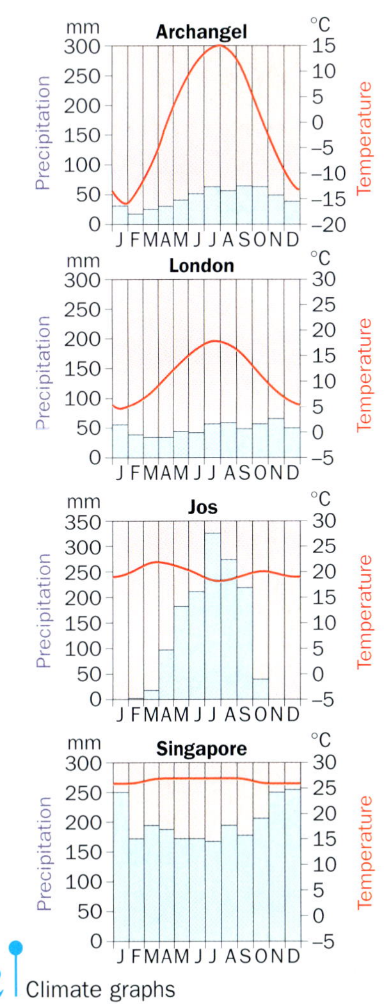

12 Climate graphs

Case study: How do farmers use natural systems?

No one understands better than farmers how natural systems work. This section focuses on the experience of two farmers working in very different environments – one in a TROPICAL area in Malaysia, the other in a temperate area in the UK. Both farmers face problems and, at first, it appears that the problems are quite different. But closer examination reveals that for both farmers the key to solving their problems is in understanding how natural systems work. There may even be things they could learn from each other.

Task

1 Work with a partner. Each of you choose one of the farmers below. Study all the information about your chosen farm and its natural system. Make notes about your farm by answering the following questions:

- What are your inputs and outputs?
- What methods do you use (processes)?
- What impact do you have on the natural environment?
- What limits does the natural environment put on your farming methods?
- What problems do you face?
- How have you tried to solve the problems?

13 Badan, a farmer in Sarawak

Badan in Malaysia

Badan is a farmer in Sarawak, in East Malaysia, part of the island of Borneo (see Map 19 on page 28). He and his family live in a traditional longhouse community in the rainforest that still covers much of Sarawak. Each family grows rice, bananas and other crops on hillsides around the village. Sometimes they hunt and gather food in the forest. Nowadays they also buy food from shops on visits to their nearest town (a half-day journey by boat on the river).

In the past, people farmed by SHIFTING CULTIVATION. They cleared a small plot of one or two hectares in the forest by chopping down trees and burning the vegetation. They grew crops for two or three years until the soil lost its fertility. Then they moved on to another plot. Now there is more pressure on the land both from a growing population and from other users of the forest. Big companies clear large areas of the forest for timber or for plantations of oil palm trees. This means less land is available for farming, so people cannot shift from one plot to another as they used to. If they grow crops on the same land for too long, yields begin to fall and they cannot produce enough food. One solution is to use chemical fertiliser to increase yields, but this is expensive. No one knows exactly what the long-term effects of chemicals on the environment will be.

14 Tom, a farmer in Wales

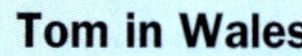

Tom in Wales

Tom is a farmer in mid-Wales. He owns a mixed farm that grows barley and produces sheep on about 100 hectares of hilly land. He uses machinery to do most of the work on the farm and employs one other full-time person. In this part of Wales trees still grow on the steepest land, but most of the best land was cleared for farming long ago.

For farmers in the UK and elsewhere in Europe the main problem is growing not too little food, but too much (see pages 255–57)! The price paid to European farmers for food is low because their farms now produce more than we need. The SUBSIDIES that farmers used to get for producing food have been cut, so many of them have been forced out of business.

Farms are able to produce such high yields with the help of artificial fertilisers and other chemicals which destroy pests and diseases. Unfortunately, the use of chemicals over many years damages the natural balance of the soil and creates other environmental problems such as river pollution. Consumers in the UK are worried about the harmful effect that all these chemicals may have both on the environment and on their bodies. They have begun to demand safer food, such as organic food, produced without the use of chemicals. Tom is considering reverting to more traditional methods in order to produce organic food.

In any ecosystem the NUTRIENTS, or food, that plants and animals need to live are continually being cycled. The NUTRIENT CYCLE consists of a number of STORES – places where the nutrients are kept – and TRANSFERS between these stores. The size of the stores and the rate at which nutrients are transferred varies from one ecosystem to another.

When farmers, anywhere in the world, remove the natural vegetation and use the soil to grow crops they break the natural cycle and take nutrients from it. The most successful farmers, who can farm sustainably, are those who understand the natural system and are able to work with it rather than against it.

SAVE AS . . .

2 Look at *either* Figure 15 *or* Figure 17:

a) Identify each of the inputs, outputs, stores and transfers in the system.

Choose from items in the box below.

- Sunlight
- Trees
- Humus (decomposed leaves)
- Soil
- Nutrients taken up by roots
- Nutrients washed from soil by rain
- Nutrients in minerals from weathered rock
- Leaves fall and decompose
- Nutrients dissolve and sink into soil

b) Draw a labelled diagram of the nutrient cycle for the farm which you chose. Use the inputs, outputs, stores and transfers that you identified in Task 2a).

c) Explain how farming would change this cycle.

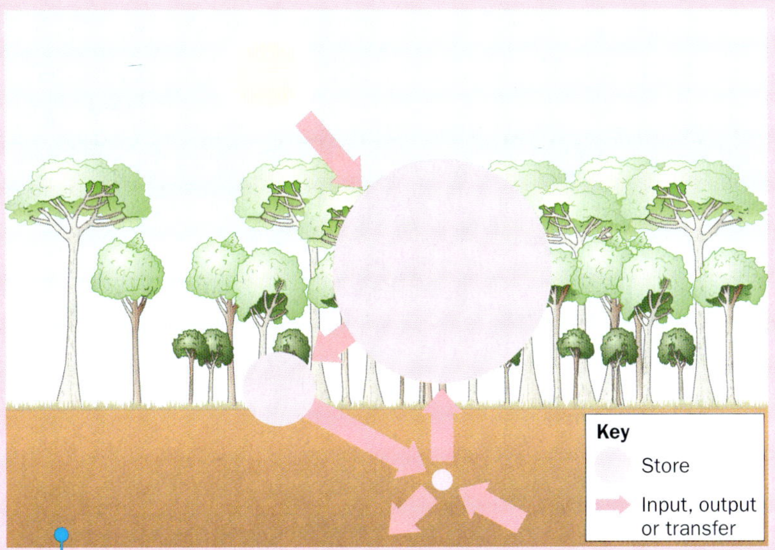

15 Nutrient cycling in tropical rainforest

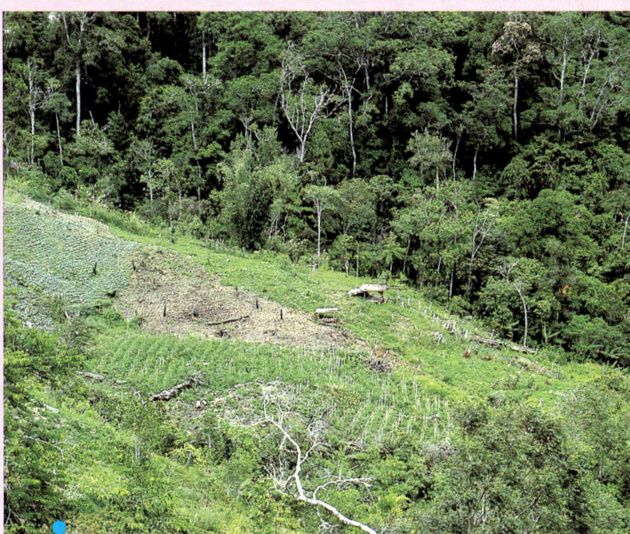

16 Shifting cultivation in Malaysia

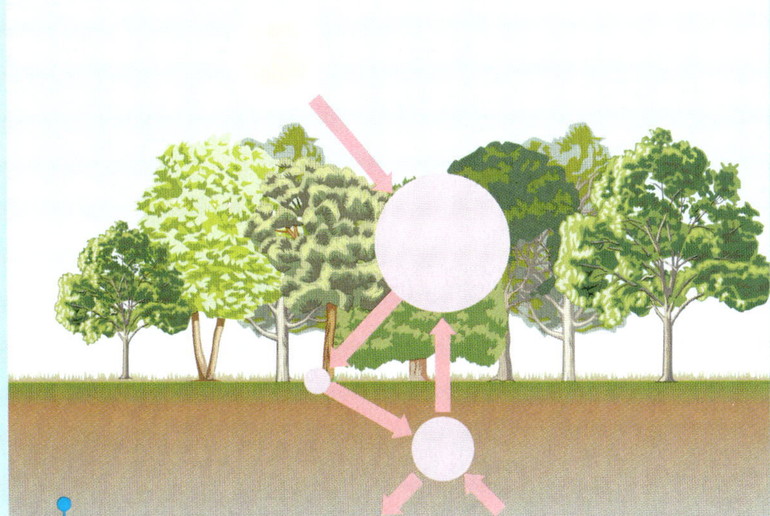

17 Nutrient cycling in temperate deciduous forest

18 Farming in Wales

The secret of soil

The secret of successful farming lies in the SOIL. This is true anywhere in the world. Soil is formed from the weathered remains of the underlying rock together with decayed organic matter, or HUMUS. Plants obtain both water and nutrients from the soil.

Soil is a system, with its own inputs and outputs. Within the soil many processes are happening, as you can see in Figure 19. While the natural ecosystem remains intact these processes are in balance. But when, for example, the forest is cleared for farmland, the whole system is disrupted. The balance of inputs and outputs changes and some processes may stop. The soil may lose its natural fertility and, in extreme cases, may disappear altogether if it is washed away by water or blown away by wind.

It is possible to study soil as a system by looking at SOIL PROFILES. These are vertical sections through the soil showing its separate layers or horizons (Figures 20 and 21).

Some farmers try to use methods that help to keep the soil fertile. CROP ROTATION helps farmers to maintain fertility without using artificial fertilisers (Figure 22).

AGROFORESTRY is a type of agriculture that mimics natural forest (Figure 23).

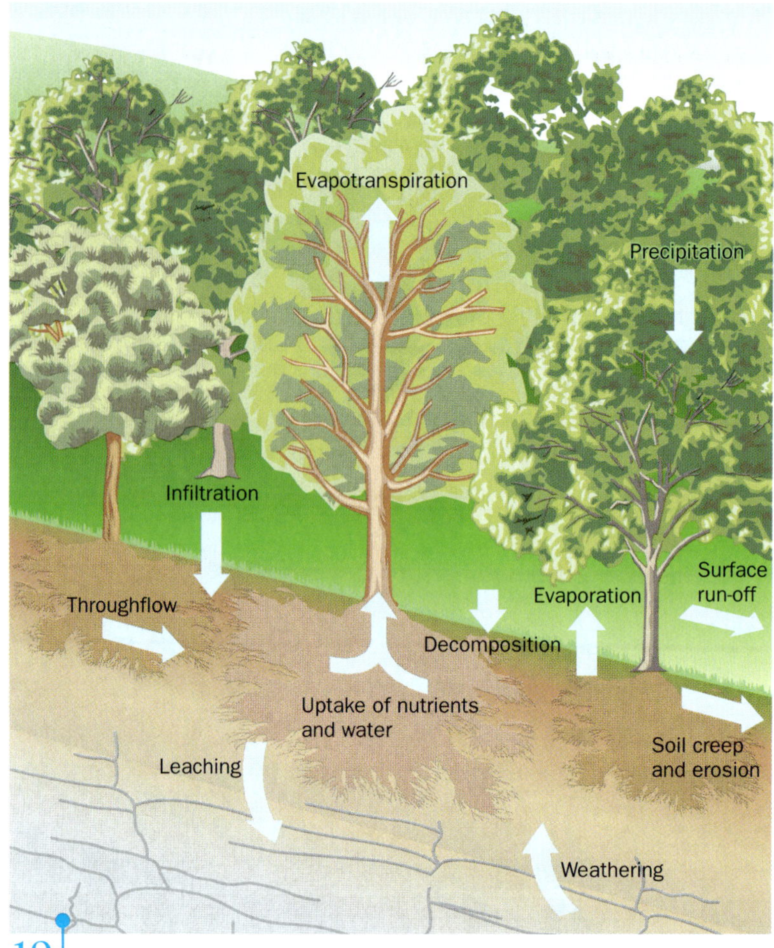

19 Processes at work to maintain the soil balance (see Glossary for all of these terms)

O

A

B

C

O horizon
Leaf litter is decomposed by bacteria and insects to form humus. This happens more quickly in a hot climate than in a cooler climate.

A horizon
Dark-coloured layer, rich in humus. Organic material gives the soil a fine crumbly structure with many air spaces.

B horizon
Pale coloured layer with less humus. Nutrients are leached (washed out) by water draining through. Iron and aluminium are hydrated by water to make soil red.

C horizon
Mainly weathered rock fragments. Weathering happens more quickly in a hot, wet climate than in a cooler one, so the soil is deeper.

20 Tropical soil profile. Tropical soil can be up to 30 metres deep!

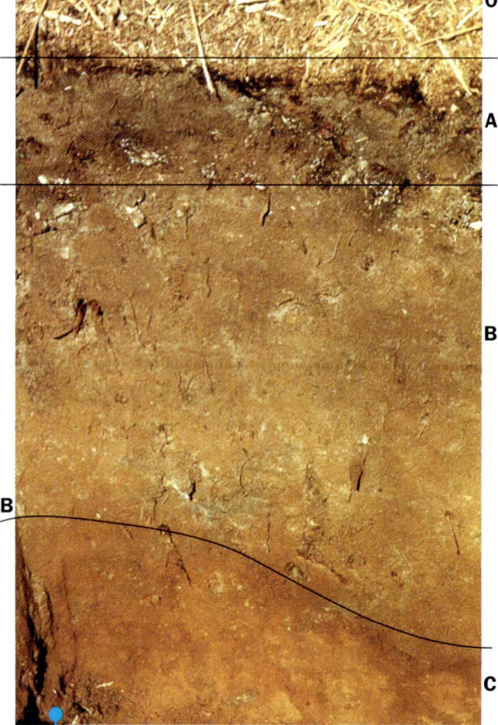

21 Temperate soil profile. This soil is about 1 metre deep

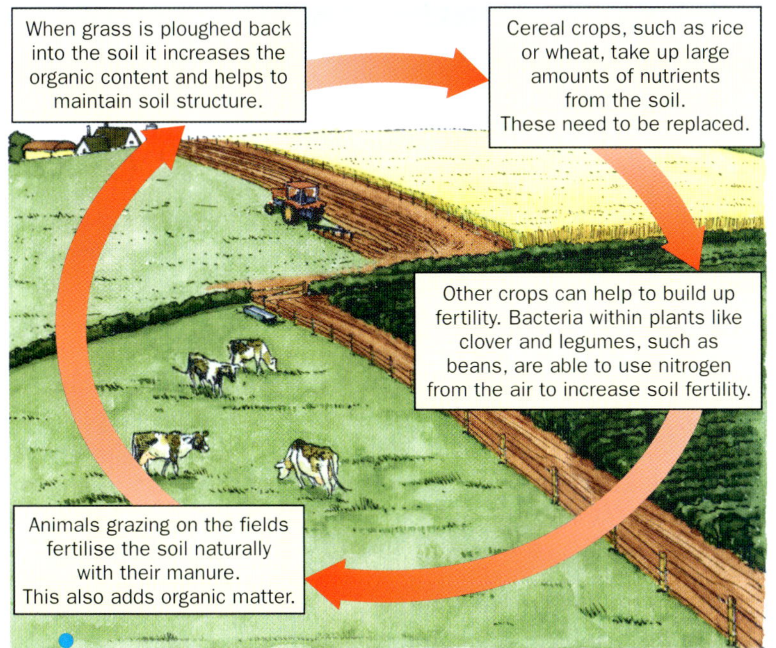

When grass is ploughed back into the soil it increases the organic content and helps to maintain soil structure.
Cereal crops, such as rice or wheat, take up large amounts of nutrients from the soil. These need to be replaced.
Other crops can help to build up fertility. Bacteria within plants like clover and legumes, such as beans, are able to use nitrogen from the air to increase soil fertility.
Animals grazing on the fields fertilise the soil naturally with their manure. This also adds organic matter.

22 Crop rotation

Trees provide shelter from heavy rain and keep the soil in place. They also help to maintain the cycle of water and nutrients.
Crops planted between the trees can be harvested without the soil losing its fertility.

23 Agroforestry

Tasks

SAVE AS . . .

1 Look at Figure 19.

a) List the processes within the soil system. Write a sentence to describe each one. Check your answers using the Glossary on pages 376–82.

b) Think about what might happen if the trees were chopped down. Suggest what impact this would have on each of the soil processes.

2 Look at *either* Figure 20 *or* Photo 21. (Instead of Photo 21 you could examine a soil profile in your school grounds – but don't dig any holes without permission!)

a) Draw a sketch of the soil profile in the part of the world where 'your' farm is (see Task 1 on page 8). Label any horizons that you can see.

b) Is there any evidence of the processes in Task 1? Name the processes and give evidence. For example, a dark horizon is evidence of decomposition.

c) Suggest how these processes might affect the soil's fertility.

3 Look at Figures 22 and 23. They show two methods that farmers have used in different parts of the world to make farming more sustainable. Would either of these methods help you on your farm? Suggest why.

Focus tasks

ICT

4 Work with the same partner as for Task 1 on page 8. Bring together the information that you have gathered from this section (pages 8–11). By now, one of you should be an expert on farming in a tropical environment, the other will be an expert on farming in a temperate environment. Now you are going to compare experiences and give advice to help solve each other's problems. Task 5 gives further reference information for research.

In real life Badan and Tom have recently met each other through an internet website for farmers. Imagine a discussion they might have via e-mail. It will help if you both have your own e-mail address and separate computers. If this is not possible you could write letters on paper or talk face to face.

a) **E-mail 1**: Write to your partner about your farm. Describe the way you farm and the problems you face. Use the notes that you made earlier to help you.

b) **E-mail 2**: Reply to your partner's message. Suggest possible solutions to their problems. Give advice from your own experience.

c) **E-mail 3**: Evaluate the advice from your partner. Do you think it would work in your environment? Write back to explain why/why not.

5 Farmers around the world face similar problems under different circumstances. You can find out more about farming in tropical and temperate environments on the internet. The National Farmers Union has a well-organised website with an educational section. It is found at: www.nfu.org.uk.

You can find farms on this site that are similar to Badan's and Tom's, for example Sitio Batal, a rice farm in the Philippines, and Merthyr Farm, a hill farm in Wales. This research will help you to do the Focus task.

1.2 *How can we manage fragile environments in the UK?*

DO YOU KNOW?

- There are currently 13 national parks in England and Wales, and 87 Areas of Outstanding Natural Beauty in the UK. Scotland has National Scenic Areas but no national parks.
- In the UK, national parks are areas where people live and work – not just places to go on holiday.
- The world's first national park was Yellowstone Park in the USA, set up in 1872. Most countries now have national parks.

AIMS

- **To understand how an ecosystem in the UK has been changed by people.**
- **To assess the potential conflict between conservation and recreation.**
- **To consider ways in which a FRAGILE ENVIRONMENT can be managed more sustainably.**

★ **FACTFILE**

Besides national parks, the UK has several other types of protected area:

- **Areas of Outstanding Natural Beauty (AONBs)** have similar aims to national parks but usually cover a smaller area. They are often more intensively farmed.
- **Heritage Coasts** are stretches of undeveloped coast that are protected in a similar way to national parks. One-third of the UK's coastline is protected.
- **Environmentally Sensitive Areas (ESAs)** are areas where conservation of the landscape depends on maintaining traditional farming methods.
- **Nature reserves** are small areas of a natural ecosystem protected from any type of development. Public access may also be restricted.
- **Sites of Special Scientific Interest (SSSIs)** are small areas that contain an important example of an ecosystem. Special permission is needed for development in these areas.

A CASE STUDY: THE SOUTH DOWNS

Almost everyone is concerned about the environment these days – or so it seems. In the UK, some areas have been singled out for special protection (see the Factfile below). Best known are the NATIONAL PARKS which are large areas of beautiful landscape that are protected from most types of development. In 1999 the government announced a plan to create another national park – the South Downs – to bring the total number to 13. Altogether, national parks cover 10 per cent of the land area in England and Wales – slightly less than the area of urban land. Do we really need another national park? Is it possible to have too much protection?

The two main aims of national parks in England and Wales are:

1 To conserve and enhance the natural beauty, wildlife and cultural heritage.

2 To promote opportunities for understanding and enjoyment of the countryside.

In addition, they have to consider the needs of the local economy and the people who live and work in the area.

Focus tasks

1 You have been called in as an environmental consultant by the government. You are going to use the next seven pages to produce a report on the need for environmental protection for the South Downs. The report should contain the following sections:

a) Why make the South Downs a national park? (Use pages 13–17.)

b) Why does chalk downland need to be protected? (Use pages 14–15.)

c) How can conservation and recreation be balanced? (Use pages 16–17.)

d) How can the South Downs be managed? (Refer to the case study on the Seven Sisters Country Park on pages 18–19).

Each section should be at least a paragraph, or up to a page. It may also contain maps and diagrams where these will help. The following websites may help you:

Council for National Parks: www.cnp.org.uk
South Downs Virtual Information Centre: www.vic.org.uk

2 Look at Photo 2.

a) Describe the distribution of national parks in England and Wales.

b) Using an atlas map, explain how the national parks compare with the distribution of **i)** mountains, and **ii)** urban areas.

c) Suggest how access to national parks differs for people in each part of England and Wales.

d) Do you think the South Downs is a good choice for the location of a new national park? Give reasons for your answer.

An Uplift for the Downs?

BARELY AN HOUR from London by road or rail, the South Downs have long provided millions of people with solitude, inspiration and gentle exercise. But while swathes of wilderness, mountain and coastline in England and Wales were granted the ultimate protection of national park status by post-war governments, the rolling chalkland of the South Downs somehow lost out.

Never mind that they accommodate more visitors than many other scenic areas – an estimated 32 million annually – while providing Britain's most popular long-distance footpath: the South Downs, in spite of intensive lobbying, have never been granted national park status. Now, the government is to make a final decision amid warnings that urgent action is needed to safeguard the 100-mile [169km] chain, already designated an Area of Outstanding Natural Beauty. No one doubts that the Downs need special status. Campaigners have listed about 30 areas damaged, or under threat, from new road schemes, shopping centres, industrial estates and even a sewage works. The dilemma facing John Prescott, in charge of the Department of Environment, is whether or not to place the Downs in the same category as the Lake District, Snowdonia, the Yorkshire Dales and the Norfolk Broads.

Not everyone is happy, because national parks are planning authorities in their own right. They, not local councils, decide how an area is developed. In the Downs most councils oppose park status. They argue that national parks do not lend themselves to 'long, thin landscapes which straddle so many administrative boundaries'.

Campaigners like Paul Millmore, author of a guide to the 100-mile [169km] South Downs Way, disagree: 'Why should someone from the South East have to get in a car or train to go to a national park in the north? There's been a nibbling-away at the edges of the Downs all the time – new houses and new developments – which add up to a steady encroachment because there is no unified plan.'

1 Extract adapted from the *Guardian*, 12 June 1999, before the announcement of the creation of the new South Downs National Park

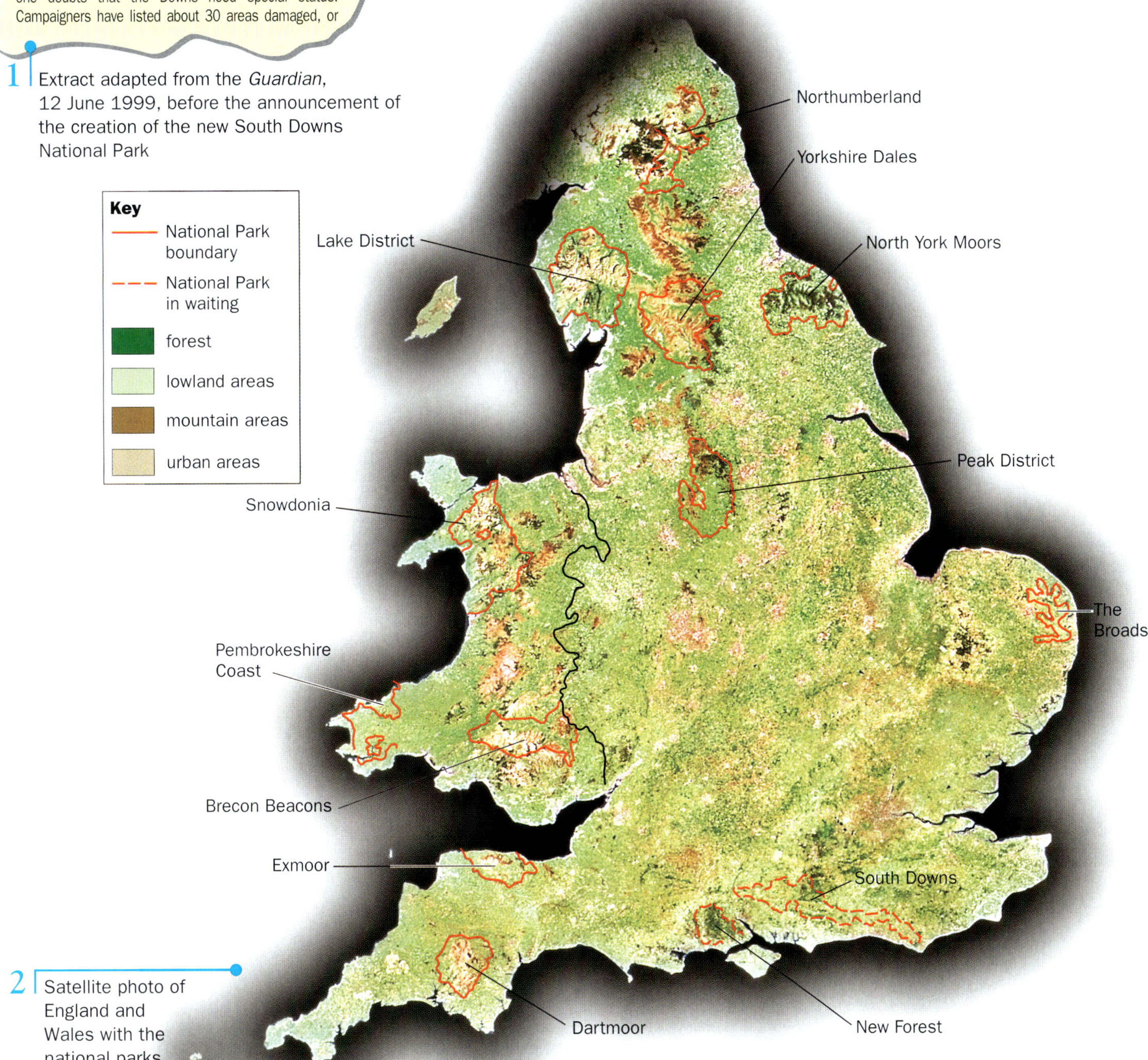

2 Satellite photo of England and Wales with the national parks superimposed

The South Downs – a managed ecosystem

3 The South Downs in Sussex, today

The view from the South Downs in Sussex is spectacular and, at first sight, entirely natural. But look carefully at Photo 3 and you may find clues that the landscape is not as natural as it first appears. In fact, it isn't really natural at all.

Most of the South Downs are CHALK DOWNLAND. This supports an ecosystem mainly comprising grasses and small flowering plants. Chalk downland is valued for its BIODIVERSITY or rich variety of species, many of which are unique to the Downs. There can be up to 45 species of flowering plants and mosses per square metre of land.

Today's landscape was created hundreds of years ago when people replaced the original natural ecosystem – beech woodland in this case – with farmland. The thin, chalky soils of the Downs lack the natural fertility needed for growing crops, but they are good for keeping animals on grass. Intensive grazing of sheep prevents the growth of larger plants, and helps to maintain the rolling green landscape. Without sheep the Downs would slowly revert, through a process of NATURAL SUCCESSION, back to woodland (see Figure 5). However, over the past 50 years, much of the traditional chalk downland ecosystem has been destroyed by modern agriculture. The most obvious change has been the replacement of pasture by arable crops. At the same time there has been an increase in field size to accommodate modern agricultural machinery, with the loss of trees and hedgerows. The resulting landscape in the 21st century is one that is less green and has far less biodiversity than it originally had.

4 The South Downs in the 1950s

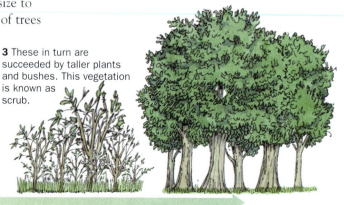

1 Regular grazing helps fine-leaved grasses like sheep's fescue to thrive.

2 When grazing stops, coarser grasses dominate and shade out the finer-leaved grasses that are typical of the Downs.

3 These in turn are succeeded by taller plants and bushes. This vegetation is known as scrub.

4 Eventually the Downs would revert to beech woodland – the natural vegetation before people began farming.

5 Succession from chalk downland to beech woodland

The changes brought about by farming on the chalk downland are not always visible. Farmers now spread fertiliser over wide areas to increase grass production. This allows them to keep more sheep, but it also has an important impact on the ecosystem.

Chalk downland is a fragile environment. Even a relatively small change such as the addition of fertiliser is enough to alter its delicate balance. The ecosystem has a natural diversity of plant and animal species, linked to each other through a complex FOOD WEB. You can see some of the species that make up this food web in Figure 6. By adding fertiliser, the farmer encourages the growth of one or two grass species so that other plant species cannot survive. This leads to the loss of animal species and reduces the biodiversity of the whole ecosystem. What remains is a simple FOOD CHAIN with just grass and sheep. Rare plants, such as orchids, and birds such as the skylark, are being lost in Sussex as the chalk downland disappears.

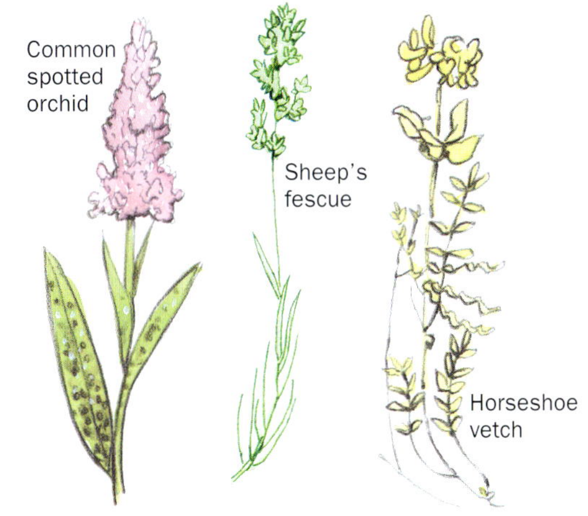

Tasks

1 Create your own chalk downland food web. Look at Figure 6. Your teacher may give you a sheet with more information about each species.

a) Identify as many possible food chains as you can. For example: earthworm → shrew → kestrel. Each chain may have two, three or four species.

b) Combine the food chains to make a food web. It will look a bit like this:

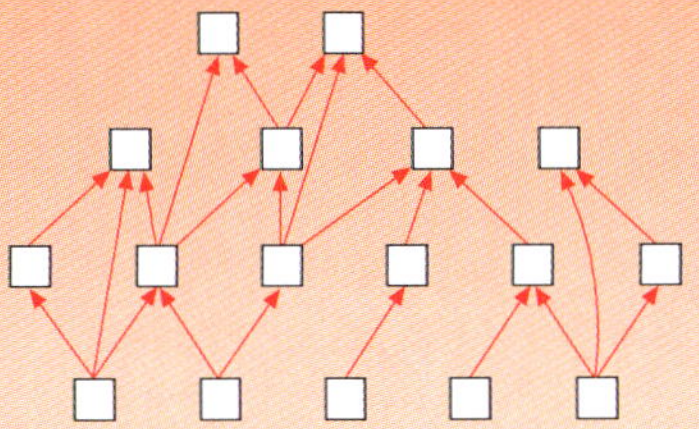

c) Explain how modern farming methods reduce biodiversity in the ecosystem.

SAVE AS...

2 Use this task to help you to write the second section of your report, 'Why does chalk downland need to be protected?' (see page 12). Study all the sources on pages 14–15.

a) 'Chalk downland is a fragile environment.' Explain what this means.

b) Explain the role of farming in:

i) maintaining the ecosystem

ii) destroying the ecosystem.

6 Species found in the chalk downland ecosystem (not to scale)

Conservation and recreation – a conflict of interest?

The South Downs, stretching from Winchester in the west to Beachy Head in the east, are regarded by many as one of the UK's finest landscapes. Despite changes in farming and the immense pressures that 32 million visitors each year bring, the Downs still retain much of their beauty and character.

You may think it is surprising that the South Downs had to wait so long before the process of designation as a national park got under way. They have always had more visitors than most of the UK's national parks in more remote, upland areas. For this reason they are also in greater need of protection. Along the south coast linear urban development has created a huge demand for land to build on. As space along the coast runs out, towns and cities like Worthing and Brighton are growing northwards, taking chunks out of the Downs as they expand. It is unlikely that such development would have been allowed had the South Downs already been a designated national park.

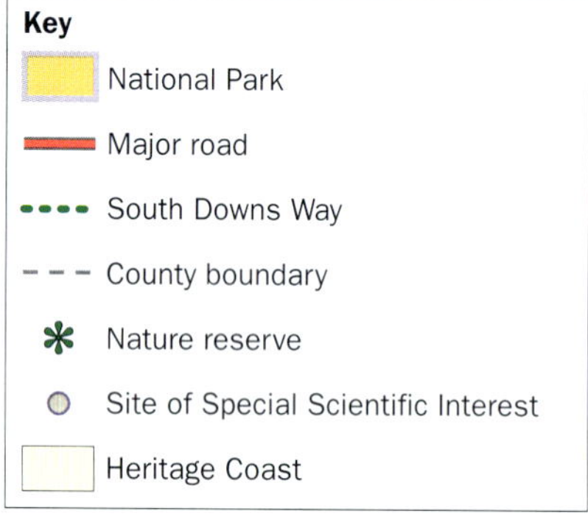

7 The planned South Downs National Park

8 The pub and information centre at Beachy Head

9 Part of the South Downs Way near the coast

The two main aims of any national park can be summarised quite simply: CONSERVATION and RECREATION. The role of the national park authority is to manage the area so that each of these aims can be achieved. The basic aims are explained in more detail in the Factfile.

The two aims can often be in conflict. The problem is that if an area is designated as a national park in order to protect it, it is likely to attract an increased number of visitors. Within an area like the South Downs it is possible to identify both PRESSURE AREAS and VULNERABLE AREAS. Pressure areas are those where there are already a large number of visitors, and any further increase in numbers is only likely to make things worse. Vulnerable areas are those that are more remote and where natural wildlife habitats are sensitive to any increase in visitor numbers.

★ FACTFILE

Conservation

National parks aim to conserve:

- the natural beauty of the landscape
- the variety of natural habitats and individual species
- a quality of remoteness
- archaeological, historical and traditional features, buildings, landscapes and townscapes.

Recreation

National park authorities aim to provide opportunities for understanding and enjoyment. To achieve this they have to:

- negotiate access for the public to open land
- maintain a network of well-signposted footpaths
- provide information centres, car parks and picnic areas
- manage tourist numbers and traffic levels
- provide educational materials and opportunities.

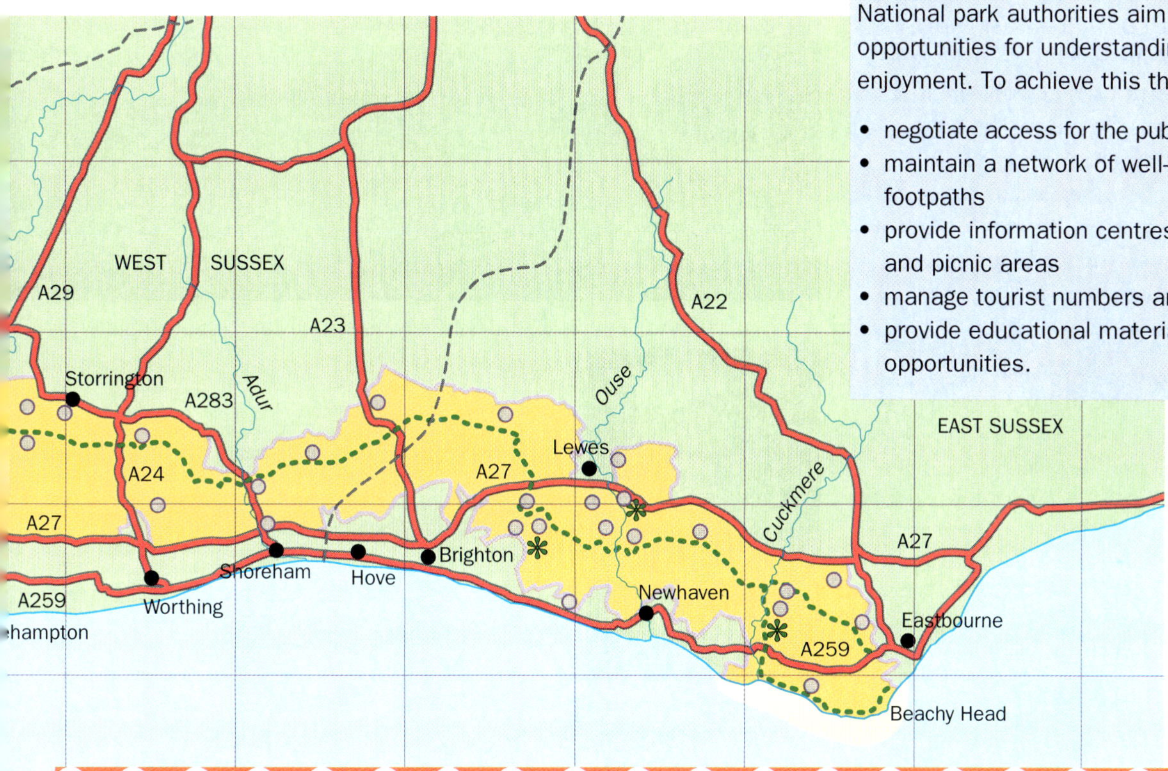

WEST SUSSEX
SUSSEX
A29
A23
A22
Storrington
A283
Adur
Ouse
EAST SUSSEX
Lewes
A24
A27
Cuckmere
A27
A27
Brighton
Shoreham
Hove
A259
Worthing
Newhaven
Eastbourne
A259
hampton
Beachy Head

Tasks

1 Look at Map 7.

a) Work out the size of the planned South Downs National Park. Use the scale on the map to measure the distance from west to east and from north to south.

b) Estimate the area of the planned national park. Each square on the map represents roughly 100 km^2.

c) Describe the shape, size and location of the planned national park. Explain how this might make it more difficult to manage.

2 Look at Photos 8 and 9.

a) Suggest how the area in each photo might be important for conservation and/or recreation (the information in the Factfile will give you ideas). What evidence can you see in each photo of any conflict between the two aims?

b) Would you describe the area in each of these photos as a *pressure area* or a *vulnerable area*? Give reasons.

SAVE AS . . .

3 Use Tasks 1 and 2 to help you to write the third section of your report, 'How can conservation and recreation be balanced?' (see page 12). Suggest how the national park authority could balance conservation and recreation throughout the whole national park scheme. Which areas should be conserved? Where should recreation be encouraged? Should there be any limits on numbers of people and what they are allowed to do?

How should the Seven Sisters Country Park be managed?

The Seven Sisters Country Park is one of the most popular places for tourists to visit in the South Downs. It attracts 450,000 visitors each year and the number is likely to increase now that it is part of a national park scheme.

The country park contains a large area of chalk downland, and smaller areas of other natural environments that people come to enjoy. The Cuckmere River cuts through the chalk, forming a valley. It has created a number of classic MEANDERS on the FLOODPLAIN. As the river flows towards the sea there is an area of SALTMARSH, formed where plants have invaded the old ESTUARY. There is also an artificial LAGOON, created to attract wading birds. Finally, at the coast there are chalk cliffs and a shingle beach, making this one of the best examples of chalk coastline in Britain. Here, the rolling Downs have been eroded by the sea to form the Seven Sisters – seven large chalk cliffs.

10 The Cuckmere valley in the Seven Sisters Country Park

The South Downs Way spans the whole length of the proposed national park and goes through the Seven Sisters Country Park. As well as being a footpath it is also a bridleway used by horse-riders and cyclists. There has been a huge increase in the amount of off-road cycling in recent years. Our company has a shop next to the country park where we hire out bikes to visitors. We warn cyclists to stick to the designated cycle routes and to keep off smaller footpaths. Now that the South Downs is to become a national park we expect an increase in the number of visitors and the amount of business that we do.

11 Roy Smith works for the Cuckmere Cycle Company

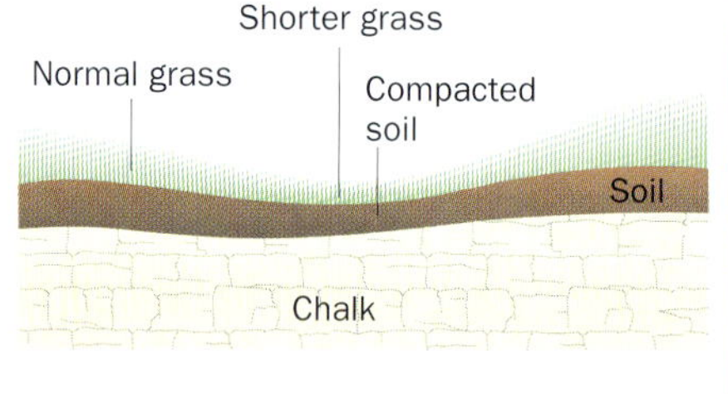

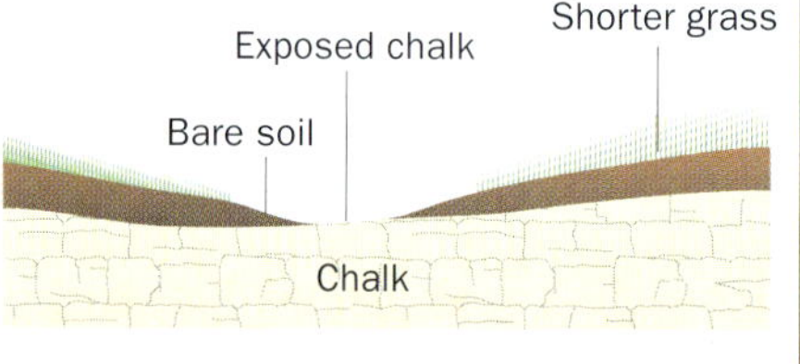

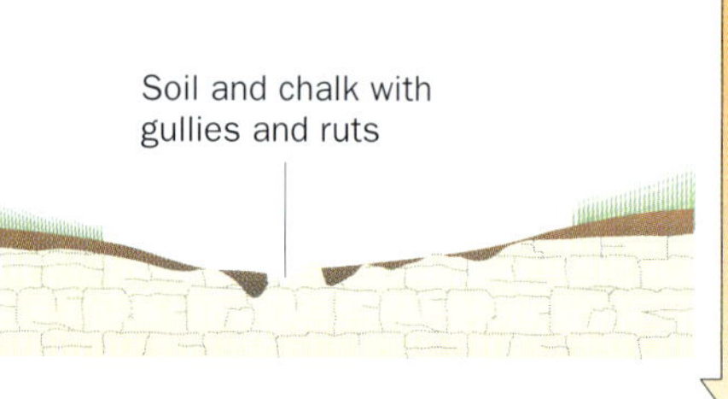

13 Stages in the erosion of a footpath on chalk downland

At the Seven Sisters Country Park we try to educate people about the value of the natural environment and how they should treat it. If the number of visitors grows, as we expect, education will be even more important. Already there is evidence that pressure of numbers is damaging the fragile chalk downland. The South Downs Way has been badly eroded in places by the number of people using it, particularly the section along the Seven Sisters. Some of the worst damage is caused by cyclists who ignore warnings and cycle on footpaths. Without education the problem is likely to get worse. People wrongly believe that in a national park they have the right to go where they want. In reality the land belongs to someone else and access has to be limited.

12 Ann Murray is a member of the countryside management team at the Seven Sisters Country Park

Tasks

Decision-making exercise

1 Work in a small group to devise a management plan for the Seven Sisters Country Park. Your main aims are to conserve the natural beauty and wildlife in the park and to encourage understanding and enjoyment.

Stage 1

Look carefully at Map 14. On a copy of the map, divide the park into three zones:

Active zone where all types of recreation, including cycling, picnicking and canoeing, would be allowed.

Buffer zone where fewer people would go and activities would be limited to walking.

Remote zone where people would be discouraged from going in order to conserve the natural environment.

You can make the three zones any size you want, depending on how you think the park should be managed. Think carefully about where each zone should be located.

Stage 2

If you are going to conserve the environment you will need to control numbers of people and access to each part of the park. Decide how you are going to do this. You may decide to keep the existing car parks, bridleways and footpaths, or you may want to change them. The following are some options that you could consider (choose at least three):

- enlarge the car parks
- reduce the car parks
- charge money for car parking
- provide more footpaths
- provide fewer footpaths
- more signposts, to encourage walking
- no signposts, to discourage walking
- provide maps for keen walkers
- ban cycling and horse-riding
- provide more bridleways for cyclists and horses.

Mark your own car parks, footpaths and bridleways on the map. Think about how they will relate to the three zones.

Stage 3

Draw a large annotated map to show your plan. Your map should show how the plan will balance the need for conservation and recreation. Notes around the map will help to explain the decisions that you took.

SAVE AS . . .

2 Use the plan you have drawn up in Task 1 to complete the final section of your report, 'How can the South Downs be managed?' (see page 12).

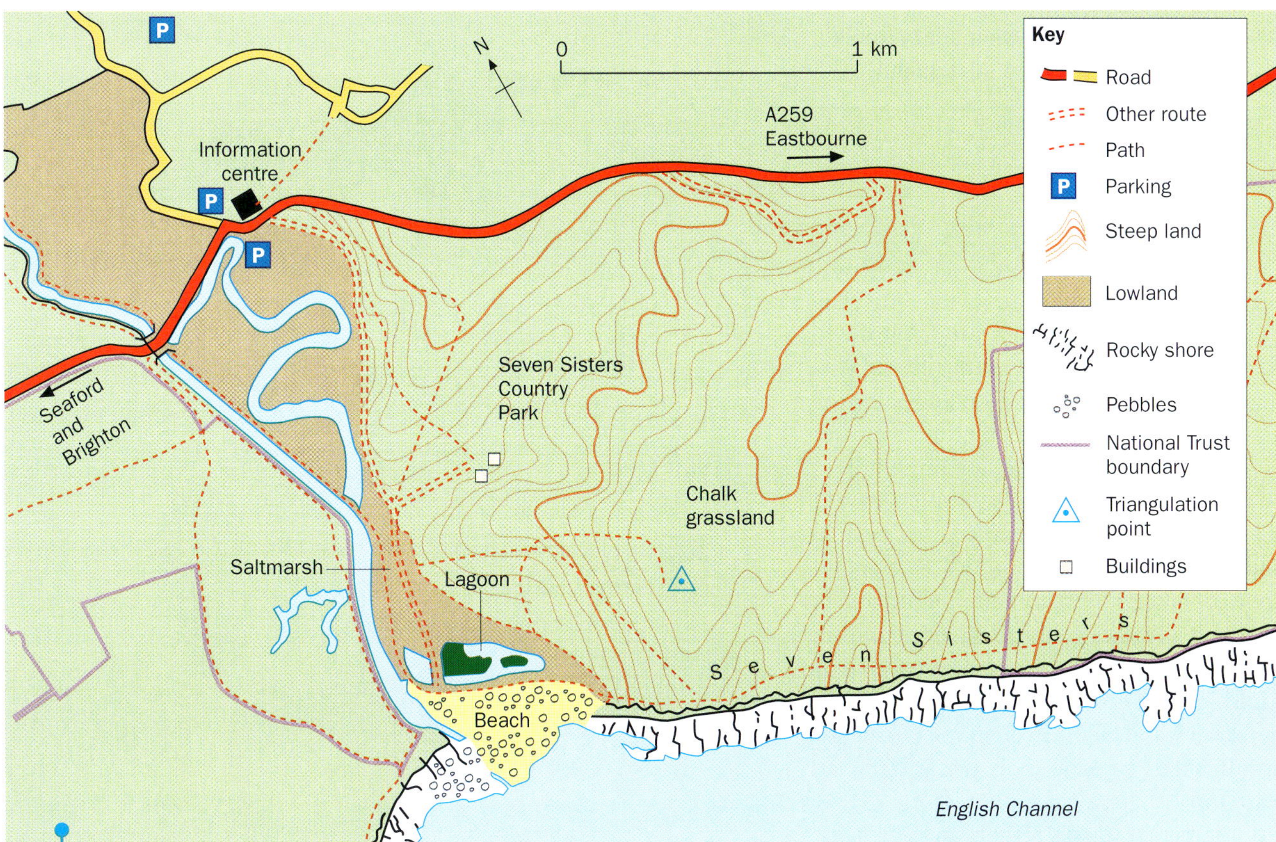

14 Seven Sisters Country Park

1.3 Is the global environment under threat?

DO YOU KNOW?

- People have been living sustainably in TROPICAL RAINFORESTS for thousands of years.
- Tropical rainforests cover about 7 per cent of the Earth's surface – an area about the size of the USA. About 18 million hectares of tropical rainforest are destroyed each year – an area about the size of England and Wales.
- Of all the world's ecosystems, tropical rainforest has the greatest biodiversity, including many species that have yet to be discovered.
- 135 million people live in areas that are severely affected by DESERTIFICATION. There is a constant threat of hunger or starvation.

AIMS

- **To understand the changes to natural ecosystems caused by human activities.**
- **To recognise the wider environmental impact of these changes.**
- **To explore ways in which ecosystems can be managed more sustainably.**

1 Burning rainforest in Indonesia: in 1997 many fires got out of control and smoke covered much of South East Asia

2 Disappearing forests and expanding deserts

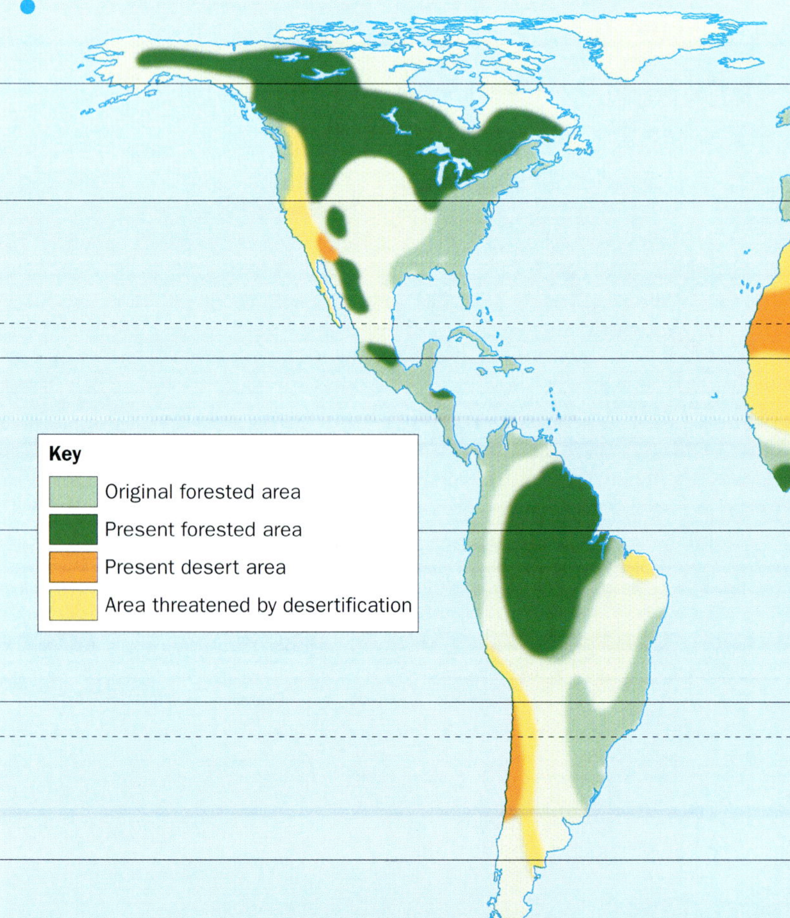

As far as we know the Earth is the only planet in the Universe that has life on it. Yet the most powerful inhabitants of the planet – the human race – are in danger of destroying the diversity of life they have inherited. Nowhere on Earth these days escapes the impact of people's activities. Natural environments that took millions of years to evolve have been lost within a few hundred years, and many more are under threat. Tropical rainforests, coral reefs and wetlands could all disappear within our lifetime. In densely populated parts of the world, natural ecosystems have been replaced by farming and cities. Even in remote areas, uninhabited by people, pollution and global warming are changing the natural environment. Human activity is affecting those areas too.

★ **FACTFILE**

Expanding deserts

Deserts currently cover about 10 per cent of the Earth's land surface and their area is expanding. People, again, are one of the main causes of this change.

Population growth puts pressure on the land to produce more food. People chop down trees and plough up grass in order to plant crops, while animal herds graze whatever natural vegetation remains. In arid areas, unreliable rainfall, together with human mismanagement, has led to vegetation loss and land degradation. This process is known as desertification. In Africa, the whole Sahel region south of the Sahara is in danger of turning into desert.

Disappearing forest

Half of the world's original forests have been chopped down, while half of the remaining areas have been altered by people so that they no longer resemble the original forests. The loss of natural forest ecosystems threatens not only the survival of thousands of individual plant and animal species but also the balance of the global environment.

Forests still cover 30 per cent of the Earth's land surface and play a vital role in the global water cycle and in maintaining global temperatures. Much of the forest cover in densely populated areas, like Europe, has already been lost. Today, DEFORESTATION is happening even more rapidly in tropical areas like South East Asia.

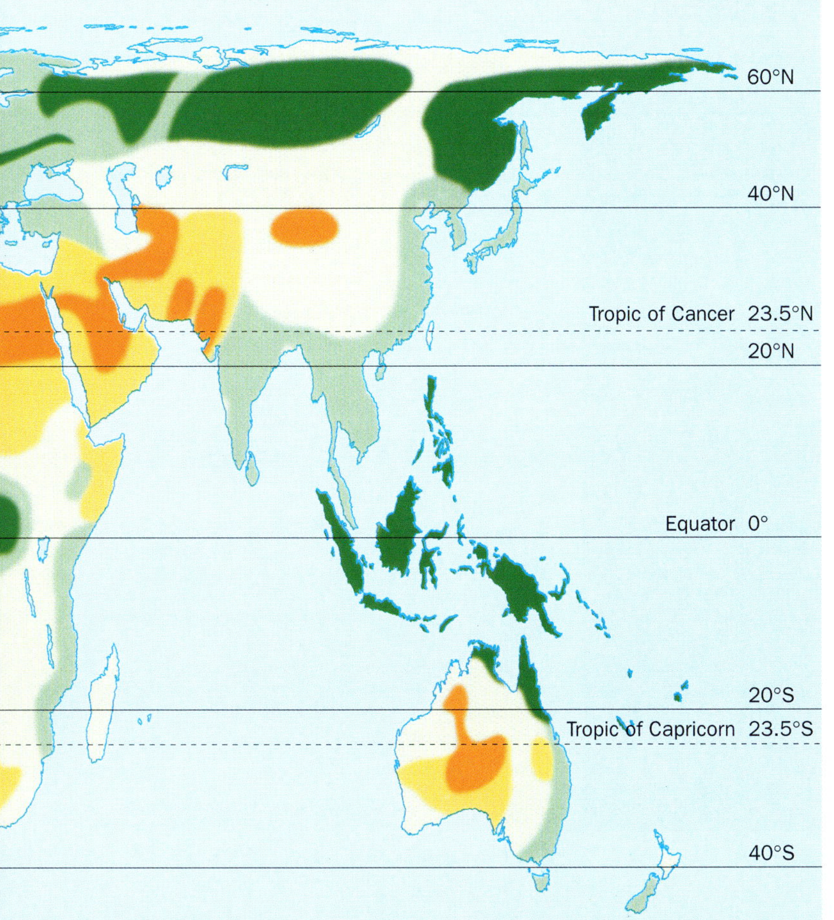

Tasks

1 Look at Map 2. You can also use an atlas if you need to.

a) Name five countries that appear to have lost most of their forest. In each case, what type of forest was this? (Look back to Map 11 on page 7.)

b) Name five countries that still appear to have a large area of original forest. In each case, what type of forest is this?

c) Overall, which type of forest appears to have been most affected by deforestation? Suggest reasons for this.

2 Look at Map 2 again. You can use an atlas to help if you need to.

a) Name five countries that are threatened by desertification.

b) Explain why these countries are most at risk of desertification. Look at maps of world climate and population growth to help you.

Focus task

3 How do you feel about the future of the global environment? Are you optimistic or pessimistic? After reading these pages you may feel that there is no hope! Indeed, there are many reasons to be worried. The danger is that we will all give up and do nothing about it (rather like the boy on page 1). But there are hopeful signs too. Both the case studies in this chapter have positive ideas about the ways in which people can protect the environment and live more sustainably.

As you work through this chapter, make a table with two columns:

Reasons for optimism	*Reasons for pessimism*

You could also collect stories about the environment, from newspapers and TV, to add to your list. (Are these more likely to be optimistic or pessimistic, do you think?) Towards the end of this chapter, you will be asked again to consider how you feel about the future of the global environment.

A CASE STUDY: CAN FARMING IN KENYA BE SUSTAINABLE?

Mukethe Mbithi has good reason to feel pleased. She is a farmer in the Machakos district of Kenya, an area that has become well known for its success in SOIL CONSERVATION. Three-quarters of Kenya is described as being ARID or SEMI-ARID – the east of the country is particularly dry. Kenya also has one of the world's fastest-growing populations, rising at almost 4 per cent each year. In combination, DROUGHT and population growth are the two major causes of desertification.

The Machakos district where I live is prone to regular periods of drought. It is also one of the most densely populated parts of Kenya. You would expect that it would be one of those places in Africa, which you often see on TV, where nothing grows and people go hungry. In fact, the opposite is true. We grow plenty of food and everybody is well fed. So what is the secret of our success, and what lessons does it provide for other parts of the world?

3 Mukethe Mbithi

4 Where Mukethe Mbithi lives

5 A fertile hillside in the Machakos district of Kenya

Life has not always been so good in Machakos. In the 1930s the district suffered from serious problems of SOIL EROSION and famine. In those days Kenya was still a British colony. A soil expert working there, after a field trip to the district, described it like this:

> *'The Machakos Reserve is an appalling example of a large area of land that has been subjected to unco-ordinated and practically uncontrolled development. The inhabitants are rapidly drifting to a state of hopeless and miserable poverty and their land to a parching desert of rocks, stones and sand.'*

	1932	**1989**
Population	240,000	1,393,000
Population density (people per km^2)		
in wet areas	80	400
in dry areas	50	150
Population growth (% per year)	2.7	3.1
Livestock population units (one unit = 1 cow or 5 goats)	214,000	333,870

6 Changes in Machakos, 1932–89

Kenya is on the edge of the Sahel – a region that runs from west to east across Africa south of the Sahara Desert. Since the 1960s this region has become drier and the frequency of drought has increased. In places, DESERT has replaced the neighbouring SAVANNA and has also taken over land that was once used for farming.

Savanna was the natural ecosystem in the Sahel and in much of Kenya. It is a mixture of tall grass, bushes and scattered trees. It grows in a semi-arid climate with wet and dry seasons, in the transition zone between wet tropical climate, where rainforest grows, and desert climate. Savanna vegetation is adapted to survive months of drought and to make best use of the rainfall from heavy storms during the wet season.

7 Satellite photo of Africa; the boxed area shows Kenya

Month	J	F	M	A	M	J	J	A	S	O	N	D
Temperature (°C)	27	27	28	27	25	25	24	24	25	26	26	27
Rainfall (mm)	38	64	125	211	158	46	15	23	31	53	109	86

8 Temperature and rainfall for Nairobi, Kenya

Tasks

SAVE AS . . .

1 Look at Photo 5 and Table 6.

a) Describe the landscape of Machakos in the photo. What evidence of human activity can you see?

b) From the table, write four sentences to describe changes in Machakos since the 1930s. In what ways would these changes put extra pressure on the environment?

c) How does the landscape today differ from the description of Machakos in the 1930s? Given the data in the table, why are these changes surprising?

2 Look at Photo 7.

a) From an atlas, draw a sketch map of Africa. Mark and label the Sahara Desert and the Sahel. Draw the outline of Kenya on your map.

b) Describe the cloud distribution over Africa. How does this help to explain desertification in the Sahel?

3 Look at Table 8 and Photo 9.

a) Draw temperature and rainfall graphs for Nairobi in Kenya. Describe the annual climate with the help of your graphs. During which months are the wet season and the dry season?

b) Suggest at which time of year Photo 9 was taken. Give reasons for your answer.

c) Explain how savanna vegetation is adapted to the climate.

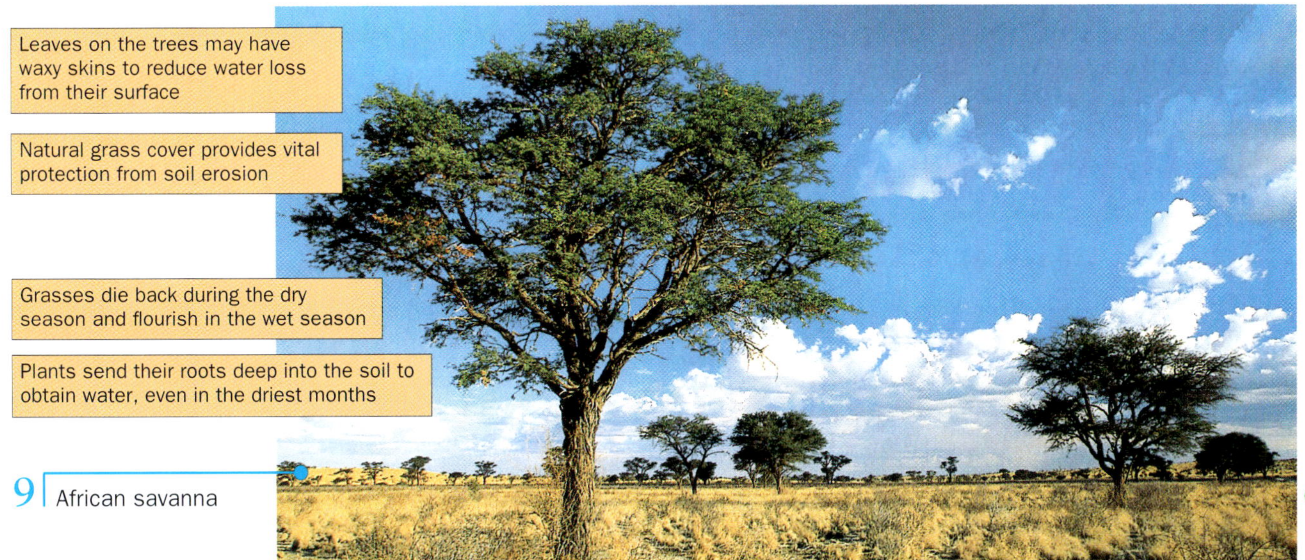
Leaves on the trees may have waxy skins to reduce water loss from their surface
Natural grass cover provides vital protection from soil erosion
Grasses die back during the dry season and flourish in the wet season
Plants send their roots deep into the soil to obtain water, even in the driest months

9 African savanna

How can balance be restored to the ecosystem?

Desertification is a process that degrades land, turning an area that was once fertile and productive into one that is barren and useless. Desertified land is quite different from the desert you see in picture postcards – an endless landscape of rolling sand dunes. It is more likely to be a brown, rocky landscape without soil or vegetation, where the most interesting feature is the occasional dead tree.

Climate change and population growth both contribute to the process of desertification. Although the savanna ecosystem is adapted to survive periods of drought, a decline in rainfall over many years may gradually turn it into desert. The process is accelerated in areas where population growth puts more pressure on the land to produce food. OVERGRAZING and OVERCULTIVATION are two of the main causes of desertification in Africa.

Soil erosion is an important part of the process of desertification. After the vegetation cover has been removed there is nothing to protect the soil that lies beneath. Deep GULLIES – like those in Photo 10 – may form as water flows quickly over the surface where vegetation has been removed.

Soil erosion is like theft. Sometimes it happens so slowly that you don't notice until it's too late. A steady flow of water or wind over bare ground can lift the small soil particles from the surface, eventually leaving behind just stones and rock. At other times, soil erosion can be a sudden, violent event that happens right in front of your eyes. Soil that has formed over thousands of years can be washed away in a few minutes. This happens on steep slopes where fast-flowing water can carve out deep channels, or gullies. The soil is quickly washed down to the nearest river.

11 Mukethe Mbithi

10 Soil erosion

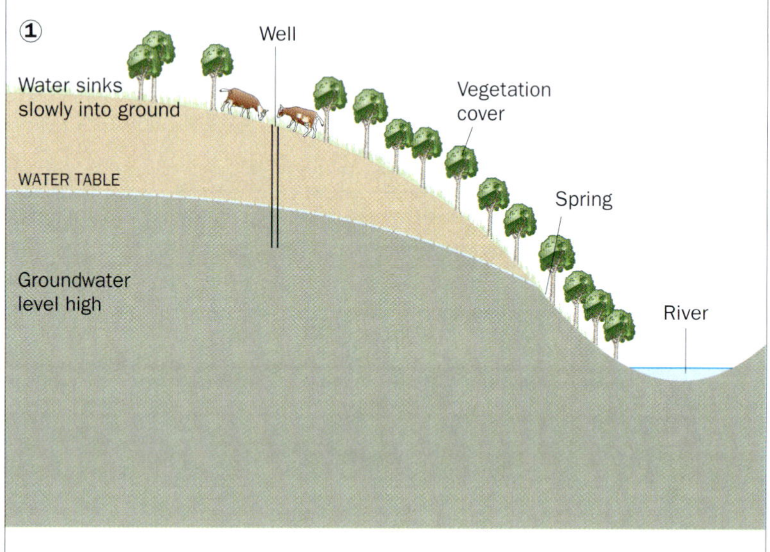

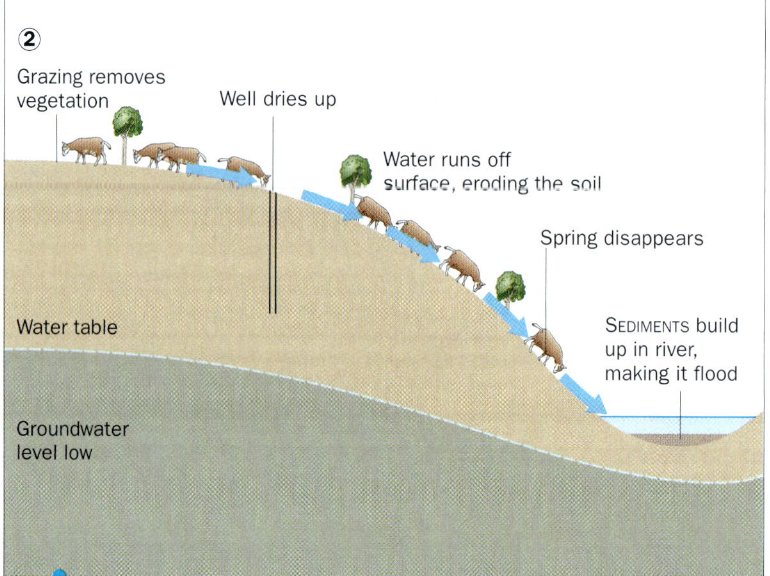

12 The effect of removing vegetation on a slope

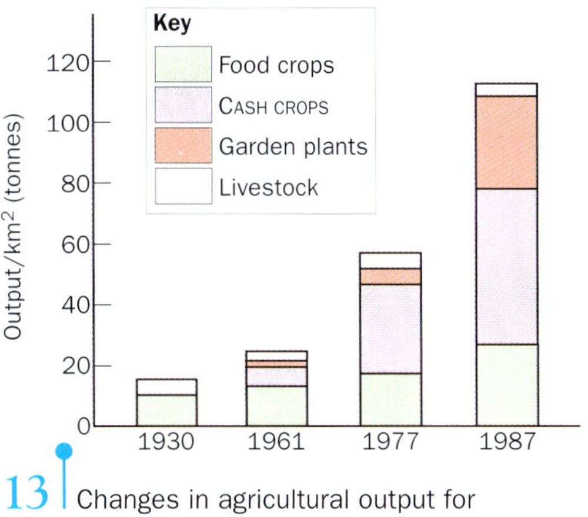
Key
Food crops
CASH CROPS
Garden plants
Livestock
Output/km² (tonnes)
0
20
40
60
80
100
120
1930
1961
1977
1987

13 Changes in agricultural output for Machakos district, 1930–87

Until 1970 the Machakos district, like many other drought-affected parts of Africa, struggled with the problem of soil erosion. As the population increased, more land was cleared for cultivation. Much of this land was on steep hillsides where the rate of soil erosion was high. Even worse, the increase in cultivation meant that less land was available for grazing. Livestock were squeezed on to smaller areas of land that became overgrazed and, therefore, prone to erosion.

Farmers in Machakos knew that this type of farming was unsustainable and was leading to environmental catastrophe. The Kenyan Government set up the National Soil and Water Conservation Project (NSWCP) with support from the Swedish International Development Agency. Machakos was chosen as a pilot district and a soil and water conservation project was launched there in 1979.

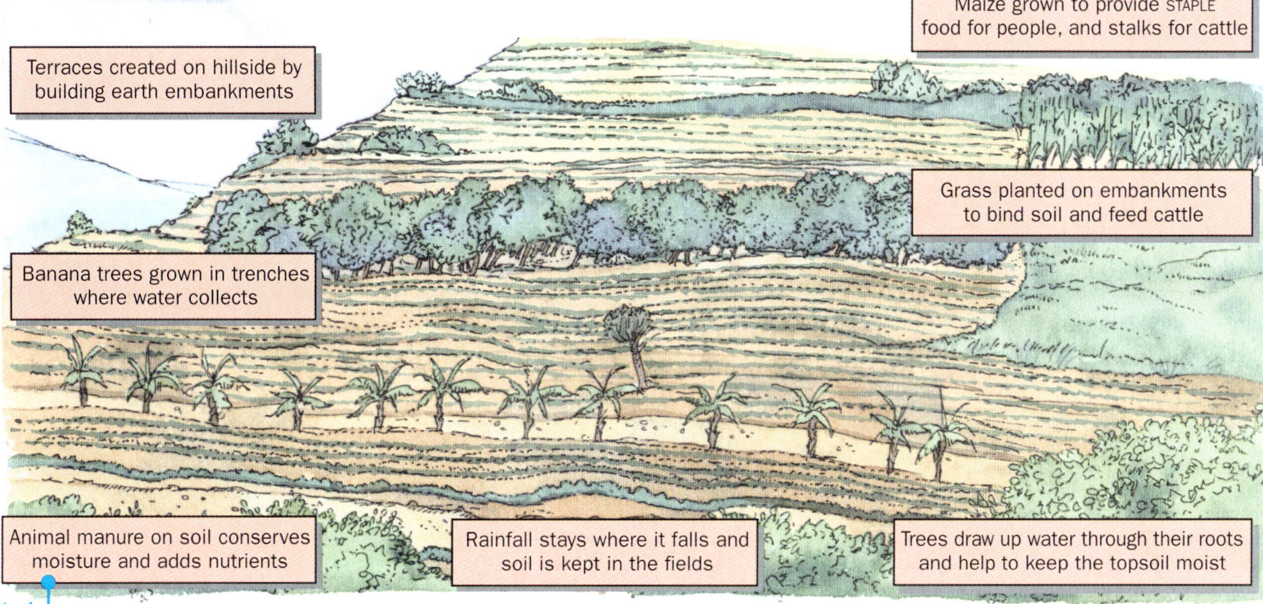
Maize grown to provide STAPLE food for people, and stalks for cattle
Terraces created on hillside by building earth embankments
Grass planted on embankments to bind soil and feed cattle
Banana trees grown in trenches where water collects
Animal manure on soil conserves moisture and adds nutrients
Rainfall stays where it falls and soil is kept in the fields
Trees draw up water through their roots and help to keep the topsoil moist

14 Soil and water conservation measures in Machakos

Tasks

SAVE AS . . .

1 Look at Photo 10 and Figure 12.

- **a)** Draw a sketch of the photo to show the impact of soil erosion on the landscape.
- **b)** Draw a second sketch to show what it might have looked like before the soil was eroded.
- **c)** Annotate your sketches to describe the process of soil erosion that produced this landscape.
- **d)** Suggest at least three reasons why you would expect the Machakos district to suffer from soil erosion.

2 Look at Figure 13.

- **a)** Describe the changes in agricultural output in Machakos since 1930.
- **b)** Try to explain these changes (look back at Table 6 on page 22 to help you). Are you surprised by them?

3 Look at Figure 14. You are going to draw a flow diagram to show how people have helped to restore balance to the ecosystem.

- **a)** Make copies of the labels on the drawing (your teacher may give you a sheet of these labels). Cut out each label.
- **b)** Place the labels on your desk. Move them around to find links between them. Each label should be linked to at least one other. Some may be linked to more than one. Stick the labels on a page in your workbook and draw arrows to show the links.
- **c)** Use the diagram you have made to explain how the changes have helped to make farming more sustainable.

What can be learned from traditional methods?

People of my age in many parts of Africa can remember a time when the land was a lot greener than it is today. The environment has changed beyond recognition in our lifetime. But this need not be a one-way process. In Machakos we have proved that the clock can be turned back. With a better understanding of nature it is possible to do good rather than harm. We don't need outside experts to come and solve our problems. Often, it is the old methods that work best. We need to educate our children to produce a new generation of farmers using these methods. That will help to reverse years of environmental deterioration.

15 Mukethe Mbithi

Mulching
Bare ground between plants is covered with a layer of organic matter, such as straw, grass or leaves. MULCHING helps to retain soil moisture, reduces weeds and improves the soil structure.

Contour barriers
Almost any material can be used to build barriers along CONTOURS, for example stones, grass or earth. Soil collects behind the barriers to form natural terraces. Trenches in front of the barriers help to catch water.

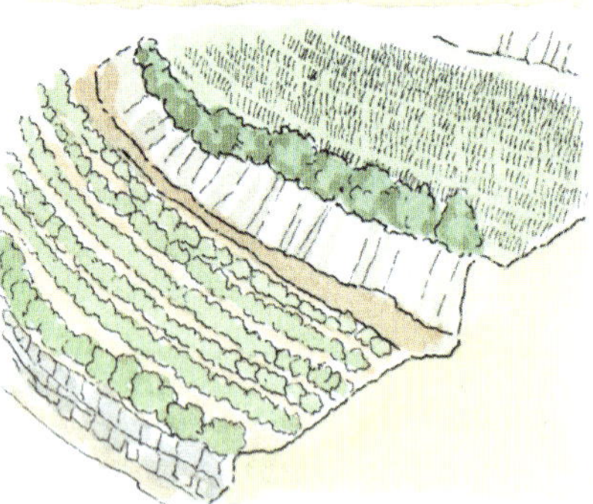

Agroforestry
Planting crops among trees helps to protect the soil from erosion, particularly after the crops are harvested. Tree roots also help to bind the soil, and the trees can provide fruit, fodder and firewood.

Rock dams
Land where gullies have begun to form can be restored by building PERMEABLE rock dams. These slow down the flow of water and prevent further soil loss.

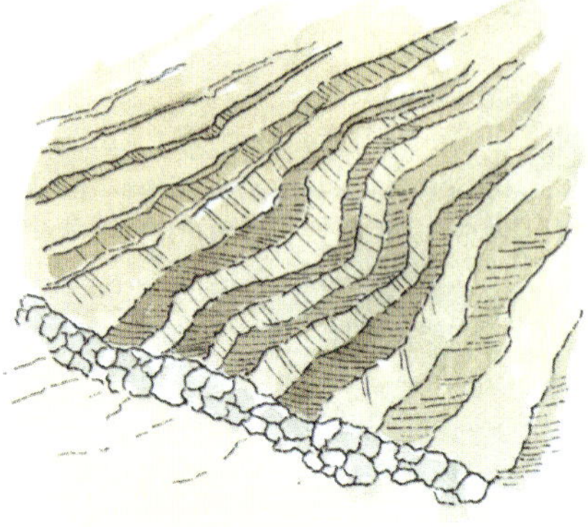

Terraces
TERRACES are built on steep slopes to create flat surfaces on which to grow crops. They remove the sloping land and so reduce soil erosion and surface water run-off. They require skill and hard work to build.

Intercropping
A variety of crops, mixed together in alternate rows or sown at different times, help to protect the soil from rain. The aim is to cover the whole area throughout the year.

Contour ploughing
Land should never be ploughed up and down a slope since this encourages erosion. Instead it should be ploughed along the contour lines, the plough lines then forming a natural barrier to water flowing down the slope.

Land reform
Farmers who do not have secure ownership of the land they farm are less likely to adopt soil conservation methods since these are often long-term projects involving a lot of work. Land should be owned by the people who farm it as they are more likely to look after it.

16 Traditional soil conservation methods used in Africa

17 Machakos women building terraces

So how were people in Machakos able to combat increasing drought and population growth? Rather than seeing themselves as part of the problem, the people of Machakos saw themselves as part of the solution.

They set up self-help groups to work on conservation projects to improve the land. They built terraces to reduce the problem of soil erosion and they also used other soil conservation techniques, like mulching and INTERCROPPING (see Figure 16 on page 26).

Since the mid-1980s an average of 1,000 kilometres of new hillside terraces have been built each year. As a result, crop yields have increased enough to feed the growing population and to provide a surplus that can be sold. More importantly, the people here have developed a farming system that seems to be environmentally sustainable.

Focus task

Work with a partner or small group. Look at Figure 18. It shows an imaginary area in an African country where desertification has become a problem. You have been appointed by the government as consultants to set up a project to conserve soil and water, and to restore the land to productive farming.

1 Assess the problems that the area faces. How has the environment been damaged? How have people contributed to the problem? Is there any evidence of ways in which people conserve the environment?
2 Using the information in this case study (pages 22–27), suggest how soil and water can be conserved and the land restored to productive farming. Your ideas may differ from one part of the area to another, depending on the type of damage. Explain why you would use these techniques.
3 Draw a labelled sketch to show what the environment would look like after your project had improved the environment.

★ THINK!

The case study of Machakos may not match your stereotype of Africa. We are used to seeing TV images of floods and droughts in Africa, with people appearing to be helpless victims. In Geography you need to be careful about these stereotypes. Of course there are floods and droughts in Africa and people are victims of hunger and starvation. But this is only part of the picture (although it happens to make dramatic news stories). There are also success stories in Africa that we don't usually see in the news, where people overcome problems rather than become victims. These successes, like the story of Machakos, may take a long time and involve hard work. So be careful that you present a balanced picture when you answer questions, and use factual case studies rather than stereotypes.

18 A desertified landscape

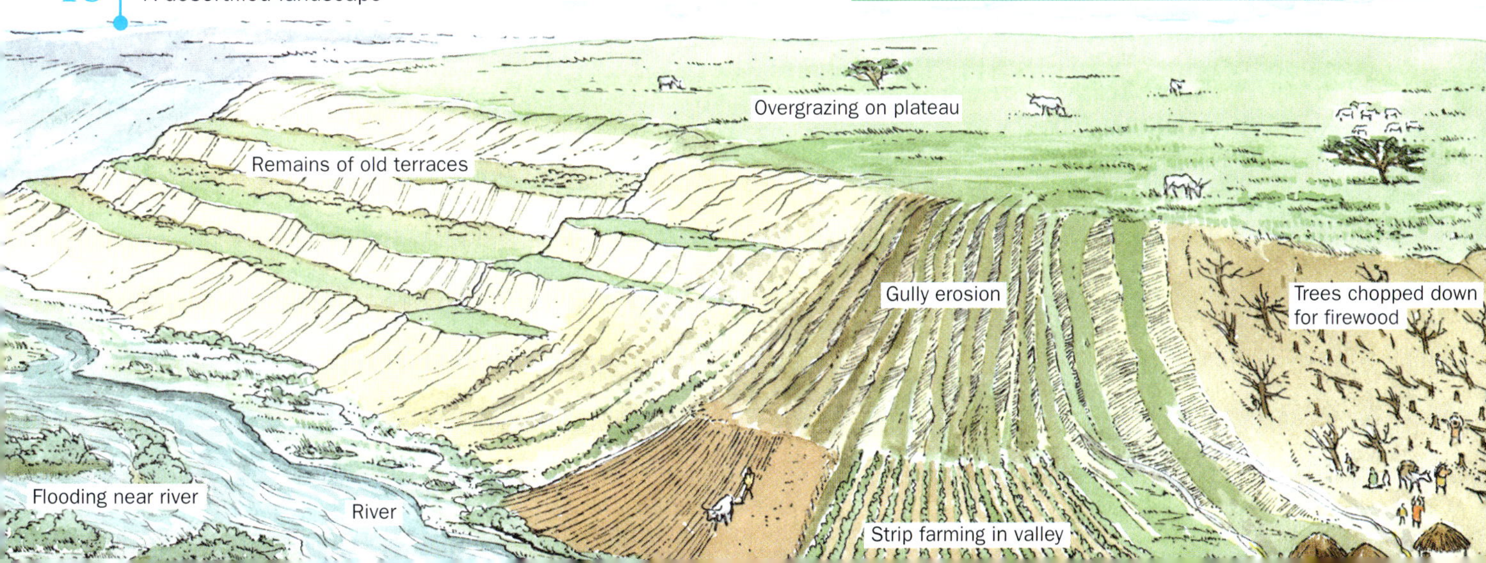

B CASE STUDY: CAN THE MALAYSIAN RAINFOREST BE MANAGED SUSTAINABLY?

Why is rainforest so important?

Tropical rainforest means different things to different people. To the tourist the rainforest might be where to go for the trip of a lifetime, to the scientist it is a living laboratory, to the hard-pressed government of a LESS ECONOMICALLY DEVELOPED COUNTRY (LEDC) it could be a valuable resource, and to the INDIGENOUS rainforest inhabitant it is home.

19 Rainforests in Malaysia

The tourist

It was another day in the rainforest. Overhead a sound like a cascading cutlery drawer indicated that something – a hornbill or an orang-utan – had collided painfully with a trellis of palm leaves. Somewhere to the right, something, possibly a bearded pig, was fidgeting through the liquorice-dark undergrowth. On the ground, scavenger ants the size of small spanners hopped over ground pock-marked by the tunnels of burrowing reptiles.

The thrill about a holiday in Borneo is that you are never quite sure what might be about to emerge from behind a strangler vine and scare you stupid, or possibly tear you limb from limb.

The scientist

Rainforests vary enormously over relatively small areas, particularly where there are changes in altitude, slope, bedrock, humidity and soil moisture. In Malaysia, the forest is dominated by numerous species of a single tree family, the dipterocarps (Photo 22). These huge trees distinguish South East Asian lowland rainforest from rainforests elsewhere in the world. The upper canopies of the trees are among the highest in the world – some reaching over 65 m. On average, the trees grow quite slowly, adding less than 1 cm to their diameter each year. Many local plants and animals are found only in this part of the world. Most species will only be represented by one or two individuals in each site. This means that any impact or event that reduces the size of habitat can cause the number of local species, and thus the biodiversity, to fall rapidly.

The government

Almost 60 per cent of Malaysia is covered by natural forests. A further 14 per cent of the land area is covered by rubber and oil palm plantations. This represents an area the size of the UK and makes Malaysia one of the most forested countries in the world. By comparison, only 10 per cent of the UK is forested. The Prime Minister of Malaysia has stated that Malaysia will ensure that at least 50 per cent of its land will remain permanently under forest cover. He has urged developed countries to join in the greening of the world.

Malaysia's policy of SUSTAINABLE FOREST MANAGEMENT distinguishes between areas of 'productive' forest and 'protection' forest. Two-thirds of the forest area is 'productive' forest where commercial logging is permitted. The remaining third is protection forest, managed to conserve biodiversity and to protect soil and water resources.

The rainforest inhabitant

We use the plants in the forest as remedies for a wide range of problems. There are 20 species that we use to treat stomach problems, 15 for skin problems, and 14 for childhood ailments. We can also treat leprosy, toothache and jaundice. Roots are most commonly used in traditional medicine, and leaves as well. In most cases only a single plant part is used, but in some cases several plants are used in combination. Most of our traditional herbalists are growing older and there is a danger that their knowledge of the forest will die with them.

Not many of us now make trips into the forest to collect food. We still eat jackfruit, durian and rambutan, which are easy to find. People do still farm, fish and hunt in the forest, but these are often part-time activities and mainly for older men and women. More people these days have money, and most of our food is bought from a shop.

20 Views on the Malaysian rainforest

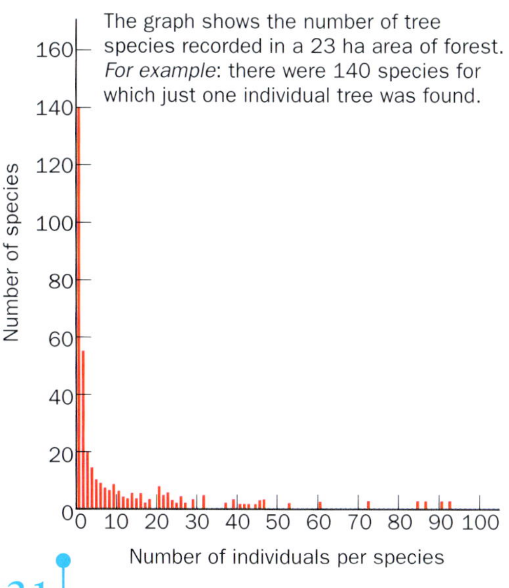
The graph shows the number of tree species recorded in a 23 ha area of forest. *For example*: there were 140 species for which just one individual tree was found.
Number of species
0
20
40
60
80
100
120
140
160
0 10 20 30 40 50 60 70 80 90 100
Number of individuals per species

21 Biodiversity in the Malaysian rainforest

22 Dipterocarp trees

23 An orang-utan, a species found only in the forests of South East Asia

Task

Fieldwork investigation: compare the biodiversity of the Malaysian rainforest with woodland in the UK

1 Carry out an investigation of a local woodland. Work with a small group under the supervision of your teacher. It is important that you never carry out fieldwork alone.

- **a)** Choose an area of woodland to investigate. It could be a small patch of woodland in a local park or part of a larger forest. If possible, choose an area of about one hectare (100 × 100 m). Divide the area of woodland into equal-sized 10 × 10 m squares. You can measure the area with a tape measure.
- **b)** Identify all the different tree species in your square. You can do this with the help of a tree identification book. Identification is easier if the trees still have their leaves.
- **c)** Count the number of individual trees belonging to each species in your square, and record your findings. Repeat this for more squares. Between all the groups in your class, try to record the species for 100 squares.
- **d)** Back in the classroom, combine your findings with those of all the other groups. Draw a graph, similar to Figure 21, to show the number of each species over the whole area. What area did you cover? How many species did you find? How many trees of each species were there?
- **e)** Write up your investigation. In your conclusion, compare your findings with Figure 21 (but note that this was based on an area of 23 hectares).
  - Which ecosystem has the greater biodiversity and why?
  - How could your findings influence the way that each ecosystem should be protected and managed?

SAVE AS . . .

2 Work in a group of four. Each person should read one of the speech bubbles in Figure 20.

- **a)** From your text, select up to five important facts about rainforests in Malaysia. Share them with your group.
- **b)** From all the facts your group has found, choose ten that you think are the most important from a geographical view. Put these facts into a sensible order, trying to link them together. Then write your own geographical account of rainforests in Malaysia.

★ THINK!

What is the point of doing your own investigation when you could just as easily get the information from a book? The answer is that there is no substitute for getting your feet muddy! You are much more likely to remember what you find out through investigation than if you just read about it. If an exam question asks you for a named example of an ecosystem you have studied, you will have one at your fingertips. (And the examiner, not having been there, probably won't be able to argue with your results!)

How is the Malaysian rainforest used?

Tropical rainforest is more than a collection of plant and animal species. When an area of rainforest is chopped down it is not just the local ecosystem that is lost. The impact is felt over a much wider area. The forest helps to enrich the soil and prevent soil erosion, it maintains a fresh water supply, and it absorbs carbon dioxide from the atmosphere, helping to keep the climate stable. If the forest is removed, the whole system is changed, with consequences that are hard to predict. These pages show some of the economic activities which impact on rainforest ecosystems.

Small-scale farming

Traditional farming practices, like shifting cultivation, are in decline in Malaysia. One reason is that as other uses of the forest, such as logging and plantations, increase, the area of land available for traditional farming is shrinking. Farmers used to farm a plot of land for two or three years until the soil lost its fertility. Then they would move on to a new area, leaving the soil to recover for many years. Now, they often farm the same land continuously, using chemicals to keep the soil fertile.

24 Small-scale farming in Malaysia

Logging

Timber is one of Malaysia's main exports. Four-fifths (80 per cent) of the country's timber is sold abroad. Malaysia has a policy of sustainable forest management in order to maintain this valuable resource. Commercial companies pay the government for the right to log areas of forest. They are only allowed to cut down trees with a trunk diameter over 50 cm. Smaller trees are left standing, and this helps the forest to regenerate more quickly. Forests are logged every 30 years. Malaysian rainforests are one of the most productive ecosystems in the world.

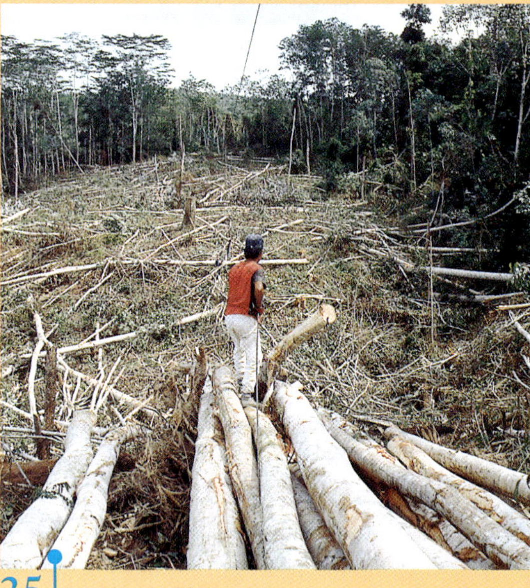

25 Logging in Malaysia

Key
Forest
Agricultural tree crops
Shrubs/secondary forest shifting cultivation
National park/wildlife sanctuary
Urban areas/other land use

0 100 km

THAILAND
Kuala Lumpur
N
SINGAPORE
BRUNEI
INDONESIA

26 Land use in Malaysia

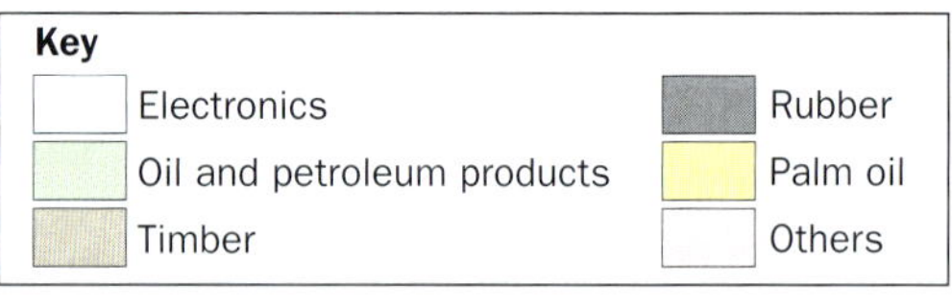

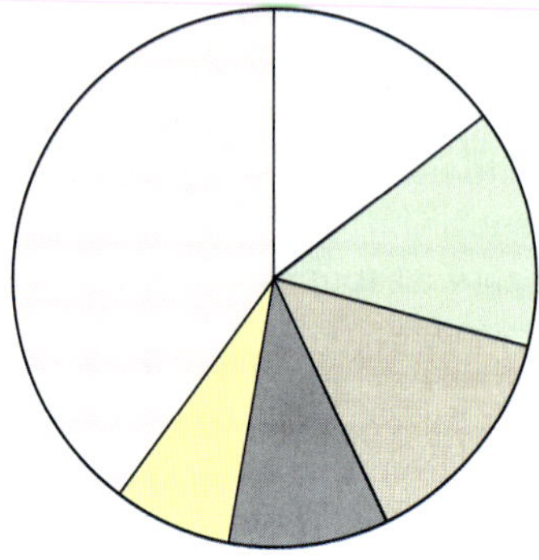

27 Malaysia's exports, 1980–92

Hydro-electric power

Malaysia has an ever-increasing demand for cheap energy to fuel industrial growth. HYDRO-ELECTRIC POWER (HEP) is a cheap (and clean) alternative to coal and oil for generating electricity. The densely forested valleys of Sarawak have some of the best sites for HEP schemes in Malaysia. Building dams across valleys creates huge RESERVOIRS that flood the forest and the homes of indigenous forest people.

28 Reservoir created by the Pergau Dam in Malaysia

Plantations

Large areas of forest have been cleared for use as PLANTATIONS (large areas producing a single tree crop). In Malaysia the two main plantation trees are oil palm and rubber. Plantations rely on chemical inputs such as fertilisers and pesticides and these find their way into the surrounding environment. While the trees are growing they provide little soil cover, so soil erosion is a problem. SILT may collect in rivers and cause flooding.

29 An oil palm plantation in Malaysia

Environmental impact assessment

Before a new development is allowed to happen, the planning authority – usually either the national government or the local planning authority – may ask for an ENVIRONMENTAL IMPACT ASSESSMENT (EIA) to be carried out. An EIA examines the impact that the proposed development would have on each part of the natural environment. An EIA can be used to identify the most suitable location for a new development or to compare the impact that different types of development would have on the same environment.

Tasks

1 Work with a partner.

a) Look at Photos 24, 25, 28 and 29 and read the information about each economic activity.

b) Complete an environmental impact assessment chart, like the one shown here, for each of the four activities. Discuss each option with your partner. Tick one box in each of the three main columns for each economic activity.

	Amount of change			Scale of change		Positive or negative changes	
Small-scale farming	No change	Some change	Complete change	Small-scale	Large-scale	Positive	Negative
River							
Soil							
Atmosphere							
Landscape							
Traditional lifestyle							
Vegetation							

2 Write a report for the Malaysian Government to recommend how the rainforest should be developed. Divide it into two sections.

a) Compare the environmental impact of the four activities including:

- the amount of change from each activity
- the scale of the change from each activity
- whether the changes would be positive or negative.

b) Consider the economic importance of the four activities. Figure 27 will help you. Which activities will help the government to achieve its goal of economic growth? How important should environmental considerations be? Balance the needs of environmental protection and economic growth to recommend how you would develop the rainforest.

Deforestation – a problem for Malaysia, a problem for us all?

Globally, each year an area of tropical rainforest the size of England and Wales is destroyed. Some of the highest rates of deforestation are found in the countries of South East Asia. However, the impact of deforestation goes beyond national boundaries. In 1997, forest fires in the Indonesian part of Borneo spread smoke over the whole region, causing an environmental and economic disaster. City residents were forced to wear masks, thousands of people suffered breathing-related illnesses, and tourist resorts had to be closed down.

On an even wider scale, global temperatures are rising at their fastest rate ever, and the frequency of natural disasters caused by climate change is on the increase. It may be no coincidence that these trends are happening at a time of rapid deforestation.

Task

1 Look at Figure 30. Use the information around the photo to explain how deforestation can lead to local flooding and to global climate change.

a) Divide a sheet of paper into 20 boxes. Write each statement from Figure 30 into a separate box.

b) Cut out the statements. Sort them into two groups – one related to local flooding and the other to global climate change. You may want to put some statements in both groups. If so, make extra copies.

c) Arrange the statements into two concept maps, both with deforestation at the centre. The maps should help to explain how deforestation can lead to **i)** local flooding and **ii)** global climate change. Draw arrows to show links between the statements. Stick the statements down to complete the maps. Give each one a title.

★ THINK!

Concept map

In Task 1 you are creating what is sometimes called a concept map. The aim is to help you hold the big picture in your mind. Everybody thinks differently, so the best concept map is the one you make yourself. It needs to show relationships and links in the way *you* see them. You will be making quite a lot of concept maps in this book.

30 The impact of deforestation on the local environment. But what will the impact be on the global environment?

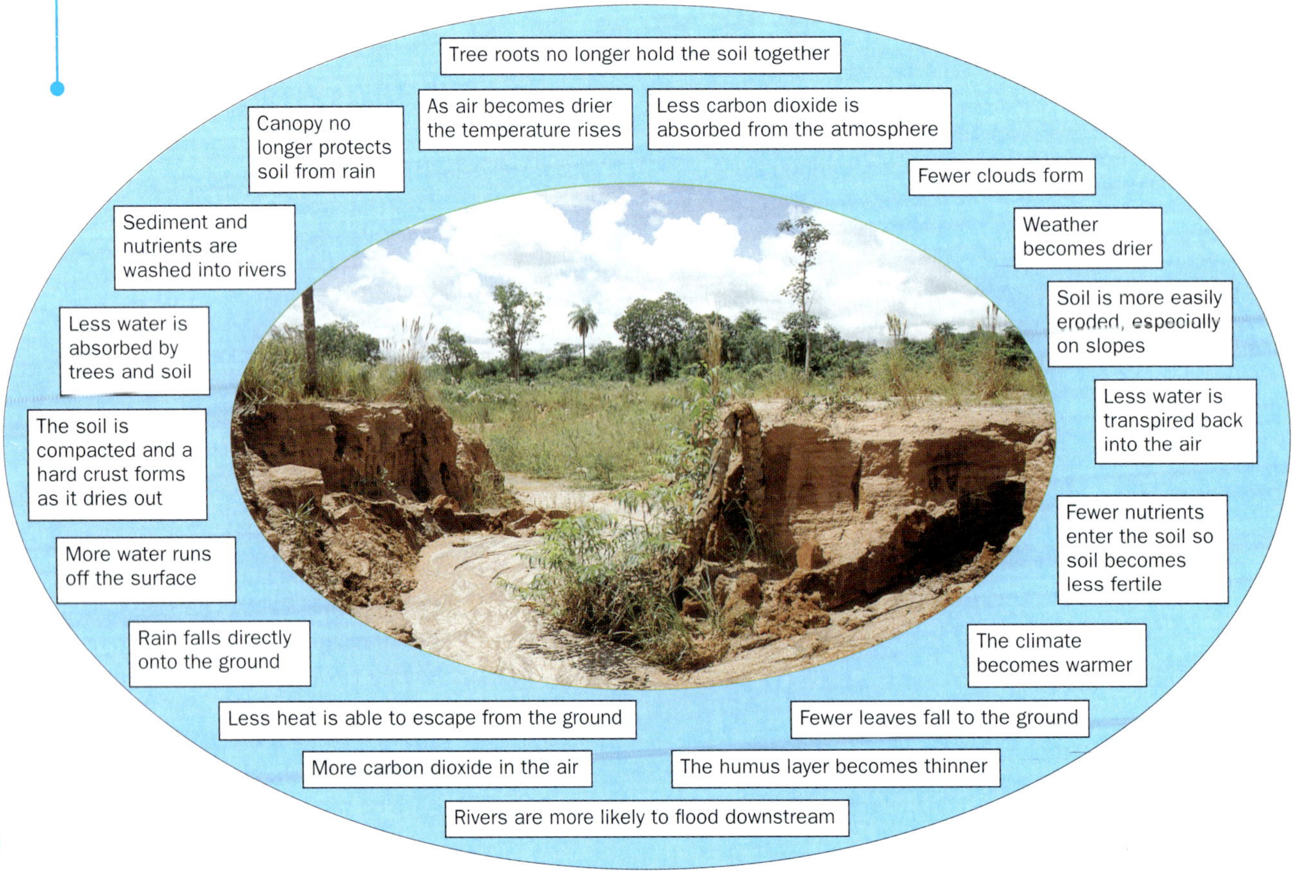

Tree roots no longer hold the soil together

Canopy no longer protects soil from rain

As air becomes drier the temperature rises

Less carbon dioxide is absorbed from the atmosphere

Fewer clouds form

Sediment and nutrients are washed into rivers

Weather becomes drier

Soil is more easily eroded, especially on slopes

Less water is absorbed by trees and soil

Less water is transpired back into the air

The soil is compacted and a hard crust forms as it dries out

Fewer nutrients enter the soil so soil becomes less fertile

More water runs off the surface

Rain falls directly onto the ground

The climate becomes warmer

Less heat is able to escape from the ground

Fewer leaves fall to the ground

More carbon dioxide in the air

The humus layer becomes thinner

Rivers are more likely to flood downstream

A Rainforest University

The Malaysian Government has set up a network of national parks and wildlife sanctuaries to protect their natural environment (see Map 26 on page 30). However, the traditional lifestyle of the indigenous rainforest people is also being threatened. Increasingly, people are leaving their communities to seek work in the cities. Sometimes this is by choice, but frequently they are forced from their ancestral land by logging, plantations and HEP schemes. Only the older people are left to continue the traditional lifestyle. As they die their intimate knowledge of the rainforest is lost.

There is now a proposal to establish a 'Rainforest University' among the Penan people of Mulu National Park. Here, the indigenous people would share their traditional skills and knowledge of the forest with visitors and, at the same time, benefit from Malaysia's economic development. Companies would pay the Penan to help their research into rainforest products, which could be used commercially. In this way, knowledge of the forest would be kept alive to benefit both indigenous people and the wider world.

31 A Penan inhabitant of the rainforest – knowledge of the forest may be lost when such older people die

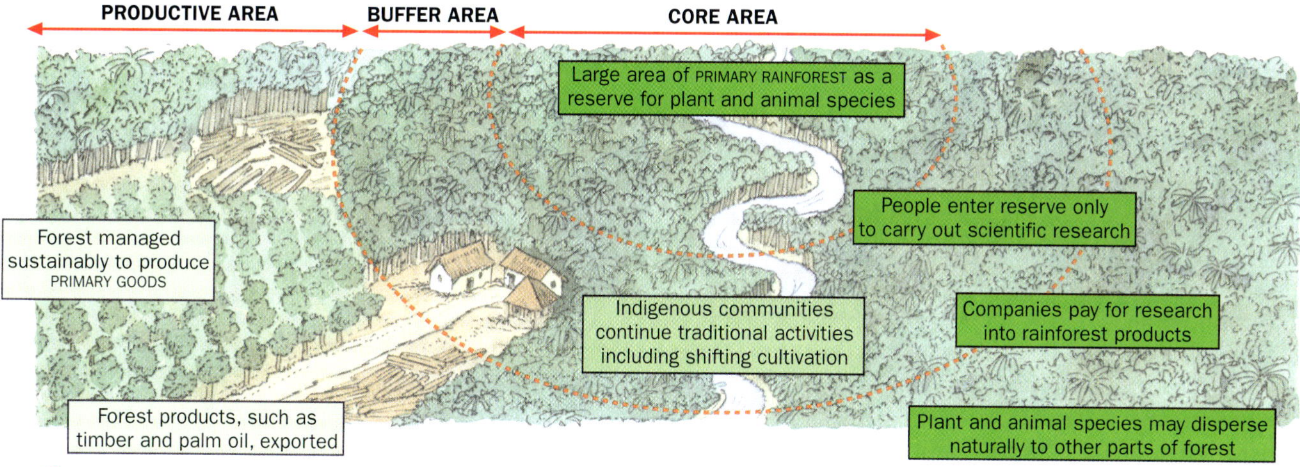
PRODUCTIVE AREA
BUFFER AREA
CORE AREA
Large area of PRIMARY RAINFOREST as a reserve for plant and animal species
People enter reserve only to carry out scientific research
Forest managed sustainably to produce PRIMARY GOODS
Indigenous communities continue traditional activities including shifting cultivation
Companies pay for research into rainforest products
Forest products, such as timber and palm oil, exported
Plant and animal species may disperse naturally to other parts of forest

32 Proposal for a Rainforest University to protect both the natural environment and its human inhabitants

Focus tasks

2 Work in a small group. Put yourselves in the role of the Penan people. You will work at the Rainforest University. You have to devise a Rainforest Curriculum for visitors to the rainforest.

Start by looking at a copy of your own school timetable.

a) Which subjects could you learn in the rainforest? How would you be able to teach them?

b) Which subjects would it be impossible to teach in the rainforest? Why?

c) Now think about subjects which are important to learn but which are not taught in your school (e.g. building or medicine). Which of those might you be able to teach in the rainforest? (Be as imaginative as you can.)

d) What exams or assessment would you set visitors at the end of their stay at the University?

e) Finally, explain how the Rainforest University would benefit:

- the rainforest ecosystem
- the Penan people
- visitors from the outside world.

3 Look back at the table that you completed for Task 3, page 21, listing reasons to be optimistic and pessimistic about the global environment. On balance, how do you feel about it now? Are you optimistic or pessimistic?

You could hold a class discussion or debate entitled 'Should we be worried about the global environment?'.

REVIEW CASE STUDY: HOW CAN THE CAPE PENINSULA BE PROTECTED?

The Cape Peninsula is a national park in South Africa. It has been proposed as a WORLD HERITAGE SITE – a status that would put it alongside some of the world's best-known landmarks, such as the Grand Canyon, the Great Wall of China and the Tower of London.

I come from South Africa, one of the most beautiful countries in the world. I live in Cape Town. Everyone who comes here says what an amazing city it is – we've got Table Mountain, stunning beaches and probably more flowers than anywhere else in the world. Every weekend our family goes somewhere different for a picnic. There are so many places to choose from, and all so close to the city. When I was little we weren't allowed to go to some of those places because we were black and the law in South Africa said we couldn't go to areas for white people. It's different now – we can go anywhere. But many people still don't know what beautiful places there are on our doorstep. We need to protect them for the future so that everyone can enjoy them.
I don't see what all the fuss is about. Sure, the Cape Peninsula is a great place to live and we've got some of the best surfing beaches. But there are so many beauty spots in South Africa – why do we need to protect this one? The real problem with Cape Town is lack of space. The city is built on a narrow strip of land between Table Mountain and the sea. TOWNSHIPS have spread over the Cape Flats – an area east of the city that is almost dry enough to be a desert. I've never been in a township but I'm told that they're dreadful places to live in, and more people arrive there each day. It seems crazy to protect all the empty land on the Cape Peninsula when there are people desperate for space.

In 1994 the years of APARTHEID that had divided black and white people in South Africa came to an end. This was an important event, not just for people but also for the environment. Until 1994 the main issue for most South Africans was justice. Because of the inequalities of apartheid, people did not have time to think about environmental issues. They were too busy struggling for decent health and education and the right to vote. With the arrival of democracy and black majority government, basic inequalities are now being tackled and the quality of life has started to improve.

For years, black South Africans were excluded from large parts of their country. They were banned from going to 'whites-only' beaches and forced to live in some of the most desolate areas. Now that they are able to travel freely, and enjoy places from which they were previously banned, people are taking more interest in environmental issues. Since the ending of apartheid, tourism to South Africa has also begun to grow. This has led to a wider appreciation of the country's natural beauty.

Finally, with the end of apartheid, South Africa was readmitted to the United Nations Organisation. The UN Educational, Scientific, and Cultural Organisation (UNESCO) grants World Heritage Site status to places that are recognised internationally for their outstanding value. Currently, there are no World Heritage Sites in South Africa. The country has applied for consideration to be given to the Cape Peninsula, which includes both Cape Town and Table Mountain.

★ FACTFILE

World Heritage Sites

World Heritage Sites are natural or cultural sites recognised internationally as being worth protecting for the benefit of future generations. Natural sites should fulfil some, or all, of the following criteria:

- An outstanding example of a major stage of the Earth's evolutionary history.
- An outstanding example of landforms or ecosystems.
- A unique, rare or superlative natural phenomenon or an area of exceptional natural beauty.
- An important example of a habitat where threatened species still survive.

The site should also meet the following conditions:

- It should have adequate long-term protection.
- It should include all the key elements and processes of the natural environment.
- It should be of sufficient size to be sustainable.

Once a site has been designated as a World Heritage Site, the country within which it is located must ensure that the site is protected, and must take responsibility for it on behalf of the international community.

33 The Cape Peninsula, where Cape Town is built, includes the world-famous landmark of Table Mountain

34 Table Mountain, with Cape Town in the foreground

35 Springtime on the Cape Peninsula: the Cape is home to over 2,000 species of flowering plants, giving it the highest species density of any region in the world.

Task

1 On page 37 you will prepare a plan for protecting the Cape. But first, look nearer to home. Work with a partner. You have to make a case for greater environmental protection for a site in your local area. Choose a site that you think has natural or cultural value and that ought to be protected.

a) Find out more about the site you have chosen. You may be able to arrange a visit. You could interview visitors and/or a manager of the site.

- Is it a natural or cultural site? Why is it important?
- What landforms, habitats, species or buildings need to be protected?
- How is it already protected? For example, is it part of a national park or an urban conservation area?
- Does it need further protection?

b) Prepare a case for protection of the site, using the information that you have found. Present your case to the rest of your class. Each pair will put forward a case for their own site. The class could vote on whether each site should be protected or not.

Why is the Cape under pressure?

The Cape Peninsula demonstrates on a small scale many of the pressures that affect the global environment.

Urban growth

Cape Town's current population, which is 2.5 million, is expected to grow to 6.2 million by the year 2020. This is due mainly to large numbers of rural South Africans migrating to the city to find work. Land around the edge of the peninsula is already fully developed. Most growth is likely to be in the townships on the Cape Flats to the east of the city. Many of these areas are informal settlements where people build their own homes on whatever vacant ground they can find. More homes need to be built in other areas if the peninsula is to be protected from developments like the one below.

Increasing demand for water

Cape Town has a Mediterranean climate. This means that summers are usually dry and annual rainfall is quite low – about 600 mm per year. With an increasing population the demand for water is growing at about 5 per cent each year. At this rate it will soon outstrip supply. Many township homes still lack running water, and government programmes to provide water will further increase demand. Currently the city depends on water from the local mountain catchment area, but this is unlikely to be able to provide sufficient water in the future.

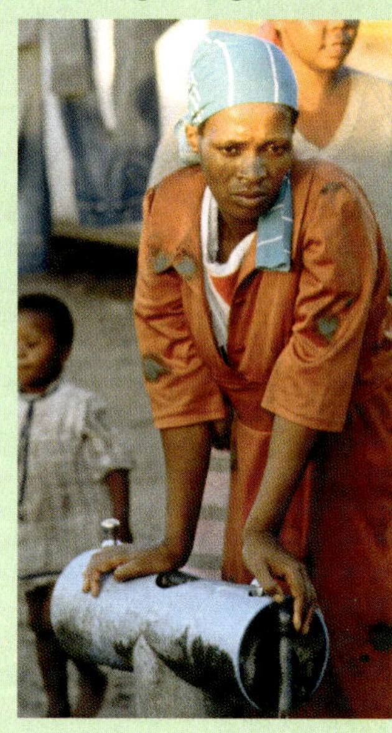

Tourism

Tourist pressure on the Cape Peninsula is growing with the increased number of visitors to South Africa, as well as greater movement of people within the country. Tourists need hotels, campsites and recreation facilities such as golf courses. All these put extra pressure on the limited space and on the water supply. Cape Town and Table Mountain are the main attractions, but many people also visit other parts of the peninsula. This can have a direct impact on the fragile ecosystem, mainly through trampling. Tourism in these areas needs to be carefully managed.

Alien species

It is not just the human invasion that is a problem for the Cape Peninsula. Alien plant species, originally brought to South Africa by travellers from Europe and Australia, have also invaded the Cape and are rapidly replacing the local plant species. There is a danger that rare species found nowhere else in the world may be lost for ever. Often the new species, particularly the fast-growing pines, eucalyptus or willow trees, are much larger than the plants they have replaced. They take up more water through their roots and this reduces the supply of groundwater that reaches rivers and reservoirs.

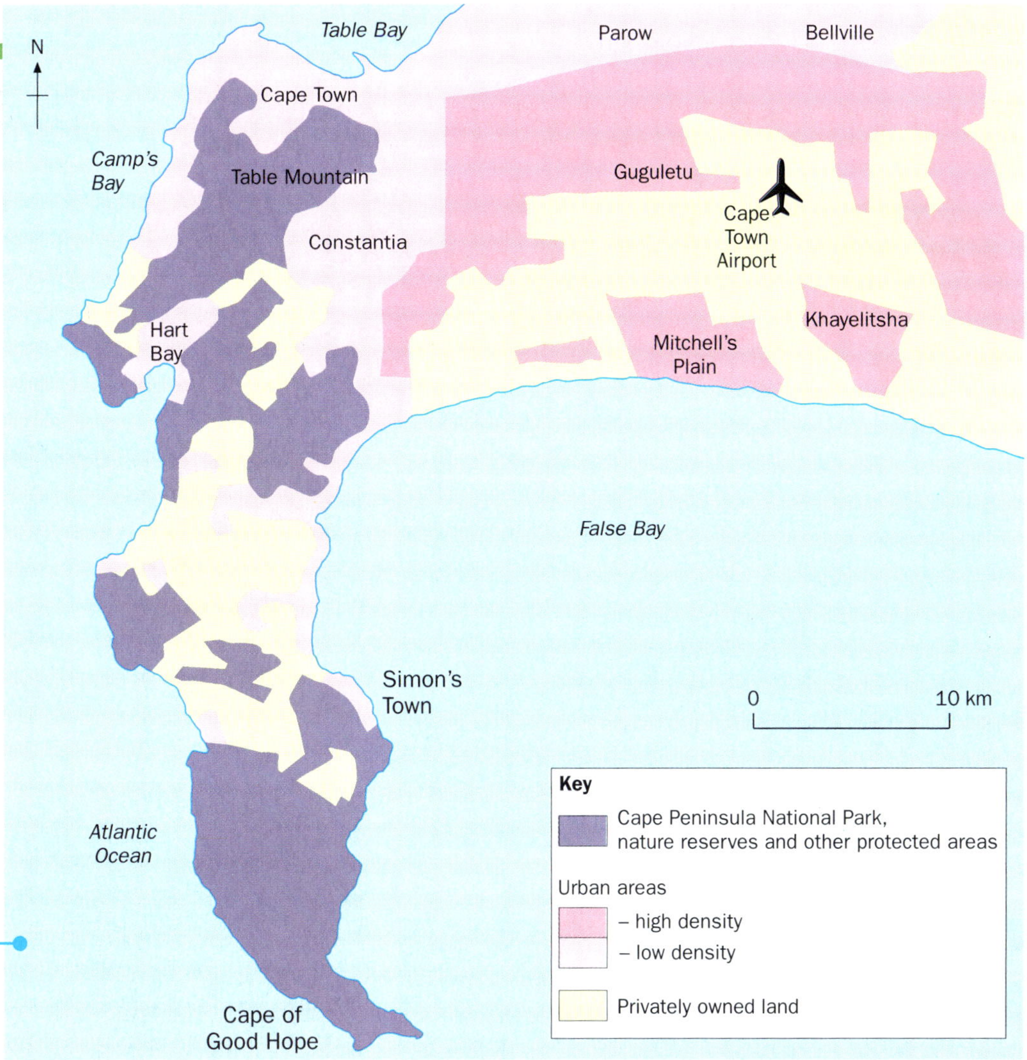

N
Table Bay
Parow
Bellville
Cape Town
Camp's Bay
Table Mountain
Guguletu
Cape Town Airport
Constantia
Hart Bay
Khayelitsha
Mitchell's Plain
False Bay
Simon's Town
0
10 km
Key
Cape Peninsula National Park, nature reserves and other protected areas
Urban areas
– high density
– low density
Privately owned land
Atlantic Ocean
Cape of Good Hope

36 Land use around Cape Town and the Cape Peninsula

Focus tasks

At the Cape, a unique fragile environment and growing population have to compete within the confined space of the peninsula.

1 Devise a sustainable strategy to protect the Cape.

a) Look at Map 36. Describe the location, size and shape of the Cape Peninsula's protected natural environment. What advantages and/or disadvantages might this have for the protection of the natural environment? On a copy of the map, show how you would protect the environment.

Highlight areas that could be *productive*, *buffer* and *core zones* (see Figure 32 on page 33).

b) Draw a diagram, like the one below, to show the main components of the Cape Peninsula system.

Urban growth
Water supply
NATURAL ENVIRONMENT
Tourism
Alien species

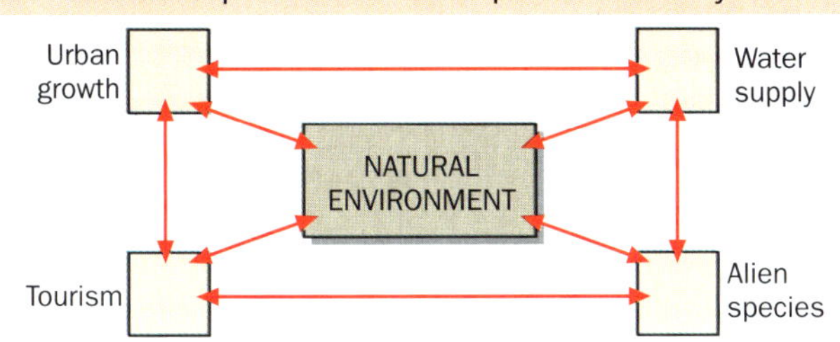

- Label the arrows to describe the links between components in the diagram. For example, on the arrow between urban growth and water supply you could write: 'Urban growth is leading to a 5 per cent increase in demand for water each year'.
- Explain why the way that people use the Cape Peninsula is unsustainable.

2 Using all the information on pages 34–37 produce a presentation to argue the case for including the Cape Peninsula as a World Heritage Site. In your presentation show how the area meets all the requirements of World Heritage Site status. Your presentation could use ICT and may include photos that you could obtain from websites such as:

City of Cape Town: www.capetown.gov.za
South African Tourism site: www.cape-town.org

Why protect the environment?

We have to protect our environment to make sure it is sustainable. It doesn't matter whether it's on our own doorstep – like the Cape Peninsula is for us – or part of the wider global environment – like tropical rainforest. The arguments are the same. People depend on the natural environment, and not just because it looks nice. If we chop down trees or pollute the air we're destroying the natural system, making it unsustainable – and we are part of that system. We need to protect the environment if we want to survive.

37 The South Downs, England

39 Machakos, Kenya

38 Rainforest, Malaysia

40 Cape Peninsula, South Africa

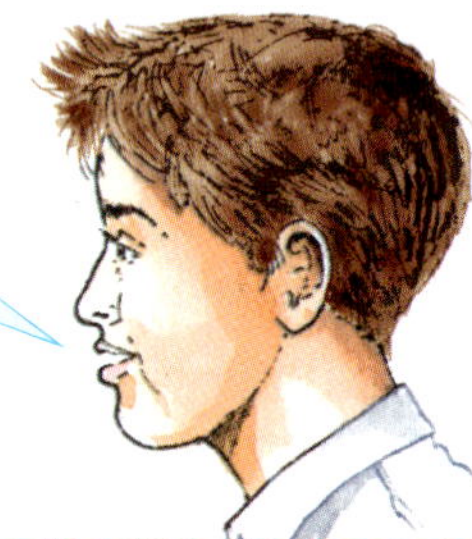

OK – now you've convinced me. But how do we convince the people who really matter – the governments and large companies that take the decisions? And, come to that, how do we convince ordinary farmers and people who use the environment just to survive?

Focus task

ICT

1 You are going to produce *either* a short ICT presentation *or* a magazine article entitled: *Why protect the environment?*

You have to persuade your audience that it is in all of our interests to protect the environment. Don't just use emotional arguments – 'Wouldn't it be awful if ...'. Think about what you have learned in this unit. List all the reasons why we should protect the environment. Tell people why the issue of sustainability is so important, and what will happen if we don't protect the environment.

Remember, you have to persuade people like farmers, company directors and prime ministers, who make decisions that affect the environment, so your arguments need to be convincing. The more that you can refer to real places and real cases, the better. Think about the case studies that you have explored in this unit, or in your local area. Make sure that your conclusion spells out the importance of environmental protection.

2 Natural Hazards

How do natural disasters affect our lives?

Earthquake measuring 7 on the Richter scale rips Indian town apart. At least 5,000 feared dead.

Flooding in UK – worst since records began

Hurricane Hannah in Belize – local people plead for help as floodwaters rise

Natural disasters affect us all, and can take us all by surprise. Our ability to survive on a restless and sometimes unpredictable planet may have more to do with the level of economic development of the country we live in than with the hazard itself.

2.1 Why do LEDCs suffer more than MEDCs from tropical storms?

DO YOU KNOW?

- HURRICANE comes from the word 'Hurican', the Carib god of evil.
- TROPICAL STORMS are called hurricanes when they reach wind speeds of more than 119 km/h.
- A mature hurricane produces energy equivalent to about half of the worldwide electricity-generating capacity.
- A STORM SURGE for a tropical storm can be from 6 to 10 m high, caused by winds pushing up the ocean surface ahead of the storm.

AIMS

- **To discover how tropical storms are formed.**
- **To find out how tropical storms affect people's lives.**
- **To investigate the reasons for differences in the impact of storms in a more economically developed country (MEDC) and a less economically developed country (LEDC).**
- **To consider the role of aid in helping people to recover from storms.**

Case study: Hurricane Mitch

As Hurricane Mitch approached the Caribbean coast of Central America at the end of October 1998, no one was aware of the death and destruction that it would cause. What followed led to the devastation of the impoverished countries of Honduras and Nicaragua, and a worldwide effort to help rebuild their shattered economies.

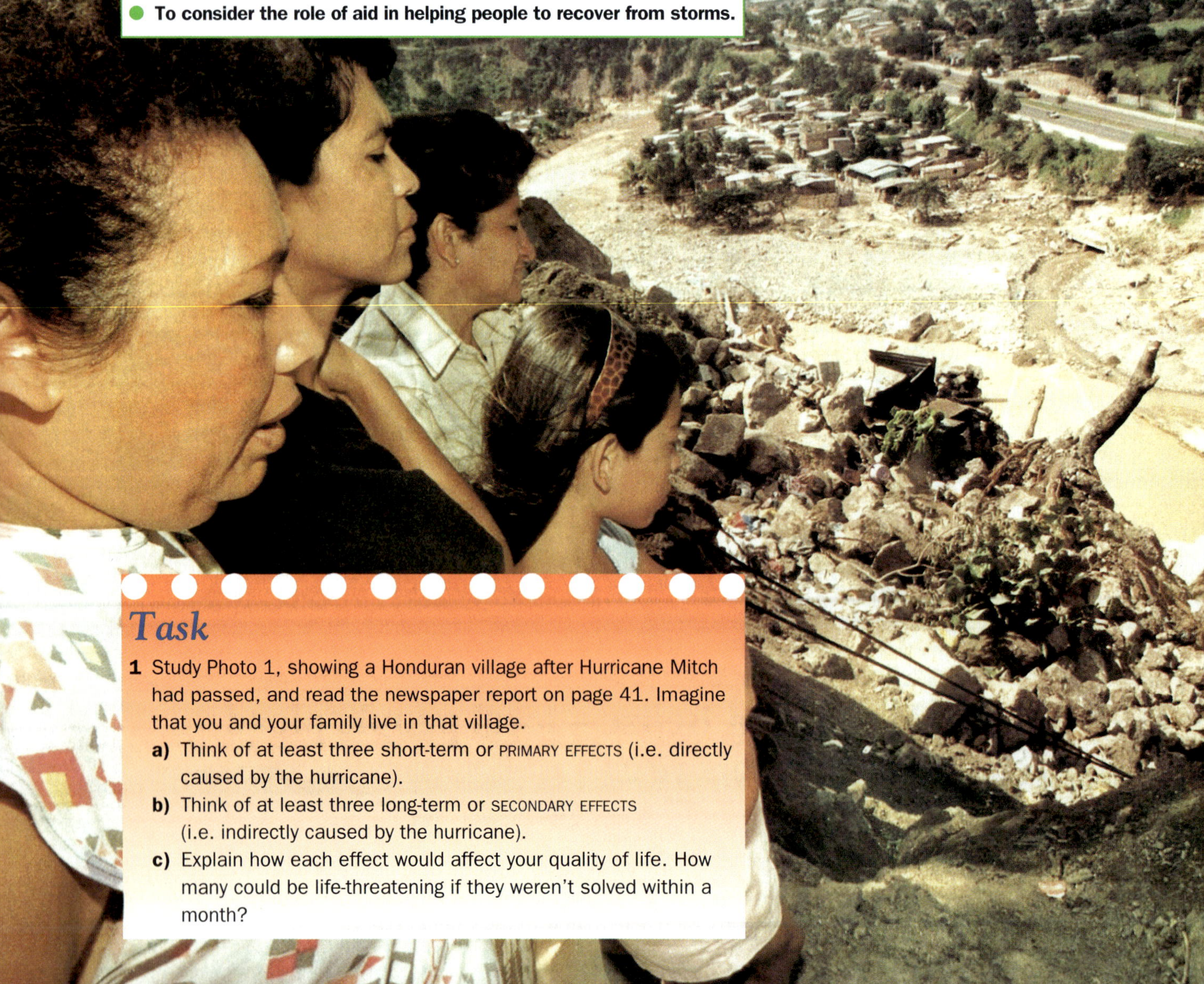

1 The effect of Hurricane Mitch in Honduras

Task

1 Study Photo 1, showing a Honduran village after Hurricane Mitch had passed, and read the newspaper report on page 41. Imagine that you and your family live in that village.

  a) Think of at least three short-term or PRIMARY EFFECTS (i.e. directly caused by the hurricane).

  b) Think of at least three long-term or SECONDARY EFFECTS (i.e. indirectly caused by the hurricane).

  c) Explain how each effect would affect your quality of life. How many could be life-threatening if they weren't solved within a month?

2 Central America – the area affected by Hurricane Mitch

3 Satellite photo of Hurricane Mitch heading for Central America

Hurricane Mitch roars in

The death-toll from the worst storm to hit Central America in the 20th century seemed likely to exceed 7,000 yesterday.

The director of the Honduran national emergency committee, Dimas Alonzo, said floods and landslides caused by tropical storm Mitch may have cost as many as 5,000 lives in Honduras alone. The Honduran president appealed for international aid and announced he was introducing measures to combat looting.

'There are corpses everywhere,' he said. 'The floods and landslides erased many villages and whole neighbourhoods from the map ... I ask the international community for help.'

The swollen Rio Choluteca has turned Tegucigalpa city centre (the capital of Honduras) into a vast lake, while the hillsides are strewn with the wreckage of shanty homes. In all, 800,000 of the country's 5 million inhabitants are reported homeless.

A dam across the Rio Choluteca, caused by a landslide, is now threatening to burst and add to the devastation in the city, where looting is rife.

In Nicaragua, rescue workers continued to pull bodies from the black volcanic mud at the scene of one of the worst disasters in the country's history.

1,000 to 1,500 people have been killed at the Casita volcano, north-west of Managua. Swollen by torrential rains caused by Mitch, the crater lake at the volcano's summit overflowed, witnesses say, causing a mudslide that wiped out communities. The mud, in places 20 ft deep, covers about 30 square miles.

'It is a giant cemetery, the inhabitants were buried; there were very few survivors', one rescue worker told us.

The Nicaraguan president, Arnoldo Aleman, declared the devastation to be the 'worst in living memory. These difficult times have brought mourning to every Nicaraguan home.'

Up to 50 bridges on main highways, along with many minor bridges, have been destroyed. These include all those on main roads in and out of the Nicaraguan capital Managua.

Along with other governments in the region, Nicaragua is now requesting foreign aid including emergency food. Honduras and Nicaragua, the two countries worst affected, are the poorest in the Americas after Haiti.

With hundreds of bodies rotting in the open air, and water supplies and other utilities disrupted, another fear is of epidemics, including malaria and cholera.

Postscript – The final death-toll was estimated at more than 11,000. Records were too poor to give a more precise figure.

4 Adapted from a newspaper report on Hurricane Mitch

★ **THINK!**

Ask your own questions!

We get used to answering questions posed by other people but it is not always the best way to investigate a topic. Asking your own questions can be better.

- You are the person most likely to ask questions that capture your interest, which will help you to fill in the gaps in your own knowledge.
- You are more likely to remember things that you find out for yourself.
- Searching for relevant information helps to develop your research skills.
- Even if you find it impossible to answer your own questions, that is helpful! It will help you to refine your enquiries, learn from earlier mistakes and become a better investigator in the future.

Tasks

Investigating Hurricane Mitch

1 You are going to start with your own questions.

What questions do you want to ask about Hurricane Mitch?

A good starting point is: 'Who? Where? Why? What? When?'

We call it the '5Ws' approach.

a) Try to think of at least three questions beginning with each W. Some questions have already been written to help you get started.

b) Record them in a table like this:

	Questions	Answers
Who?	1. *Who was affected by the hurricane?* 2. 3.	
Where?	1. *Where did Mitch hit land?* 2. 3.	
Why?	1. *Why were communications damaged?* 2. 3.	
What?	1. *What caused most of the deaths?* 2. 3.	
When?	1. *When did the hurricane happen?* 2. 3.	

c) Read through the information on pages 40 and 41 and try to answer the questions that you have posed in your 5Ws table.

d) With a partner, group or class, pool your findings. Note any similarities between the questions that you each asked and the answers that you arrived at. You will need them for Task 2.

2 Now you are going to begin an investigation: 'Why did so many people die as a result of Hurricane Mitch?'

The answer may sound obvious – because it was a bad storm – but think again. When similar storms hit the USA that summer hardly anyone died. So it is interesting to ask: 'Why did so many die this time?' Look carefully for things that are specific to this storm, or specific to Honduras and Nicaragua. Work like this:

a) Look back at your questions and answers to Task 1. Are any of them going to be useful for this enquiry?

b) Consider what additional questions you need to ask. You will return to the questions at the end of the chapter and see how many of them you can answer.

ICT

3 You are a reporter for an international newspaper. You are covering the impact of Hurricane Mitch on Honduras. Your newspaper wants a photograph for the front page.

You have to e-mail a photo and a description to your head office in London. The description must not be more than 100 words, so you need to choose your words carefully.

a) Use the internet to find a suitable photo. Try this website: www.hurricanemitch.com

b) Paste the photo into a word-processing package such as Word and write your description. Then attach it to an e-mail and send it to the address your teacher will give you. Good luck! Your editor will not give you another assignment unless you get this one right.

Focus task

Why was Cornelio Guevara jailed in Tijuana?

4 Cornelio Guevara, a Honduran from the coastal town of Choluteca, was jailed in Tijuana, Mexico on 1 December 1998 for 'loitering'. He had been picked up on the outskirts of town. He would not say why he was loitering. You need to solve this mystery to explain why.

a) Read the statements in the boxes below. Your teacher may give you a copy of these. If so, cut out the statements and lay them out on your desk.

b) To help you to understand them, it is useful to group or categorise them under headings. You can use any headings you like, but simple ones that cover several facts are often the most useful. Between three and five headings is about right. For example: Impact of Hurricane Mitch, Cornelio, Choluteca, Honduras economy, US immigration. You should find Map 2 on page 41 useful.

c) When you have worked out why Cornelio Guevara was jailed, write up his story. It should begin with the earliest events and end with his jailing.

mystery statements

1 The volcanic soils around the Choluteca River were easily washed into it.

2 Hurricane Mitch hit Honduras on 29/10/98 and brought torrential rain for over seven days.

3 Honduras owes other countries US$4.45 billion. It needs US$1.5 million/day to pay the interest on its loans.

4 After Hurricane Mitch, aid was slow to arrive in Choluteca.

5 Cornelio had invested his life savings to buy his boat. He could just meet the repayments.

6 Tijuana is a Mexican town on the border with the USA.

7 The Choluteca River flooded as soon as the storm arrived.

8 It is illegal to cross into the USA unless you have a passport and/or a visa. Cornelio did not have a visa.

9 The banana and melon (export crops) farms near Choluteca were destroyed. 50 per cent of workers lost their jobs.

10 The damage to Honduras is estimated at US$2.7 billion.

11 Cornelio has relatives in San Diego in the USA.

12 Floods smashed barrels of pesticides, contaminating the rivers and fish stocks.

13 Cornelio's prawn boat was washed out to sea and his nets were ruined.

14 A 300 m section of the Pan-American highway at Choluteca bridge was washed away.

15 The Mexican police arrest anyone they suspect is trying to cross illegally from Mexico into the USA.

16 Prawns, coffee and sugar are all export crops for Honduras.

17 Cornelio spent three days in the Choluteca River before he was rescued.

18 Choluteca is a thriving town in a fertile farming area on the Choluteca River, near the west coast of Honduras.

19 Emergency services found it hard to get to the survivors.

20 Cornelio lived in a tent for two months after the flood.

★ FACTFILE

Tropical storms

Hurricane Mitch was a tropical storm. Tropical storms are areas of intense low pressure. They occur in tropical oceans. They are called by different names in different parts of the world: hurricanes, CYCLONES and TYPHOONS (see Map 5).

They are formed when prolonged heating by the Sun (solar radiation) raises the surface temperature of the sea to 27 °C. The warm seawater heats the air lying immediately above it to form an unstable warm air mass. This is less dense than the surrounding air, so it rises and produces a LOW PRESSURE zone. This pulls in cooler air (at surface level) from HIGH PRESSURE zones nearby, creating wind. The greater the pressure difference, the faster the winds travel.

The winds move in an anticlockwise direction in the northern hemisphere, but in a clockwise direction in storms in the southern hemisphere.

Rapidly rising moist air cools. Water vapour in the air condenses to form masses of dense cloud and heavy rain. The winds and rain increase in intensity the nearer they are to the centre of the storm. However, right at the centre of the storm is the EYE. Within the eye, descending air warms and gives clear skies and no wind, although the seas remain very violent. The eye averages 24 km in diameter but gets smaller as the hurricane speed increases.

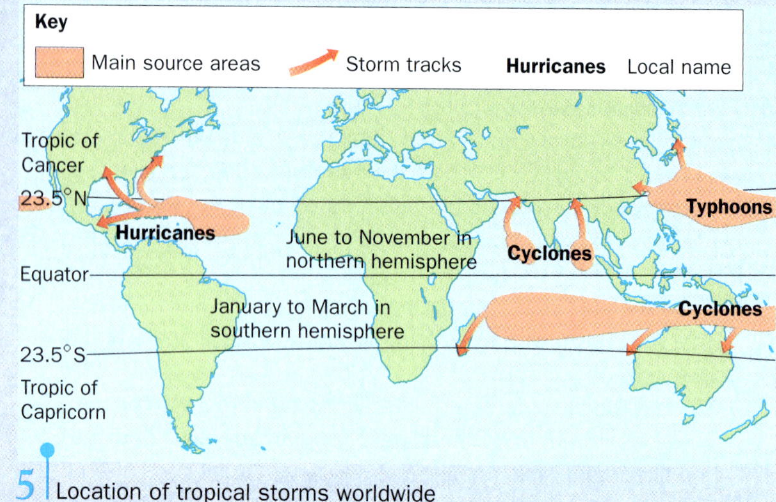

5 Location of tropical storms worldwide

The path of a tropical storm

Map 6 shows the path of Hurricane Mitch in detail. In an average year, more than ten hurricanes form over the Atlantic and usually head west towards the coasts of Central America and southern USA. They increase in speed as energy and moisture are transferred from the warm sea by evaporation. As storms approach the coast their paths become difficult to predict as pressure systems over the land start to affect the direction of the storm.

Hurricanes get less violent as they lose their energy source, the warm tropical sea. Then the hurricane is reduced to the status of a storm and eventually to a DEPRESSION.

The dangers of tropical storms

The main problems caused by tropical storms such as Hurricane Mitch are the intense winds that can gust to over 280 km/h, destroying houses and communications. These winds also drive the sea forward, 'piling up' the waves. The low pressure at the eye of the storm can cause the sea surface to rise up, leading to storm surges that can overwhelm coastal defences. Torrential rain feeds already swollen rivers, causing flooding, and surface run-off starts landslides.

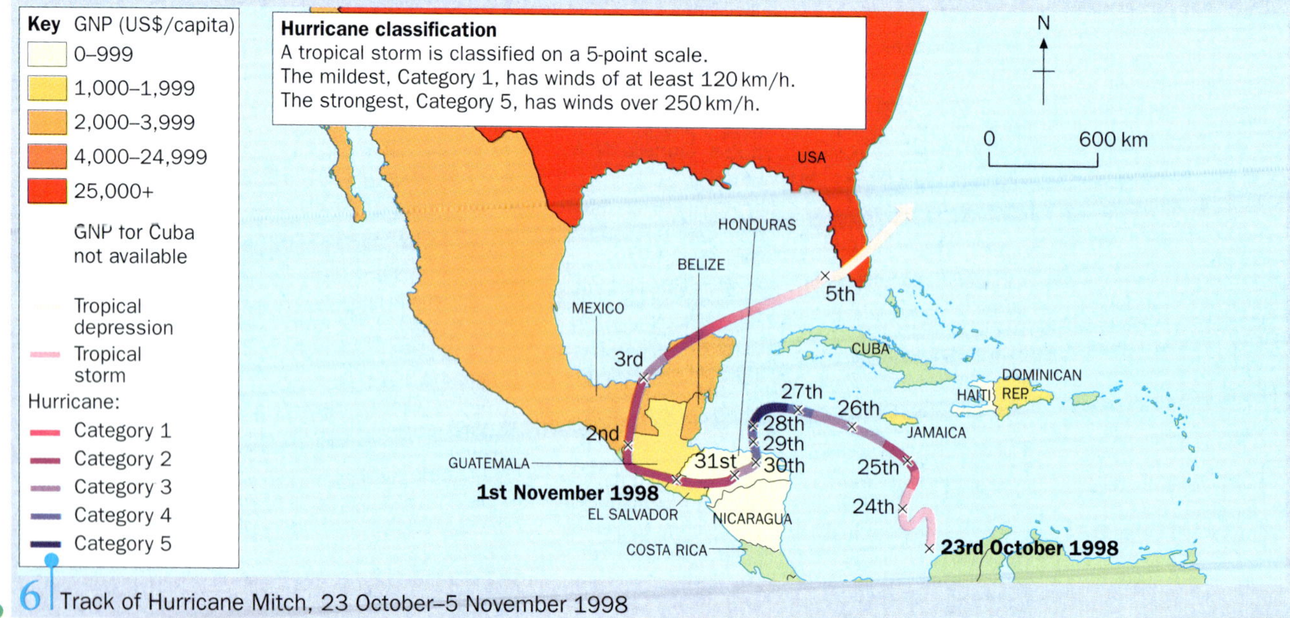

6 Track of Hurricane Mitch, 23 October–5 November 1998

Tasks

1 Using the information on pages 40 to 44, construct a timeline to summarise the events and effects of Hurricane Mitch. On one side of the line show events, on the other side show effects.

Start with the earliest information on the formation of the hurricane. Continue to the events after the hurricane had passed. You may need to estimate dates and, to keep it simple, group some effects.

2 Discuss: Did the most damage occur during the hurricane, or after it had passed?

SAVE AS . . .

3 Write a paragraph or draw a series of labelled sketches to explain the origins and tracks of tropical storms. Include the following key words and phrases:

solar radiation, late summer, 27 °C, warm air, uplift, low pressure, evaporation, anticlockwise winds, westwards.

4 Draw a Venn diagram like the one below, to compare the experiences of Honduras (Hurricane Mitch) and Texas, USA (Hurricane Gilbert). The diagram has been started for you.

Experiences unique to Honduras	Common to both	Experiences unique to Texas, USA
Mudslides	*Businesses lost money*	*Cities evacuated*

5 a) Discuss: What are the most important differences between the two countries' experiences?

b) Suggest reasons for these differences.

SAVE AS . . .

6 a) Draw a simple copy of Figure 7.

b) Categorise the information for Texas into Prediction and Precaution.

c) Write three sentences about which of these you would expect to find wholly or partially implemented in Honduras.

How do MEDCs defend themselves against tropical storms?

Predicting the exact path and speed of tropical storms is very difficult and expensive. Only rich countries can afford all the technology needed to do it.

The USA, for example, has a co-ordinated system for tracking hurricanes. Radar, aircraft, sea-based recording devices, weather satellites and other devices supply data to the National Hurricane Center in Florida. Data are collected on wind speed and direction, the location and size of the eye, air pressure and temperature within the storm. With this information, both very precise tracking of every hurricane's path and more reliable predictions of what the storm will do next are possible.

With adequate warning, communities can take precautions against storm damage. When Hurricane Gilbert (1988 – category 5) hit the Mexico/US border near Monterrey, large sections of the Texan coastline had been evacuated at an estimated cost of $30 million for 300 km of coastline. A combination of well-established evacuation routes, good roads, storm wardens who visit people, rest centres away from the most vulnerable places, high levels of private car ownership and excellent communications ensured a safe exodus for the majority of the population. The inconvenience and loss of business was enormous, but a very low death-toll (3 killed in the USA) was astounding considering that Gilbert was the strongest hurricane to hit the western hemisphere in the 20th century.

The USA can also afford longer-term precautions. To reduce the risk of flooding and damage from future storms, seawalls have been built in coastal towns. Many houses are built on stilts so that floodwaters can pass below them.

In LEDCs, despite access to some of the information from technology, the situation is very different, as you have seen in Honduras.

7 Hurricane prediction and precaution, in Texas, USA

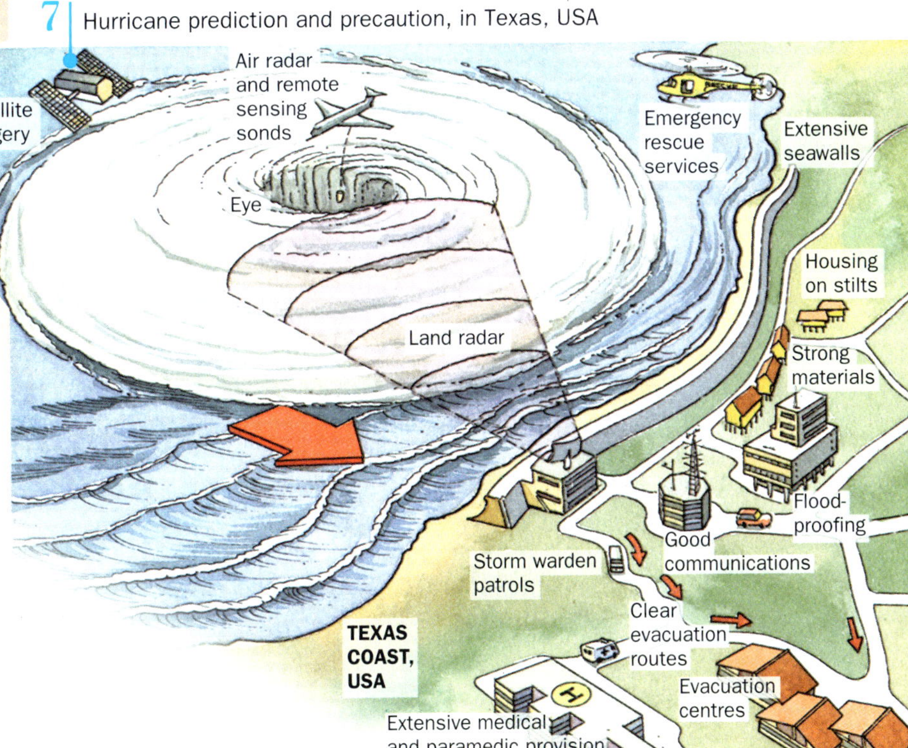

Why does poverty make hurricane danger worse?

How much damage there is and how well a country recovers from the effects of a hurricane depends on many factors (Figure 8) and many are interrelated. But by far the biggest factor is the wealth or poverty of the country.

- **Before a storm:** a poor country will be less able to afford to take precautions to protect people and property; less able to invest in the technology to predict storms; less able to inform its citizens, many of whom live in remote villages, about the impending disaster; and poor infrastructure, with low vehicle ownership and unsurfaced roads, means that its citizens will be less able to take action to defend themselves even if they know a storm is coming.

- **After a storm:** a poor country is less able to rebuild its damaged infrastructure, industry or agriculture; and less able to train emergency services. If it borrows money to do all these things it risks sinking deeper into debt.

FACTORS AFFECTING DAMAGE AND RECOVERY

(a) CLIMATE, VEGETATION AND TERRAIN

Honduras	USA (Texas)
• Rainfall heavy on humid northern coast (2,500 mm, summer high). • Vegetation ranges from tropical mangrove swamps on coast to pine forests in interior. • Main central plateau is cut by deep valleys and has volcanic mountain ranges (2,800+ m).	• Humid subtropical climate zone near coast (1,300 mm, summer high). • Dense forests in the east to deserts in the south-west – mainly productive farming land. • West Gulf coastal plain: this is a low, flat area ranging in height from sea-level to about 90 m.

(b) POPULATION AND ECONOMY

	Honduras	USA
GNP (US$/capita)	600	26,980
Population density/km^2	40	26
Population growth (% pa)	3.6	1.0
Life expectancy (years)	64	75
Urban population (%)	42	74
Population/doctor	1,510	470

COMMUNICATIONS/INFRASTRUCTURE

Honduras	USA
Railways – 620 km Airports – 30 local Roads – 18,500 km; 12% paved, incl. PanAmerican Highway (160 km in Honduras)	Railways – 283,099 km Airports – 5,100 Public roads – 68,449 km of National Interstate Highways

EMERGENCY SERVICES

Honduras	USA
• Limited emergency facilities with army and police providing main back-up. • Volunteer help essential and aid organisations important.	• Well-equipped emergency services. • Emergency plans and contingency funds (e.g. California $1 billion fund). • National Guard.

STORM FORCE

Honduras (Mitch)	USA (Gilbert)
4th strongest to hit Caribbean in 20th century. Winds up to 280 km/h.	Strongest to hit Caribbean in 20th century – category 5. Winds up to 360 km/h.

8 Factors affecting damage and recovery from hurricanes: Honduras and the USA

Tasks

1 a) Study Figure 8. Make your own simplified version using just the (blue) boxed headings around the edge of a piece of paper. Make sure you allow plenty of room to add labels to your diagram.

b) Draw lines between factors which you think are linked, and label the line to explain the link. For example, you might link infrastructure and emergency services because the poor infrastructure makes it hard for emergency services to reach devastated villages.

c) Decide which factors or problems Honduras can most easily do something about and which are more difficult to change. Rank them on a scale of 1 to 3, and then colour code your diagram as follows:

- easiest to do something about → green
- possible but hard to do something about → blue
- impossible to do something about → red.

SAVE AS . . .

2 Look again at the factors or problems that you think are easiest to do something about. Write a paragraph to explain why poverty might make it difficult for Honduras to tackle even these problems.

Can aid help recovery from tropical storm damage?

The damage inflicted by frequent storms may bring financial ruin to countries that are already poor and don't have wealth-creating industry or agriculture to provide the money for rebuilding.

When the damage inflicted by a tropical storm is too great for an LEDC to cope with, the country may ask for international help (see Figure 9). AID may be sent as money, food, medical supplies, materials or technological expertise. The donors range from private individuals to governments and world banking organisations. Aid and LOANS from the last two groups are often linked to conditions such as improved economic performance and/or trade agreements. These solutions are seen by some as creating problems of their own. LEDCs struggle to meet interest payments and may lose control of their own economies.

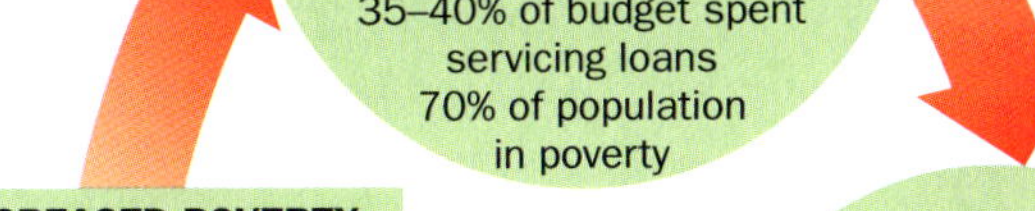

POVERTY
Honduras before the storm
Debt: US$ 4.45 billion (multilateral organisations 60%, The Paris Club 26%, private 5%, others 9%)
Annual repayment: US$ 564 million
35–40% of budget spent servicing loans
70% of population in poverty

STORM DAMAGE
5,657 dead; 12,272 injured; 8,058 disappeared; 80,000 homeless.
US$ 2.76 billion direct losses (60% of GNP).
70% of infrastructure ruined; industry and agriculture severely disrupted.

INSUFFICIENT AID
Inability to rebuild infrastructure/economy or improve precautions ready for next storm. Existing debt repayment unlikely.

INCREASED POVERTY
'Honduras has no revenue to pay for its debt and reconstruction. Not because of bad policies or because it doesn't wish to, but because that capacity was crushed by the storm.'
Roberto Flores Bermudez, Honduran Ambassador to London

SUFFICIENT AID
Ability to rebuild infrastructure/economy and improve precautions. Future storm damage potential reduced. Debt increased, but more chance of repayment.

CIRCLE BROKEN

FORMS OF AID

Disaster relief
Raise money from private and government donations. Aid not 'tied', usually short-term, and amount generated depends on public reaction.

Voluntary aid
Raise money from private donations, e.g. Oxfam, Cafod, Save the Children, etc. Aid not 'tied'; usually involved in smaller projects.

Bilateral aid
Usually from one government to another. Often 'tied' to trade arrangements; may be denied to countries seen as 'unfriendly' or with poor human-rights records.

Multilateral aid
MEDCs give money to international organisations which then distribute it to poorer nations. In theory not 'tied', e.g. International Monetary Fund (IMF) and The Paris Club. The Paris Club is an organisation of about 20 MEDCs which are collectively owed £300 billion by LEDCs. They can suspend loan repayment (debt moratorium), write off some of a country's debt, or provide debt relief for countries that adopt a strict economic policy.

9 Can aid break the vicious circle of poverty?

3 Study Figure 9.
For each of the following people write two statements explaining what they might suggest as the best short-term measure and the best long-term measure to help Honduras recover from Hurricane Mitch:

- relief worker
- flood victim
- civil engineer specialising in infrastructure
- member of the Honduran Government
- member of The Paris Club.

You could get ideas from the UN website at: www.un.hn/mitch/entrada.htm

Focus task

Why did so many people die as a result of Hurricane Mitch?

4 This is the question posed on page 42. You now have a lot more information to use to answer it. You have found out about how Honduras's *physical geography* and its *poverty* led to a very high death toll.

a) Write up your answer using the above two points as headings.
b) Explain which you think was the more important.
c) Add other points if you wish.
d) Hurricane Mitch was so catastrophic that it attracted a lot of attention and there are still many live websites about it. One is mentioned in Task 3. You can use a web search to add to your answer.

2.2 What are the causes and effects of flooding?

DO YOU KNOW?

- As world population increases, more people live in areas that are liable to flood.
- The Mississippi basin and the DELTA in Bangladesh are both densely populated areas because past flooding has left fertile sediments behind.
- The Mississippi floods in 1993 covered 4 million hectares, destroyed 40,000 buildings, killed 45 people and cost an estimated $12 billion.
- Bangladesh suffered floods in 1987 and 1988 which covered 40 per cent and 57 per cent of the whole country. The death-toll probably ran into hundreds of thousands of people.
- Most scientists agree that global warming will bring a rise in sea level, leading to increased coastal flooding, and will probably cause an increase in storms leading to more river floods.

AIMS

- **To understand the causes and consequences of flooding.**
- **To make decisions about managing the flood risk.**
- **To compare two different types of floods – coastal and river – in LEDCs and MEDCs.**

1 Surveying the flood damage

A CASE STUDY: THE MALTON FLOOD – EASTER 1999

'We know this area can flood but we cannot understand why no one could do anything to stop it becoming so bad,' said one resident of Malton in Yorkshire as he mopped out his home when the floods finally subsided. 'We had plenty of warnings, but the water just kept on rising and everyone seemed helpless. You do not expect that in a developed country at the end of the 20th century.'

DIARY OF THE FLOOD

Tuesday 2nd March *Heavy snowfall on North York Moors.*

Wednesday 3rd *Snowfall continues. Whitby completely cut off.*

Thursday 4th *Temperature rises. Rainfall replaces snow.*

Friday 5th *The ENVIRONMENT AGENCY issues a yellow flood alert and Ryedale Council workers start to distribute sandbags to vulnerable properties. Emergency services prepare to deal with flooding.*

Saturday 6th *The alert is stepped up to amber. More sandbags are supplied. The River Derwent bursts its banks in Pickering town centre.*

Sunday 7th *The alert becomes red. Malton and Norton are flooded. The railway is cut. Telephones out of action.*

Monday 8th *All road links between Malton and Norton are cut. Large areas of the Derwent FLOODPLAIN are under several feet of water. Emergency services use dinghies to rescue people. Stamford Bridge centre is flooded.*

Tuesday 9th *The river is at an all-time record level of 19 metres in Malton. The flooded area is still spreading in Norton. More than 140 homes in Norton and Malton are flooded. The fire brigade works throughout the day, even though the fire station is almost cut off by floodwater.*

Wednesday 10th *The flood stops rising.*

Thursday 11th *The flood finally starts to recede from Malton and Norton. Health alert issued because of the possible contamination of drinking water and spread of raw sewage across flooded area. (No recorded instances of disease spreading, though.)*

Friday 12th *The clean-up begins. Insurance companies set up a 'shanty town of builders, plasterers and decorators' on spare land, at point M on Photo 8 (page 51).*

. . . Friday 12th November *(eight months later) Most repair work (replacing floorboards and replastering) is finished. Much redecorating has been completed. In some cases, however, plastering and decorating needs to be done again, because it was attempted before the houses had dried out (properly).*

A total of 200 homes and businesses in the North Yorkshire towns of Malton and Norton were swamped by the flood of 1999. Estimates of the financial cost of the damage varied, but it was probably around £10 million. The psychological and emotional cost of the experience was immeasurable. One resident said, 'When you have seen one flood steadily overwhelm your house, you fear that you will never feel safe there again.' In 2000 and 2001 they were flooded yet again!

2 At the height of the flood, 160 cubic metres of water per second flowed down the river near Malton. This was like emptying the water from a 25-metre swimming pool into the river *every two and a half seconds.*

3 Waist-high floodwater

5 Coping with the floods

4 A government minister visits flood-damaged Stamford Bridge

Task

1 Look at the people in the photos on these two pages. Choose two people who might have contrasting views of the flood, and try to put yourself in their place. Describe your feelings about the flood.

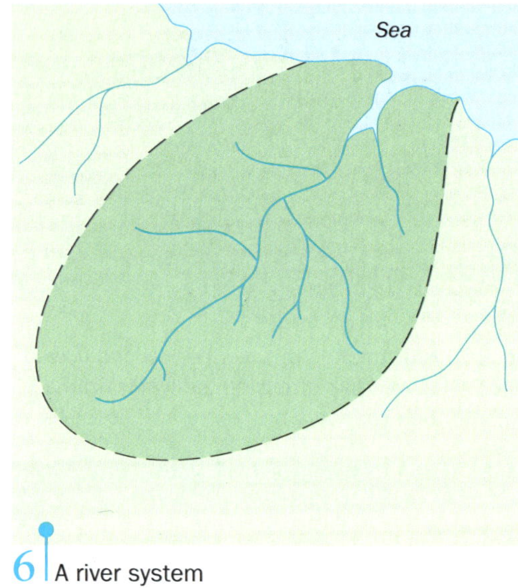

6 A river system

Tasks

SAVE AS . . .

1 On a copy of Figure 6, label the following features:

a) the SOURCE of the main river

b) a TRIBUTARY which flows into the main river

c) a CONFLUENCE where a tributary meets the main river

d) the WATERSHED which is the line drawn round the river system to separate it from neighbouring river systems

e) the CATCHMENT AREA of the river – which is the whole area drained by the river and its tributaries – also known as a DRAINAGE BASIN

f) the MOUTH, where the river meets the sea.

★ THINK!

Use the correct terms!

In an exam you will gain marks for 'correct use of technical terms'. Learn the capitalised technical terms in this task and try to use them regularly in your work on rivers and flooding.

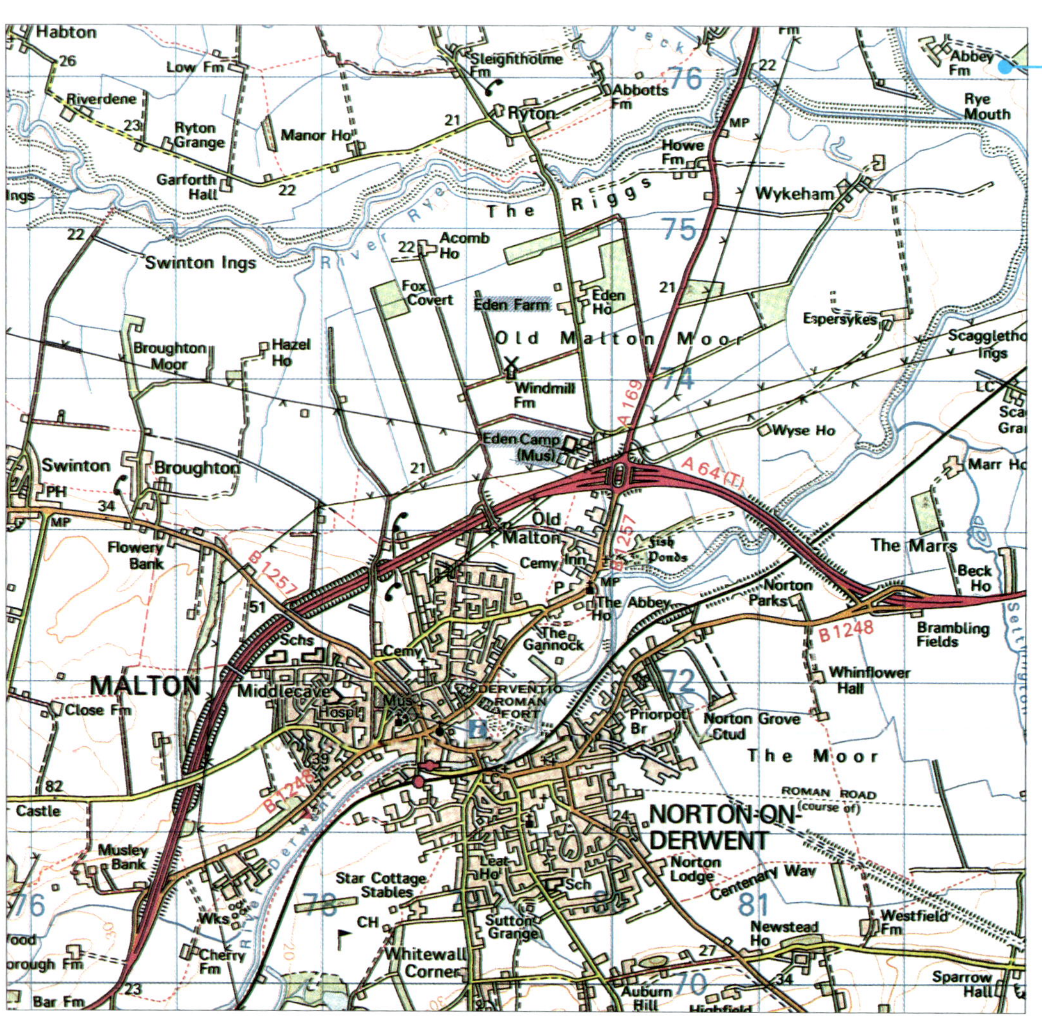

7 Malton. Reproduced from the 1997 1:50,000 Ordnance Survey map by permission of the Controller of HMSO © Crown Copyright.

8 An aerial view of the flooding in Malton and Norton

2 Photo 8 was taken from a plane flying over grid reference 791717 on the Ordnance Survey map extract (Map 7). The camera was pointing south. Place a sheet of tracing paper over the photograph. Then, referring to Map 7, mark on the paper:

a) the normal course of the River Derwent

b) the course of the stream that flows from the small lake at A on the photo and then enters the Derwent through a CULVERT at B on the photo

c) the railway line

d) the main road B1248

e) the total flooded area – shade this light blue

f) flooded housing – shade this a darker blue

g) flooded industry and shops, etc. – shade this area very dark blue.

3 Try to estimate the size of the area in the photograph that has been flooded.

4 In the rest of this chapter, whenever a building or area is named, try to locate it on Photo 8 and mark it on your tracing paper overlay.

A river as a system

A river can be thought of as a system (see page 3). Systems are a useful way to understand many geographical ideas. A system is anything that is made up of different parts which work together. Geographical systems all have four things in common: inputs, transfers, stores and outputs.

In a RIVER SYSTEM these are:

- Inputs – PRECIPITATION, melting snow, eroded rock, sewage and waste, etc.
- Transfers – river flow, THROUGHFLOW, piped water for human use, etc.
- Stores – lying snow, soil moisture, GROUNDWATER, lakes, reservoirs, etc.
- Outputs – flow to the sea, EVAPOTRANSPIRATION, water used by farms and industry, etc.

You are now going to apply this systems way of thinking to understanding the Malton flood (see Focus task).

Focus task

What caused the Malton flood?

Many different causes contributed to the Malton flood. Some are natural causes, and some have been brought about by human actions. Some causes start way back in history. Others are recent. Some can be easily controlled, but others are probably out of people's control.

1 Your task is to find out the main cause, or causes, of the Malton flood. Read pages 52–58, and each time you come across something that you think is a cause of the flood, list it under one of these river system headings:

- Inputs
- Transfers
- Stores
- Outputs

This will give you a classification of the causes of the flood.

The history behind the Malton flood

To understand the causes of the flood you are going to have to start way back in history.

★ Stage 1: the Ice Ages

KEY QUESTION: How did the Ice Ages increase the flow of water through the narrow gap in the hills at Malton?

Study Maps 9, 10 and 11, which show the Vale of Pickering and the North York Moors before, during and after the last ICE AGE. (The Ice Ages started about 2 million years ago and the last one ended about 10,000 years ago.) Then complete Tasks 2–6.

Note that all place names and locations in Maps 9–11 are modern.

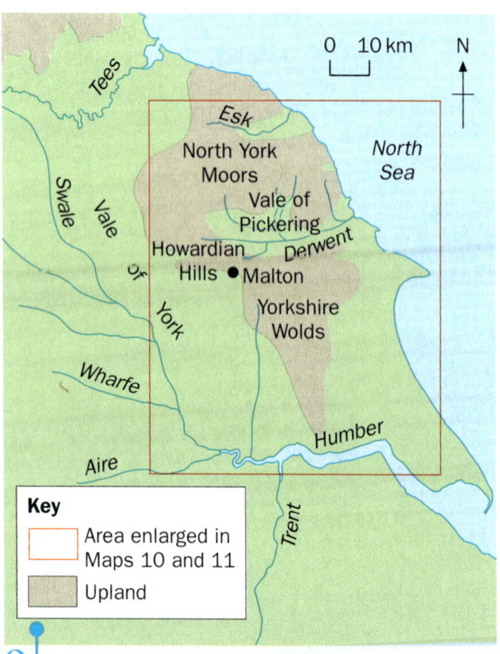

9 East Yorkshire before the last Ice Age

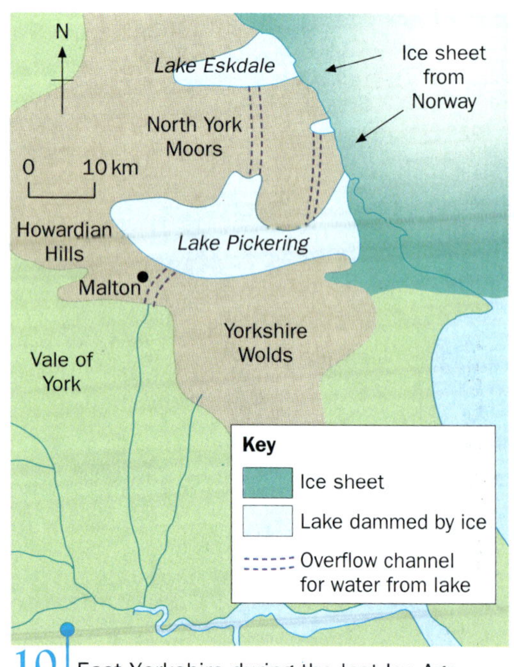

10 East Yorkshire during the last Ice Age

Tasks

Stage 1

2 Describe the drainage pattern of the Vale of Pickering (shown on Map 9) before the Ice Age.

3 When ice started to melt at the end of the Ice Age, water from the Vale of Pickering could not follow its old course to the North Sea.

a) Why not?

b) What happened to the meltwater at first?

c) How did the water eventually escape to the sea?

4 When the ice sheet finally melted, water from Lake Pickering still could not follow its old course to the North Sea. Why not?

5 The floor of Lake Pickering was covered with a layer of fine SEDIMENT deposited in the bottom of the lake. Suggest how this might have affected farming in the Vale of Pickering in recent years.

6 Point A on Map 11 shows one of the sources of the River Derwent.

a) How far does the river have to flow from its source until it reaches the sea?

b) How far does it drop between its source and Malton?

c) What does this suggest about the probable speed of flow of the river?

d) How will this affect the likelihood of the river flooding at Malton?

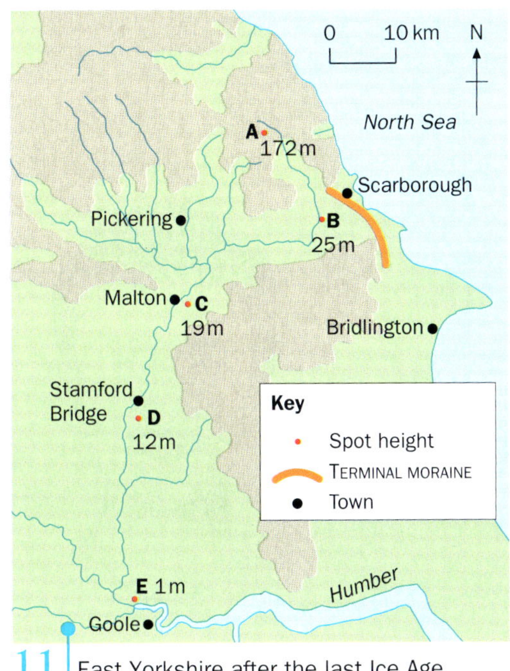

11 East Yorkshire after the last Ice Age

★ Stage 2: 10,000 BC to the present

KEY QUESTION: How have land use changes in this period speeded up the flow of the Derwent?

Over the last 10,000 years the use of land in the Derwent basin has been steadily increasing the flood risk lower down the valley. Large parts of the North York Moors used to be covered with a deep layer of PEAT soil. This acted as a huge sponge which soaked up rainfall and slowed down RUN-OFF from even the worst of storms. Other parts of the moors were forested, and the forest also helped to slow down the run-off.

When the rivers left the moors they crossed the gently sloping Vale of Pickering. They flowed slowly across the vale, with many meanders, and marshes. In periods of flood the water spread out across the floodplain. Natural storage stopped the water from rushing down to Malton.

From the Bronze Age onwards settlers altered this natural system, sometimes accidentally, sometimes on purpose:

- Bronze Age farmers cut down the forests on the hills to use the land for shifting cultivation. When they abandoned patches of land the soil was quickly eroded away.
- In the Middle Ages, farmers drained the lowland around the edges of the vale. They built banks along the rivers to try to stop them from flooding, and then pumped water off their fields to allow crops to grow.
- In the 19th century people began to use the moors for grouse shooting. Old vegetation was burnt to stimulate new growth to feed the birds. This burning often led to the destruction of peat.
- After the Second World War there was a rush to increase food production. The British Government and, later, the European Union paid grants to farmers to drain land. Cultivation spread onto the floodplain of the Derwent. Great efforts were made to get rainfall off the land and into the river as quickly as possible.

Tasks

Stage 2

7 Study Map 7 on page 50, particularly the area around Old Malton Moor to the north of Malton. Find at least three pieces of evidence to show that people have made changes to the drainage which will get the water off the land more quickly. Look at the following areas in particular: 7874, 8175 and 8274.

Hints:

- In square 8274 a little stream runs alongside the main River Derwent. It cannot join the main river because of the embankment. What must happen to allow this water to drain into the Derwent?
- *Ings* is a local dialect word which dates back many centuries. It means 'newly reclaimed fields'. In this area it is used to describe land that was once marsh.

8 Suggest how the building of the Malton by-pass in the 1980s will have affected drainage.

★ Stage 3: the past 2000 years

KEY QUESTION: Why did a town develop at Malton, close to the floodplain?

There has been a settlement on the site of Malton since Roman times. Work through the following tasks to understand why it was built here despite the risk of flooding.

Tasks

Study Map 7 on page 50.

1 a) Find two pieces of evidence to show that the Romans occupied this area.
b) What do you think was their main reason for building 'Derventio' on this site?

2 To the west of Malton the Roman road ran through Broughton, Swinton and other villages at the same height above sea-level.
a) Suggest why this was a good route for a road.
b) Suggest why these were good sites for small farming settlements.

3 Before the INDUSTRIAL REVOLUTION several mills were built in Malton around grid reference 793715.
a) Suggest how these mills were powered.
b) The mills ground corn, sawed wood and also produced woollen cloth. Suggest where the raw materials for these mills came from.

4 The mid-19th century is sometimes called 'the railway building age'. A railway line was built from York (south of Malton) to Scarborough (to the north-east).
a) Give two reasons why it was built through Malton. You should be able to work out both reasons from the map and from thinking about your previous answers.
b) Study the contour pattern on the map very carefully. Explain why the railway line was built on the Norton side of the river, rather than on the Malton side.

5 In the late 19th century several small houses for workers were built around grid reference 792713.
a) Suggest where the people who lived in these houses worked.
b) Suggest why this land was used for low-cost housing, and not for more expensive housing.

6 In the 20th century, planners put restrictions on building on the floodplain. Suggest why.

★ Stage 4: developments since the 1960s

KEY QUESTION: How have recent changes increased the flood risk?

Marie Stewart and her son Ian (aged 14) live in Spring Field Garth in Norton (marked S on Photo 8 on page 51). Their house was built in the 1960s. Marie has lived there for 16 years and it never flooded before 1999. Here are some of their comments on the flood.

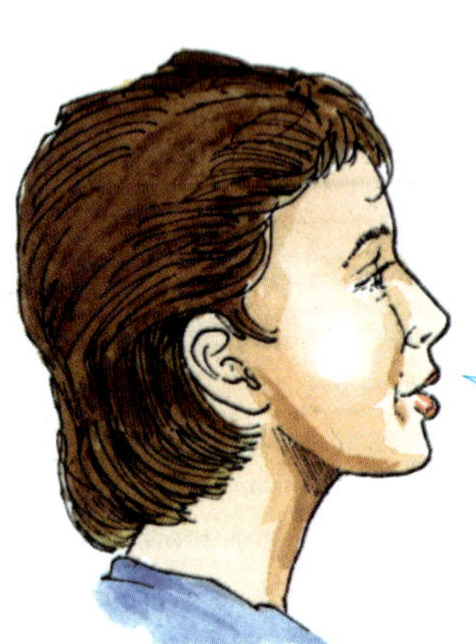

On Sunday morning the river burst its banks. It flooded the railway line and some of the houses near the river, but we felt quite safe. Our house had never flooded before. The worst floods always happened above the bridge. It's very old, with narrow arches. They get blocked by trees and rubbish swept down the river, and it forms a kind of dam, flooding the area between Malton and Old Malton. You can see on Map 7 that not much has been built there.

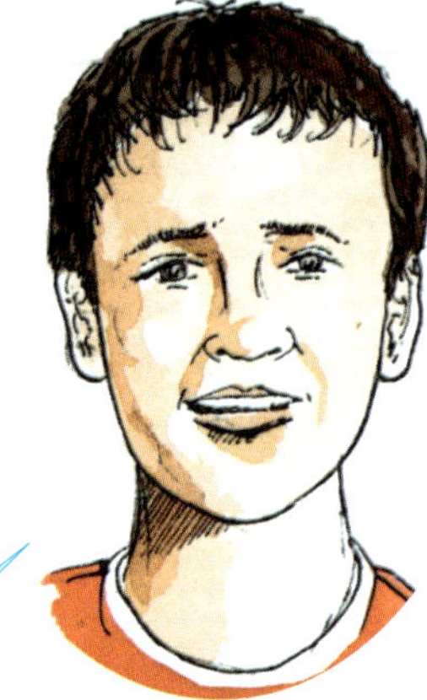

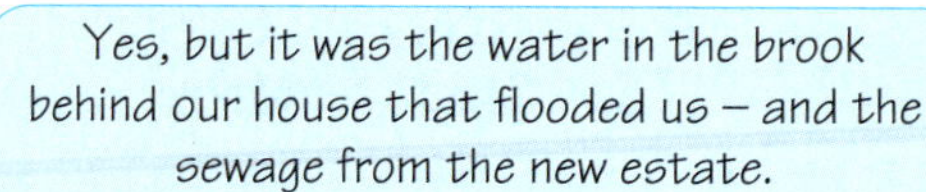

Yes, but it was the water in the brook behind our house that flooded us – and the sewage from the new estate.

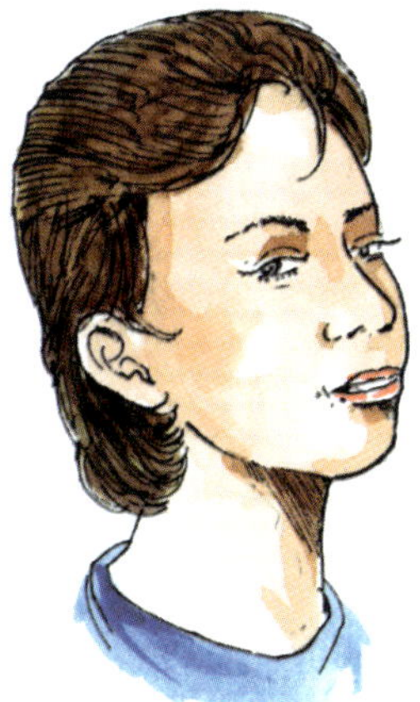

Let me explain. The brook used to power the sawmill and some of the other mills. As more and more building took place the stream was eventually put into a pipe under the buildings and roads. You can see how it just disappears on Map 7. The pipe could not take all the floodwater. It also got blocked with leaves and branches as it hadn't been cleared out properly. Now, since the flood, they're always coming and raking it out.

And the sewage from the new estate flooded up into people's houses.

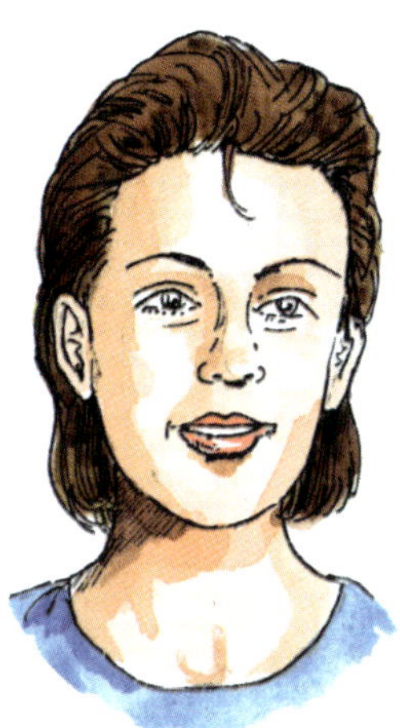

Yes. We were all worried when they built the new estate, about five years ago, that they didn't put any new sewers in. They just added the waste pipes to our sewage pipes. The floodwater blocked the sewers, and raw sewage came back up the pipes into the houses. On Monday afternoon, when we realised we were going to get flooded, we started to move all our furniture and carpets upstairs, so most of that was saved.

My sister came round to help – but all she was worried about were the fish in our garden pond. She said they'd get swept away and lost in the river. Poisoned by sewage and pollution washed from petrol stations into the river, more likely.

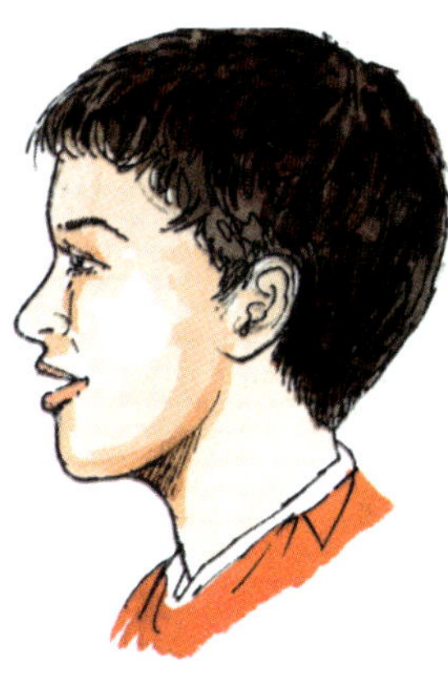

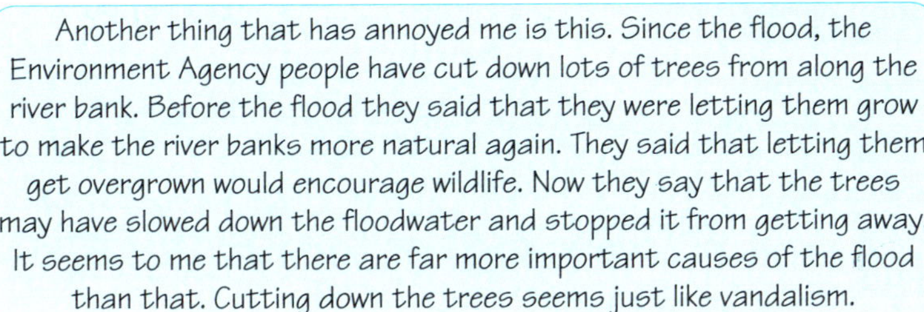

Another thing that has annoyed me is this. Since the flood, the Environment Agency people have cut down lots of trees from along the river bank. Before the flood they said that they were letting them grow to make the river banks more natural again. They said that letting them get overgrown would encourage wildlife. Now they say that the trees may have slowed down the floodwater and stopped it from getting away. It seems to me that there are far more important causes of the flood than that. Cutting down the trees seems just like vandalism.

Tasks

7 Explain how each of the following helped to cause the flood in the Stewarts' house:

- **a)** the amount of water flowing in the Derwent
- **b)** the fact that the old brook had been put in a pipe
- **c)** the fact that the pipe was not kept clean
- **d)** the sewerage system in the new estate
- **e)** the fact that all new buildings tend to make the surface of the land IMPERMEABLE, stopping water from soaking into the soil and increasing surface run-off.

8 What other causes can you find in Marie and Ian's comments?

9 **a)** Describe two sources of pollution that added to the flood problem.

- **b)** Which of the two do you think poses most danger? Explain your answer.

★ Stage 5: February 1999

KEY QUESTION: How did the weather affect the flood risk?

In February 1999, a strong blocking ANTICYCLONE built up over the European continent. This anticyclone was an area of cold, high pressure air. It was known as a 'blocking' anticyclone because it blocked the path of any depressions that formed over the Atlantic and moved towards Europe (see page 86).

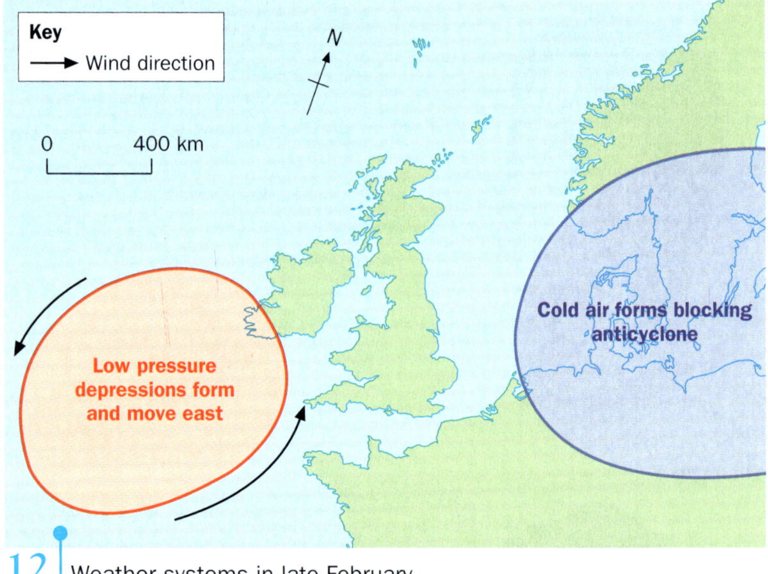

12 Weather systems in late February

Towards the end of February a series of low pressure depressions formed over the Atlantic. They contained large amounts of WATER VAPOUR because they had passed over the ocean. Air circulated around them in an anticlockwise direction. Once they had developed they started to cross Britain and Ireland.

One of these depressions crossed Britain fairly quickly, bringing snow to much of northern England. When it was blocked by the anticyclone it came to rest over the North Sea. It caused a northerly air stream to flow down the east coast of England. When this met the coast of North Yorkshire the air was forced to rise, causing cooling, and CONDENSATION of water vapour. This brought very heavy snow to the North York Moors. In fact it was so heavy that all roads to Whitby were totally blocked for 24 hours.

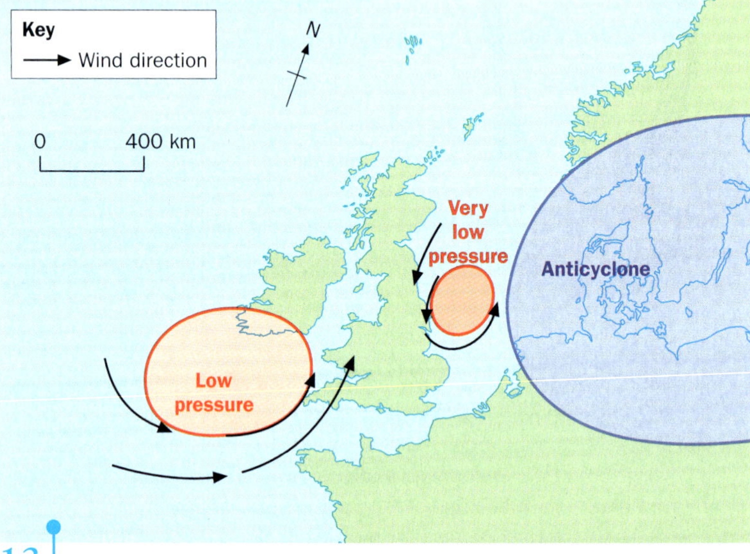

13 Weather systems in early March

Then, on 4 March, the snow turned to rain. A FRONT associated with a second depression lay across North Yorkshire. The anticyclone stopped it from moving eastward as fronts normally do. Instead the anticyclone acted like a conveyor belt, bringing warm, moist air across the country, lifting it up over the cold front and pouring more and more rain on to the land.

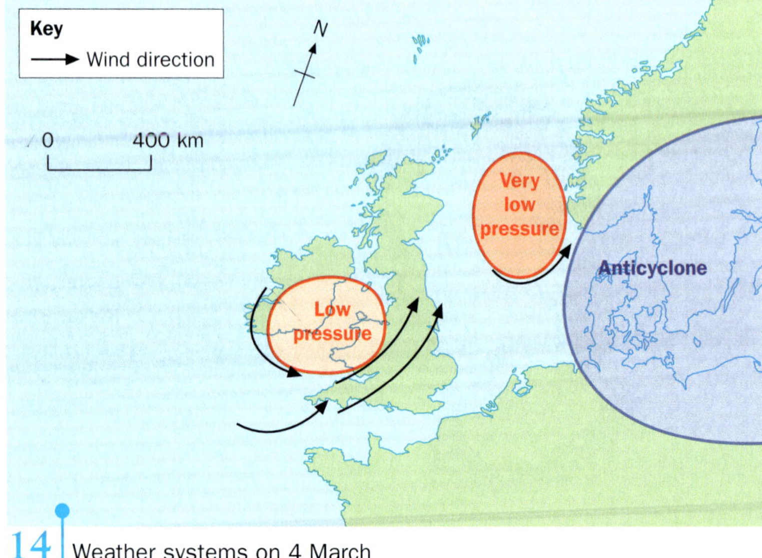

14 Weather systems on 4 March

One local man described the rain like this:

All my life I've worked on the moors cutting peat. On the Friday morning I set out to drive up on to the moors. I knew that the rain would have melted the snow. With the rain and the meltwater I knew that the surface would be totally saturated. I needed to check that my peat stacks were draining reasonably well and not being washed away.

I had my windscreen wipers on at full speed but they couldn't cope with the rain. I could only crawl along, peering forward to see. Then I came to the hill and I was shocked. You sometimes see something like a sheet of water flowing down a road when there is heavy rain. This was more than that. The main road was quite literally a river. It was frightening.

I knew then that there was nothing that anyone could do to stop the flooding. No one could plan for rain like that. I managed to turn the car around (I was worried about turning broadside on to the force of the water – I've seen boats capsize in situations like that) and went home to make sure that my house was protected. I rang to tell my friends down the valley to start moving things upstairs. I knew that things were going to get really bad.

The Environment Agency (EA) is the organisation set up by the government to plan and manage river basins. Its duties include monitoring river flow, trying to prevent flooding, and conserving the natural environment. It is also responsible for giving out flood warnings. When EA staff received reports of the rainfall that day, they asked the person sending the statistics to check his figures. These were so much higher than anything they had known before that they did not believe they could be true!

In fact, 125 mm of rain fell in 48 hours on parts of the North York Moors. This compares with the normal total for the whole of March of 87 mm.

People will ask whether such WEATHER is likely to happen again. No one knows for certain. This was the heaviest period of rainfall in the area for 100 years. However, Peter Holmes, regional flood defence manager for the EA, said, 'What we're talking about here is climate change. Scientists are telling us that we will have warmer summers and wetter winters. I'm afraid that flooding will happen again, but I can't predict when, nor how severe it will be.'

Tasks

1 The weather that led to the flood came in four main stages. Write a summary of what happened:

a) in February

b) towards the end of February

c) when the first depression crossed England

d) on 4 March.

2 Explain why so much rain ran across the land as surface run-off, rather than soaking into the surface to be stored in the soil.

3 Lying snow can be described as a 'store' in the river system. What happened to this store to help to cause the flood?

4 The work of peat-cutters over the centuries may have helped to cause the flood, by removing the deep peat soil from the moors. Explain why this increased the rate of surface run-off.

5 This was the heaviest rain for 100 years. Some people say we are unlikely to have such heavy rain again for another 100 years, so we should not worry about it. Do you agree? Explain your answer.

6 Complete a timeline diagram for the period from 27 February to 12 March to show:

a) the weather conditions in North Yorkshire

b) the state of the flood in Malton.

The consultant's view of the causes

After the flood the EA commissioned an enquiry into its causes. The enquiry also had to consider how the EA had managed the flooding. It did not have to look into the wider historical and geographical causes as you have been asked to do! The main conclusion of the preliminary enquiry stated:

> *'The flooding of property was inevitable given the snow cover and the rainfall in the upper river basin. Extensive research will be needed to show whether other influences created or contributed to the severity of the flooding.'*

The report also said that willow trees growing along the banks and bed of the river may have impeded its flow to a certain extent. The trees trapped other debris flowing downstream, causing dams to build up. The river had been declared a site of special scientific interest (SSSI) about twelve years before. Since then, the trees had been allowed to grow and spread, to encourage the development of a more natural ecosystem. An EA spokesman said:

> *'The lower Derwent Valley is an internationally important wildlife site and our maintenance work needs to take into account both conservation and flood defence interests.'*

A local resident, who had been driven out of her house by flooding, complained, 'The EA seems to think that water rats are more important than human beings.'

When the spokesman was asked why the EA did not dredge the whole river regularly to improve the flow, he said that large-scale dredging would be far too expensive to carry out. He added:

> *'The silt that lies on the river bed is a natural breeding ground for fish and insects. They also need willows to shelter under.'*

Tasks

1 Over the last seven pages you have listed the causes of the Malton flood and classified them. By now you should have a long list organised into five categories. You might feel ready to answer the big question now: 'What caused the Malton flood?' But there is another important stage needed: to think about the relative importance of these causes.

a) Pick out about 7–10 of the causes which you think are important.

b) Write each of these causes on a separate piece of card.

c) Arrange your cards into groups in order of importance. Try arranging them in a pyramid shape so that you have:

1 top cause

2 or 3 very important causes

3 or 4 quite important causes

Some other causes that should be considered

d) You can do your first arrangement quickly, but then think hard about the order and rearrange your cards if necessary. Keep moving the cards around until you are happy with the order.

2 Compare your arrangement with those of other students. Most members of the class will have different arrangements, but there should be certain causes that feature in most arrangements. (Any list of causes that did not have 'very heavy rainfall' near the top would seem a bit odd.)

3 After comparing arrangements you may decide to change your own order slightly – but do not feel you have to. At the end of the exercise, though, you must be able to provide evidence and sound arguments for your order of the causes.

SAVE AS . . .

4 'What caused the Malton flood?' Now you can answer this question, dealing with the causes in order of importance.

a) Write detailed explanations of the most important cause or causes. Make brief references to the minor causes.

b) Illustrate your answer with a simple, clear sketch map of the area. In your explanations and on your map remember to use the correct technical terms which were described on pages 50 and 52.

★ THINK!

One way to learn and revise a case study is to learn the map of the area first. Try it here. Carefully study the sketch map you drew in Task 4. Then try to draw it again from memory.

Check your new map against your original. Note where you went wrong. Then draw the map again – making it even better than your first try.

How can the flood risk be managed?

The Environment Agency has to manage the river on behalf of the government and the people who live in the region. It has to balance several responsibilities and meet the needs of a number of different interest groups, including:

- farmers
- residents of towns and villages on the floodplain
- taxpayers who have to meet the cost of river management, but who might also have to meet the cost of clearing up after floods
- environmentalists and tourists
- motorists and railway travellers
- business people who have premises on the floodplain
- insurance companies who have to pay some of the costs of clearing up after floods.

STOP PRESS

During the winter of 2000/2001 Malton was flooded again. Rainfall was not as heavy as on 4 March 1999, but several periods of heavy rain between November and February resulted in saturated soil and filled stores.

Tasks

5 Each person in the class should take on the role of one of the groups listed above.

a) Write one demand that you think your group might make to the EA as an *immediate* reaction to the floods.

b) Try to write several more points that they might raise when they have had time to think more carefully about the floods. Remember that these groups overlap. For instance, farmers are also residents and taxpayers.

6 Read the following Factfile report from the EA. Has it started to deal with your worries, and those of the other groups? Write a response to the report. You should congratulate the EA if you think it has acted sensibly so far, then go on to suggest what it should do next. If you think it has not acted sensibly, make sure you give reasons for your view.

★ **FACTFILE**

EA Report

The EA considered that it was important to let the people of Malton know how it had responded to their concerns in the period after the flooding of March 1999. This is what it said:

In the six months since the flood the EA has:

- set up a feasibility study into the provision of flood defences for Malton, Norton and Pickering
- extended an automatic telephone warning service that rings residents in danger areas as soon as a flood risk is forecast
- improved the flood forecasting service by putting more and better automatic flow monitoring stations in the upper parts of the rivers
- developed a better monitoring system for the state of repair of present flood defences
- improved liaison with farmers and other landowners to improve river maintenance on the River Derwent
- started to develop better computer models to predict how the River Derwent will react to various flood events
- cut and trimmed many of the willows and other trees growing on the banks and bed of the stream
- installed a valve where the culvert from Norton joins the Derwent, to stop floodwater from flowing back up the culvert
- set up stricter rules for people applying for planning permission to build or extend properties on the floodplain.

(The EA is not responsible for planning applications but it does have to advise the local authority on whether or not to approve applications.)

Tasks

Experiment

1 There are two reasons for the change in planning permission rules (see the Factfile on page 59). First, the EA wants to stop people from putting themselves in obvious danger. The second is more complex. This simple experiment will help to explain it.

- **a)** Pour some water into a shallow container to represent a flood on lowland.
- **b)** Now put something into the water to represent a new building (e.g. a child's toy house or a brick).
- **c)** What happens to the water level in your container? What is the lesson of this experiment for the EA?

Planning decisions

The EA tries to ensure that any new building on the floodplain makes allowances for the water it displaces. Somewhere has to be provided for the displaced water to go.

2 The landlord of the pub at X on Photo 8 (page 51) was renovating his pub. He needed to build a new ladies' toilet at the back of the pub. This would be on the floodplain. In the event of a flood, where could any displaced water go?

The landlord made a simple suggestion, which the EA approved, and planning permission was granted. Think carefully about the layout of pubs and make your own suggestion.

3 The owner of the car showroom at Y on Photo 8 wanted to build an extension, to store his cars in, on the spare land at the side of his premises. (*Note:* During the flood he had to move his cars across the field at the back of the showroom to avoid the flood.) When asked where the displaced water would go he could not offer a solution. Should planning permission be granted?

4 The owner of the clothing factory at Z on Photo 8 wanted to build a small extension in his car park at the back of the factory. This would be used for extra machines to make more clothes. He could not satisfy the EA that there was somewhere else for the displaced water to go, but the council approved the application anyway. Suggest what reason was given.

5 The town's bowls club had a wooden pavilion which was destroyed by the flood. The EA suggested three possible options for a replacement building:

- **A** replace the old pavilion with a new one in exactly the same place
- **B** build a new, bigger pavilion with a storage tank underneath for floodwater
- **C** build a new, bigger pavilion on legs to raise it above the flood level.

- **a)** Suggest why the EA thought each of these was acceptable.
- **b)** The bowls club decided that there was only one they could afford. Which do you think it was? Why?

(You will find answers to the four planning decisions on page 62.)

A plan to make Malton 'flood proof'

Malton and Norton could be made safe from floods as bad as the one in 1999 if the authorities were to do the following:

- Spend about £15 million on flood defences along the river banks in the town to stop the water spilling out of the river. This would mainly involve building concrete banks right through the town between grid references 795716 and 783713 (see Map 7 on page 50). This is known as HARD ENGINEERING. It uses man-made materials to stop the floodwater from taking its natural course (see Photo 15).
- Provide somewhere else for floodwater to go. It could be allowed to spill on to farmland further upstream by lowering river banks or removing natural barriers to flooding. Then the higher banks downstream would contain it until the flood risk had passed. This might damage crops but it would be less costly than the damage caused when houses and business premises are flooded.
- Do far more management work on the river banks downstream from Malton. In particular, cut back and remove a lot of the vegetation to speed up the flow of the river. This is known as SOFT ENGINEERING. It uses natural solutions to keep the floodwater in its natural course.

15 'Hard engineering' flood defences

Tasks

6 Draw up a table with two columns, headed 'For the plan' and 'Against the plan'. Using information in this chapter, list all the arguments you can in favour of, and against, each part of the plan described above.

7 Decide whether you think each part of the plan should be accepted. Justify your decision.

ICT

8 You are a reporter for a local paper and you have been asked to write a front-page story about the Malton flood plan. Write your story and design the front-page layout using an appropriate program. You could include:

- information from this book
- statistical information from the Centre for Ecology and Hydrology website at: www.nwl.ac.uk
- information from the BBC archive site: http://news.bbc.co.uk (when you find this site, select *archive* and search for 'Malton floods').

Ⓑ A CONTRASTING CASE STUDY: ORISSA – NOVEMBER 1999

In November 1999, soon after the Malton flood, the coast of Orissa in India was hit by a devastating cyclone. This caused coastal flooding on a much greater scale than the Malton flood. Floods are common in both areas, but 1999 was a very bad year in both places. It is interesting for geographers to compare the causes and the effects of the two floods. Note, though, that it is far easier to find out details about the flood in Yorkshire than about the flood in India, even though the Indian flood was much more serious and caused a far greater loss of life and property.

Background

Orissa state lies in the east of India, south of Calcutta, bordering the Indian Ocean (Map 16). It is one of the poorest states in India. Reasons for the poverty include the following:

- It is isolated from the main areas of development in the Ganges valley, in the west around Mumbai (which used to be called Bombay) and in the south around Chennai (previously Madras) and Bangalore.
- Rainfall totals are only moderate, and they are unreliable.
- Much of the state has poor soils because it has steeply sloping land which erodes easily.
- The GREEN REVOLUTION (see pages 266–68) had little effect in this area because most farmers lacked capital, so they were unable to buy seed; they were also poorly educated and conservative and so were unwilling to risk anything new.

The most densely populated part of Orissa lies along the coast and in the delta of the River Mahanadi (see Map 19). Here the soils are more fertile and people can supplement their food supply by fishing. In many ways the conditions are similar to those in the coastal areas of Bangladesh, which also suffer from many cyclones, although the Mahanadi delta is not as densely populated as the Ganges delta.

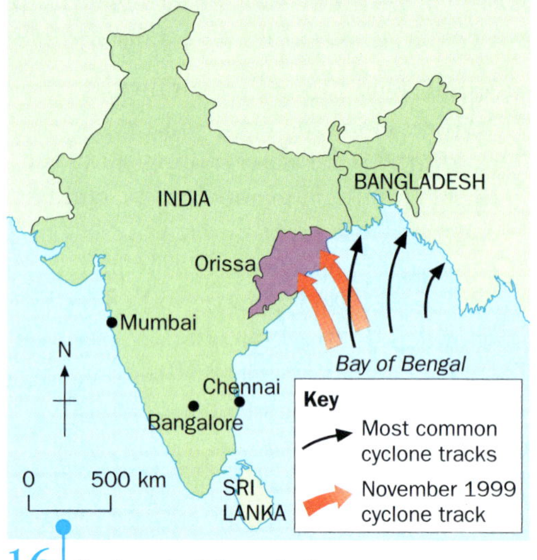

16 Cyclone in Orissa, India

17 Satellite image of a cyclone in the Bay of Bengal

18 Flood damage caused by the cyclone

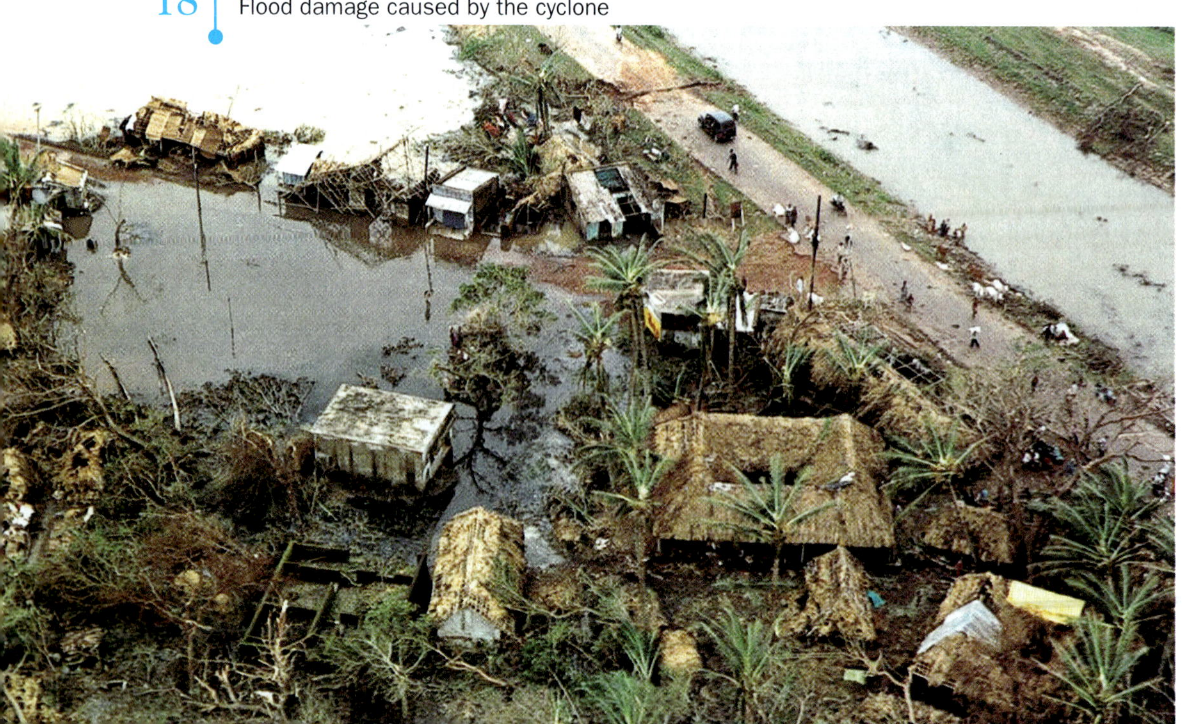

Answers to Malton 'Planning decisions' tasks on page 60

2 The landlord suggested that he would not try to keep water out of his cellar if the river flooded. The cellar could be used as a storage for extra floodwater. Easy!

3 Planning permission was not granted because the building would have added to the flood risk on the edge of the floodplain.

4 Even though this might increase the flood risk they could not turn away the extra jobs that this would bring to the town.

5 **A** was chosen. It didn't add to the displacement problems that had applied before the flood. **B**: it would have cost too much to excavate the tank. **C**: it would have cost too much to make the foundations that would have supported the legs.

★ **FACTFILE**

The Orissa flood

- **Causes** A cyclone developed over the Bay of Bengal. Most cyclones travel further north and strike the coasts of Bangladesh and West Bengal. This one swung west and hit land in the Mahanadi delta area.
  The area was flooded by a combination of:
  - huge waves blown by the wind
  - a storm surge caused as the low pressure at the centre of the cyclone sucked up the surface of the sea and raised water-levels by almost 2 metres
  - torrential rain caused as the rain-bearing winds from the cyclone were forced to rise over the coast and the hills of the Eastern Ghats.
- **Extent** Thousands of square kilometres of land were flooded. The low-lying delta land was very easily flooded.
- **Death-toll** This will probably never be known because records of the population were very poor, especially in rural areas. Five days after the flood a figure of 'at least 3,000 dead' was widely reported, but the true figure may have been 10,000 or more.
- **Homes damaged** At least half a million people were made homeless. The coastal cities of Puri and Berhampur were very badly damaged. Some coastal villages were completely destroyed.
- **Warning** There was none! The Indian cyclone tracking and prediction service is not as sophisticated as that in Bangladesh. Even if there had been adequate warnings it would have been very difficult to broadcast the news to the people in rural areas because communications there are poor.
- **Protection** There was little in place. Bangladesh has built seawalls and banks along the rivers because cyclones are so common. They also have concrete storm shelters for the people. Neither of these is so well provided in Orissa. There are plans to build storm shelters, but they were not ready for the 1999 cyclone. The area's poverty means that it will probably be many years before adequate shelters are built. Evacuation of large numbers of people was almost impossible because of poor communications. Roads were poor and sufficient vehicles were just not available.
- **Effect on water supply** Supplies of clean, purified water are limited and the sewage disposal system is poorly developed. When floods occur in an area like Orissa, the supply of clean drinking water is lost. Tens of thousands of people had only contaminated water to drink. Two weeks after the disaster, disease was starting to spread as a result of the flooding.
- **Effect on food supply** The immediate effect was that people lost their food stores when their houses were destroyed. It took several days for the government to organise food drops into the flooded areas. Even then, much of the food dropped from the air was lost in the water.
- **Long-term effects** The long-term effects mean that it will take several years for people to recover. Fields were contaminated with salty water or with pollution from sewers and flooded industry. Fishing boats were destroyed. The poor road system in the area was made even worse. International relief efforts were slow to start, because organisations were having to cope with disasters in several other parts of the world at the same time.

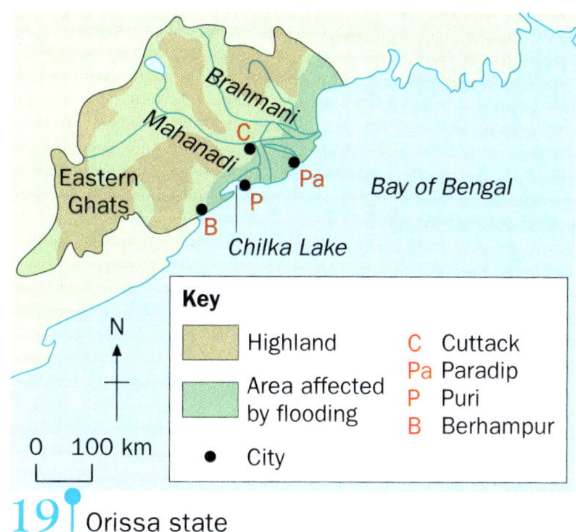

19 Orissa state

Focus tasks

1 Draw up a table with two columns to compare the Orissa flood with the Malton flood.

Head the first column 'Orissa' and the other column 'Malton'. Then write the subheadings from the Orissa Factfile down the side of your table.

Pick out the main points from the Orissa Factfile, then look back at the Malton case study to make a comparison between the two floods.

2 Using your table to help you, write two summaries to compare: the causes of coastal floods and river floods; and the effects of floods in LEDCs and floods in MEDCs.

A *summary* should be fairly brief but should express the main ideas clearly and strongly in a clear prose. Each summary should pick out several important points and compare how each point applied in Orissa (which is in an LEDC) and in Malton (which is in an MEDC). You are not expected to repeat everything that you wrote in your table of comparisons.

2.3 *How do plate movements cause hazards?*

DO YOU KNOW?

- Most EARTHQUAKES and VOLCANOES are located at PLATE margins, but some do occur within plates.
- Earthquakes may last only a few seconds but their effects can last for decades.
- Over 15,000 earthquakes happen every year but fewer than 100 usually lead to serious damage or human deaths.
- There are over 550 active volcanoes in the world and 15 are expected to have life-threatening eruptions within the next ten years.

AIMS

- **To understand that earthquakes and volcanoes are caused by plate movements.**
- **To investigate the threat posed by earthquakes and volcanoes.**
- **To contrast the differing experiences of MEDCs and LEDCs to similar hazard events.**
- **To analyse why countries respond in very different ways to earthquakes and volcanoes.**

Ⓐ EARTHQUAKES

1 Aftermath of the earthquake in Kobe, Japan

2 Earthquake debris at the Hanshin Expressway, Kobe

Case study: Kobe, Japan – an earthquake in an MEDC

Just before dawn on 17 January 1995, an earthquake measuring 7.2 on the RICHTER SCALE struck Kobe in Japan. It ruptured the Earth's crust beneath densely populated areas and caused terrible damage (see Photos 1 and 2). Kobe suffered 5,500 deaths, as well as massive destruction of buildings and INFRASTRUCTURE such as roads and railways. Japan has more than 1,000 substantial earthquakes a year. It devotes much money and expertise to limiting the effects of earthquakes. How could a developed, technologically advanced country suffer such an appalling loss of life and so much damage? The Kobe earthquake illustrates the combination of factors that determine the impact of a hazard event on people and property.

The earthquake lasted only 20 seconds but the shock waves devastated homes and offices and brought transport systems to a halt. Many of the worst-affected areas were in the bay area (see Map 3), where over 100,000 buildings (one in five) failed. These areas, e.g. Sannomiya, had a lot of older, concrete-framed housing which was constructed before new earthquake-proof building laws were enforced in 1981. Later steel-framed buildings fared much better and many more survived the earthquake.

More than 150 fires occurred in Kobe and the surrounding areas in the hours after the earthquake. Firefighters could not tackle them because streets were blocked by collapsed buildings, debris and traffic. There were also severe problems with water supply which further hindered the firefighters. Over 2,000 fractures in the water system were blamed for initial delays in restoring supplies.

The port of Kobe was devastated by widespread LIQUEFACTION (when the ground moves in waves) which destroyed more than 90 per cent of the port.

Elevated structures such as the Hanshin Expressway and the Shinkansen ('bullet train') track collapsed.

The Japanese Government did not appreciate the scale of the disaster to start with. It took a week to declare the city a disaster zone.

Recovering from the earthquake

The bill to repair all the damage caused by the earthquake was estimated at US$147 billion. However, the Japanese Government was able to set up an emergency fund of around US$10 billion to rebuild the infrastructure. This subsidised up to 90 per cent of the cost of repairing public facilities. Relief grants, low-cost loans and tax breaks for victims helped the city back on the road to recovery. Kobe estimated that it would take five years to build 80,000 houses and flats for those made homeless. But after just one year there were only 1,000 victims still living in tents, and the town won an international award for the speed at which it got back to business.

Task

1 Study Photos 1 and 2, and Map 3, and consider the problems faced by Kobe in the aftermath of the earthquake.

Draw a concept map (see page 32) to show the links between the effects of the earthquake. Use these headings in your concept map:

- Fires
- Bridges
- Expressways
- Pre-1981 buildings
- Rescue workers

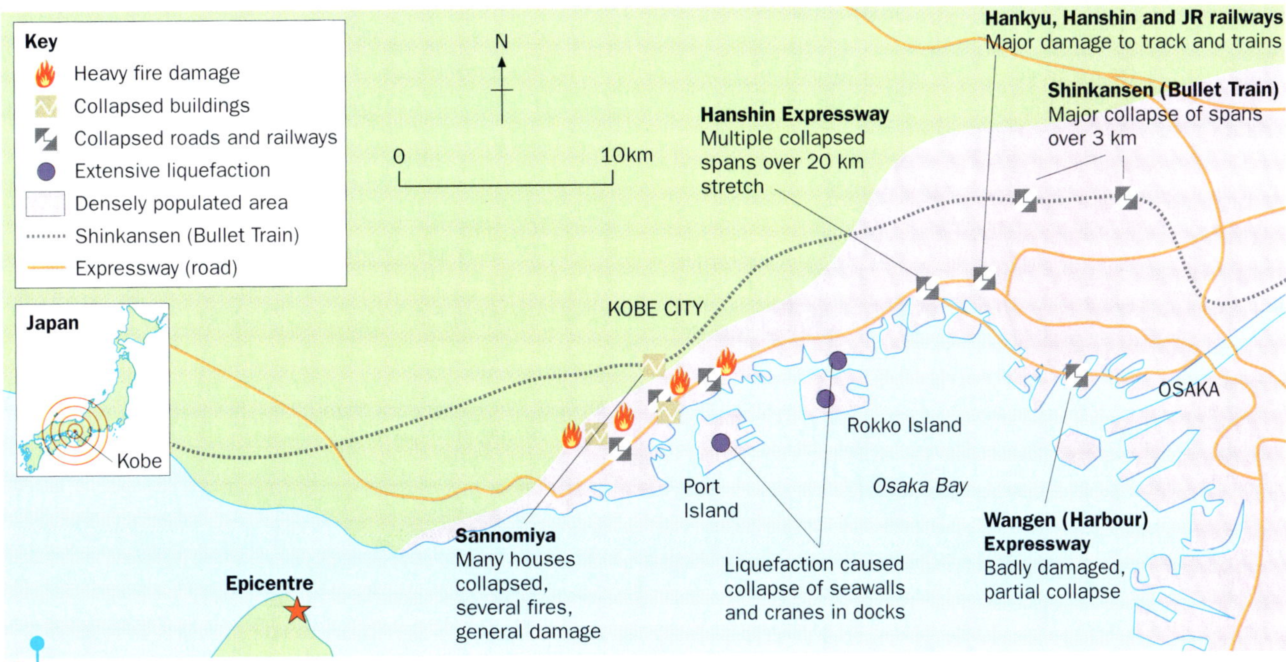

3 Damage in Kobe

What causes earthquakes and volcanic eruptions?

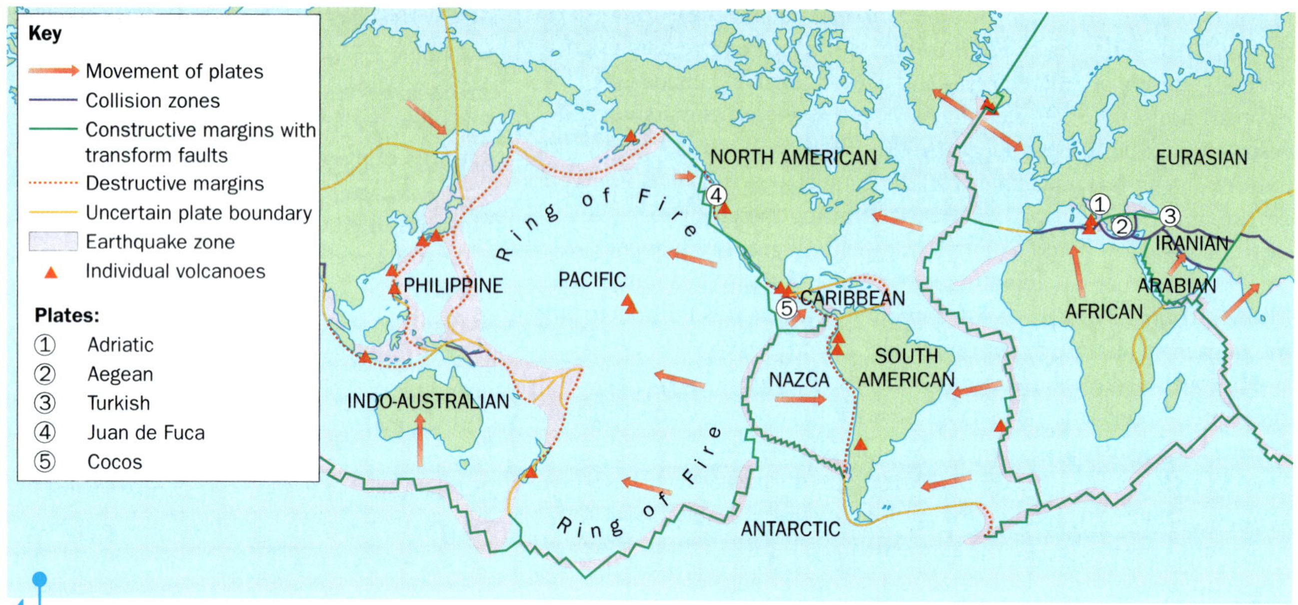

4 The world's plate boundaries

Earthquakes are caused by movements in the Earth's CRUST. The crust is divided up into several major plates (Map 4), which move as a result of CONVECTION CURRENTS in the UPPER MANTLE (Figure 5).

The major landmasses consist mainly of thick CONTINENTAL CRUST. The thinner, younger OCEANIC CRUST pushes the continental crust around. At plate margins, where plates collide, pull apart or slide past each other, tremendous forces are generated. Most earthquakes therefore occur at the edges of plates. Figure 6 shows what happens at different types of plate margins.

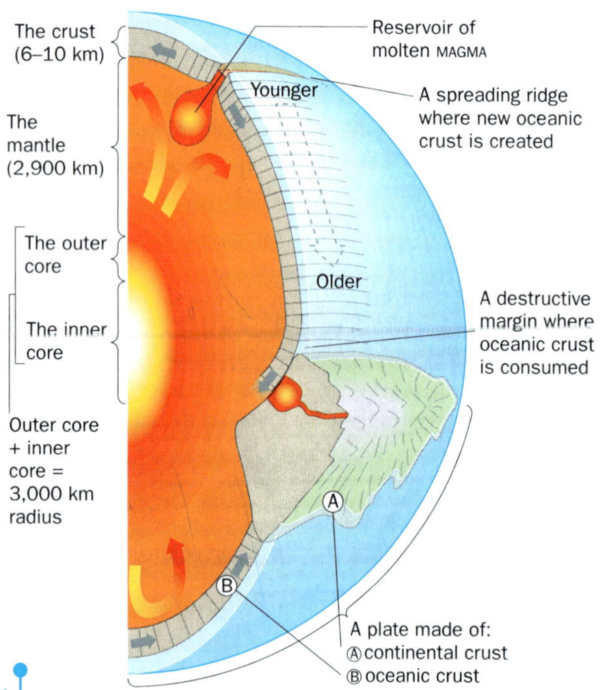

5 The structure of the Earth

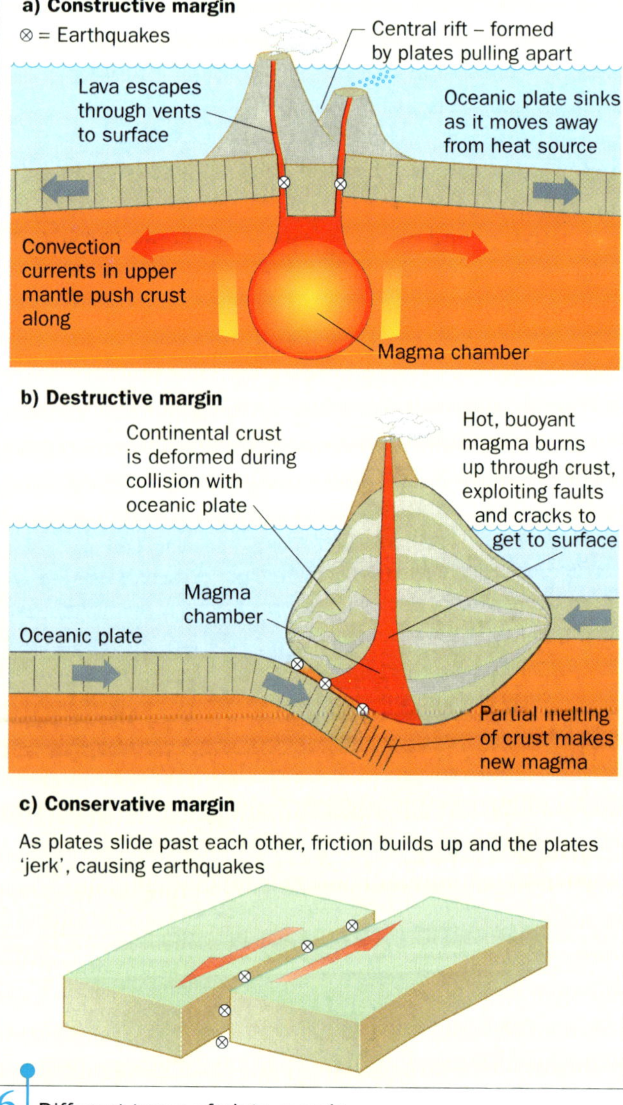

6 Different types of plate margin

A sudden release of pressure along fault lines in the crust causes violent movements, especially at the EPICENTRE, the ground directly above the *focus* of the earthquake (Figure 7). The amount of energy released at the focus determines the size or *magnitude* of the earthquake measured on the Richter scale (Figure 8). Effects can range from minor tremors to shaking so extreme that liquefaction may occur and the ground moves in undulating waves.

AFTERSHOCKS may follow in the days and weeks after the main earthquake as the Earth settles after the initial disturbance. These can cause a further hazard to rescuers and may hinder reconstruction.

The Kobe earthquake was caused by SUBDUCTION. The Philippines oceanic plate was sinking beneath the Eurasian plate. This created stresses that were relieved by a sudden movement along a huge fault line that runs through Osaka Bay to the south of Kobe (Map 9). The sediments and reclaimed land on which Kobe is built increased the effects of seismic waves, and the shaking of buildings was much worse in these areas than elsewhere.

You can see from Figure 6 that the same forces that produce earthquakes are also responsible for volcanoes. You will investigate volcanoes on pages 74–78.

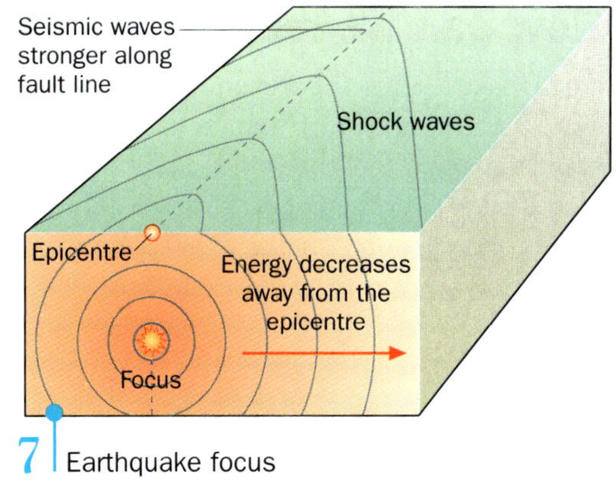

7 Earthquake focus

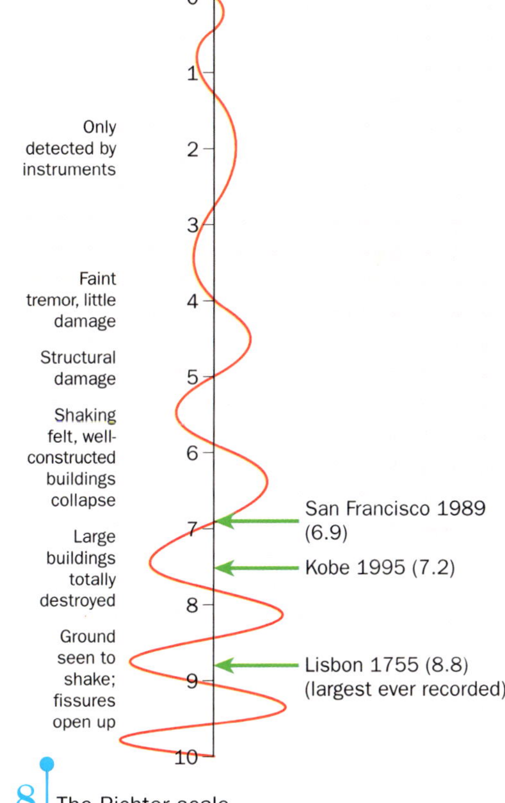

8 The Richter scale

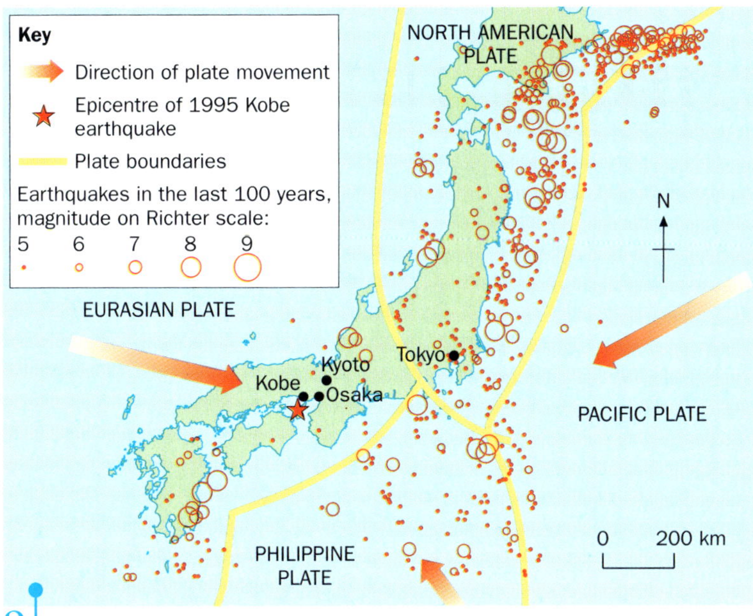

9 Movement of plate boundaries around Japan

Tasks

SAVE AS . . .

1 Study Map 9, and other information on the causes of the Kobe earthquake.

a) Explain the physical causes of the Kobe earthquake, using simple diagrams to help you where necessary.

Key terms to use: oceanic/continental crust, plates, Philippines plate, Eurasian plate, subduction, friction, pressure, focus, epicentre, seismic waves, energy, Richter scale.

b) Study the information on the effects of the earthquake on pages 64 and 65 and above. Which human factors contributed to the scale of the disaster? For each factor below, explain how it made the damage worse:

older buildings, water and gas mains, government confusion, reclaimed land, building on fault lines.

How can we manage earthquakes?

Earthquakes cannot be prevented. They occur very suddenly with little or no warning and are a 'natural hazard' affecting some of the most populated regions of the world. The average annual death-toll from earthquakes over the last century was 10,000. That is like the pupils of ten secondary schools being wiped out every year!

Worldwide, 10,000 tremors and earthquakes occur each year, but most are too small to cause great concern. But 20–50 do cause real damage. Those in areas of high population can be devastating.

Attempts to reduce the impact of earthquakes have largely concentrated on predicting when they will happen and on taking precautions to prepare people for an earthquake and its aftermath. Prediction may already be saving 2,000 lives per year.

Earthquake prediction

Methods of predicting earthquakes range from the observation of erratic animal behaviour (which triggered a mass evacuation in Haicheng, China, just before the earthquake of 1975) to the high-tech solutions in, say, California or Japan (Figure 10).

Seismically quiet periods often precede major shocks as pressure builds up. Minor tremors may signal that a big earthquake is imminent. However, even with the most advanced technology, the closest prediction is to within one year – hardly a sound basis on which to evacuate a city such as San Francisco!

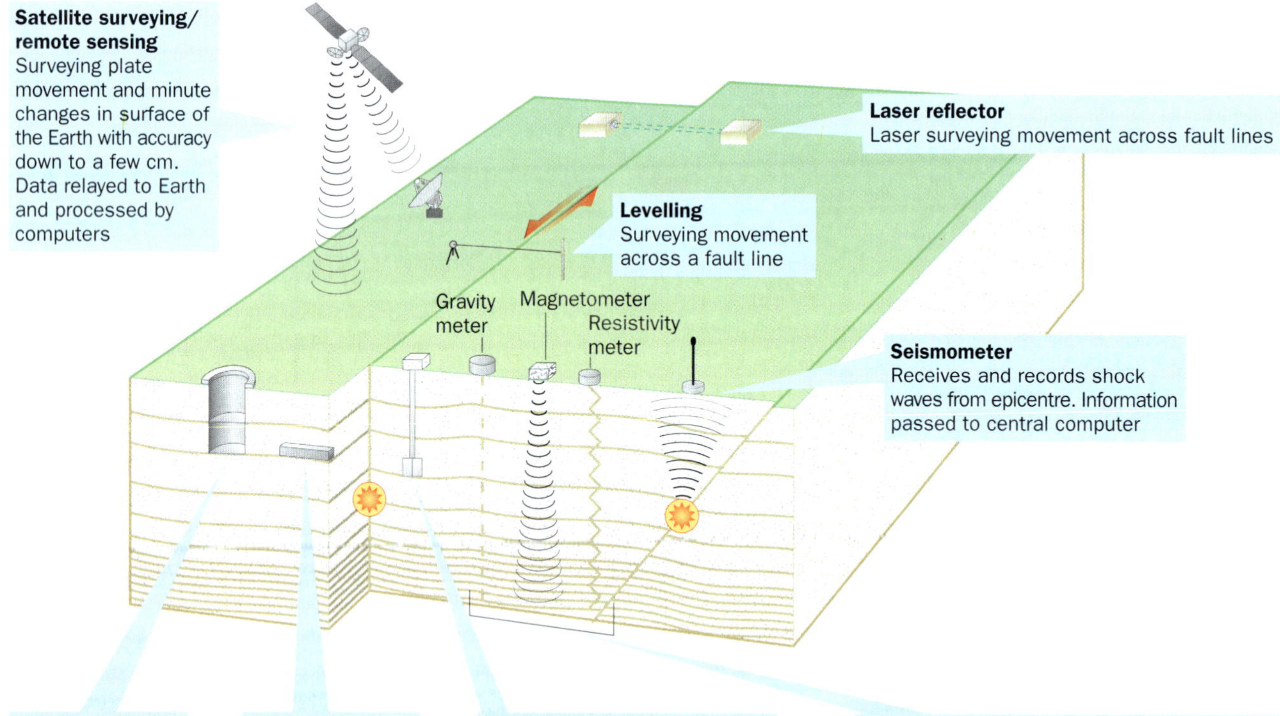
Satellite surveying/ remote sensing
Surveying plate movement and minute changes in surface of the Earth with accuracy down to a few cm. Data relayed to Earth and processed by computers
Laser reflector
Laser surveying movement across fault lines
Levelling
Surveying movement across a fault line
Gravity meter
Magnetometer
Resistivity meter
Seismometer
Receives and records shock waves from epicentre. Information passed to central computer
Water table level
The level of water in wells rises and falls as the ground is stressed before an earthquake
Strain meter
Stretching and compression of the crust is recorded automatically
Radon gas sensor
The inert gas radon is released from within minerals when they are cracked as pressure increases during the build-up to an earthquake
Gravity meter, magnetometer and electrical resistivity meter
Gravity, magnetic properties of rocks and resistance to electrical current all change as stresses increase during the build-up to an earthquake

Earthquake focus

10 Predicting earthquakes

Earthquake precautions

1. Structural planning

During the San Francisco earthquake of 1989, many buildings in the Marina district collapsed. The land had been reclaimed from the sea, and its loose structure amplified the shock waves, causing many buildings to fail. Planning regulations now try to ensure that new buildings are either in areas with solid bedrock or have foundations that are deep enough to reach solid strata.

Old brick buildings proved to be especially vulnerable in San Francisco, and large parts of the older freeway system collapsed. Strengthening older buildings with reinforcing rods is effective but expensive. More often, carbon-fibre textiles are 'glued' around supporting columns to help to reduce the likelihood of collapse.

As rigid buildings are prone to failure, new buildings such as the Transamerica skyscraper are designed with steel frames that permit some movement. Computer-controlled counterbalances (Figure 11) move heavy weights at the top of tower blocks in the opposite direction to that of the shock wave to give buildings greater stability.

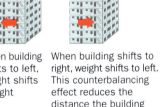

a) Weight on top of the building

b) When building shifts to left, weight shifts to right

c) When building shifts to right, weight shifts to left. This counterbalancing effect reduces the distance the building moves. The shaking stops more quickly

11 Gyroscopic counterbalances (drawing not to scale)

Fire-resistant materials, open spaces and 'BARRIER BUILDINGS', such as the Shirahige-Higashi apartments in Tokyo, can restrict the fires that follow many earthquakes.

2. Emergency drill

In Japan, earthquake drills and signposted evacuation routes are meant to prepare the public for disasters. However, they assume that there will be good warnings and time to organise – things which are often absent in real life.

Earthquake recovery

The long-term effects of an earthquake can be considerable. These include loss of housing, damage to hospitals, loss of drinking water and unsanitary conditions in camps. Until the basic infrastructure is restored, it can be difficult to deliver help. Aid sent to regions with poor communications and weak administration may not reach the real victims. Generally, wealthier countries have the resources to recover faster than poor ones. California has an emergency fund of over US$1 billion for emergency aid and recovery. However, wealth is not the only factor in recovery; organisation is just as important. In 1978 an earthquake measuring 7.8 on the Richter scale struck Tangshan in northern China and killed around 250,000 people. Careful planning meant that a complete reconstruction of the urban area has now been achieved, and all new buildings can withstand tremors up to 8.0. Similarly, the two earthquakes that hit Iran in June 1990 led to the mobilisation of 100,000 volunteers and 10,000 Red Crescent workers. Within 48 hours they had set up care and shelter for REFUGEES, so there was no health crisis.

Tasks

1 Look at Figure 10. Choose three of the methods of earthquake prediction and explain how they might help Japan to anticipate an earthquake event. Complete a table like the one below. (An example has been done for you.)

Method of prediction	How it works	How it could help	Problems
Seismograph	Measures seismic (energy) waves in the ground.	The pattern of seismic waves can only help forecast a major earthquake.	Even the best predictions are only accurate to within one year. Expensive to set up and monitor.

2 Prepare a storyboard for an advertisement for any earthquake precaution method mentioned in the text. Explain how it works and why it should be used, e.g. 'Strengthen older buildings with reinforcing rods' or 'Practise earthquake drills'.

Here is an example:

By checking bridges now we can identify . . .

Damaged girders can be . . .

Although it's expensive, the road . . .

This means rescue vehicles carrying . . .

Case study: Armenia, Colombia – an earthquake in an LEDC

12 Aftermath of the Armenia earthquake in Colombia

Over 1,000 people were killed and many thousands were injured when an earthquake ripped through the central coffee-growing region of Colombia on 25 January 1999. The earthquake, measuring 6.3 on the Richter scale, had its epicentre 170 km west of the capital, Bogotá. It was the worst earthquake in the country since 1983 when 300 people were killed. Scientists said that the earthquake was less than 32 km below the surface – far less than normal. It disrupted telephone communications in the mountainous region and brought buildings crashing down in at least five provinces. Fires raged throughout Pereira, after gas pipes were fractured. Tall buildings swayed in Bogotá. The disaster zone included 20 towns and villages in five provinces, and eight aftershocks triggered landslides that cut off many areas. Armenia, the city that took the brunt of the earthquake, suffered an aftershock measuring 4.2 on the Richter scale, which further hindered rescue attempts.

Even as rescuers fought to clear the rubble, the scale of the effect on the nation became apparent. Colombia had relatively little warning of the earthquake, and few buildings were designed to withstand earthquakes. Within hours a state of national emergency was declared. In towns lacking water and food, looters began to take the law into their own hands. Aid organisations and human rights workers had to battle against paramilitaries seeking to exploit the confusion for their own political reasons. Worries about the damaged coffee crop meant that prices soared on world markets.

13 Location of the Armenia earthquake

14 Earthquake survivor Jayson, aged 12, is rescued by the Red Cross

Tale of two sisters

Saturday 30 January 1999

Carmen Zambrano, aged 76, had been washing clothes on the rooftop terrace when the earthquake struck on Monday. As the building collapsed beneath her, she hit her head on the concrete washtub and was knocked unconscious. Miraculously she survived a four-storey fall, because no rubble landed on top of her, but a head injury required 60 stitches. She was taken away to get treatment soon after the disaster.

Her sister Martha, aged 78, had been watching television on the second floor of the house they had shared for many years and was buried under the debris. 'My legs, chest and arms were pinned down under concrete blocks,' she said. 'I only had a small space around my mouth which allowed me to breathe.' Relatives of the sisters spent Monday night desperately searching for Martha but with no success. 'I could hear them crying my name and moving around above, but I was very weak. They could not hear my calls for help,' she said.

Eventually rescue workers heard Martha's muffled cries and dug for six hours to free her, talking to her constantly. Martha, who was in shock but suffered only cuts and bruises, was taken to hospital.

Carmen, in another city where she had been flown for treatment, had no idea that her sister had been found. So when relatives returned that night from visiting Carmen, they resumed their search for Martha. Her 42-year-old niece, Ruth, said, 'Of course we found nothing. We were beginning to lose hope.' The family even discussed how they would break the news to Carmen who was coming home that day. But at 3 pm on Thursday the sisters were reunited. 'I don't know how or why we are both still here,' Martha said. 'It feels like a miracle.'

Source adapted from an article in the *Guardian*, 30 January 1999.

Tasks

1 Read the article 'Tale of two sisters' and study Photos 12 and 14.

If you had been in the building with the Zambrano sisters and found yourself the next day in the street, how much do you think your life would have changed?

Draw a table like the one below to categorise the different impacts that the earthquake would have on your life.

Immediate impact (during earthquake and hours after)	**Short-term impact** (next 3 days)	**Long-term impact** (weeks and months after)
Violent shaking . . .	Water shortage . . .	Rebuilding home . . .

2 As Earth science reporter for *News 24*, you have to produce an A4-page summary of the causes of the earthquake, entitled 'Why was there an earthquake in Colombia?', to inform viewers and to be posted as a web page.

Your summary must include text, diagrams and location map(s), and should start with a brief description of plate movements in the area.

Look back at pages 66–67 for background information then forward to Figure 21 on page 76 for specific information on plate movements in the region.

Key terms to use: oceanic/continental crust, plates, Nazca plate, South American plate, subduction, friction, pressure, focus, epicentre, seismic waves, energy, Richter scale.

3 Look back at Task 1 on page 69, and the earthquake prediction methods you included in your table.

a) Would your chosen methods and the data they could provide have helped an LEDC like Colombia?

b) Which earthquake prediction methods do you think Colombia might have had access to, and which do you suppose would have been unavailable? Give reasons for your answers.

AT LEAST 250 people were killed and 3,000 injured when an earthquake ripped through the central coffee-growing region of Colombia yesterday. The earthquake was the worst in the country since 1983.
26 January 1999

Emergency declared in 'quake zone
President Pastrana is announcing a state of emergency in the aftermath of the earthquake.
29 January 1999

Earthquake in Colombia
A powerful earthquake has struck western Colombia.
25 January 1999

Colombia starts to rebuild after 'quake
A week after an earthquake struck western Colombia, killing nearly a thousand people, the authorities have started reconstruction.
1 February 1999

UN appeal for Colombia
The United Nations has launched an appeal for 3 million dollars to help victims of last month's earthquake.
5 February 1999

Coffee prices soar after earthquake
Coffee prices rise sharply following the earthquake in Colombia which damaged the country's biggest coffee-growing region.
26 January 1999

Spain approves 50-million-dollar loan to Colombia
The Spanish Government has approved a loan to help earthquake victims, ahead of a visit to Spain by President Pastrana next week.
12 March 1999

UK rescuers fly to Colombia
The UK is sending specialist rescue workers to Colombia following the earthquake that has claimed 2,000 lives.
27 January 1999

Death Toll from Colombian Earthquake Tops 1,000; Nation in Chaos
27 January 1999

Destruction of a proud city
Teenage soldiers enforce a curfew as thousands in Armenia face up to the loss of their relatives and their homes.
27 January 1999

Colombian paramilitaries strike in 'quake aftermath
The aid agency, Oxfam, has accused right-wing paramilitary groups in Colombia of exploiting the devastation and confusion following the earthquake there, to attack human rights workers.
3 February 1999

Looting frenzy in 'quake city
Violence erupts after mobs of hungry earthquake survivors looted shops.
28 January 1999

Colombia relief effort blighted
Four days after the Colombian city of Armenia was devastated by an earthquake, concern is growing that food and medical supplies are not getting through quickly enough.
29 January 1999

15 The Armenia earthquake story unfolds

Tasks

1 Use the headlines in Figure 15 to draw a timeline to show how the events unfolded in the aftermath of the Colombian earthquake. Put events that are specifically Colombian on one side of the timeline, and events involving the wider world on the other, as shown below.

	Colombia		World
Powerful earthquake hits western Colombia		25	
		26	Coffee prices soar
		27	
		28	
		January 1999	

Colombia – the aftermath

2 You have looked at the causes and effects of the Colombian earthquake, but what about this question:

'Why was it unlikely that Colombia would have the resources to cope with this disaster?'

a) The Factfile on page 73 shows natural disasters affecting Colombia from late 1984 up to the January 1999 earthquake. Why is this recent history relevant? What could be the effect of a series of disasters on a country's ability to recover?

b) Table 16 shows the effects on two cities nearest to the epicentre. Which parts of this information could help you to explain the size or nature of the problem?

c) Table 17 shows development criteria for Colombia. Which of these facts help to explain why recovery from an earthquake might be slow?

Now answer the question above in less than one page, using your answers in the order a) – c). Try to relate each point back to the question.

Service	Armenia (population 290,000)	Pereira (population 695,000)
Financial system	Total collapse.	Many financial institutions functioning normally.
Water supply system	Available to a very low percentage of neighbourhoods.	Almost completely restored to entire city.
Sewerage system	Major damage.	Significant damage.
Power supply system	Operating at 50 per cent of normal capacity due to damaged Regivit substation.	Operating at 100 per cent of normal capacity.
Hospital system	City's only hospital and large percentage of health centres affected.	Minor damage.
Human cost	543 people dead, 1,700 injured, and 60 per cent of homes affected.	44 dead, 650 injured and 390 buildings destroyed.

16 Impact on cities near the epicentre, 27 January 1999

Development criteria	Colombia	Japan
Population (million)	30.1	122.4
GNP (US$/capita)	1,240	15,760
Population/doctor	1,190	663
% agricultural workers	30	8
Population density/km^2	28	330
Debt (% of GNP)	41	0

17 Development criteria for Colombia and Japan

★ THINK!

Colombia is classed as an LEDC, but what does this mean and how does it affect a country's ability to recover from disaster? When you answer a question like the one in Task 2, examiners will expect you to give specific data to support your point and to use different sources of evidence to give a full answer to the question.

★ FACTFILE

Recent natural disasters in Colombia

Earthquake	January 1999
Floods	July 1996
Earthquake	March 1995
Earthquake	February 1995
Earthquake	January 1995
Earthquake and mudslides	June 1994
Floods	February 1994
Floods	April 1993
Eruption of Galeras volcano	January 1993
Earthquake and volcanic activity	October 1992
Earthquake	November 1991
Fire	February 1991
Volcanic eruption	September 1989
Eruption of Galeras volcano	April 1989
Floods	August 1988
Landslide	September 1987
Floods	July 1987
Floods	July 1986
Volcanic eruption	May 1986
Volcanic eruption	January 1986
Volcanic eruption	November 1985
Floods	November 1984

Focus task

3 'Why do LEDCs suffer greater damage from natural hazards than MEDCs?' Write a report (1–2 pages), using evidence from pages 64–73 and your own research (e.g. school library and websites) to answer the question.

★ THINK!

When answering a question like Task 3, you can gain a lot of marks by referring to one type of natural hazard in depth. Pages 64–73 have equipped you to use earthquakes as your main example but, if you wish to show that you can apply your wider knowledge, you can use other hazards which you study, such as storms (pages 40–47), floods (pages 48–63) and volcanoes (pages 74–78), to support your arguments.

B VOLCANOES

Photo 18 shows a FISSURE ERUPTION in Iceland where the Earth's crust is being pulled apart. It is dangerous and dramatic – but not typical: most volcanic eruptions occur under the sea or far away from where people live.

Volcanoes can be classified as:

- **Active** – those that have had eruptions within living memory.
- **Dormant** (asleep) – those that have had eruptions recorded in history.
- **Extinct** – those that will not erupt again.

There are estimated to be nearly 550 active volcanoes in the world, of which 15 are classed as 'decade' volcanoes – that is, capable of a major life-endangering eruption within the next ten years. Most of these are at plate margins, but they can develop within plates at hot spots such as Mauna Loa in Hawaii, and where the crust has weaknesses such as in the Kenyan Rift Valley in Africa.

Effects of volcanic eruptions

Volcanoes present dangers to nearby populations as violent eruptions may occur with little warning, and they can be unpredictable in the way they erupt. However, as well as the immediate local dangers posed by molten LAVA, ASH and volcanic bombs (PYROCLASTICS), and mudflows and LAHARS (mixtures of ash and water) can travel hundreds of kilometres away from the source volcano. In Cameroon in 1986, volcanic carbon dioxide burst out from beneath lake muds and suffocated 1,700 people over an area of up to 50 km. The effects can be even more widespread. Eruptions often trigger rainfall, and the ejection of vast amounts of debris into the atmosphere may have a global impact. When Mt Pinatubo in the Philippines erupted in 1991 it sent ash more than 30 km into the atmosphere, partly blocking out sunlight and, some scientists believe, cooling the atmosphere.

18 A volcano in Iceland

What is it like when a volcano erupts?

IN THE PATH OF AN ASH CLOUD

When Mt St Helens erupted in May 1980 in Washinton, USA, two people camping outside the danger zone found that a dense, hot cloud of ash had covered their camp. Jenny Cheney explains what happened next.

'My boyfriend shouted "Get out of the tent", then . . . panic, I found I couldn't breathe and every time I gasped for air I gulped in the ash. I guess I'd fallen to the floor but I can remember that the sky had gone purple-red, though I couldn't make out anything clearly, y'know? I had to dig the ash out of my throat using my fingers – it made me retch but that probably helped. Amazingly, I had the presence of mind to try to take my contacts out – I think they were starting to melt on my eyes because it was so hot. I can't believe I had the time to put them in their case while I was nearly choking to death! It's hard to describe, but imagine having an asthma attack inside a vacuum cleaner bag while it's on!'

Many also believe that the sulphates in the ash could have reduced OZONE levels, resulting in more dangerous ultraviolet light hitting the Earth.

What are the attractions of volcanic regions?

Active volcanic regions attract people for a variety of reasons, such as:

- Tourism – frequent eruptions of Mt Etna in Sicily and Kilauea in Hawaii fill local hotel rooms, and dormant volcanoes with their spectacular scenery are attractive to tourists.
- Fertile soils – volcanic soils are extremely fertile and can produce several crops each year. These high yields mean that a small island like Java can support a population of over 100 million.
- Rocks and rare minerals – volcanic regions contain resources such as basalt, used in construction, and sulphur which has many industrial uses (e.g. cement).
- GEOTHERMAL heat – cheap electricity can be generated from groundwater heated by volcanoes, e.g. Wairakei in New Zealand, and Reykjavik in Iceland where 80 per cent of houses are warmed by geothermal heat. In fact, despite the sub-Arctic climate of Iceland, bananas, peppers and avocados can be grown there in heated greenhouses. Places like Hveragerdi, the 'green' town about 25 km from Reykjavik and built on a particularly active geothermal area, make Iceland Europe's biggest producer of bananas!

Case study: Nevado del Ruiz, Colombia – a volcanic eruption in an LEDC

Nevado del Ruiz is a large ice-capped volcano in the Andes. It is 5,200 m high – about four times the height of Ben Nevis, the highest mountain in the UK. It had twice before erupted and devastated the area around Armero in Colombia (Map 19). In 1595, and again in 1845, the combination of ash/lava eruptions and melting ice had sent lahars, or mudflows, cascading down the steep flanks of the volcano. The eruptions of 1985 began relatively quietly but eventually led to the highest death-toll caused by a volcanic eruption during the 20th century. This is how events unfolded.

19 Major volcanoes in Colombia

11 September 1985 *Gas/steam eruptions began. These are often a sign of a bigger eruption to follow.*

Late September–early November *Increased melting of glaciers produced a mudflow that travelled nearly 30 km down the Languinillas valley.*

13 November
At around 8.30 pm an explosive eruption ripped through the summit and 20 million m^3 of pyroclastics were strewn across the snow-covered glacier. Clouds of gas and hot ash helped to melt the ice, and hot lahars soon formed, moving at speeds of up to 50 km per hour. These picked up debris and swamped the mountain valleys to a depth of 40 m in places (Photo 20).

At around 11.00 pm the town of Armero, 74 km away, was hit by a lahar whilst most of its inhabitants were asleep. Final estimates put the death toll at 23,000 (over 90 per cent of the population of Armero). Most people were drowned in the 3–5 m layer of mud.

20 Valley filled with lahars, Colombia

The Colombian authorities found it hard to locate and rescue survivors, as roads were lost under the debris, and helicopters could not easily land on the unstable lahars. Other problems caused by the eruption were the contamination of water supplies and the loss of the best-quality farmland in the valleys.

What caused the eruption at Nevado del Ruiz?

Nevado del Ruiz is part of the Andean volcanic mountain range, which itself is part of the much larger 'Ring of Fire' (see Map 4 on page 66). Subduction of the oceanic crust at plate boundaries at the edge of the Pacific Ocean gives rise to a belt of volcanoes which helps to form islands and mountain ranges in a broken circle around the ocean margins.

As the oceanic Nazca plate sinks beneath the continental crust of the South American plate, friction leads to partial melting of the descending plate (Figure 21). This generates MAGMAS which rise through the continental rocks above to form ANDESITIC volcanoes, which are prone to violent ash and lava eruptions (Figure 22).

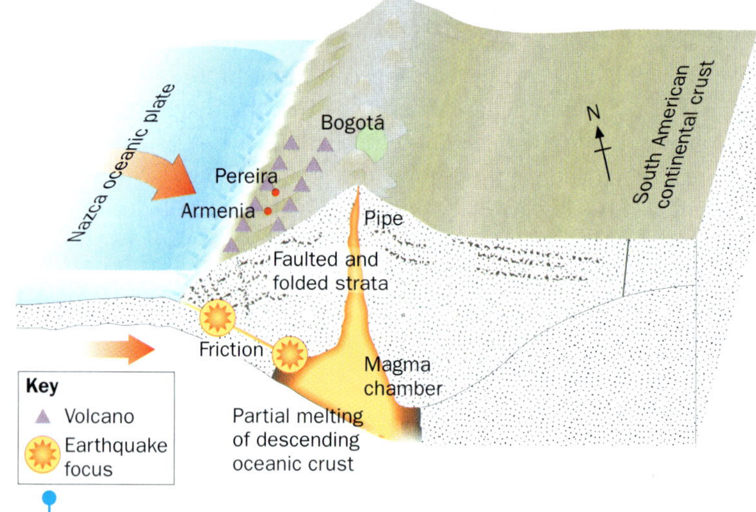

21 The Colombian plate boundary

Tasks

1 Draw copies of the simple diagrams below. Annotate the diagrams using the key words to explain why Nevado del Ruiz erupted.

Key words: Nazca plate, oceanic crust, subduction, partial melting, magma chamber, volcano, eruption, South American plate, continental crust, Andes mountains.

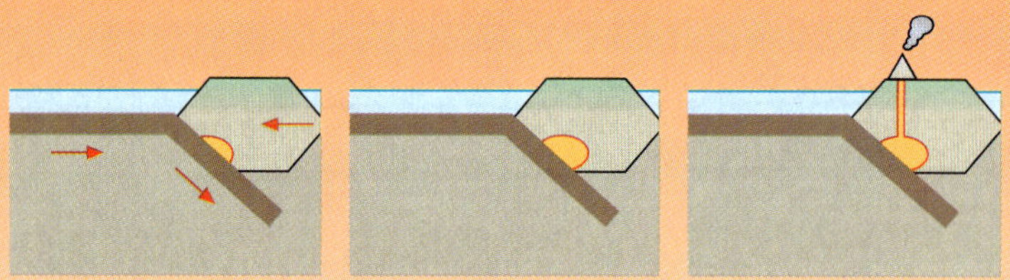

2 Study Map 19, Photo 20 and the text describing the effects of the Nevado del Ruiz eruption.

Try to find three *physical* and three *human* reasons for the high death-toll.

ICT

3 Pages 75–76 look at Nevado del Ruiz. There is a website called Volcano World which will give you more information on this and other volcanoes. You can find it at:

http://volcano.und.nodak.edu/

a) Choose either Nevado del Ruiz or another volcano and select two photos from the website which you think give information about:
- causes of eruption
- details of the eruption
- impact on the local area.

b) Label features on each photo, using the information in this book or from the internet.

The first part of the eruption of andesitic volcanoes produces ash clouds as the volcanic gases explode through the ash layers

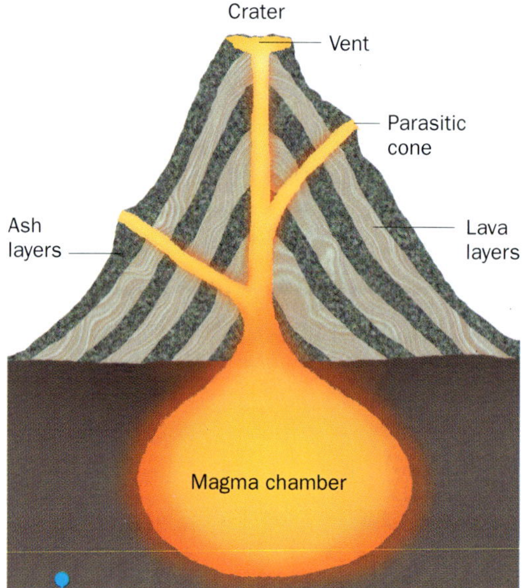

22 Cross-section through an andesitic volcano

How can we manage volcanic eruptions?

Volcanic eruptions: predictions

Figure 23 shows some methods used in the prediction of volcanic eruptions:

- Tiltmeters record shifts in the ground as magma rises towards the vent.
- Seismometers relay shock-wave data from the tremors associated with changes in the magma chambers.
- Analysis of gases from the vent gives information about the nature of the magma.
- REMOTE SENSING (global positioning system (GPS)/satellite data) records changes in the form (morphology) of the volcano.

The eruption of Mt St Helens in the USA in May 1980 was successfully predicted by careful monitoring of:

- earth tremors, created as magma rose prior to the eruption and made the ground shake
- changes in gas emissions, as gases changed in composition near to the eruption
- the physical swelling of the mountain as the magma chamber neared the surface and gas pressure increased.

All of the above were measured just before the final eruption, and the area was evacuated. This was largely successful, but the scale of the explosion was unexpected. The northwards direction of the blast led to the destruction of 550 km^2 of forest. The campers in the article on page 74 were well outside the evacuated area.

At Nevado del Ruiz (see pages 75–76), seismic stations near the summit sent information about the volcano's activity to Colombian emergency-response co-ordinators. However, a storm prevented accurate observation of the eruption itself. Warnings which were meant to alert the public to the danger were issued too late. This may have contributed to the final death-toll.

23 Methods of predicting volcanic eruptions

Volcanic eruptions: precautions

A range of methods can be used to try to control the effect of eruptions to some extent (Figure 24).

Loss of life can be minimised if the most risky locations are evacuated. At Nevado del Ruiz, areas known to be in the pathways of lahars had been identified by volcanic hazard mapping. However, this work was completed barely a month before the main eruption, so there was little time to take proper precautions against the threat of a major eruption.

Lava from the 1973 eruption of Heimaey in Iceland threatened to engulf a harbour, until ships sprayed seawater on the lava to cool and eventually halt the flow. In 1992, helicopters diverted lava flows from Mt Etna (Sicily) on to safer courses by dropping concrete blocks into the main lava channel. Earlier attempts at slowing the flow by controlled explosions had failed, and previous attempts at constructing concrete barriers had met with little success.

Despite people's success in predicting and to some extent controlling eruptions, technology doesn't always provide the solution. In many endangered areas a fatalistic attitude prevails – in the event of an eruption people evacuate the area, then return to rebuild their lives after the eruption.

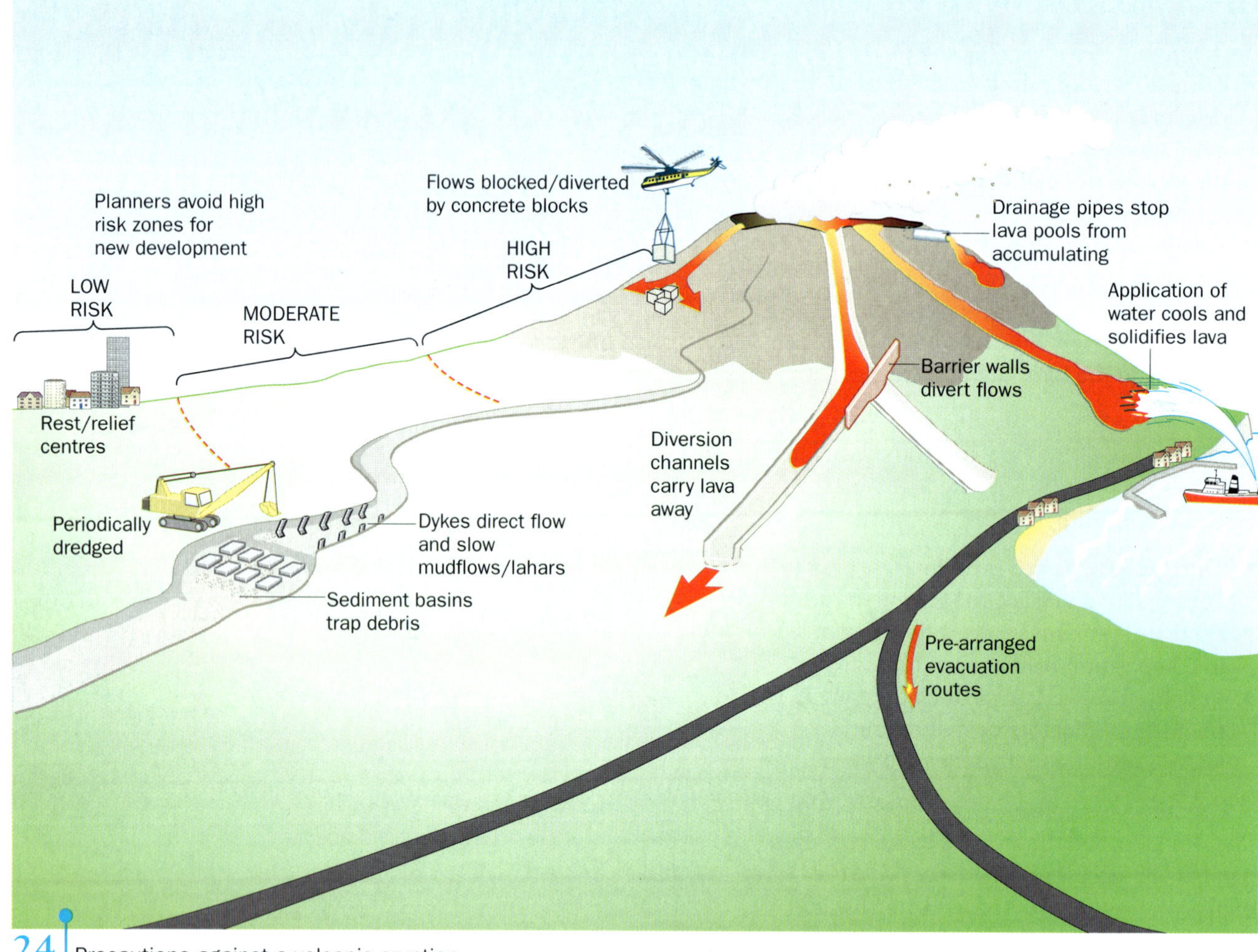

24 Precautions against a volcanic eruption

Task

SAVE AS . . .

1 We know how dangerous eruptions can be, yet many people live in hazardous volcanic environments. This can't just be because there is a shortage of land. You may well be asked in your exam to explain what the benefits are.

Draw a table like the one below to explain why people want to live in dangerous volcanic environments. Fill it in using the information on page 74. One section has been done for you.

Feature	Country and/ or volcano	Benefits
Weathering of new lavas releases nutrients	*Java*	*Crops grow well, supporting larger populations*

Focus task

2 Write a one-page report to discuss the following hypothesis:

Volcanoes cannot be controlled, but the damage they cause can be reduced.

In your report you will need to:

- say what volcanoes are and where they occur
- give specific examples of the damage they can do
- explain how they can be controlled, using real examples (methods and places)
- conclude with limitations of the methods, using real examples from different locations.

Physical Environment

Should we try to stop environmental change or simply go with the flow?

Holbeck Hall Hotel near Scarborough collapsed into the sea in 1993 because the waves eroded the land beneath it.

As it began to tumble into the sea, people asked: 'Why is it not being defended?'

The Norfolk coast at Sea Palling is being defended. It is protected by a series of specially built offshore reefs that produce small bays behind them. These reduce the power of the waves. A sand and shingle beach protects the cliffs from excessive erosion. As soon as the scheme was proposed, people asked: 'Is this really worth the trouble and money it costs?'

These two parts of England's eroding coastline have received very different treatment. Which do you think is right?

As global warming increases, sea-levels will rise, and changes to the coastline will happen faster than ever, over a wider area. More places will face the tough questions raised by any attempt to hold back environmental change:

- Why should an environment be saved?
- Who should pay to save it?
- Who will really benefit and for how long?
- Is it really worth it?
- Should we try to stop the changes or let nature take its course?

3.1 What affects our weather and climate?

Who's worried about the weather?

The weather is a constantly changing part of our lives, affecting our work and leisure time. Although variable, it is increasingly predictable thanks to a century of scientific and technological advances that have given us a clearer picture of weather and climate patterns.

Tasks

At 6 am on 3 September a low pressure zone is approaching Britain and Ireland at 50 km/h, bringing unsettled weather. As a consultant for Weather Warnings, your job is to make accurate local predictions based on data from the Meteorological Office and other weather data organisations.

By 7 am you have received three requests for weather predictions (see Figure 1). Each is for a different location and different timescale.

This is a tricky task but you have to do your best. If your customers feel that they are getting reliable advice they will use your service again.

You could do all three reports yourself, or work in groups of three doing one each.

1 Figure 2 shows the weather system at 6 am. On your own copy of the map from Figure 2 plot the position of the fronts at three-hourly intervals. This will be quite a complex map so use a different colour for each interval. For example, mark 9 am in red, 12 pm in green, and so on.
2 Mark on your map the location of the three customers.
3 Think about the progress of the weather over their site. When will it be raining? When will it be fine? Then use your own judgement to decide when each customer should experience the best weather. You might think that some of them should call off their activity altogether!
4 Choose one customer and compile a report for them explaining the best timing for their activity and why this is the best time. Mention rain, cloud cover, wind, temperature and any other details you think are relevant.
5 Write reports for the other two customers in the same way.

Cereal Group Inc.
Need a two-day weather window to harvest wheat in East Anglia. One dry day means only completing half of the task. The recent cool weather means that the harvest is behind schedule, and money is lost when the combines are idle.

Cornish Air Tours
They have two hour-long pleasure flights scheduled for lunchtime and have a coastal photography project flight going out at 5 pm. For both trips they need a relatively clear sky and good visibility. Very strong winds are uncomfortable for passengers.

Liverpool Pro-Celebrity Tennis Tournament
The organisers plan to start the tournament at noon and need dry conditions for three hours. Strong winds might be a problem. They can change the start time by two hours or delay for 24 hours if better conditions are promised.

1 Who's worried about the weather?

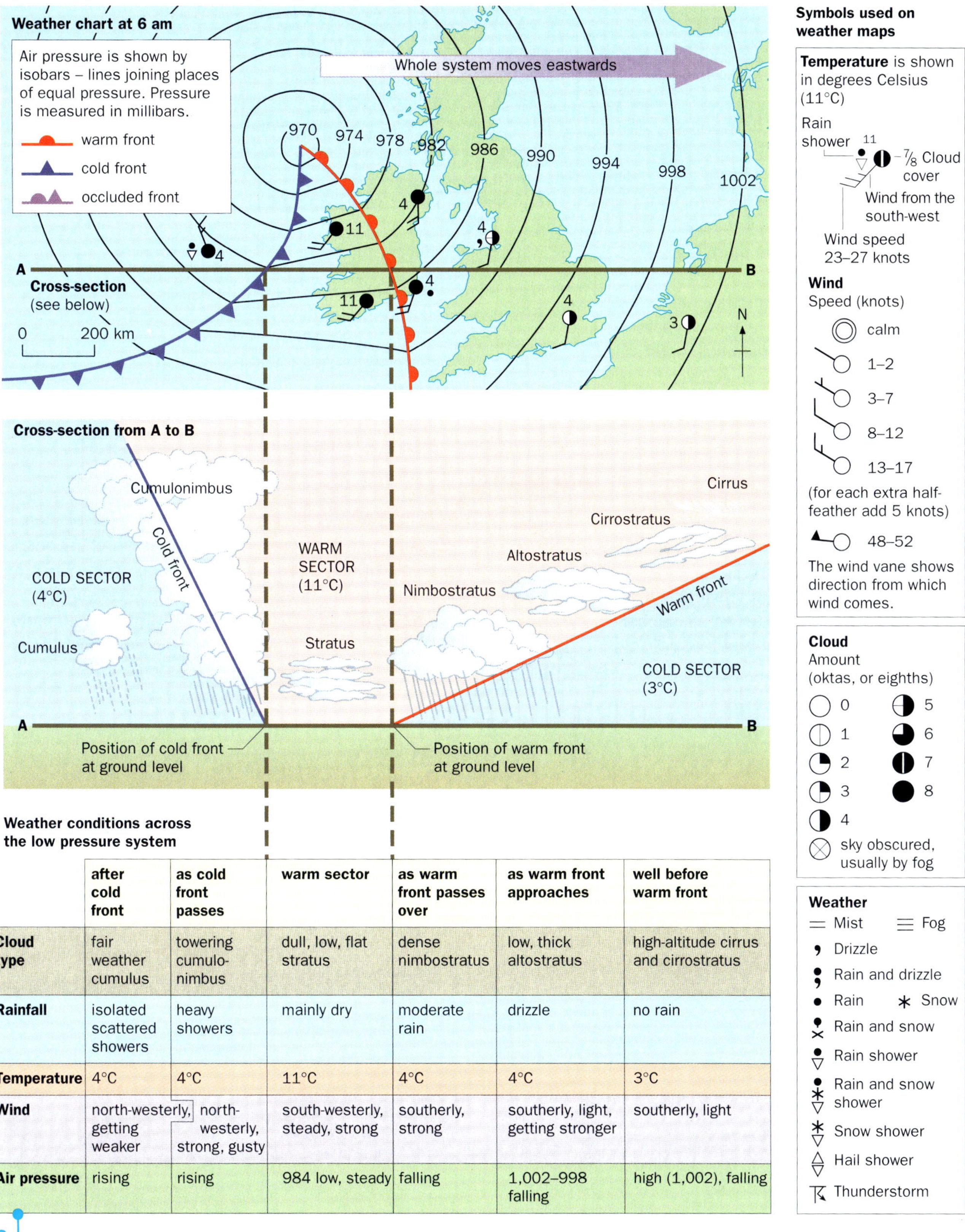

Weather conditions across the low pressure system

	after cold front	as cold front passes	warm sector	as warm front passes over	as warm front approaches	well before warm front
Cloud type	fair weather cumulus	towering cumulo-nimbus	dull, low, flat stratus	dense nimbostratus	low, thick altostratus	high-altitude cirrus and cirrostratus
Rainfall	isolated scattered showers	heavy showers	mainly dry	moderate rain	drizzle	no rain
Temperature	4°C	4°C	11°C	4°C	4°C	3°C
Wind	north-westerly, getting weaker	north-westerly, strong, gusty	south-westerly, steady, strong	southerly, strong	southerly, light, getting stronger	southerly, light
Air pressure	rising	rising	984 low, steady	falling	1,002–998 falling	high (1,002), falling

2 The passing of a front

DO YOU KNOW?

- *Weather* is the day-to-day condition of the atmosphere. *Climate* is the average weather for a region.
- Human activities may have serious long-term impacts on weather and climate.
- Global warming could lead to a 3 °C fall in winter temperatures of Britain and Ireland, because it may change the direction of flow of the warm NORTH ATLANTIC DRIFT current.
- The eruption of large volcanoes throws ash into the atmosphere and this can have noticeable effects on the whole world's climate for years after the eruption.

AIMS

- **To explain how weather affects our daily lives.**
- **To understand the way in which basic weather systems work.**
- **To recognise the ways in which we alter both our own MICROCLIMATES and global climate patterns.**

A WEATHER AND CLIMATE

The first tasks in this chapter gave you a lot of information. You may already know some of it, but other details will be new to you.

The MODEL of a low pressure system passing over Britain and Ireland (Figure 2) was based on several examples. This means that it is *typical*, but every weather system is slightly different. Forecasters therefore use *averages* to help to predict the weather.

Think about all the weather variables you considered before making your decision about when the customers should carry out their activities: the wind speed, the rainfall, the cloud cover…

- How might these factors vary by the hour, day or month?
- What additional information would you have found useful in reaching your decision and where might it have come from?
- How disastrous would it have been if your predictions for your clients had been wrong – life threatening, or merely expensive?

Weather and climate in Britain and Ireland

The weather in Britain and Ireland is changeable because of their location, but the climate has characteristic features which means it is classified as temperate or mild. Despite rapid changes from sunshine to showers, Britain and Ireland rarely experience the extremes of heat and cold, rain and drought that are common in other climates. Maps 3, 4 and 6 show how the climate varies from one season to another, and from one region to another.

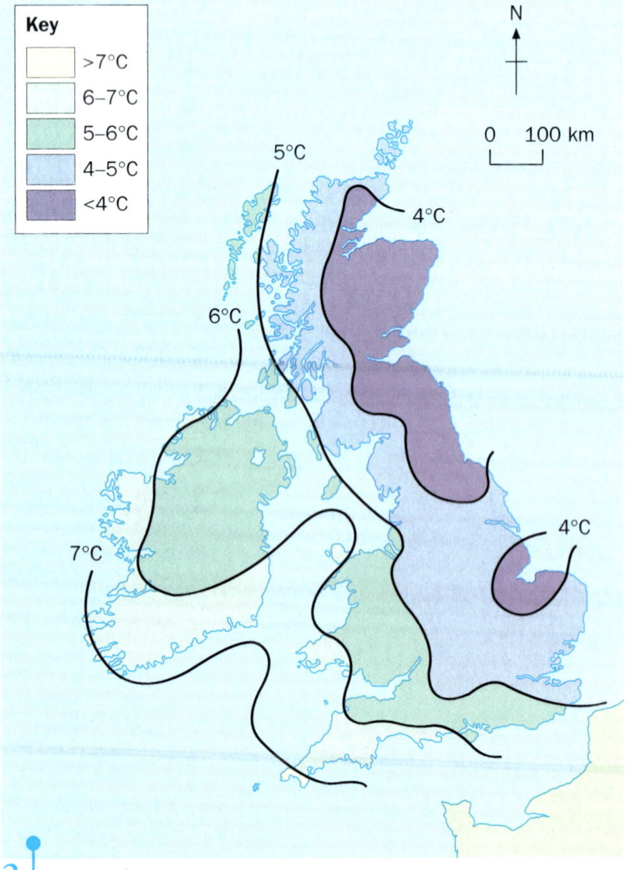

3 Winter temperatures in Britain and Ireland

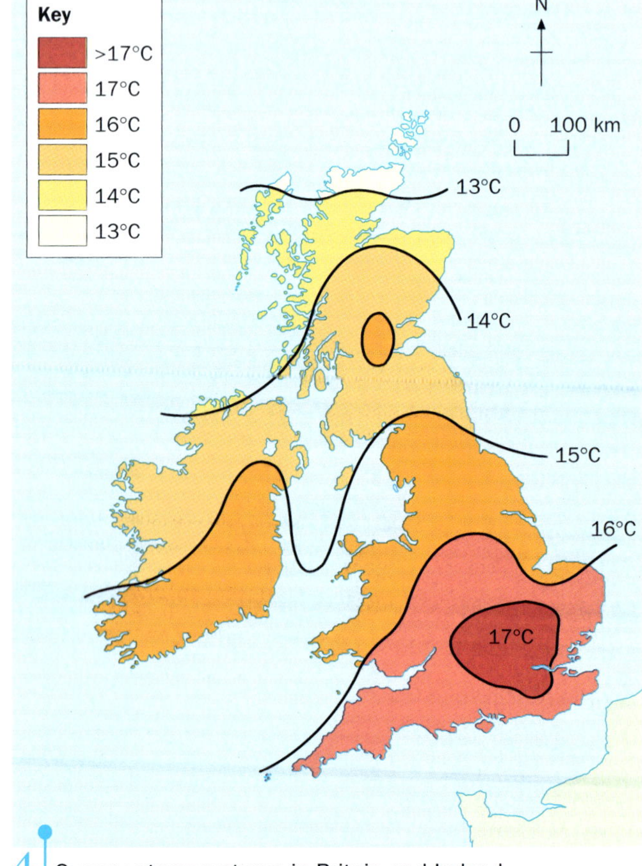

4 Summer temperatures in Britain and Ireland

The ISOTHERMS (lines joining places of equal temperature) show that in summer it is warmer in the south than in the north, with the south-east experiencing the highest average temperatures. It is cooler nearer the coasts. During the winter it is warmer in the west (especially the south-west), with temperatures falling to the east. The sea is warmer than the land.

Map 6 shows that the north and west receive greater amounts of rainfall during a year, whilst the south and east are significantly drier.

Map 7 summarises the temperature and rainfall patterns, showing how the climate of Britain and Ireland can be broken up into quadrants. However, a generalised picture such as this can be misleading. Read the appropriate description for your area, and consider how accurate it is. What could cause variation within one sector?

Factors affecting temperature

Variations in temperature result from the interaction of many factors.

- In Britain and Ireland it is LATITUDE that largely explains why temperatures are lower as you go north (Figure 5). The Sun's rays are concentrated near the EQUATOR. Nearer to the POLES they hit the Earth at an angle, so energy and heat are spread over a wider area. Also, they pass through a greater depth of the ATMOSPHERE, which increases the chances of energy being reflected by dust and other particles. This results in areas further away from the Equator receiving less SOLAR RADIATION, so the north of Britain is cooler than the south.

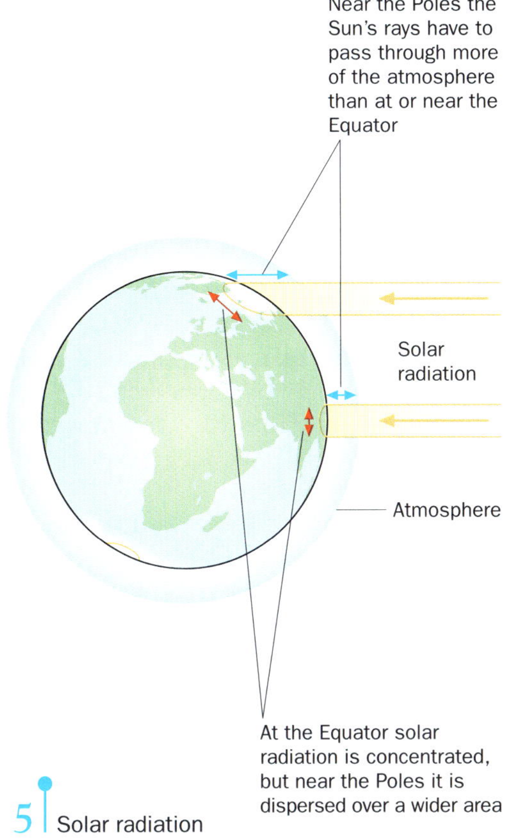

5 Solar radiation

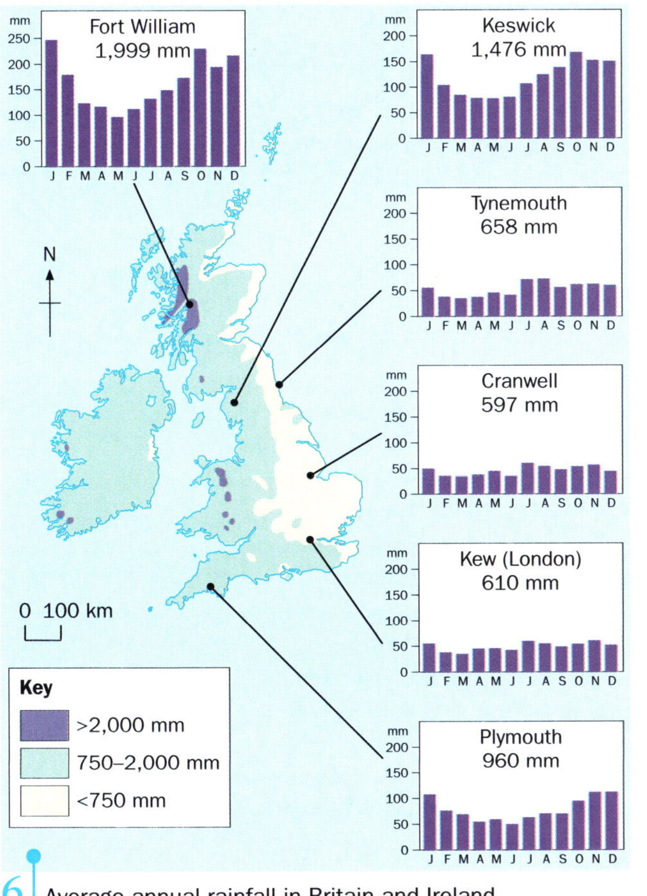

6 Average annual rainfall in Britain and Ireland

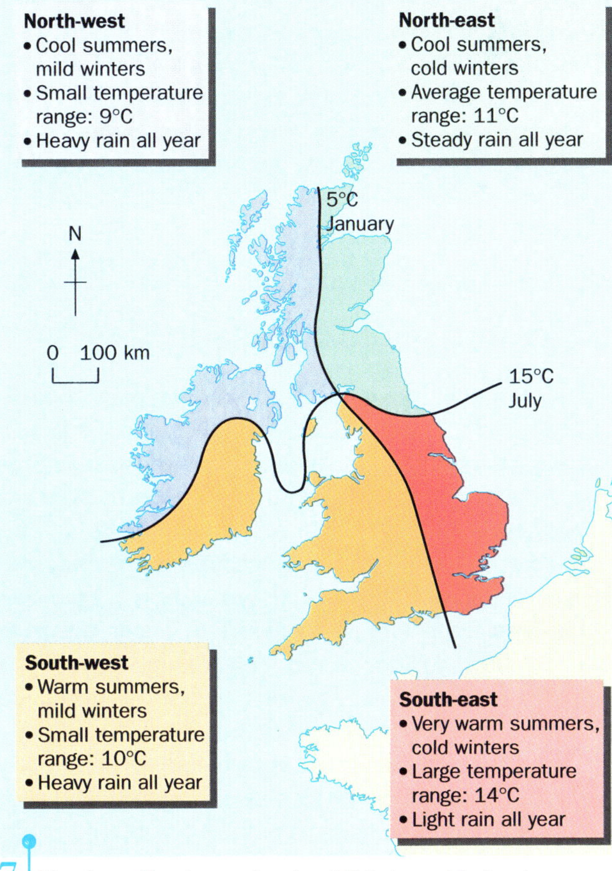

7 The four climate quadrants of Britain and Ireland

- The altitude/elevation of the land also affects temperature, which falls by about 1°C for each 100m in height. The top of Scafell Pike (974m) in the Lake District is generally about 10 °C cooler than St Bees on the Cumbrian coast, even though it is at approximately the same latitude.
- Distance from the sea has a significant effect on temperature, because it takes more energy to heat the sea. It heats up more slowly than the land so it is relatively cooler in summer. It also cools down more slowly so is relatively warmer than the land in winter (Figure 8). Places near the coast are cooled by the sea in summer and warmed by it in winter.
- The PREVAILING WINDS are the dominant winds for any area. If they blow in from the sea, the cooling (summer) and warming (winter) effects explained above are evident. This is true for Britain and Ireland, as the most common winds come from the south-west, across the mid-Atlantic Ocean. Five main air masses dominate the climate of Britain and Ireland, and each has its characteristic weather patterns (Map 9).
- 'ASPECT' refers to the direction the land faces. In Britain and Ireland land facing south receives more sunshine and is warmer than north-facing slopes. This affects farming and building design.
- Ocean currents bring air masses that reflect their properties. The North Atlantic Drift brings warm Gulf Stream waters to the western coasts of the UK, raising air temperatures so that palm trees can grow in some places along the coasts of Wales and Scotland.

Season	Sea	West coast	Land	East coast	Sea	Season
Winter	Warm	Warm wind	Cold	Cold wind	Warm	Winter
Summer	Cool	Cool wind	Warm	Warm wind	Cool	Summer

8 How the sea affects temperature

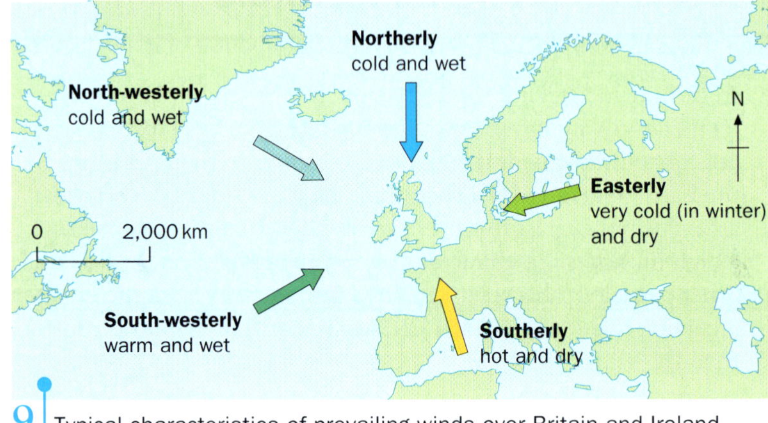

9 Typical characteristics of prevailing winds over Britain and Ireland

Types of rainfall

All rainfall is caused by warm, moist air being forced to rise and cool until the 'DEW POINT' is reached and condensation begins (water vapour becomes water droplets). The three main types of rainfall affecting Britain are shown in Figure 10:

- Relief rainfall (orographic) occurs when moist air is blown over Britain by the prevailing winds and is forced to rise up by higher land. As the rising air cools it leads to rainfall. This type of rainfall is especially common in the west and north. Heavy precipitation over the high peaks means there is less moisture on the leeward side of the high land (the RAINSHADOW effect). For example, Scafell Pike in the Lake District receives about 3,000 mm per annum, whereas Tynemouth, in the east, records only 658 mm per annum.
- Frontal rainfall occurs when warmer air is pushed up over a wedge of cooler air as two air masses meet in a frontal system. This commonly occurs when warmer tropical air is undercut by cooler polar air. As the rising warm air cools, cloud forms and precipitation falls over a wide area. The low pressure systems (depressions) that bring this rainfall usually arrive from the west and are more frequent in winter.
- Convectional rainfall happens when the ground is heated over a long period, especially during the summer. The air above is warmed, and becomes less dense, so it rises. When dew point is reached, condensation occurs but the air may continue to rise, forming tall clouds that often result in thunderstorms. This type of rain is largely responsible for the summer rainfall in the warmer south and east of Britain.

Tasks

SAVE AS . . .

1 Study Maps 3, 4, 6, 7 and 10. Make a copy of the table below, then try to identify the reasons for the variations in climate of the places listed, and fill in your table.

Climatic variations

Place	Latitude	Distance from sea	Relief/elevation	Prevailing winds	Other?
Keswick	North of England – lower temperatures and less sunny overall. More chance of rain.	Close to west coast – more chance of rain as moist air comes in off the sea.	Surrounded by hills up to over 800 m – high rainfall.	Mainly from west – frontal rainfall as winds blow off Atlantic Ocean and Irish Sea.	
Plymouth					
Cranwell					
Tynemouth					
Your town					

2 Study the information on 'Types of rainfall' in Figure 10. Draw flow diagrams to explain how relief, frontal and convectional rainfall occur. For example, for convectional rainfall the flow diagram might start off like this:

Sun's rays heat the ground → Air above ground is warmed → Lower-density warm air begins to rise

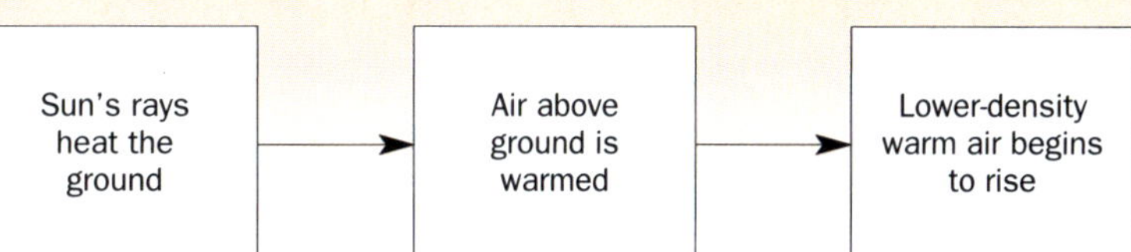

10 Types of rainfall in Britain

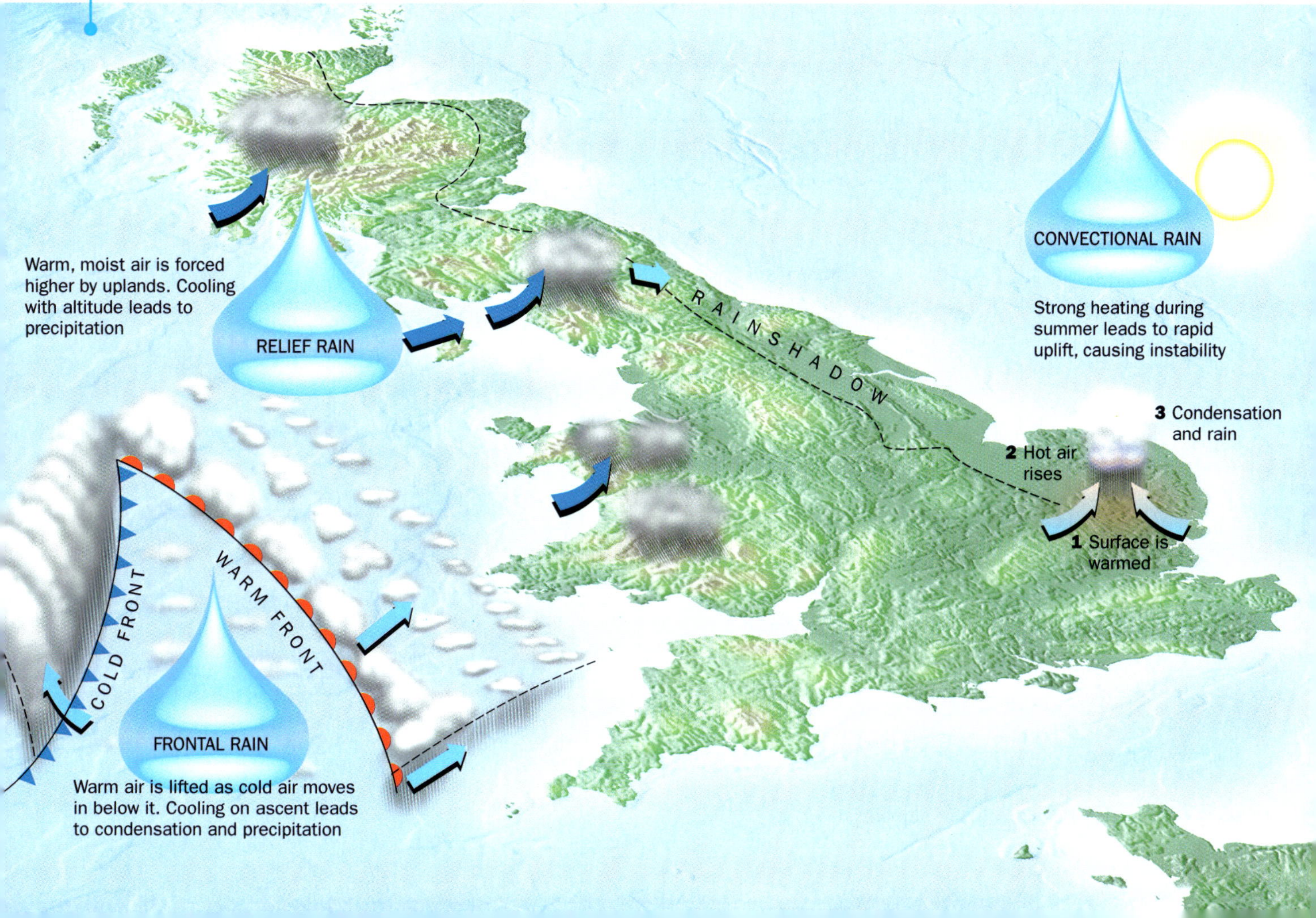

Warm, moist air is forced higher by uplands. Cooling with altitude leads to precipitation
RELIEF RAIN
RAINSHADOW
CONVECTIONAL RAIN
Strong heating during summer leads to rapid uplift, causing instability
3 Condensation and rain
2 Hot air rises
1 Surface is warmed
COLD FRONT
WARM FRONT
FRONTAL RAIN
Warm air is lifted as cold air moves in below it. Cooling on ascent leads to condensation and precipitation

Depressions . . .

The tasks on page 80 looked at the effects of a low pressure system – a depression – passing over Britain and Ireland. Depressions bring weather that is dominated by cloud, rain and wind, and they are very common in Britain and Ireland for much of the year.

Depressions form in the Atlantic where lighter, warmer air from the south (tropical maritime) meets denser, cooler air from the north (polar maritime). The differences in density mean that they don't really mix, and warmer air is forced to rise as the denser cool air undercuts it – see Figure 11. As the whole system moves eastwards, the low pressure air in the warm central sector is pushed up over the wedge of colder air in its path. As it rises it cools, leading to condensation, cloud formation and eventually precipitation. The gentle gradient of the WARM FRONT (where the cold air stops and the warm air begins) means that cloud and rain affect large areas as the depression approaches.

Air rushing into this low pressure zone creates strong winds, spiralling anticlockwise (in the northern hemisphere) into the centre. The lower the pressure the stronger the uplift in the warm sector and thus the stronger the winds.

Inside the warm sector, conditions improve slightly with cloud and rain being weaker and sunshine breaking through at times. The COLD FRONT travels faster than the warm front and, with a steeper gradient, undercuts the rear of the warmer air. This forces moist air into a rapid ascent, leading to towering cloud formations, usually accompanied by heavy rainfall. As the cold front passes, pressure rises, temperatures fall, cloud cover begins to break up and precipitation dies away.

. . . and anticyclones

Anticyclones occur when high pressure dominates the weather. Cooler descending air creates higher pressure and more stable conditions over a wide area. As the air descends it warms, enabling it to hold more moisture without condensation occurring. Because of this, high pressure zones are associated with settled, dry and bright conditions. The ISOBARS are far apart, so the winds tend to blow weakly in a clockwise direction or may even stop, leaving calm conditions. Map 12 shows a SYNOPTIC CHART which features both a cyclone and an anticyclone. The anticyclone is centred over the coast of northern France but also covers most of south east England.

Although anticyclones are rarer than depressions in Britain and Ireland, they bring very distinctive weather, depending on the time of year. In summer cloudless skies as shown in Photo 13, and strong sunshine bring rising temperatures and fine weather during the day. This is followed by rapid cooling at night as a result of radiation heat loss. In winter the rapid night-time cooling often leads to the formation of fog as the ground cools the air above it. While strong sunshine quickly evaporates the summer mists, the weaker, low-angled rays of the winter sun cannot disperse the frost and fog, and so they may persist during the day.

Large high pressure systems over Europe also help to determine the paths of depressions over Britain and Ireland as the low pressure systems are deflected by the stable anticyclones (see page 56). When the European highs are weak, low pressure systems track further south across Britain and Ireland and Europe, and can lead to storms and flooding.

Cumulonimbus
Warm front
Warm
Cirrus
Cold
Cold front
Stratus
Cold
Rain
1

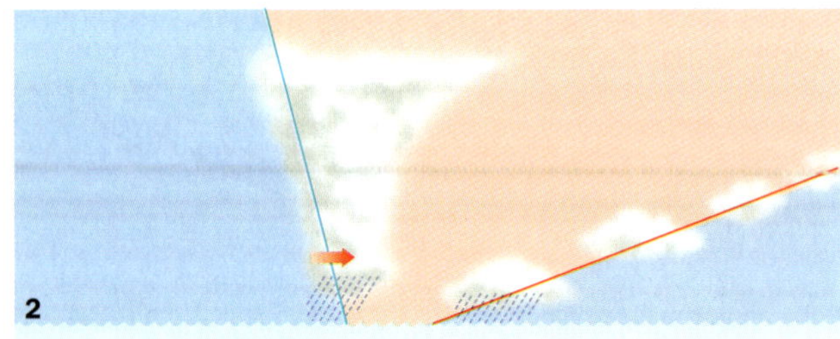

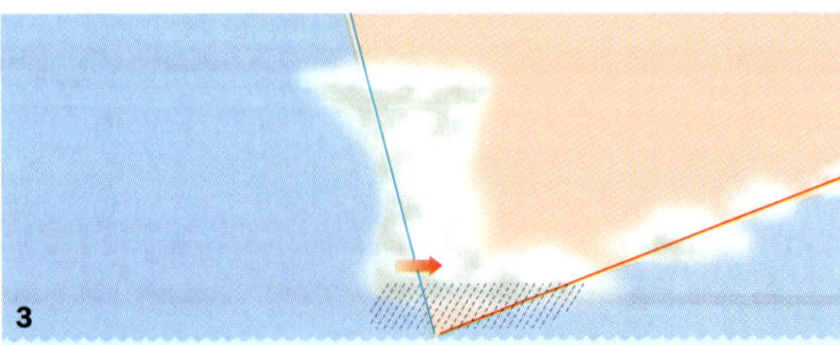

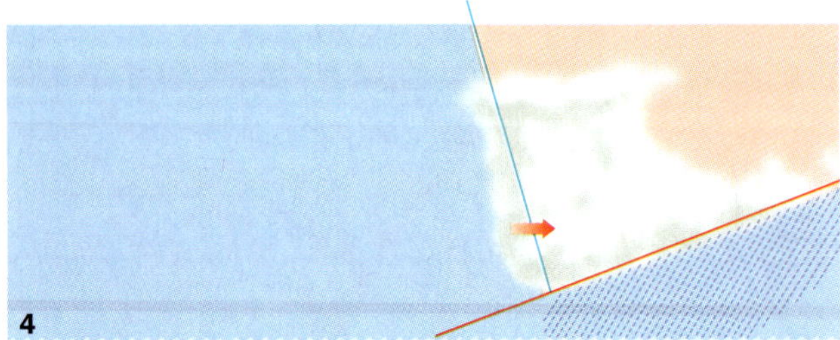

The warm front is at a lower angle (**1**). The steeper-angled cold front travels faster (**2**) until it catches up with the warm front (**3**) and rides up the warm front to form an occlusion (**4**).

11 The development of a depression

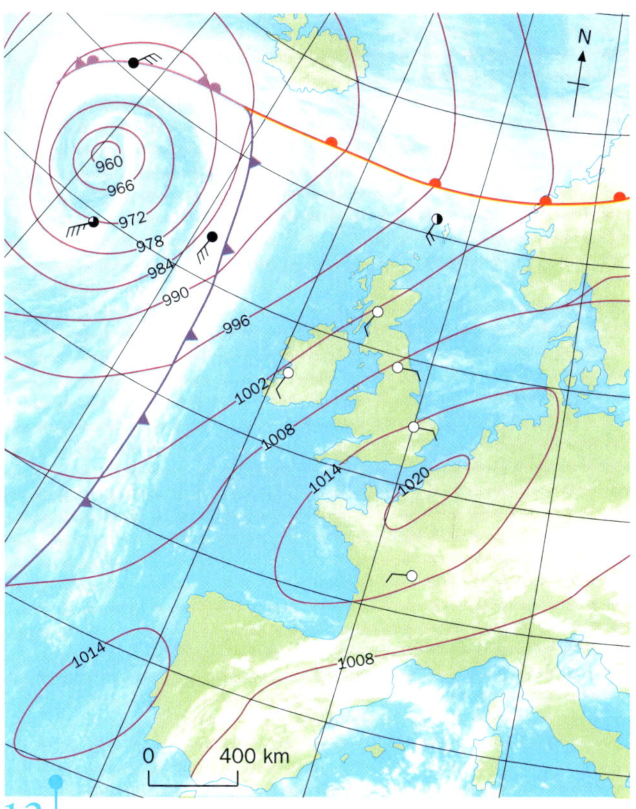

12 Synoptic chart drawn from Photo 13

13 Satellite image showing a depression, or cyclone, over the Atlantic, whilst Britain and Ireland lie beneath an area of high pressure, or anticyclone

Tasks

1 Study Map 12. It shows the high and low pressure zones you can see in Photo 13.

- **a)** Look carefully at the anticyclone on both the map and the photo. To what extent does the map explain the photo?
- **b)** Does the photo contain any information that the map doesn't?
- **c)** What does the map tell you that the photo doesn't?
- **d)** Look at the low pressure area over the North Atlantic. Repeat tasks a), b) and c) for this area.

2 Study Figure 11 (1).

- **a)** Arrange the following weather statements into a timeline sequence to describe the development of a depression. The first two have been done for you.

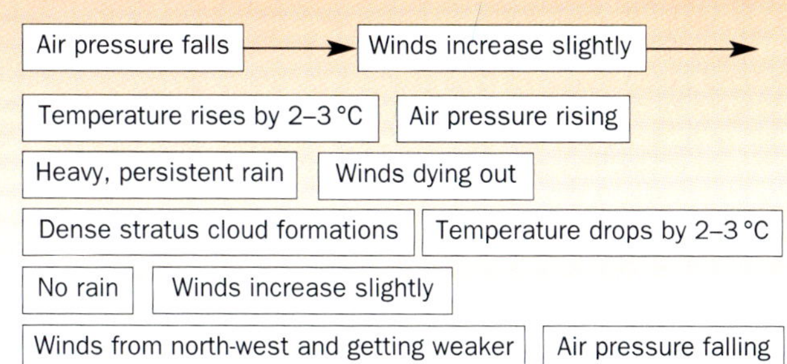

- **b)** Write a single-page report to explain why the weather sequence you produced in task 2a) might be expected when a depression passes over Britain and Ireland.

ICT

3 Visit the Meteorological Office website:

www.met-office.gov.uk

- **a)** Download a recent satellite image and related synoptic chart for the same date (or find a synoptic chart in a newspaper).
- **b)** Decide whether the weather system is one of high pressure or low pressure.
- **c)** Find out what happens to this system by downloading successive images and charts and following its development over the next few days. Describe the changing events.

★ FACTFILE

Weather forecasting

Weather prediction or forecasting is based on identifying patterns of atmospheric behaviour. By comparing current conditions with past records, a reasonably accurate prediction can be made as to the likely weather conditions in the next 24 hours. Accuracy rates drop rapidly as the time scale of prediction is lengthened because the movement of air masses is affected by so many variables.

Synoptic charts

- 'Synopsis' means summary, and synoptic charts sum up the weather data on a map. They can show several different weather conditions at one time, e.g. as on Map 12 on page 87.
- Usually temperatures, cloud cover, precipitation, wind speed and direction are recorded using symbols (see Figure 2 on page 81), whilst air pressure variations are shown by isobars (lines joining places of equal pressure).
- Data are collated from weather stations, ships, aeroplanes, balloons and satellites. Forecasters are able to visualise weather formations across large areas.

Satellite images

The satellite image in Photo 13 on page 87 is an infrared photograph which highlights moisture in the atmosphere and shows the cloud formations associated with low pressure.

Two particular satellites are used for obtaining information about the weather, and for prediction:

- **Meteosat** is a satellite that stays over the Equator, orbiting at the same speed as the Earth rotates. You will find it is good for images of low pressure cloud formations, and shows the clear skies during high pressure phases.
- **NOAA** orbits from pole to pole and is closer to the Earth than Meteosat. It is good for more detailed and varied analysis of weather conditions.

You can access images from these satellites via the internet, as follows.

Meteosat images from: www.nottingham.ac.uk/meteosat
NOAA images from: www.noaa.gov

Tasks

Mind movie

Stage 1

1 You have 60 seconds to construct a movie in your head called 'The Perfect Summer Holiday'. It must have scenes dealing with eating, clothes, accommodation, weather and activities.

2 When your 'movie' is complete, jot down around the edges of a piece of paper all the features of your 'mind movie'.

3 Write **Weather** at the centre.

4 Draw links between the elements of your perfect summer holiday and the weather. Annotate the links to explain how these factors may be related to the weather.

Stage 2

5 a) Read the statement below by a Tourism Liaison Officer for a summer holiday resort.

b) In pairs or small groups, discuss the similarities between the holiday in your 'mind movie' and the Tourism Officer's view.

- Do they describe similar weather?
- Do they have positive or negative views of the effects of the summer weather?
- Could they be the same place?

The high summer temperatures and prolonged drought mean that local crops fail and cannot support tourist demands. This has led to high levels of imported food and overuse of IRRIGATION systems so that the WATER TABLE has fallen, emptying village wells. In addition, rapid evaporation has led to SALINISATION of the soil, which in parts is becoming too salty to farm. Tourists suffer, as the sewers and drains become very smelly, and in the height of summer on average there are two deaths related directly to sunstroke every year. The tourists don't dress sensibly and are often sunbathing at midday when locals take a 'siesta'. Summer thunderstorms catch many visitors off guard, and flash floods have washed away inexperienced drivers on mountain roads.

SAVE AS . . .

6 Write a paragraph explaining how 'good' weather may bring advantages and disadvantages.

B HOW CAN HUMAN ACTIVITY CHANGE OUR WEATHER AND CLIMATE?

Microclimates

The climate that exists over a large area describes a general pattern. Smaller variations can produce local differences called microclimates. Natural features such as hills and valleys cause temperature, wind and precipitation conditions that may differ from the general pattern. Human activity can also create or affect microclimates. In some cities tall buildings can create a 'canyon' effect as winds are concentrated at major junctions. For example, winds blowing off Lake Michigan produce powerful gusts along Chicago's grid-pattern streets.

Heat islands

Most commonly, however, industrialised urban areas generate 'heat islands' where temperatures can be more than 6 °C higher than in surrounding rural areas (Figure 14).

Higher temperatures in urban areas are caused by:

- buildings storing heat during the day and releasing it at night
- central heating/energy leaks from houses and factories
- wind speed being restricted by tall buildings so air gets warmer
- pollutants in the atmosphere which reduce energy loss by long-wave radiation, scattering it back to Earth instead
- reduced vegetation in urban areas, which means that less energy is used up in evapotranspiration, so it is available to the heat island.

Precipitation and cloud cover are also affected in urban areas as a result of extra energy in the heat island and pollutants acting as nuclei for condensation. This leads to:

- 10 per cent more cloud cover
- reduction in sunshine hours
- increased rainfall (heavier and for longer periods)
- increased hail, snow, thunder and lightning.

Areas up to 100 km downwind of large cities also suffer climatic change as the air modified by the city is carried by the wind.

SAVE AS . . .

Urban microclimates

7 Complete the following sentences:

- Three reasons why it is warmer in the city than the countryside are . . .
- We get a lot more hail downwind of the city because . . .
- The winds gust to high speeds downtown when . . .
- It tends to be cloudier in cities because . . .

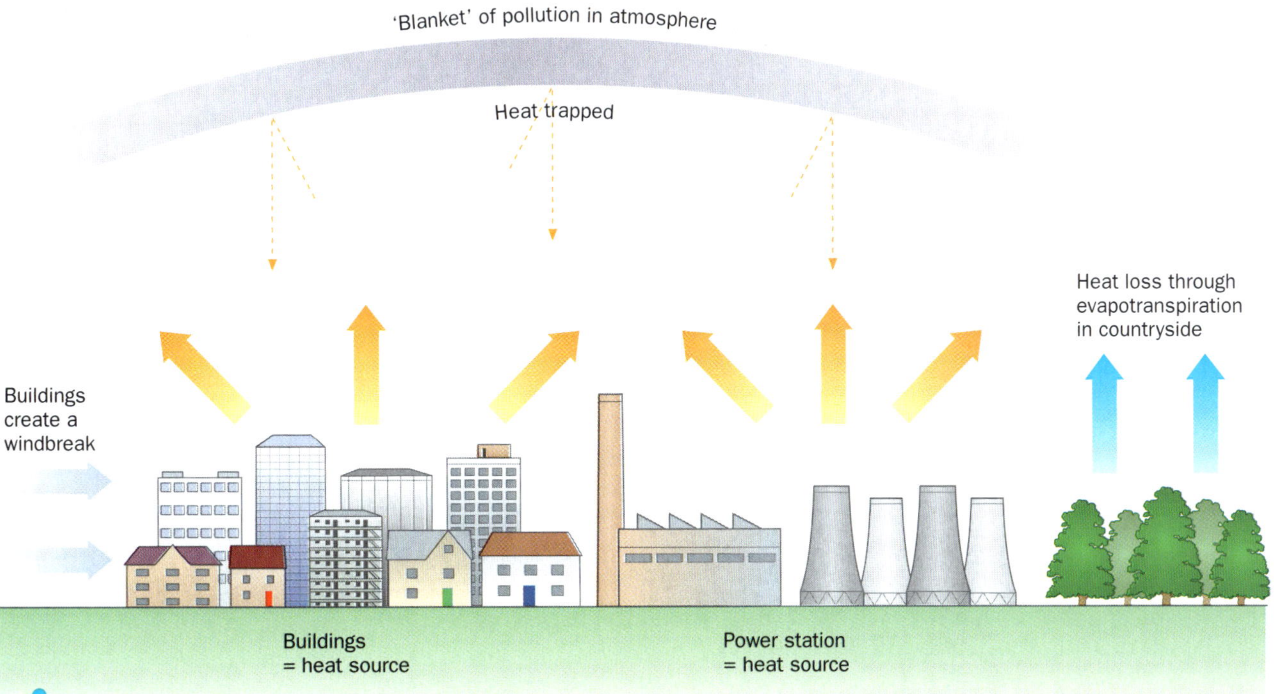

14 Urban microclimates

Temperature inversion

The way in which physical and human conditions interact is shown by the TEMPERATURE INVERSIONS experienced in Los Angeles, California, USA (Figure 15). A layer of descending air warms because it gets compressed. This traps air from the city and stops it from rising. Pollution from buildings and cars cannot escape. A poisonous layer of smog builds up on many days of the year, creating major health problems for people.

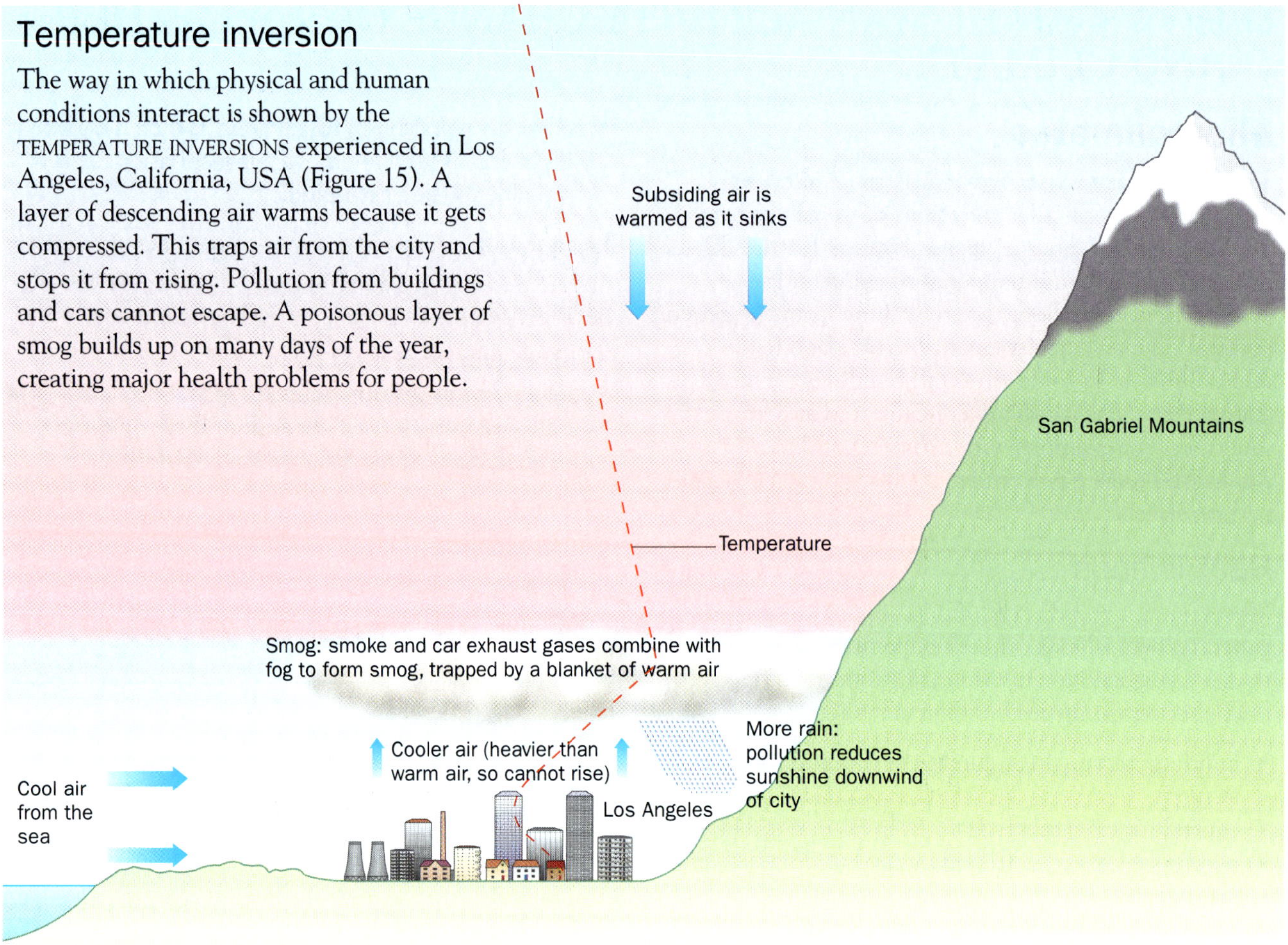

15 Los Angeles: temperature inversion and urban pollution

Task

SAVE AS . . .

1 a) Study Figure 15. Then copy this graph which shows air temperature and altitude.

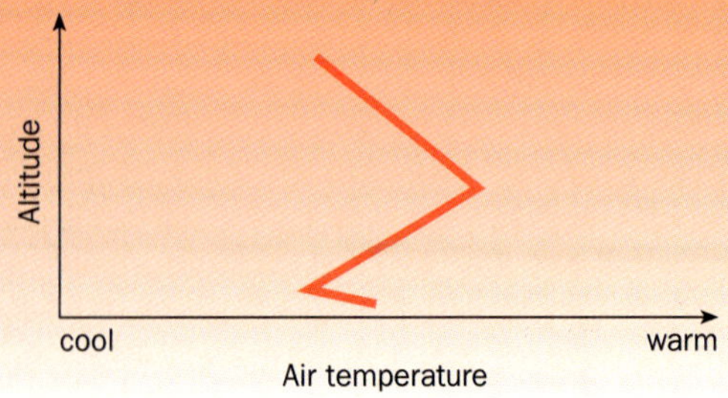

b) Add the statements below in the correct place to produce an annotated graph.

- Air from sea cools air over city
- Subsiding air stops smog rising
- Condensation at altitude
- Smoke particles lead to creation of smog
- Warm air cannot rise past overlying cold air

c) Give your graph a title.

Global warming

Human activity can unintentionally change our weather and climate. The previous section looked at how it could affect local climates. This section looks at how such activity can affect global climates and gives you the opportunity to explore one of the most important climate topics facing the world today. As you work through the section, look for evidence both for and against the view that global warming is a genuine danger, and consider who is responsible for solving the problem. This will help you with the tasks on page 93.

The average temperature of the Earth has fluctuated significantly over the past 10,000 years. The current debate over whether the recent rises in global temperature are a natural variation or a problem caused by people involves industry, conservationists and governments. However, there is increasing consensus amongst scientists that GLOBAL WARMING, an accelerated heating of the atmosphere resulting from human activities, is an alarming reality. Graph 16 shows the increase in average global temperatures since 1860.

The Sun has always warmed the atmosphere, producing the GREENHOUSE EFFECT – Figure 17. Short-wave radiation from the Sun penetrates the atmosphere and warms the surface of the Earth. The subsequent loss of heat by radiation from the Earth's surface is partially checked at night because the longer-wave outgoing radiation cannot easily penetrate the clouds and dust particles. The scattering of energy back to Earth helps to maintain atmospheric temperatures at a level humans can tolerate. It is worth noting that during the last Ice Age the Earth was less than 5°C cooler than it is today – not a big difference, but without the natural greenhouse effect it would have been 33°C cooler.

Oxygen and nitrogen make up 99 per cent of the Earth's atmosphere, but it is the variations in the remaining 1 per cent of gases, particularly carbon dioxide and ozone, that cause most concern, because these retain more heat. In addition, pollutants such as sulphur dioxide, nitrogen dioxide and methane are considered to be responsible for additional heat being trapped in the atmosphere (Figure 18). Increases in these so-called 'GREENHOUSE GASES' are believed to have increased the heat-retaining properties of the atmosphere, leading to global warming.

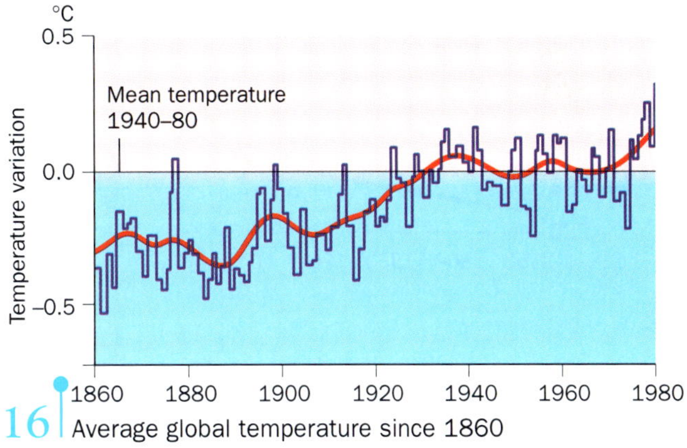

16 Average global temperature since 1860

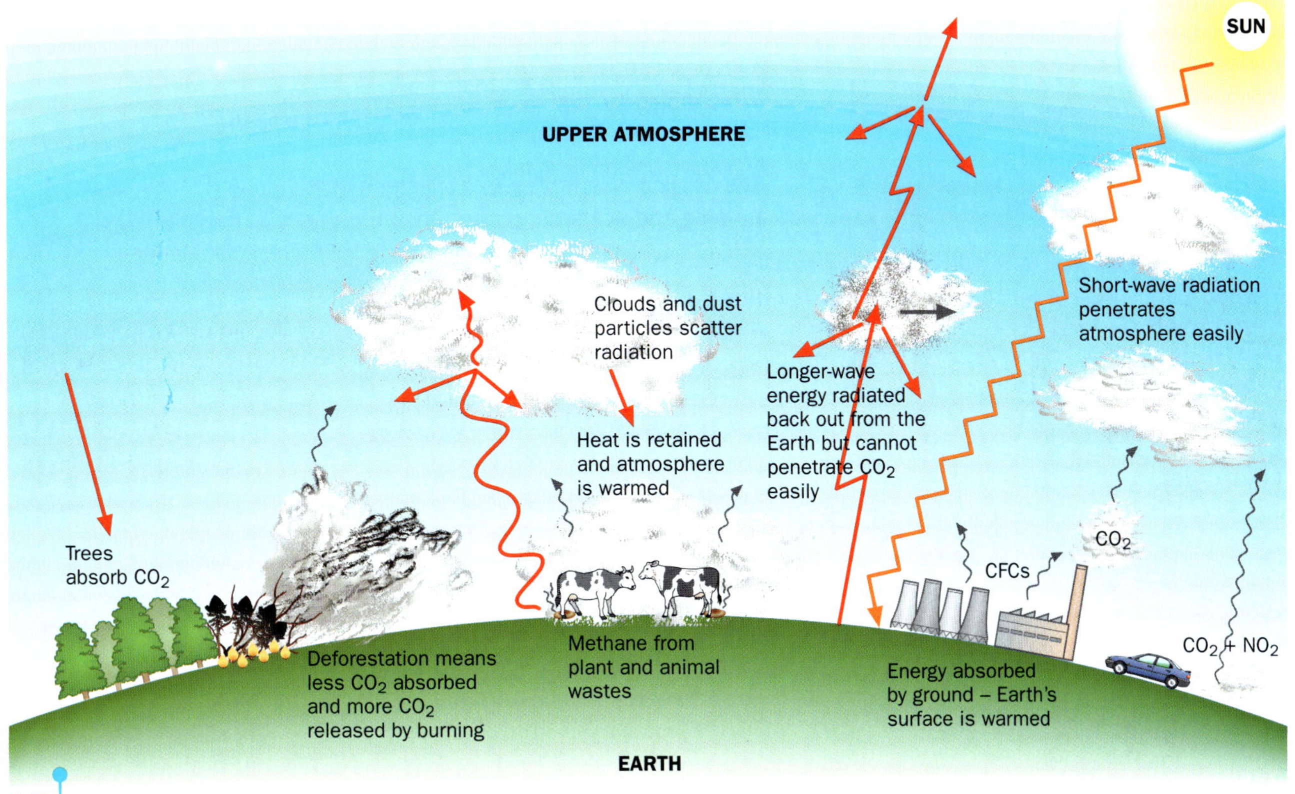

17 Global warming

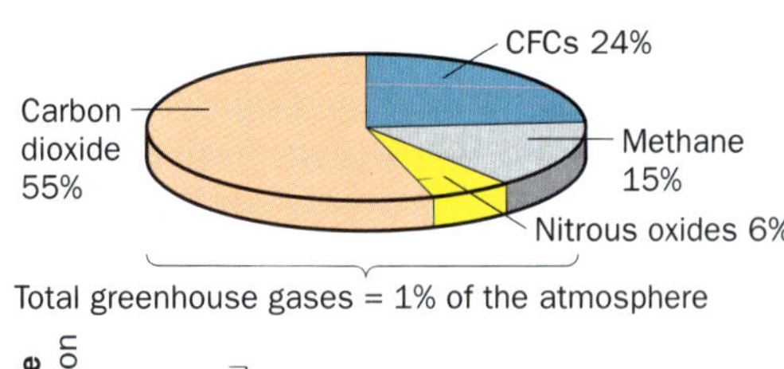

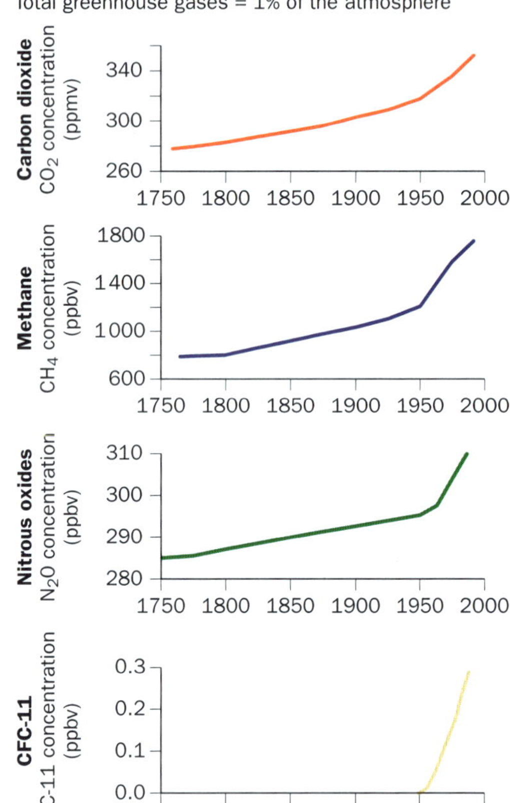

18 Greenhouse gases

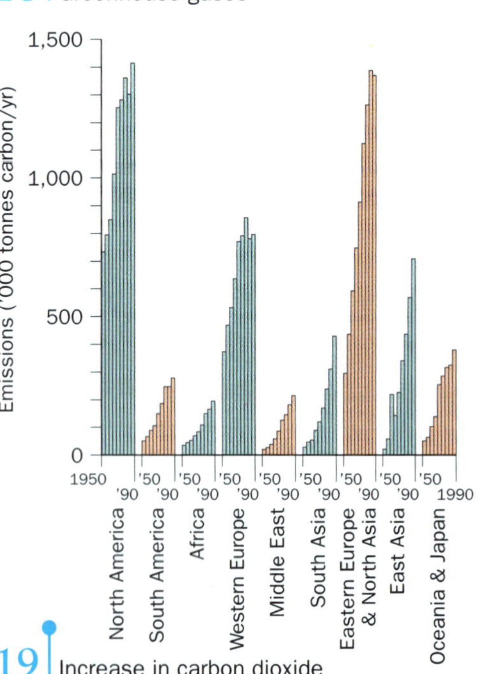

19 Increase in carbon dioxide emissions, 1950–90

★ FACTFILE: GLOBAL WARMING

Causes

- The burning of FOSSIL FUELS releasing carbon dioxide into the atmosphere is the biggest single contributor to global warming, with MEDCs being the worst culprits (Graph 19).
- Deforestation, particularly in LEDCs but often encouraged by demand in MEDCs, is another major contributor to the problem.
- The late-20th-century boom in plastics manufacturing accounts for the sudden rise in chlorofluorocarbons (CFCs).
- The mistakes by the earlier industrialised countries (e.g. the UK) are being reproduced to a greater degree by newly industrialising countries eager for economic growth. MEDCs such as the USA and the UK continue to consume vast amounts of fossil fuels, but developing countries such as China are now increasingly seen as the main problem for the 21st century. They need cheap energy sources to sustain industrial growth and are likely to follow the lead of MEDCs, using huge quantities of fossil fuels and so making the problem worse.

Consequences

Whilst the main culprits of global warming can be identified at a local scale, the effects will be global. Scientists believe that a 1°C increase in world temperature is all that the world can tolerate before climatic chaos sets in. If the 0.6°C rise in global temperature over the last century is repeated over the 21st century, some scientists predict the following results:

- A rise in sea-level of 0.3–1.5 m as ocean temperatures increase and seawater expands.
- Accelerated melting of ice caps leading to a rise in sea-level of more than 5 m.
- Increased frequency and intensity of storms and hurricanes.
- A decrease in the total amount of precipitation, and greater variability in rainfall over continental areas, because warmer air holds more moisture.

The implications of these changes on flooding, hurricanes, drought, farming and water supply are shown in Map 20. See also pages 276–77 to see how global warming could affect Bangladesh.

Solutions

Since the Kyoto conference of 1997, world governments have agreed to reduce overall greenhouse gas emissions by 5.2 per cent from 1990 levels. Each country has its own target to contribute to this overall target. Some scientists called for 60 per cent reductions but this would mean ending the use of fossil fuels altogether. Fossil fuels are relatively cheap. To stop using them would mean a huge rise in the cost of energy, which would raise the costs of industrial production and would threaten jobs: a frightening prospect for any government.

The USA's target was 7 per cent. Ironically, whilst encouraging developing nations to clean up their act, in March 2001 President Bush withdrew the USA from all its Kyoto commitments.

1999 the hottest year of the millennium

Anthony Browne, Environment Correspondent
Sunday 14 November 1999

This year is set to be the hottest of the millennium. With only six weeks to go to the New Year, the weather in Britain has been so much warmer than usual that only an unusually long cold snap in December would bring the annual average temperature below the previous record, according to one of Britain's top climate scientists.

And the Meteorological Office predicts December will be 'warmer than normal'.

The prediction is a dramatic reinforcement of evidence that the whole world is warming up at an increasingly rapid rate. Britain's wildlife has begun to be affected, with spring arriving earlier and autumn later. Researchers warn that 4,000 km^2 of East Anglia is now at such severe risk of flooding that new building should be banned.

Dr Mike Hulme, of the Climatic Research Unit at the University of East Anglia, said: 'There's a good chance 1999 will break the record.'

Britain has accurate weather records back to 1659, and Hulme said: 'From what we know of the climate before then, if it is the hottest year since 1659, you can argue it will be the hottest of the millennium.'

The likely record for 1999 is particularly shocking since this year has not had a notable heatwave. Not one month in 1999 has broken the record, but every month, apart from June, has been consistently a degree or two warmer than average.

Almost all scientists now consider the evidence for global warming to be 'incontrovertible', and many believe it is accelerating at an alarming rate. Last year was the hottest of the last millennium for the world as a whole, although it was not a record for Britain.

Scientists now predict that global temperatures could rise by as much as 4 °C by the year 2100, causing severe storms, the melting of the Antarctic ice cap, and rising sea-levels.

The Labour Party committed itself at the election to reducing greenhouse emissions by 20 per cent but has since relegated this to a 'goal'.

From the *Guardian*, 14 November 1999

Tasks

1 Write sentences to answer each of these questions.

a) Which 'greenhouse gases' are responsible for global warming?

b) What human activity is the biggest single contributor to global warming?

c) How does the increase in greenhouse gases cause global warming?

2 a) Where will the worst effects of global warming be felt – in MEDCs or LEDCs?

b) Suggest one way in which the MEDCs can help the LEDCs to cope with the problems of global warming.

Focus task

Should we worry about global warming?

3 In groups, research the effects of global warming. You are to give a brief talk (3–5 minutes) supported by an ICT/flipchart/OHP presentation. Use the information on pages 91–93 and websites to prepare your talk. Include the following sections:

- a definition of the term 'global warming'
- evidence that global warming is taking place
- the link with greenhouse gases
- the possible impacts of global warming
- your conclusion about whether individuals, governments or manufacturers should do anything about it, and if so what.

You should include graphs of atmospheric gas compositions and maps showing possible effects of global warming. Here are some possible websites:

www.environmentaldefense.org
http://globalwarming.enviroweb.org
www.sierraclub.org/globalwarming

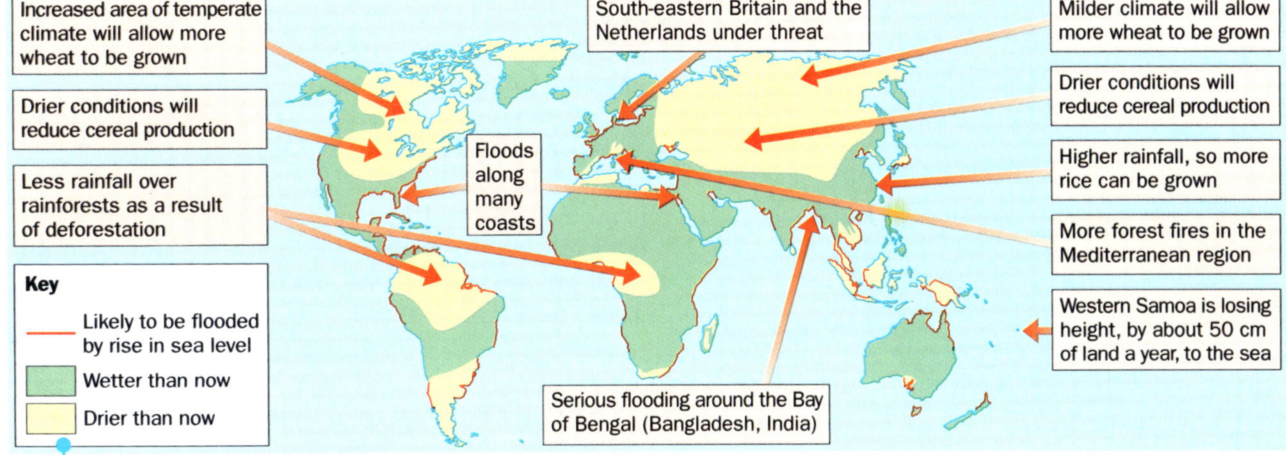
Increased area of temperate climate will allow more wheat to be grown
Drier conditions will reduce cereal production
Less rainfall over rainforests as a result of deforestation
Key
Likely to be flooded by rise in sea level
Wetter than now
Drier than now
Floods along many coasts
South-eastern Britain and the Netherlands under threat
Serious flooding around the Bay of Bengal (Bangladesh, India)
Milder climate will allow more wheat to be grown
Drier conditions will reduce cereal production
Higher rainfall, so more rice can be grown
More forest fires in the Mediterranean region
Western Samoa is losing height, by about 50 cm of land a year, to the sea

20 Impact of global warming

Ⓒ *HOW AND WHY DOES CLIMATE VARY AROUND THE WORLD?*

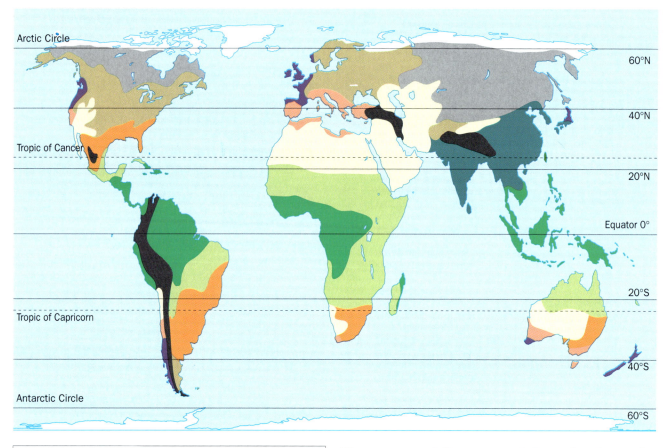

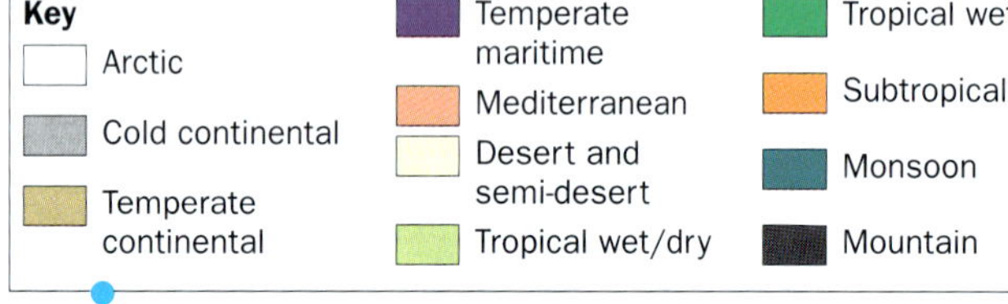

21 World climate zones

There are many different types of climate found around the world. Map 21 shows the major climate zones of the Earth as distinct regions, but it is important to remember that these zones actually merge into each other without precise boundaries.

Atmospheric circulation

The pattern of climate zones can be explained by the way air circulates in the atmosphere (Figure 22). Air at the Equator is heated and rises (A). This creates a low pressure zone (B) and causes heavy convectional rainfalls. Cooler air is dragged in at ground level to replace the rising air (C). Air sinks back to the surface at about 30° N and 30° S (D). As it sinks this air is warmed. It brings hot, dry conditions typical of desert climates. Some air from this high pressure zone flows back towards the Equator (C). Figure 22 shows that this produces two large cells of circulating air. They are called the HADLEY CELLS.

Some of the air from the high pressure regions flows towards the poles (E). This gives warm winds which pick up moisture from the ocean. When these winds meet cooler polar air they produce the depressions which often dominate the weather of Britain and Ireland.

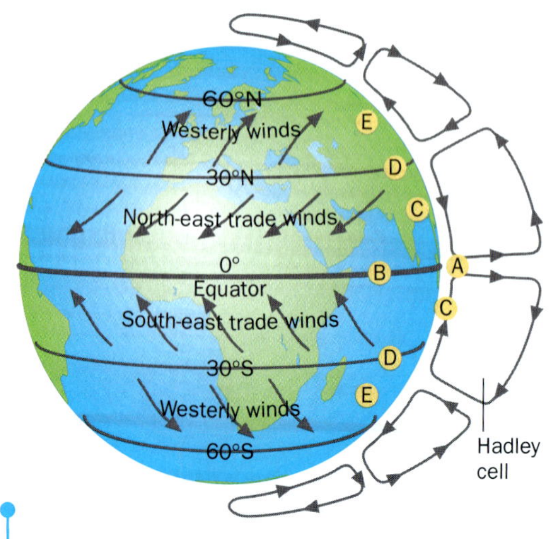

22 Atmospheric circulation

The tilt of the Earth leads to an apparent seasonal shift in the position of the overhead Sun. The Sun is directly overhead at the Tropic of Cancer (23.5°N) in summer, on 21 June, so the low pressure zone moves northwards towards this location, dragging the prevailing wind patterns with it. This situation is reversed at the winter solstice on 21 December when the equatorial low and its associated winds move south.

Ocean currents

The passage of the prevailing winds over the surface of the oceans creates frictional drag, which in turn generates ocean currents. The main ocean currents follow a roughly circular pattern: anticlockwise in the southern hemisphere and clockwise in the northern hemisphere (Map 23).

Those currents that move water from tropical areas polewards, such as the North Atlantic Drift, are said to be warm currents, while those that move cooler waters from nearer the poles back towards the Equator, such as the Canaries Current, are said to be cold currents. Their effect on climates is profound. It is the warming effect of the North Atlantic Drift that brings much of the rainfall to Britain and Ireland, because it warms south-westerly air masses enough to hold moisture. The cool Canaries Current restricts the uptake of moisture, and so it keeps the coastal areas of north-west Africa surprisingly dry.

Task

1 Using Map 21, Figure 22, Map 23 and an atlas, draw up a table like the one below to explain the climate in **three** of the following locations:

- Sahara (Nouakchott)
- Peru (Lima)
- South Africa (Cape Town)
- Germany (Berlin)
- Jamaica (Kingston)
- Malaysia (Kuala Lumpur)
- China (Beijing)

Place		
Climate region		
Latitude		
Altitude		
Distance to coast		
Prevailing winds		
Ocean current		

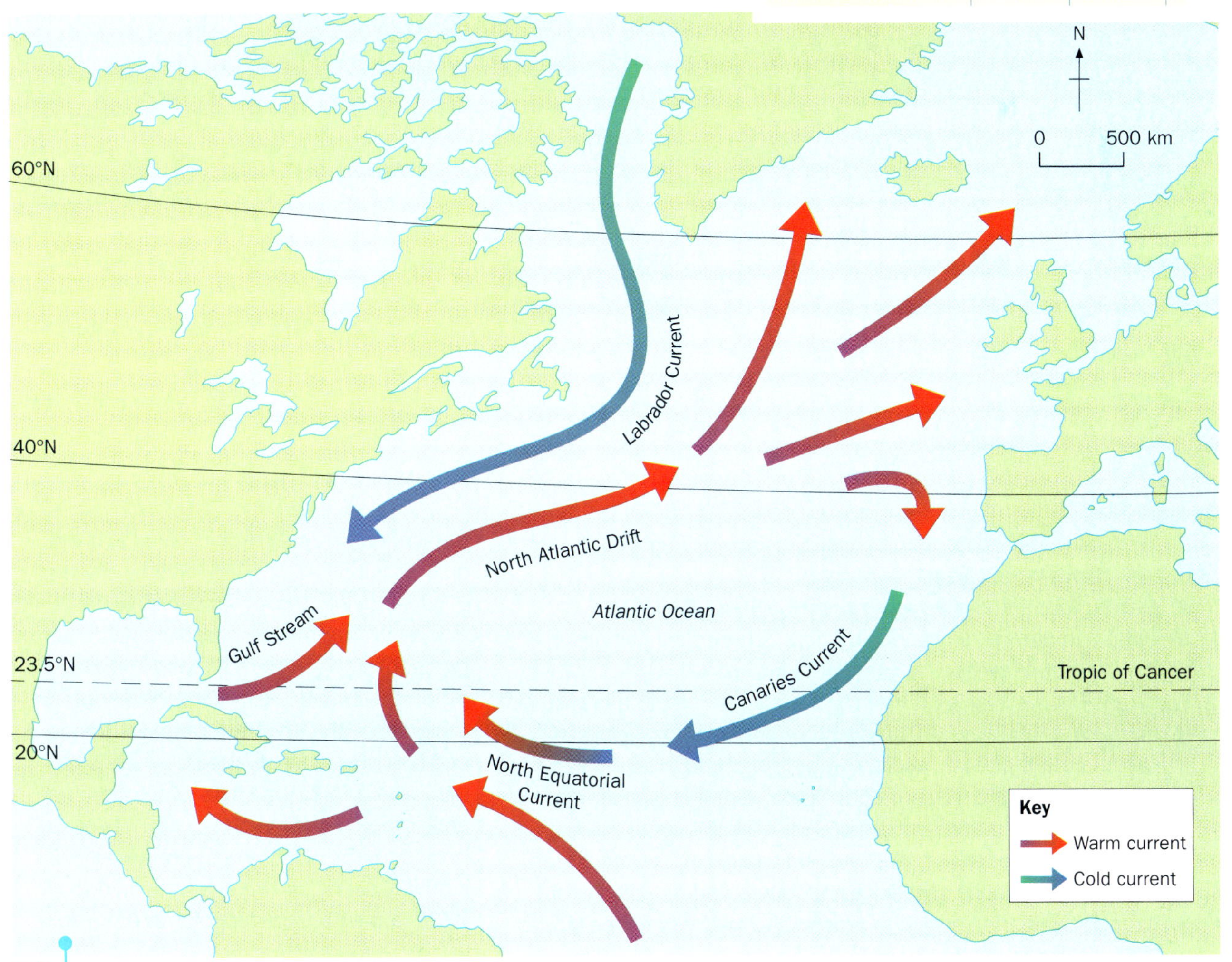

23 The major Atlantic Ocean currents

Connecting climate and ecosystems

Look at Map 24, which shows the major world ecosystems. Compare it with the world climate map, Map 21. How do the main climate zones link with the ecosystems? What are the reasons for these links? In many cases you will find that the climate zones and ecosystems do not match exactly, because in reality climate and vegetation zones rarely have precise boundaries but merge into each other.

In addition, the effect of altitude, prevailing winds, ocean currents, soil/rock types and human activity all have their part to play in determining the extent of ecosystems.

The link between climate and ecosystems is clearly demonstrated in tropical regions where the combination of high temperatures and humidity generates rapid vegetation growth.

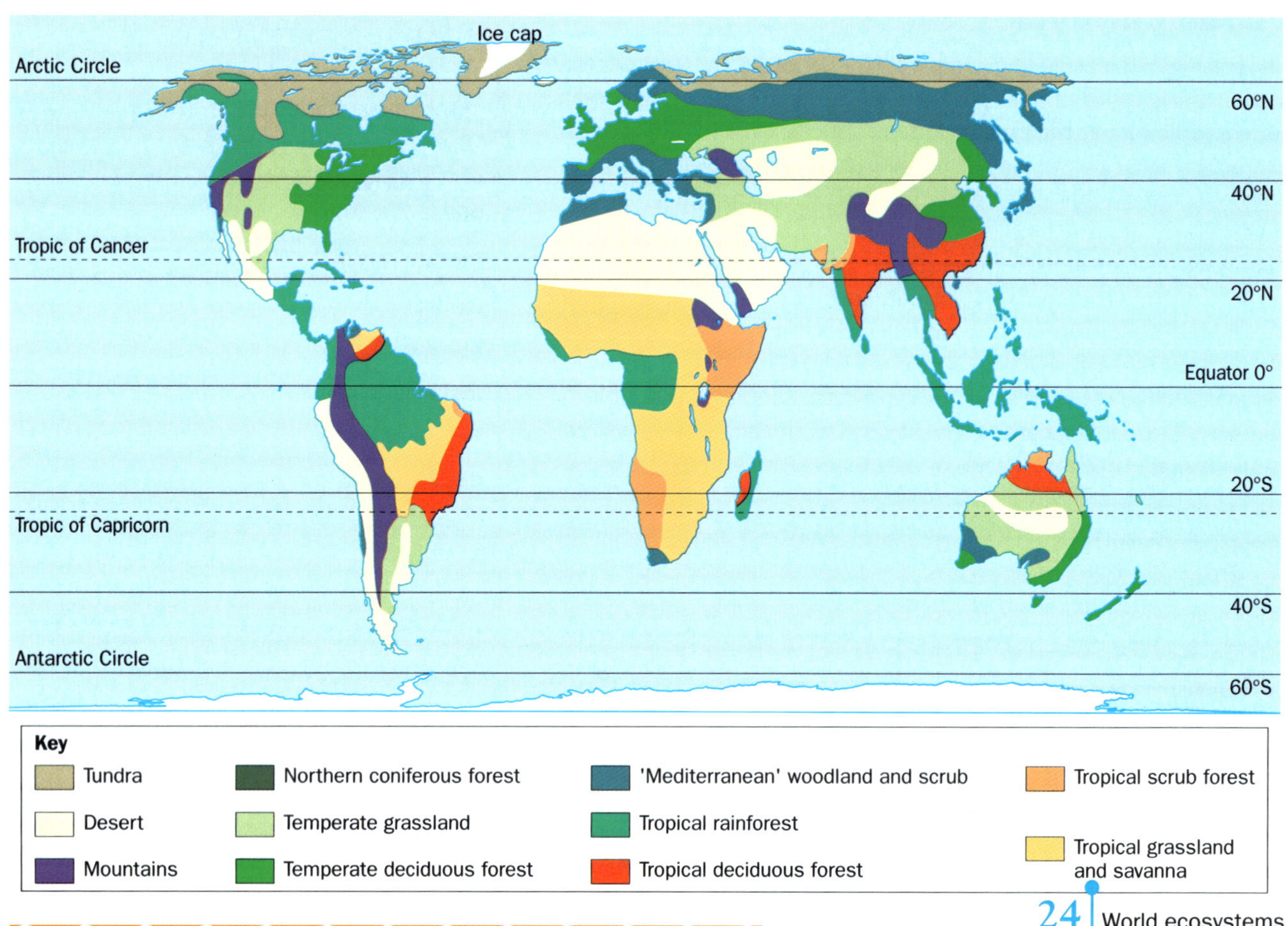

24 World ecosystems

Tasks

1 Study Map 24.
For each of the three locations chosen in Task 1 on page 95, find out the characteristic vegetation type.
- **a)** Record your findings as an extra row in your table.
- **b)** Write a sentence to describe the link between the climate and the vegetation types.

2 The Trans-Pacific Banking Corporation wishes to set up a branch in Kuala Lumpur in Malaysia. The directors are keen to be able to explain to their European workers what living conditions might be like, and to make them aware of seasonal variations in climate.
Study the information and sources on page 97. Write a two-page report for the bank, describing and explaining the Malaysian climate. Include graphs showing temperature and rainfall. Finish your report by offering advice to prospective visitors on lifestyle, clothing, timing of holidays, etc.

★ THINK!

Climate graphs are often included in exam questions but are not well used by students.

When using climate graphs in your answers, always quote specific figures from the graphs, e.g. maximum temperatures/minimum rainfall, using the units shown on the graph axes. Describe seasonal trends and patterns, quoting specific months of the year for highs and lows of temperature and rainfall. For example, to say 'Malaysia is hot all year round' would not gain many marks. Quoting an average temperature of 27 °C and an annual range of 2 °C will gain you more marks.

Case study – Malaysia

The humid tropical climate of the Malaysian peninsula shown in Map 25 is characterised by heavy rainfall and high temperatures (Figure 26). The rainfall pattern is dominated by annual MONSOONS (see Source 27). The south-west monsoon results from low pressure north of the Himalayas. Later in the year the north-east monsoon occurs as air rushes the other way towards the low pressure area over Australia.

These factors combine to produce the widespread tropical rainforest ecosystem. Rainforest soils are poor in nutrients even though the rainforest vegetation above them is abundant. The rainforest is a fragile ecosystem which may prove unsustainable if deforestation continues at current rates.

25 Malaysia

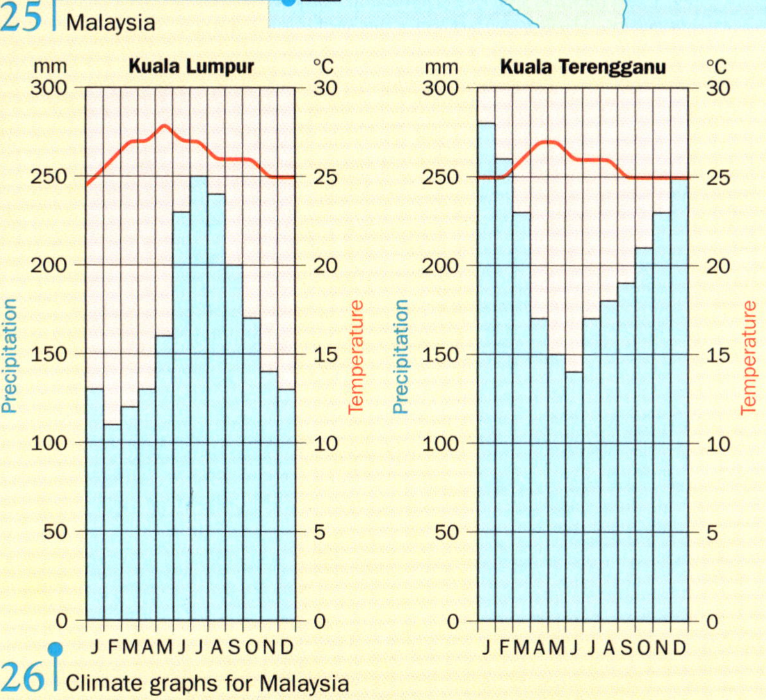

26 Climate graphs for Malaysia

Malaysia is subjected to maritime influences and the interplay of wind systems which blow from the Indian Ocean and the South China Sea. The year is commonly divided into the south-west and north-east monsoon seasons. The north-east monsoon blows from October to February and brings rain to the east coast of the peninsula and to Sabah and Sarawak. The south-west monsoon is from mid-May to September, bringing rain to the west coast. At all times the heat of the Sun tends to bring convective rainstorms. Average annual rainfall is between 2,032 and 2,540 mm. The average daily temperature throughout Malaysia varies from 21°C to 32°C, though in higher areas temperatures only reach 12–26°C. Relative humidity is generally high, with readings of over 80%.

From *Malaysia in Brief* published by Information Division, Ministry of Foreign Affairs

27 The climate in Malaysia

Task

3 Source 27 is a very bland account of Malaysia's weather and it wouldn't really tell someone intending to travel to Malaysia what the weather might be like. Use the information on this page and your own research to write two paragraphs describing what weather might be like in Kuala Lumpur

a) during the north-east monsoon season

b) during the south-west monsoon season.

3.2 *Drainage basins and river landforms – does nature know best?*

DO YOU KNOW?

- A drainage basin is the whole area drained by a river and its tributaries.
- The dividing lines between drainage basins are called watersheds.
- The world's largest drainage basin, that of the Nile in Egypt, is over 3 million km^2.
- The UK's longest river is the River Severn (about 300 km). The world's longest river is the Nile (6,990 km).

AIMS

- **To explain how the hydrological cycle leads to variations in discharge.**
- **To consider why our different uses of drainage basins may be in conflict with each other and with the natural system.**
- **To understand how rivers shape the landscape.**
- **To assess the success of attempts to control pollution of the Rhine.**

Stephen Evans, Catchment Area Manager for North West Water (NWW), is responsible for the River Ribble drainage basin. Figure 1 is his diary for one workday.

1 Stephen Evans' diary

30/3/2000

8.30 am Overcast day. Check our computer printout of discharge levels from remote upland river measuring stations. These give us the first indication of how the basin is responding to heavy rainfall. Discuss forecasts with Mel from meteorological section. We decide that Flood Level Warning should remain at the lowest level but agree to monitor precipitation levels closely as situation remains unsettled.

9.00 am Meeting with Water Quality Objectives team. Target three firms who discharge waste water into the river system for thorough pollution checks following laboratory reports from Barnoldswick area. Familiarise new team members with the European Community Sensitive Area regulations. These rules protect areas in danger of pollution damage, and fines are incurred if they are broken.

11.00 am Visit from neighbouring water authority to discuss possibilities of water transfers out of Ribble catchment to basins further east where water is needed for industry and homes. Try to balance potential physical impacts of losing water and effects on local farmers and factories who take water from the Ribble with increased income from selling our water.

12.30 pm Drive to Stocks Reservoir on River Hodder, an important tributary, to meet with Carl Radford. The Hodder reservoir supplies drinking water. Carl reports on soil flow and surface wash data for the Bowland Fells, an important environment in the drainage basin. He feels he can prove that forested areas are better protected than deforested areas after heavy rain. Agree to mention findings at next meeting and leave to inspect the tree-planting schemes.

3.00 pm Meeting with Caroline Ellison of Yorkshire Dales National Park Authority in Settle. Discuss impact of the drainage basin on farming in Upper Ribble. We are monitoring impact of new low-fertiliser methods at selected farms.

5.00 pm Attend opening of newly renovated section of River Douglas embankment. Hand out awards to local volunteer groups for planning and helping restoration.

7.30 pm Talk in Preston at meeting to discuss the future for the Ribble catchment area. Very lively meeting. Canoeists, anglers and bird-watchers who use the Ribble concerned about environmental quality and future water levels. Demands ranged from constructing new reservoirs (hotly contested!) to extending the water-sharing 'Lancashire Conjunctive Use' schemes, pumping water from one place to another – very expensive! Best comment of the night came from a Y11 student who responded to a farmer insisting that more water should be pumped up from groundwater with the observation that increasing the output would do nothing about the input. She then asked some very good questions about future rainfall predictions, in the light of recent variability. Her point about our demands for water needing to be based on a more sustainable level of use did not find favour with some of the audience!

Focus task

1 Read Figure 1. Identify at least two:

- groups of people
- human activities
- environments

that depend directly on the River Ribble drainage basin. Keep your lists. You will return to the River Ribble in the case study on pages 104–105.

2 A heavily engineered river

Task

2 Look at Photos 2 and 3, which show two different approaches to river management.

a) Draw a Venn diagram with features that are unique to Photo 2 on one side, features unique to Photo 3 on the other side, and features common to both in the overlapping section. Only include those features that are related to the river.

b) What will these rivers be like in ten years' time if nothing else is done to manage them?

c) To what extent do these two photos show sustainable use and management of our rivers?

3 An unengineered river

Ⓐ UNDERSTANDING AND MANAGING DRAINAGE BASINS

The hydrological cycle is a CLOSED SYSTEM, i.e. the same water circulates without significant loss or gain. The drainage basin is part of this system as you can see from Figure 4. But the drainage basin is an OPEN SYSTEM – there are significant variations in *inputs* and *outputs* at different times. These can have a big effect on a drainage basin system, and sometimes on our lives.

Task

1 Before you read through pages 100–111, think of at least five questions that you can ask to help you to investigate the possible impacts that changes in the amount of water in a drainage basin might have on people, on activities and on environments. Avoid questions that can be answered with 'Yes/No'. Examples of useful questions are:

- How can water be lost or gained from a drainage basin?
- How are people affected by flooding?

Add your own questions. These will help to focus your reading. As you read, note down possible answers. If your ideas change as you read more, don't be surprised! By modifying or adding to your initial ideas you will end up with a deeper understanding of the topic.

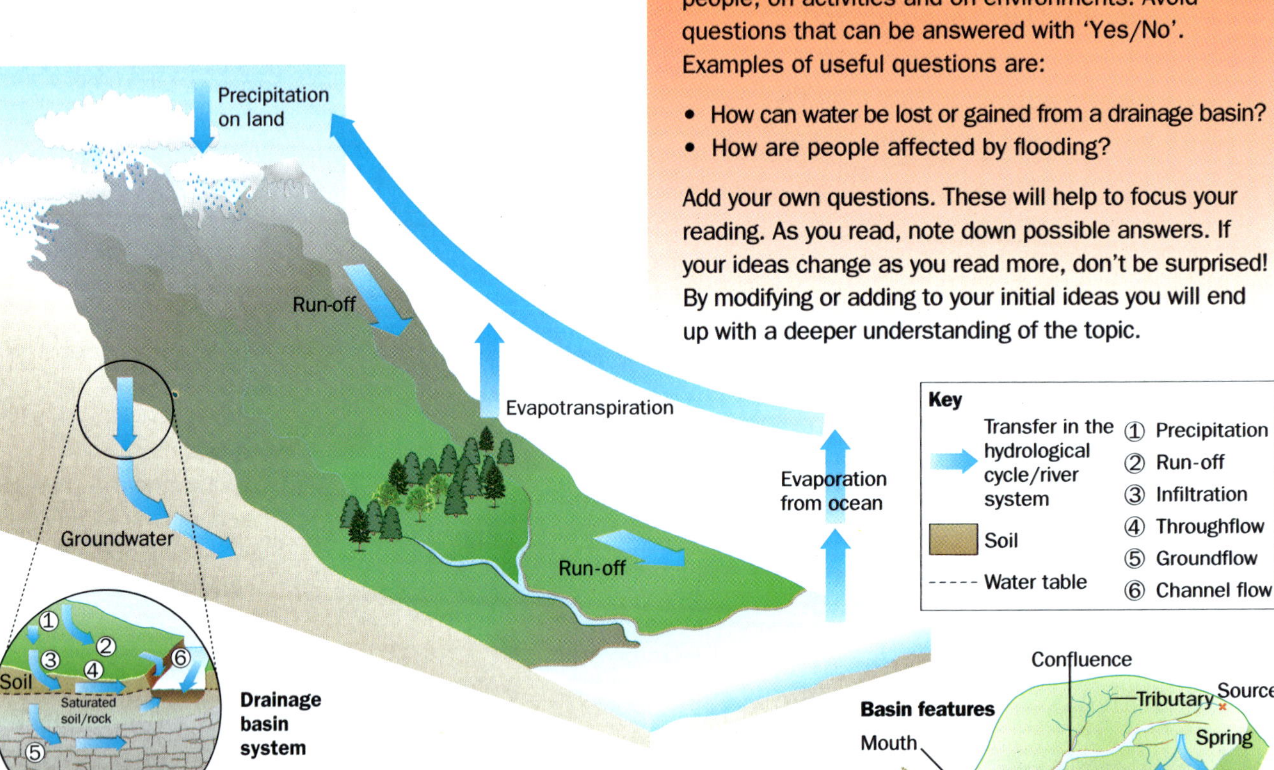

4 The hydrological cycle

The drainage basin system

Drainage basins are open systems, i.e. inputs and outputs change, and they have *stores* and *transfers*. Inputs of water to a drainage basin arrive mainly as *precipitation* in the form of rain and snow. The level of input varies over time (see Figure 5) and also from place to place (see Figure 7 on page 102).

When it rains, some of the water will be INTERCEPTED by vegetation. Then plants will return moisture to the atmosphere from their leaves. (The creation of water vapour in this way is known as *evapotranspiration*).

The path the water takes once it gets to the ground is dependent on many factors (Figure 4). Most of the water will sink in (INFILTRATION), but if the ground is saturated or impermeable, or the rainfall is too heavy, then it will run off over the surface. If it has infiltrated the soil, the pull of gravity and the slope of the land will cause the water to move through the soil as throughflow. If water percolates down to the water table (the top level of saturated rock), downward movement may continue as GROUNDFLOW, which is much slower and may take water deep into the rocks. The level of the water table will fall during dry periods and rise during wetter periods, and this can have a major effect on the level of DISCHARGE of the river.

The type of rock beneath the drainage basin affects the way it stores and transfers water. Some rocks are *permeable* and allow water to pass through them easily. Rocks such as limestones and sandstones may be termed AQUIFERS, as they store high levels of water even at dry times of the year.

The processes described here operate at variable rates at different times of the year. In summer, when foliage is densest, interception is at its greatest, whereas in winter frozen soils may hinder infiltration. URBANISATION of previously farmed or natural areas also reduces infiltration, increases levels of run-off and alters the ability of a basin to retain water.

How does precipitation affect river discharge?

A river is the main output of the drainage basin system. A river's discharge is calculated by multiplying the velocity of the river by its volume at a given point and time. It is measured in cubic metres/second (cumecs).

Hydrologists plot the response of drainage basins to rainfall events on flood HYDROGRAPHS (Figure 5). There is a gap between peak rainfall and peak discharge. This is called LAG TIME. By analysing flood hydrographs, hydrologists can build up a picture of how specific rivers react to variations in precipitation. The data are vital tools in flood prediction (see Chapter 2.2), as the lag time is critical in determining the likelihood of a flood. Rivers with a short lag time and high peak discharge are more likely to flood than rivers with a lengthy lag time and low peak discharge. Figure 6 shows how various factors can affect the flood hydrograph. As you know from Chapter 2.2, human activity can affect discharge quite markedly. In urban areas, there is little lag time. Increased flooding risk is a feature associated with urbanisation (see Figure 7).

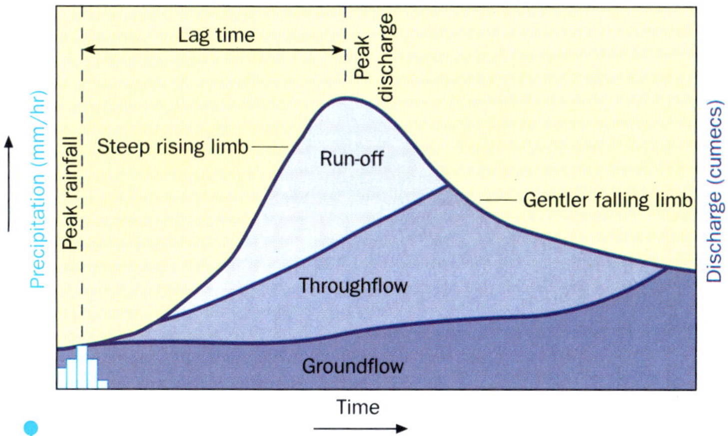

5 A typical flood hydrograph

Nature of drainage network
Drainage density = $\frac{\text{Total length of streams (km)}}{\text{Total drainage basin area (km}^2)}$
Lower figure = longer lag time

Vegetation cover
More vegetation = longer lag time, as vegetation intercepts rainfall

Angle of slope
Increased angle = shorter lag time

Stormflow
Makes up most of the peak discharge. It consists of water that flows over the surface or through the soil during the storm. It reaches the river quickly because during a storm there is too much water for it all to infiltrate into the soil

FACTORS AFFECTING A FLOOD HYDROGRAPH

Rainfall intensity
More, or more intense, rainfall = shorter lag time

Rock and soil types
Impermeable rocks and soil = shorter lag time

Seasonality
In summer, drier soil = longer lag time as rain infiltrates.
In winter, soil is frozen or saturated = shorter lag time, because water cannot infiltrate and so runs off quickly

6 Factors affecting a flood hydrograph

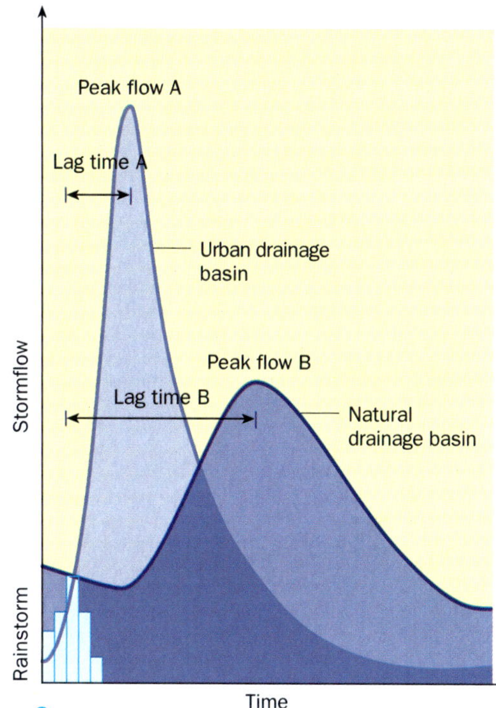

7 Hydrographs of natural and urban drainage basins

Tasks

Pages 100–101 use a lot of technical terms. Do you know what they all mean?

2 You are going to play the game **Taboo** to find out. Work in pairs. Decide who is going to go first (A) and who is second (B).

- **A** Explain to your partner how the *hydrological cycle* works, but without using any technical words (this will make sure you understand their real meaning). For this game a technical word is anything in *italics* or CAPITAL LETTERS on pages 100–101.
- **B** Now it is your turn to explain the *drainage basin system*, again without using the proper geographical words.

SAVE AS...

3 Based on your work in Task 2, agree between you on non-technical explanations for both the hydrological cycle and the drainage basin system, and write them down, this time adding the correct geographical terms in brackets.

Tasks

Understanding a hydrograph

1 a) Annotate a copy of the hydrograph in Figure 8 by completing the following statements (where necessary), and then putting them in the appropriate places.

- The discharge hasn't yet responded to increased precipitation.
- It takes ___ hours for discharge to return from peak flow to pre-storm levels.
- The steeper rising limb reaches peak flow in ___ hours.
- Base flow is ___ .
- The lag time is ___ hours.
- The storm flow peak is ___ cumecs.
- Flooding is most likely at this point.
- The peak of the storm occurs at ___ hours.

b) Add the following statements in a different colour:

- Time to issue a flood warning.
- Put warning signs on local roads.
- Put sandbags around doorways in villages.
- Phone insurance company.
- Mop out flooded homes.
- Have fire service and army rescue dinghies at the ready.

The 1991 Hodder flood

Now that you are familiar with the terms used on hydrographs, you can use these terms and ideas to describe and explain a flood event which took place in England in 1991.

2 The hydrograph in Figure 9 shows precipitation and discharge on the River Hodder in the week beginning 19 December 1991.

a) Describe the relationship between precipitation and discharge shown on the hydrograph.

b) Explain the similarities and differences you see.

c) Calculate the approximate lag time in hours between each rainfall peak and discharge peak, and find the average lag time.

d) Do your findings suggest that the Hodder has a fast or a slow response to rainfall? Why do you think it responds in this way? Refer to maps 10 and 11 to help you.

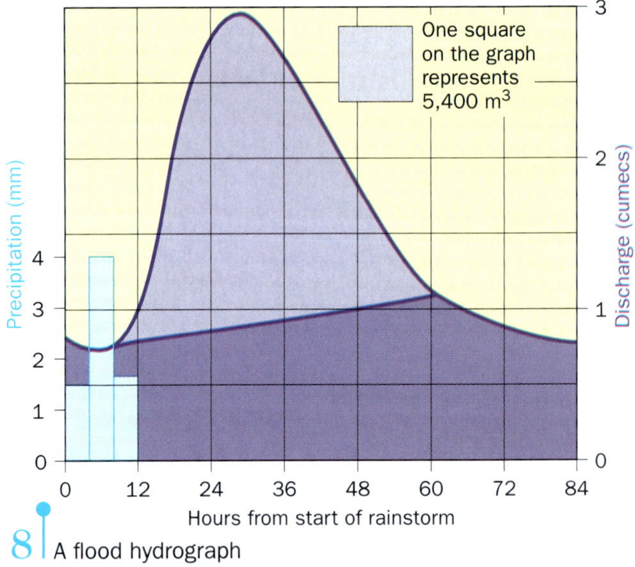

8 A flood hydrograph

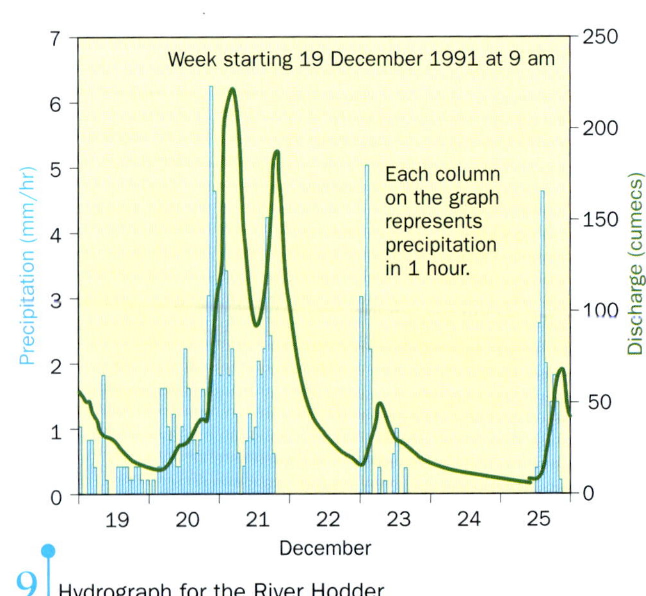

9 Hydrograph for the River Hodder

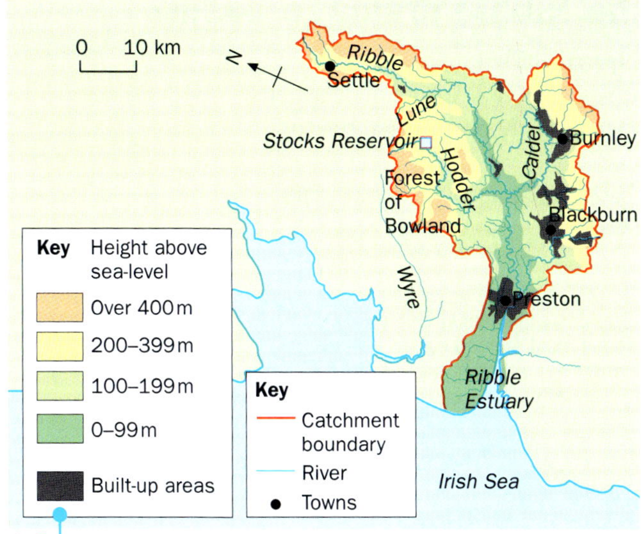

10 River Ribble catchment: relief and built-up areas

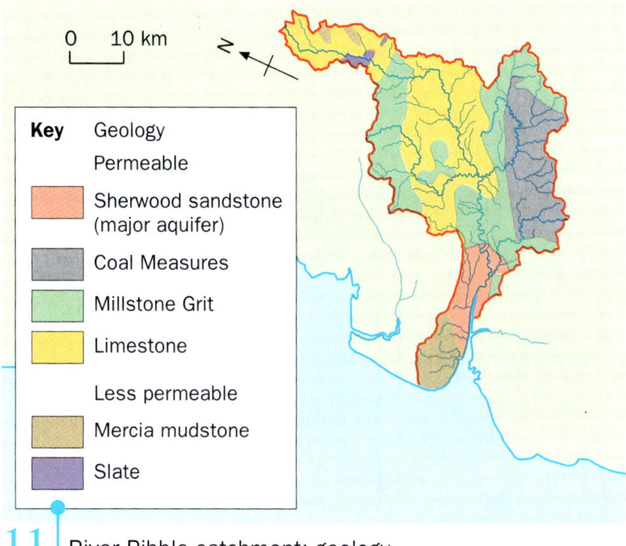

11 River Ribble catchment: geology

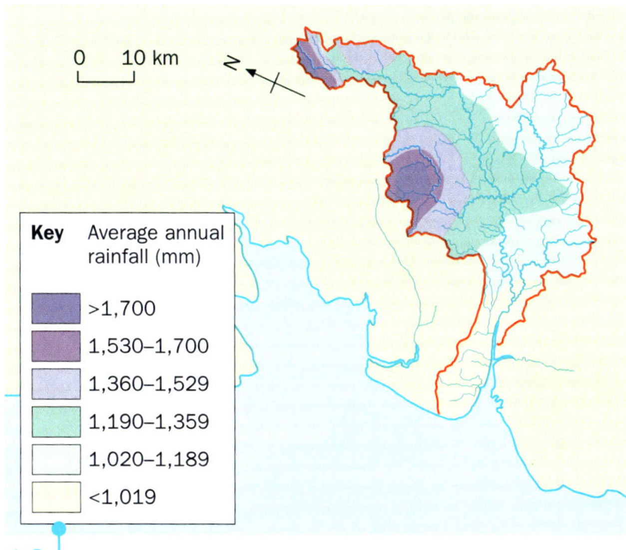

12 River Ribble catchment: rainfall

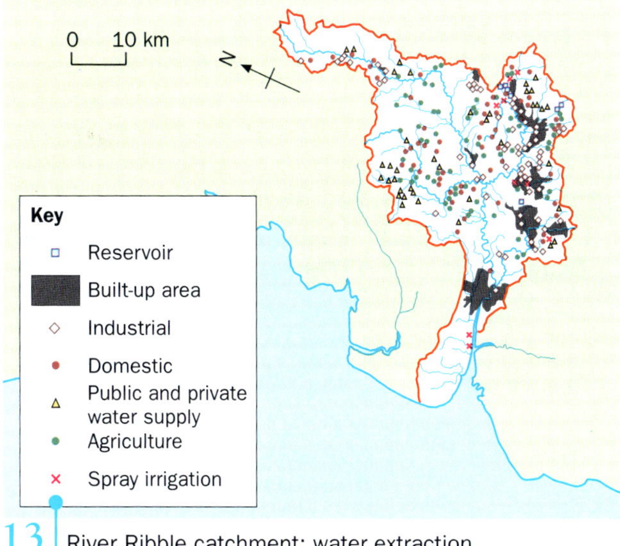

13 River Ribble catchment: water extraction

Case study: How can we manage the Ribble drainage basin?

Drainage basins can have a dramatic impact on human life, particularly in extreme conditions when they have too much water and so flood, or when there is insufficient water during a prolonged drought. The drainage basin can supply water for houses, agriculture or industry. It is also used for recreation and waste disposal.

These factors lead to conflicts of interest and a need for careful management (see Figure 14). Figure 1 on page 98 described some of the measures already taken to manage the Ribble.

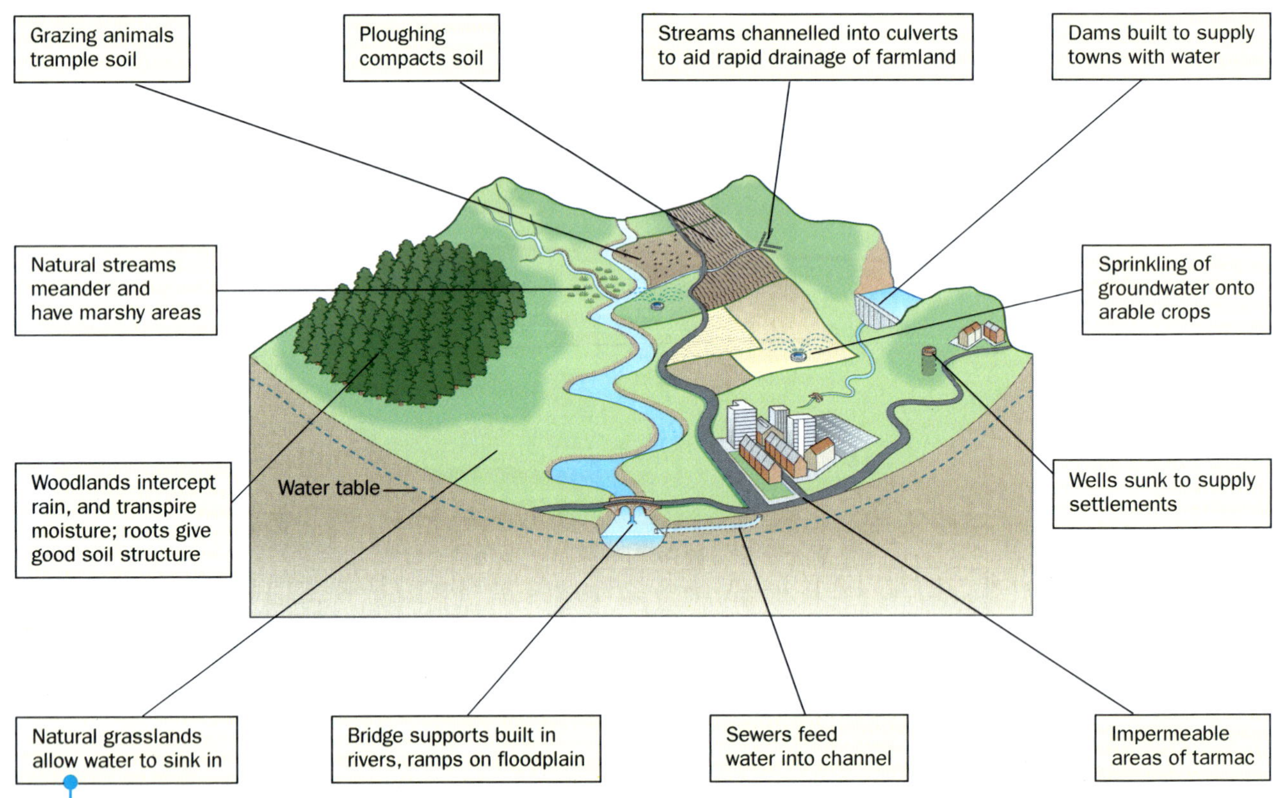
Grazing animals trample soil
Ploughing compacts soil
Streams channelled into culverts to aid rapid drainage of farmland
Dams built to supply towns with water
Natural streams meander and have marshy areas
Sprinkling of groundwater onto arable crops
Woodlands intercept rain, and transpire moisture; roots give good soil structure
Water table
Wells sunk to supply settlements
Natural grasslands allow water to sink in
Bridge supports built in rivers, ramps on floodplain
Sewers feed water into channel
Impermeable areas of tarmac

14 The effect of human activities on a drainage basin

Task

1 Using the River Ribble as a case study, look carefully at Maps 10, 11 and 12. Use the maps to complete a copy of the table below. This will help you to answer the question 'Why do some parts of a drainage basin have more water than they need, whereas others do not have enough?'. The south-west (Preston area) has been done for you.

2 Write a paragraph to explain the reasons shown in your table.

Discharge (cumecs): 0 10 20 30 40 50 60 70 80 90 100
J F M A M J J A S O N D
Months

15 River Ribble discharge, year averages

Factors / Area	South-west (Preston area)	South-east (Burnley area)	North-west (River Hodder area)	North (Settle area)
Precipitation	*Low, mainly less than 1,000 mm/yr*			
Geology	*Permeable rocks and a major aquifer*			
Urban areas	*Preston is largest town in catchment: high demand for water probable*			

The Ribble shows how water quality in a drainage basin can vary over a relatively small area. One main tributary, the River Hodder in the Forest of Bowland, is an exceptionally clean river of high water quality, making it a very important source of water supply for central Lancashire. In contrast, the other main tributary, the River Calder to the south-west of the Ribble (Map 10), provides a valuable resource for industry and has much poorer water quality as a result. A long industrial heritage in the area means that there is a lot of residual cleaning up to be done.

North West Water's plans to improve the quality of the environment across the whole basin are strictly monitored by the Environment Agency (see Figure 16). They must also meet EU rules.

Water supply

One particular issue is water supply. Here are the problems:

- Water supply constantly fluctuates.
- Water is removed from surface sources for a variety of reasons (see Map 13).
- Population density varies.
- Water quality varies.

At the beginning of the 20th century, dams and reservoirs were seen as the best way to ensure water supply. Later technological advances allowed schemes such as the *Lancashire Conjunctive Use Scheme* to combine water from rivers, reservoirs and boreholes (via pumping schemes) and to distribute water to areas of high demand. Water is pumped via pipeline from the River Lune (to the north) across the River Wyre and then to the Preston area. Groundwater is also pumped out of the sandstone aquifer north of Preston to add to the total water supply.

Today, attention is focusing on managing demand rather than on just increasing supply, i.e. educating the public or businesses in efficient use of water, or charging more for the service. The aim of water supply management is to find a sustainable way of meeting people's needs. You will return to the issue of water supply in Chapter 3.3.

Environmental concerns facing the River Ribble in the late 1990s

1 Businessman fined £8,000 for dumping waste in Ribble estuary at Lytham.
10 June 1999

2 Court orders paper mill to pay more than £50,000 for Lancashire river pollution.
27 September 1999

3 EA announces prosecution following River Darwen fish deaths.
28 August 1998

4 EA gives people a say in the future of the environment in Douglas catchment.
25 November 1998

5 EA's 'hall of shame' points the finger at guilty corporate polluters.
18 March 1999

6 *Environment Action*, issue 15 (August/September 1998) – chief hits out at 'appalling' record of pollution by some factories in the drainage basin.
14 August 1998

7 EA consults on tough new measures to save threatened salmon stocks.
20 July 1998

8 NIPA Laboratories Ltd ordered to pay £132,500 after river pollution incident.
13 May 1999

16 Extracts from Environment Agency reports

Focus tasks

3 Here are some aims of drainage basin management:

- to protect the natural environment
- to ensure clean drinking water for the population
- to ensure reliable water supply for industrial users
- to warn the public of possible floods
- to control pollution.

a) Describe one measure taken in the Ribble Valley to help to achieve one of these aims. You will need to use pages 104–105.

b) Explain how it helps to meet that aim. It might meet more than one aim – if so, explain how.

4 Each of the objectives is important but sometimes managers need to prioritise.

a) Summarise each objective on a different card.

b) Using the text on pages 104–105 to help you, add other objectives of drainage basin management that you have come across. Summarise each on a separate card.

c) With a partner agree a priority order for these objectives.

d) Discuss whether you think Stephen Evans (Figure 1 on page 98) would agree with you.

5 Using Figures 11–16 write a report for North West Water called 'Facing the Future', on the probable changes that will occur in the next 50 years as the Ribble basin becomes increasingly urbanised. Consider the needs of industry, farming and residential areas. Think about the effects of possible climate change on future supply and of changing levels of demand.

B RIVER LANDFORMS

The work of rivers

Rivers shape the landscape by the processes of EROSION and DEPOSITION. Most of the energy of a river is used to overcome the friction of the flow of the water over the bed and banks. As streams or rivers fill up after periods of heavy rainfall (bankfull), a lower percentage of the water is then in contact with the bed and banks (the wetted perimeter). The faster-moving water is able to transport its sediment load and any excess energy erodes the channel.

River erosion

HYDRAULIC PRESSURE – the sheer force of flowing water – can erode the river channel. However, more often it is the transported material that leads to erosion by:

- ABRASION – sediment is rubbed along the bed and banks, wearing them away.
- ATTRITION – larger particles are broken into smaller pieces as they collide.
- CORROSION – weak acid in the river dissolves rocks that form the bed and banks, e.g. limestone.

River transportation

A river transports its load in different ways (Figure 17): in SOLUTION, and by SUSPENSION, SALTATION and TRACTION. Finer material is easily transported but the heaviest load may be moved only when the channel is near bankfull or during extreme flood events. The increase in energy of the river at such times means that movement of sediment may lead to catastrophes such as the one at Lynmouth in 1952 when boulders that had lain upstream of the village for over 200 years were suddenly rolled along the channel, with devastating results (Photo 18).

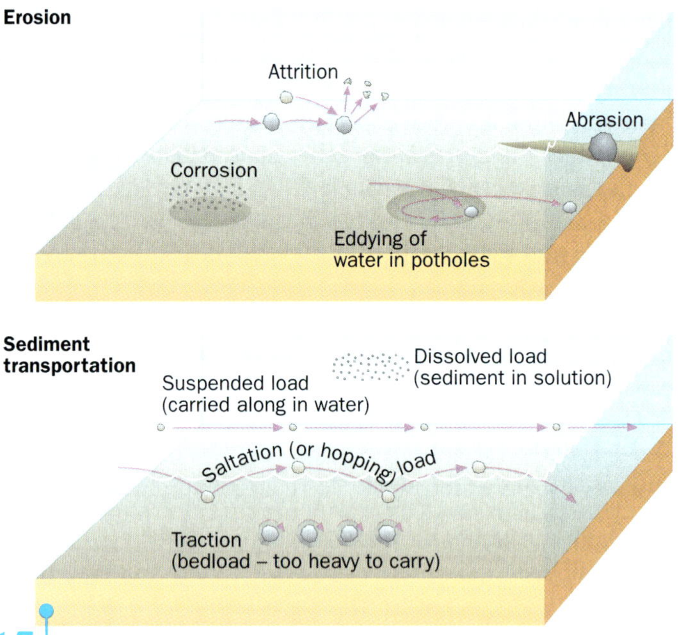

17 River erosion and transportation

18 Flood destruction at Lynmouth, Devon, August 1952

19 High Force waterfall, Teesdale

Erosional features

- Waterfalls may form when rivers run from harder rocks on to softer ones. As the less resistant rocks erode more rapidly, the overlying harder rocks will be undercut until they collapse. Repetition of the process and removal of the eroded materials from the PLUNGE POOL will cause the line of the waterfall to retreat upstream (Photo 19 and Figure 20), leaving a steep-sided *gorge*.

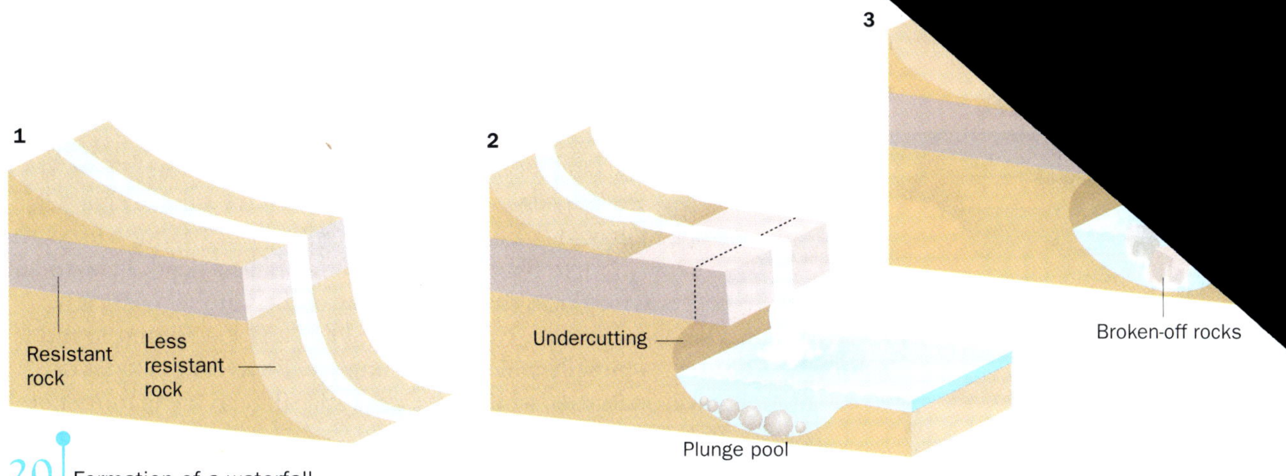

20 Formation of a waterfall

- RAPIDS may form where a river falls more gradually over layers of hard and soft rock, producing a series of smaller steps.
- In upland areas where water has a long way to fall to sea-level, V-SHAPED VALLEYS are common. Vertical erosion is relatively strong, and steep-sided V-shaped valleys are cut as the channel removes debris from the valley slopes.
- As the stream winds through the hillsides, it forms INTERLOCKING SPURS (see Photo 25a on page 109).
- At periods of low flow it may be possible to see POTHOLES in rocky river beds where the swirling action of pebbles has scoured out the channel floor (Figure 17).
- In lowland areas the river carries a greater volume of water. Sideways (lateral) erosion is greater than vertical erosion and the river tends to meander. The outside bank of the meander is eroded as the water flows fastest on the outside of the bend, cutting a deeper channel with steeper sides (Figure 21). The water flows more slowly on the inside of the bend, leading to deposition. One feature of meandering is the formation of OX-BOW LAKES, explained in Figure 22.

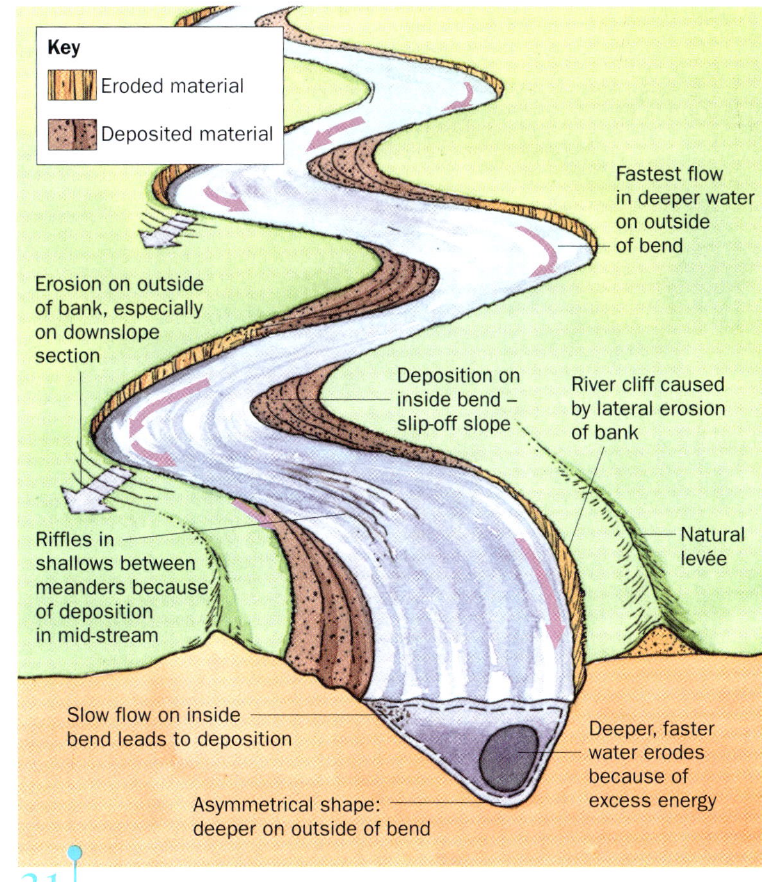

21 Formation of a meander

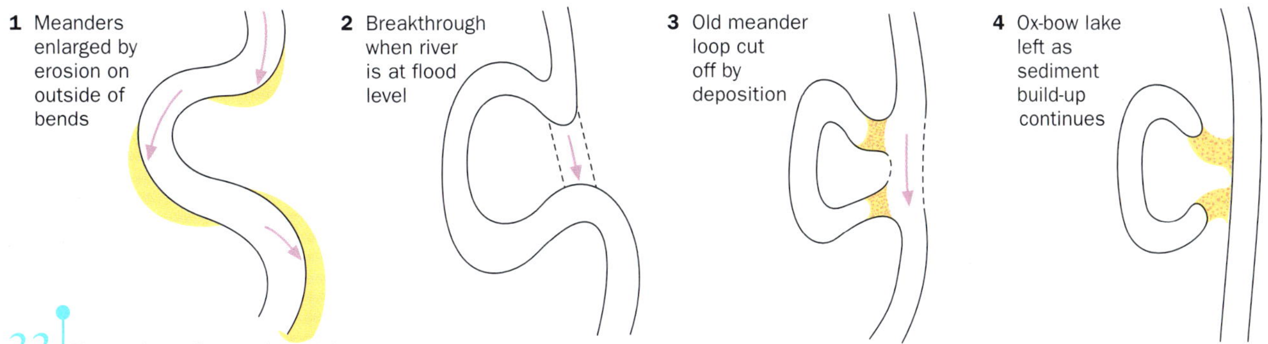

22 Formation of an ox-bow lake

Depositional features

- FLOODPLAINS When a river floods, spilling over its banks, it slows down and deposits its suspended load. Repeated flooding will build up layers of nutrient-rich ALLUVIAL SEDIMENT, forming a floodplain with fertile land for farming and flat land for building.
- LEVÉES are low ridges which develop along the sides of channels prone to flooding. As water bursts over the banks during a flood, the increase in frictional drag caused by the shallow depth of water and vegetation reduces the speed of flow of the water dramatically. The heaviest materials are dropped first and stay nearest to the bank, eventually building up alongside the river as mounds of gravel and sands, whilst finer clays and silts can travel further out on to the floodplain (Figure 23).
- Deltas Where rivers carrying large amounts of sediment enter still waters such as a lake or a sheltered sea, a delta may form.

  When a fast-moving river meets the sea or a lake, it is like hitting an immovable wall. It slows down suddenly and deposits all but the finest sediment, so blocking its own channel. The blocked channel subdivides into several smaller, more efficient channels (DISTRIBUTARIES) which fan out into the sea. The shape of the delta depends upon the balance between the volume of the sediment (more sediment pushes the delta outwards) and the strength of the waves (which wear away the delta). This is demonstrated by the Nile delta (Photo 24), which is retreating because less sediment reaches the Mediterranean each year as a result of the effect of the Aswan dam. When this dam was built to control river flow and generate hydro-electricity the effects on sediment transportation were not fully realised. Silt is trapped by the still waters of Lake Nasser and so much less now reaches the delta. This means that the waves have enough energy to erode the delta.

24 Satellite view of the Nile delta, Egypt

Deltas provide fertile, if unstable, agricultural land and in many countries (for example the Ganges–Brahmaputra delta, Bangladesh) support large populations in hazardous locations.

Task

1 a) Work in pairs to produce definition sheets for the erosion and deposition features described on pages 106–108. One person should work on erosion features, the other on deposition features.

Type the name of each feature using a large font (one term per sheet of paper). Type each definition on a separate piece of paper (as shown opposite).

b) Now match your partner's definitions to the correct features.

c) When you have matched them correctly, mount them for a classroom display.

Feature:
levées

Definition:
Low ridges which develop along the sides of channels prone to flooding

25 River Ribble: from source to mouth

SAVE AS . . .

2 Use the information from pages 106–108 to help you to analyse Photos 25a–e. These were taken at increasing distances from the source of the Ribble. Draw a simple sketch of each photo and annotate it to show river features.

★ THINK!

You are often encouraged to provide a simple annotated sketch map or diagram in exam questions. This allows you to include specific features. Be sure to include only information that is really *relevant* (rather than just related) to your answer. By remembering one feature on a simple diagram you are more likely to remember others.

POLLUTION – THE HAZARDOUS JOURNEY

One particular issue in managing drainage basins is controlling pollution. Pollution has many causes and may come from a single 'point' *source* (e.g. one factory), from an area (e.g. an industrial zone), or a whole region. The *product* may be a gas (e.g. carbon dioxide), a particulate (e.g. smoke), or dissolved in water (e.g. nitrates). These pollution products travel along *pathways*, including the atmosphere, streams or through the soil, to *sinks* such as seas, groundwater, lakes and soil. Here they may build up to reach dangerous levels of concentration, entering food chains that pass the contaminants from plants to humans.

Think of the whole process as a hazardous journey ending in disaster! So should we clean up the mess at the crash site, intercept the travellers bent on destruction, or stop them from leaving home in the first place? Do we pass laws banning these activities, jail those who are responsible, or appeal to their wish to survive?

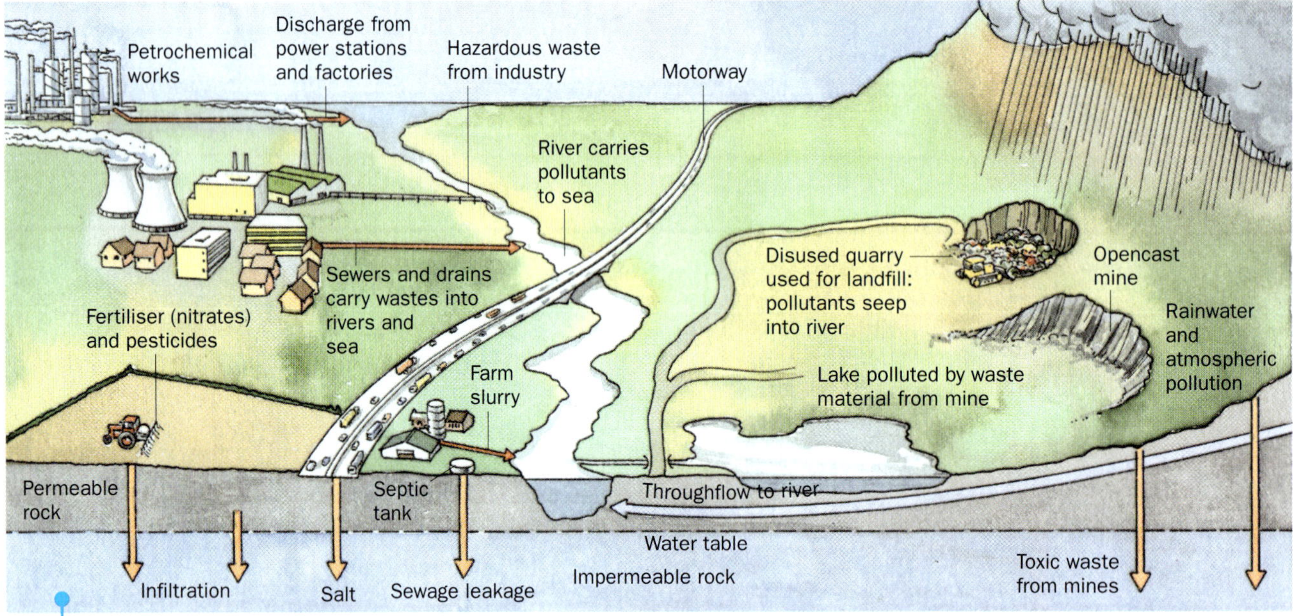

26 Causes of water pollution

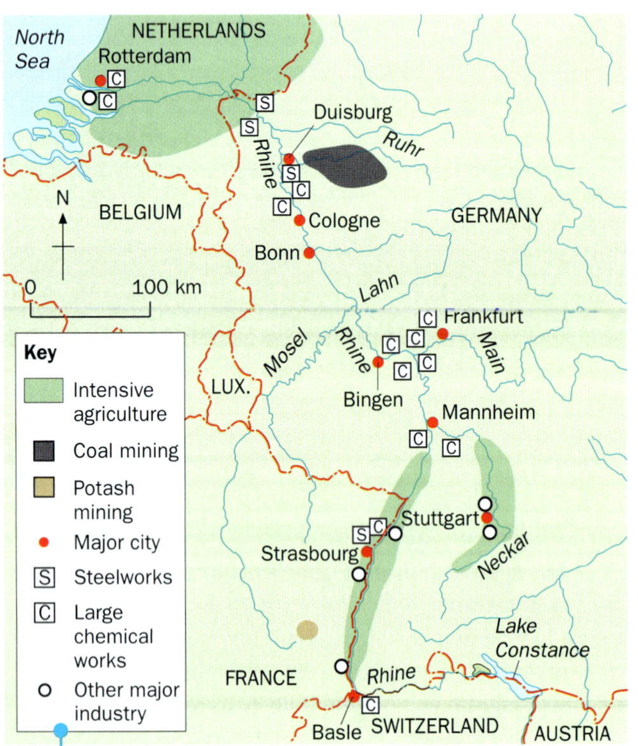

27 Sources of water pollution on the Rhine

Case study – the Rhine Action Plan

The Rhine drainage basin has 50 million people living in an area of 225,000 km^2. It has one of the highest concentrations of industry in Europe. The 1,300 km long river acts as a transport artery from Basle on the Swiss border to the North Sea. Large quantities of the region's imports and exports are carried on the 35,000 vessels that use the river each year.

During the 19th and 20th centuries the river acted as a waste-disposal system for the mines, steelworks, chemical factories and farms in the Rhine basin (Figure 27).

By 1970 the river was almost biologically dead, with 70 per cent of German industrial pollution ending up in the river. The river around Cologne was termed a 'Danger Zone'. At the same time the river provided drinking water for 20 million people! Action was desperately needed. Since the Rhine runs through several countries it needed internationally agreed action. Between 1950 and 1987 £40 billion was spent by countries along the Rhine in an effort to reduce pollution – with limited success.

The need for more effective action was shown in November 1986 when the Sandoz Company, a chemicals producer near Basle, had a huge warehouse fire. It showed how difficult it is to deal with river pollution.

The effects

- Fire-fighting washed 1,300 tonnes of hazardous substances into the Rhine, including 200 kg of deadly mercury.
- The river was pronounced 'biologically dead' for 300 km downstream.
- 500,000 fish and most eels were killed.
- Evidence suggests that other firms used the disaster as a convenient cloak to dump hazardous chemicals. The company BASF released 1,100 kg of pesticides.
- The main 'plug' of pollution took ten days to reach the North Sea.

The solution

- 25,000 tonnes of soil were removed from the contaminated site to reduce groundwater pollution.
- Nearly £20 million was spent in settling insurance claims.
- Sandoz built new, safe basins to retain chemicals during any fires and stopped using over 100 dangerous chemicals, including mercury.
- Sandoz financed environmental research with its 'Rhine Fund'.

Effective internationally agreed environmental action really started with the Rhine Action Plan (1987). (See clipboard opposite.)

The Rhine Action Plan (1987)

Aims

- Regenerate the ecosystem, with the return of salmon as the key indicator of success.
- Improve water quality to ensure clean drinking water.

Methods

- Public Relations campaign to convince industry of the benefits of a clean river.
- Introduce rigid safety procedures and river monitoring.
- Advice given to industry and agriculture on how to reduce pollution.
- Courts try polluters for first offences. If they offend again the courts can close down the offender's factory or farm.
- Target the 45 worst pollutants and reduce them by 50 per cent by 1995.

The **results** of international co-operation are impressive:

- A 70 per cent reduction in mercury, lead and dioxin levels.
- A doubling of oxygen levels in the water.
- A rise in species from 27 to 100+.
- First salmon caught in 1990 and spawning by 1995.
- Nitrates from intensive farming have been reduced but still represent one-third of all North Sea input.

You can find out more about the plan at these websites:
www.rri.org/envatlas/europe/germany/de-index.html
www.iksr.org/icpr/index.htm

Tasks

1 On a table like this, record the case study of the Sandoz fire described in the text above.

Types of water pollution	Source	Pathway	Sink

2 Study the information about the Rhine Action Plan. Sort the measures taken into 'carrots' and 'sticks'.

Focus task

3 Use your work on the River Ribble and River Rhine as the basis to answer the following question:

How can our use of rivers be made more sustainable?

Support your answer with evidence in the form of facts, figures, graphs and maps.

3.3 *Can we manage our scarce water resources?*

DO YOU KNOW?

- One billion people have no access to safe water.
- The equivalent of 12 jumbo jets full of children die every day from diseases caused by unclean water.
- England and Wales alone require around 55 million litres of drinking water a day.
- 71 per cent of the world's surface is covered in water, but 97 per cent of that is salt water.
- Of the 3 per cent that is available as fresh water, three-quarters of it is trapped as ice.

AIMS

- **To explain why some parts of Britain and Ireland have more water than others.**
- **To understand how water supply is managed.**
- **To be aware of the problems as well as the benefits of building reservoirs in Britain and India.**
- **To compare the effects of drought in Africa and the USA.**

Tasks

1 Work in pairs. One person reads out items **A** to **C** below, giving the other person time to write down their choice, and their reason for making that choice each time.

> You are shipwrecked far from home and are washed up, alone, on a small, dry desert island.
>
> **A** A friendly inhabitant appears and offers you a bucket of diamonds or a bucket of water – which would you choose?
>
> **B** The person returns again (unexpectedly) after ten minutes and offers the same choice. What is your choice this time?
>
> **C** The person comes back in another ten minutes. At the third time of asking what would you choose?
>
> You never see that person again ...

Discuss:

2 a) Did you make the right choice?

b) Did your choices change between **A** and **C**?

c) Did you need more information before choosing?

SAVE AS . . .

3 Write down how you think each of the following factors affects the way you view the importance of water resources. Give an example for each factor:

- supply
- current needs
- future demand
- climate
- water purity
- past shortages.

The water–diamond paradox (or puzzle) in Task 1 raises some important questions about the value we place on water. Most people switch their choice to diamonds by the time they reach decision C. Maybe you were smarter than most! Without water we die within three or four days. Water is seen as an abundant commodity in many parts of the world and we often use it wastefully. In times of shortage we think it's only temporary – someone will sort out the problem. There'll be another bucket along in ten minutes! But what if there were no more buckets coming along?

Just as the price of diamonds may fall if the supply increases, so the value we put on water may change if it becomes scarce and the demand for it is greater than the supply.

For England the year 2000 was a wet one. First it experienced a very wet spring, after years of drought worries in places such as Yorkshire and East Anglia. Then in the autumn and winter many parts of the country experienced the worst floods in recorded history. In the same year, Rajasthan and Gujarat in India suffered a severe drought as the failure of the monsoon rains led to famine conditions. As climate change seems likely to bring more changeable weather (see pages 91–93), the problem of ensuring an adequate water supply is one that most countries face today or will have to face in the near future.

1 People collecting water from a specially transported tank in England, during a drought in 1987

2 People on their daily visit to collect water from a well in Sudan

Tasks

4 Study Photos 1 and 2.

a) List ways in which these photos are similar and ways in which they are different.

b) How might each country's ability to cope with water shortage be different? Explain your answer.

5 Study the newspaper article in Figure 3 on the drought in Rajasthan and Gujarat in India. Discuss:

a) What are the human and physical causes of the problem?

b) What are the human and physical effects of the problem?

c) Who is responsible for sorting out the problem?

d) What are the short-term solutions to the problem?

e) What are the longer-term solutions to the problem?

Indian drought: Government under attack for slow response to crisis

Sanjeev Miglani in Delhi

The Indian government, under fire for a slow response to a drought overtaking its western states which is said to be the worst this century, has pledged to commit all its resources to averting a looming national crisis that raises the spectre of famine and affects an estimated 50 million people.

The agriculture minister, Sunderlal Patwa, said 23,406 villages in the desert state of Rajasthan and 9,421 villages in neighbouring Gujarat were affected by 'conditions of scarcity'. But he said newspaper reports of people fleeing the two states were wrong.

The prime minister, Atal Bihari Vajpayee, made a televised appeal at the weekend for Indians to provide financial help to the afflicted states.

The government said it had released 100,000 tonnes of wheat and rice each for Rajasthan and Gujarat which would be distributed in the drought-hit areas at subsidised prices, and Mr Vajpayee told parliament that train-loads of animal fodder were being rushed there.

Meanwhile his Bharatiya Janata party said: 'We appeal to all political parties to join in this endeavour and not politicise it, otherwise it would divert the nation's attention from the problem confronting the people.'

The opposition Congress party accused the government of acting too late to help a rural population already hit by high debt, rising fertiliser prices and falling groundwater levels.

The drought is hitting Rajasthan and Gujarat because the annual monsoon rains have failed in some areas over the past two years.

From the *Guardian*, 26 April 2000

PAKISTAN
Rajasthan
INDIA
Gujarat
N
0 400 km

3 Drought in western India

Supply and demand for water in the UK

Maps 4 and 5 show the distribution of rainfall and population in the UK and Ireland. In the west of Britain and Ireland precipitation provides enough clean water to provide for the population most of the time. However, in the drier east, with the highest population, the most industry and the most intensive agriculture, demand often outstrips supply, as shown in Map 6.

The current water shortage in Britain isn't as severe as in other places such as Africa, where 25 countries will lack sufficient fresh water by 2025. However, providing adequate clean water absorbs a massive amount of time and money. We still have to make careful use of what we have.

Most of Britain's water comes from surface water in rivers and lakes, but as more is needed it is increasingly taken from aquifers below the ground (see page 100).

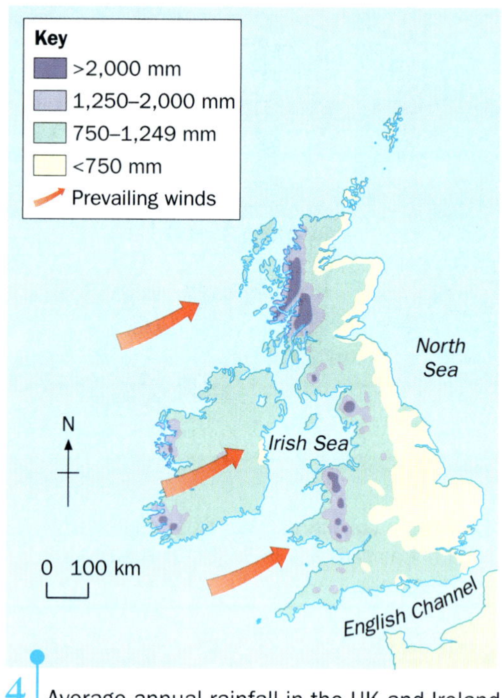

4 Average annual rainfall in the UK and Ireland

Whose water is it anyway?

The simple solution to any water supply problems in Britain would be to share out the water more fairly. Since the 19th century, valleys have been flooded, in Wales and the Lake District (areas with high rainfall) in particular, to supply cities in England. Map 7 shows the major reservoirs in Britain and Ireland. It is difficult to imagine new large reservoirs being built in those areas today. Regional water authorities looking after the needs of their own areas, and the privatised suppliers concerned about the interests of their shareholders, seem to have little incentive to help other regions unless they are prepared to pay handsomely for the water.

Political parties such as Plaid Cymru in Wales question whether it is right to give so much of their water away to another country. Environmental objections to new reservoirs make it very hard to achieve planning permission.

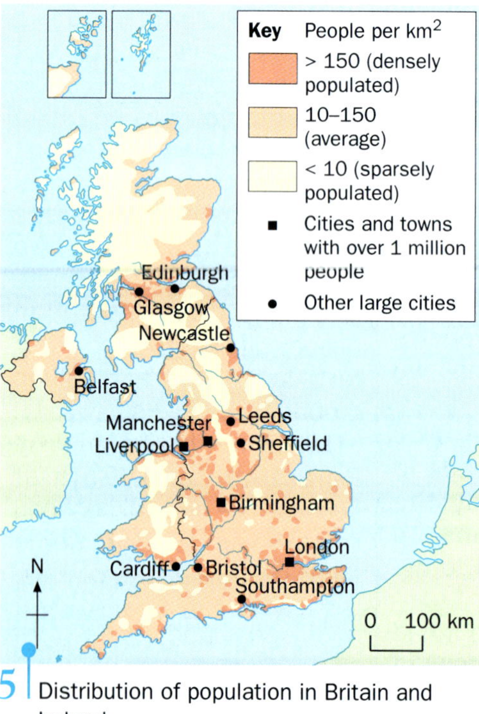

5 Distribution of population in Britain and Ireland

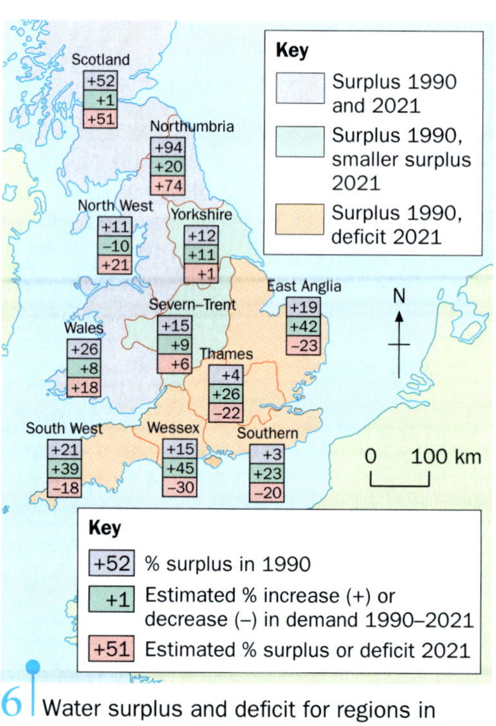

6 Water surplus and deficit for regions in Britain

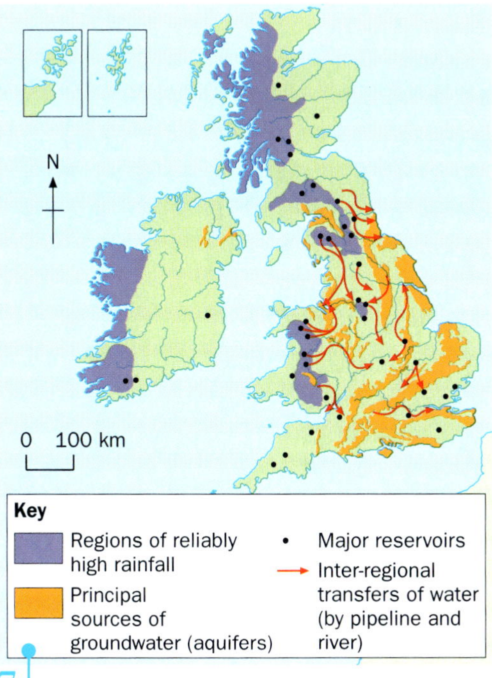

7 Water supply in Britain and Ireland

Water shortage in LEDCs

In many LEDCs people suffer great water shortages almost all of the time.

- In Delhi, the capital city of India, many SHANTY TOWN dwellers have to share one tap in the street among 200 people. Some people with access to clean water can make a good living by selling bowls of water to people in neighbouring shanty towns.
- In north-east Kenya, the women in some villages have to spend up to three hours every day travelling to wells and streams to collect water. They may carry a pot or can full of water, weighing up to 20 kg, back to their village on their head, to use for drinking, cooking, washing, and for their animals.
- The Ganges delta region of Bangladesh regularly suffers from flooding. However, the river water is contaminated with sewage, industrial waste and fertiliser washed from fields upstream. Charities like Oxfam and ActionAid provide drilling equipment so that people in some villages can drill down over a hundred metres to aquifers below the delta. Unfortunately even this water is sometimes contaminated with heavy metals, washed out of the rocks above.
- Nomadic tribes in the Sahel region of Africa have learnt to live with scarce water resources. The eldest people are held in great respect. Only they have enough knowledge and experience of travelling through the region to know where the few tiny supplies of water can be found in the dry season.

So next time you hear that some people in Britain might be stopped from using hosepipes to wash the car, or that sprinklers on the greens of the golf course might be banned, just remember how this compares with the situation in other parts of the world.

Task

1 Study the diagram below, which shows some uses of water in the north-east of England.

Who needs water in the north-east?

- The people of Tyne & Wear need 100 million litres a year for domestic purposes
- S & N Breweries 450 million litres a year
- Crowtree Leisure Pool holds 150,000 litres of water
- Nissan car manufacturers 525 million litres a year
- MetroCentre retail & entertainment complex needs 175 million litres a year

a) Draw a similar diagram for shops, industries and leisure facilities in your area. You won't be able to give precise quantities (unless you do some careful research) so just indicate the main users.

b) Colour code your diagram to show where clean water is vital (e.g. drinking water) and where it might not be really necessary.

c) Look at the number of times you have highlighted clean water as important in your own examples. What does this mean in terms of water supply and water treatment?

SAVE AS . . .

2 'Why do some parts of England and Wales have less water than they need?' Write a three-paragraph answer to this question.

a) Explain the physical and climatic reasons why some regions receive a lot of water.

b) Explain the human reasons why some regions require a lot of water.

c) Explain how the imbalance between supply and demand can lead to water shortages in some areas and surpluses in others.

★ THINK!

Examiners often say that students don't give precise details and don't add annotated sketches to support their answers.

So, when answering Task 2, add a base map of England and Wales to explain the *physical* and *human* causes of the problem. Annotate the map using Maps 4–7 to supply detail. Quote actual figures – you don't need *all* of them! Be selective in order to prove the point you wish to make.

Solving the problem – increasing supply?

Figure 8 shows how water is used in England and Wales. Rising urban population, improvements in living standards and increased needs in agriculture, industry and electricity generation have all helped to push up demand.

There are only two ways of dealing with the problem: increasing supply or reducing demand. The Factfile shows some of the ways of increasing water supply – and the estimated cost. Some of these schemes currently exist; others present great physical problems (see the Ribble case study in Chapter 3.2). For example, ABSTRACTION of surface water and groundwater leads to low flow in some rivers and to wetlands drying out because water is removed before it reaches them.

★ FACTFILE

Ways of increasing water supply	Estimated cost per million litres per day
• **Reduce waste from leaking pipes** underground. It is estimated that in some areas 33 per cent of water is lost before reaching its destination.	**£0.5 m**
• **Abstract more water** from rivers and underground aquifers in water surplus areas and send it to where it's needed using existing rivers and pumping stations.	**£0.5 m**
• **Re-use effluent** once it has been cleaned in water treatment plants, by putting it either directly back into the mains or back into rivers. This already happens in the Thames region.	**£2.5 m**
• **Build new reservoirs** to store more water, probably in upland regions in the west and north (see Kielder case study opposite).	**£2.5 m**
• **Inter-regional transfers** to pipe water from surplus to deficit areas, possibly over long distances to fill up underground aquifers near to large cities.	**£5 m**
• **Develop a national water grid** to serve all areas of England and Wales, using existing rivers, new pipelines, aqueducts and pumping stations to transfer water from surplus to deficit areas.	**£6 m**
• **Desalinate seawater** – trial plants are working in Norfolk and in Saudi Arabia – quantities of over 800 million litres per day are possible.	**£6 m**
• **Transfer by boat from Europe**, e.g. using oil tankers and storage depots at coastal locations in Britain and abroad.	**£10 m+**
• **Tow icebergs to Britain** – it is believed that the ice would reach Britain largely intact.	**£10 m+**

Task

1 a) Study the Factfile which gives different options for increasing water supply in Britain. Copy and complete a table like the one below to summarise the possible advantages and disadvantages of each option. Refer to the information on pages 114–17 and to the case study of drainage basin management in the Ribble Valley (pages 104–105). You should also use your wider knowledge to think about the possible environmental and social implications of each option.

b) Fill out your scores for each option.

c) Choose two options which you think are the best on balance and write a paragraph to explain your choice.

Method	Advantages	Score A 1 to 5	Disadvantages	Score B −1 to −5	Combined A and B	Rank
Fix leaking pipes	*Could save 33% of all water. Does not damage ecosystems in rivers, etc. Least-cost solution.*	*+5*	*Will take a long time to complete. Digging up roads will disrupt traffic. Some leaks may not be accessible.*	*−2*	*−3*	

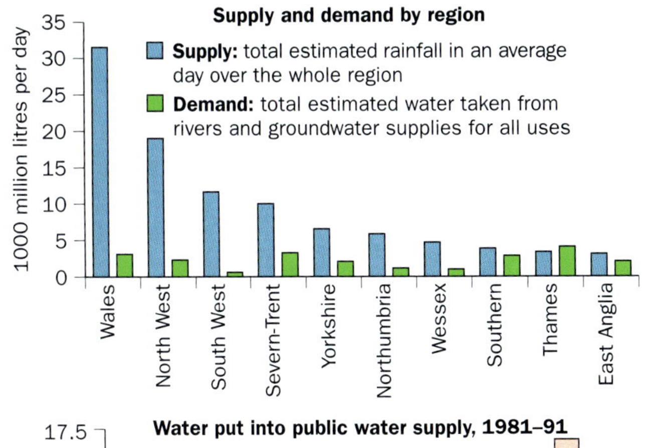

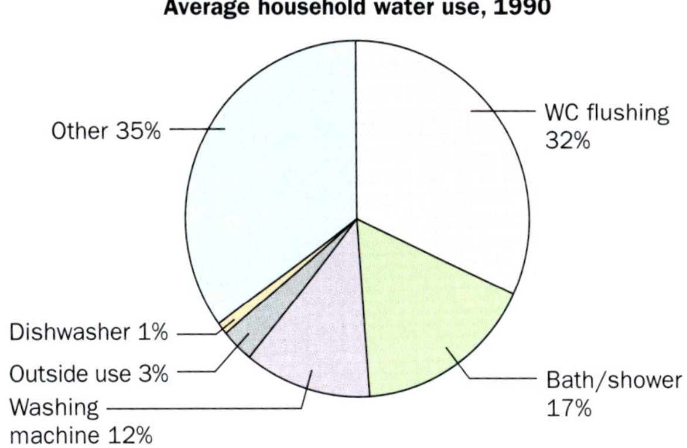

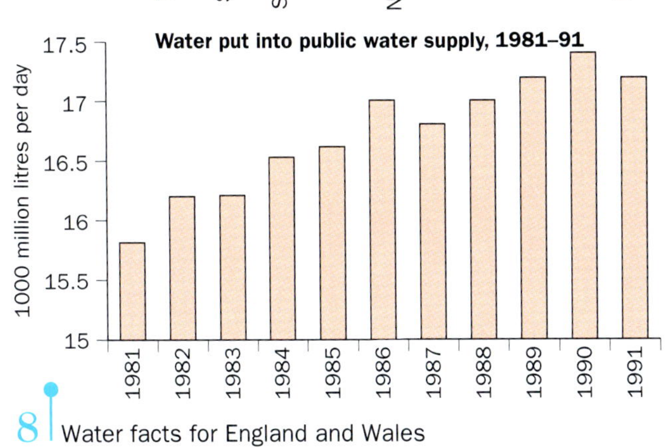

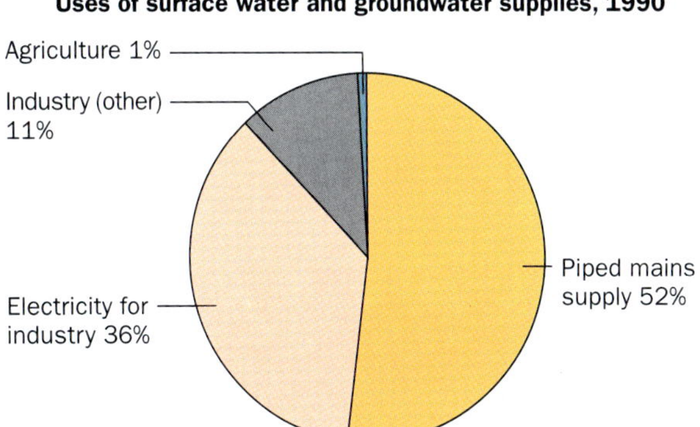

8 Water facts for England and Wales

Case Study: Kielder Water – white elephant or golden goose?

Kielder Water Reservoir in Northumberland was started in 1975 and completed in 1982 (Figures 9 and 10). It cost £167 million. The aim was to meet the projected increase in demand for water resulting from a rising population and the growth of HEAVY INDUSTRY in the north-east. A total of 1,084 ha of former forest and an attractive but sparsely populated valley were flooded to create a huge reservoir bigger than Ullswater in the Lake District. One and a half million trees had to be cut down to make way for the 12 km-long artificial lake.

The need for water overruled many concerns about the impact on the local environment at the time, but the doubts resurfaced when demand failed to grow as much as had been anticipated.

Supporter

Kielder not only protects us from future shortage but the twin hydro-electric power generators create £1 million profit each year. The project was always much more than just a local reservoir – it serves the whole north-east, with water being pumped as far away as the River Tees for cities like Middlesbrough. If global warming means the south and east get drier, Northumbrian Water will be able to sell water to new locations. Tourist and recreational facilities have really taken off with visitor numbers so high that new accommodation was opened in summer 2000. Local businesses benefit from the money the visitors spend, and the financial gains from selling surplus water mean that this could be a real golden goose for the economy of the area.

Critic

The decline in heavy industry in the north-east during and since the 1980s has meant that Kielder has never been less than 90 per cent full. We have all the water we need and more. Was it really worth flooding a huge part of an Area of Outstanding Natural Beauty, even if it means we can export water to Yorkshire or Europe in the future? Tourism alone can't justify it. We had plenty of other attractions in the area already. At the end of the day, I look at the money spent and wonder if we could have better spent it on fixing leaking pipes and reducing waste rather than building yet another white elephant!

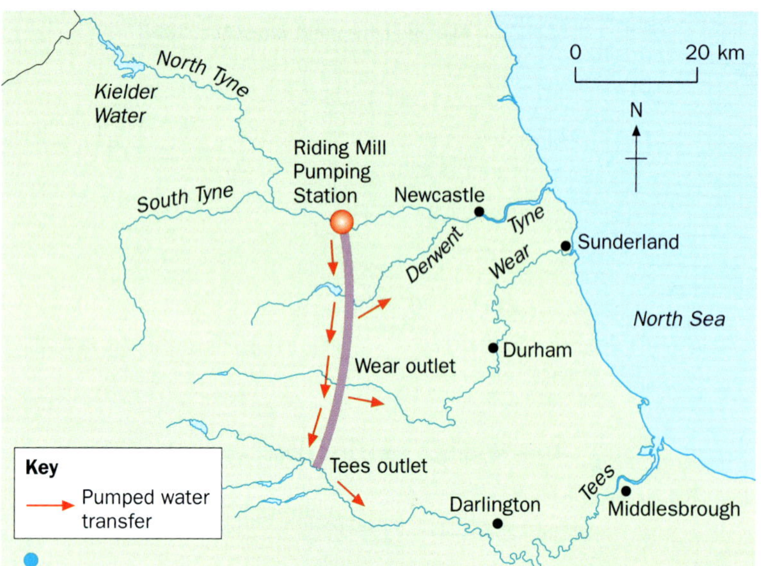

9 How Kielder Water serves the North-East with water

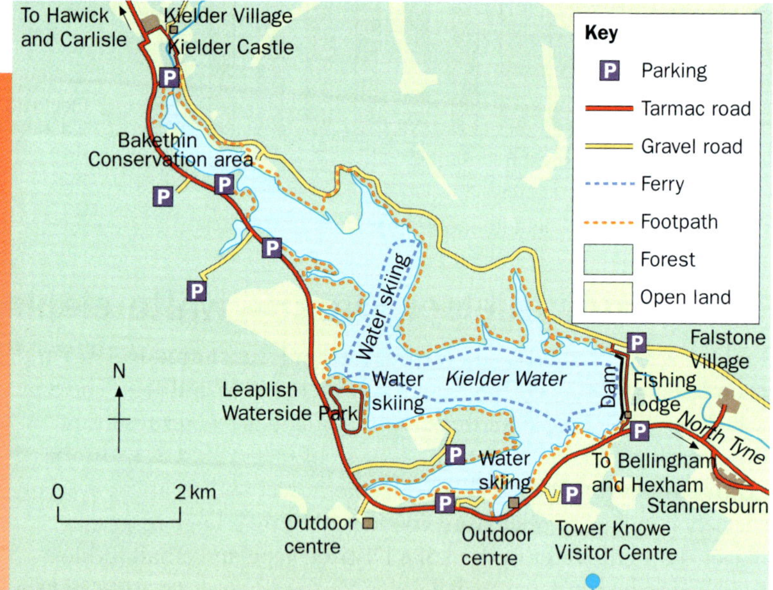

10 Kielder Water and features

Tasks

1 Using the information in Map 9, explain how Kielder Water provides water for the north-east of England.

2 Study Map 10. List the features shown that are related to:

a) recreation or tourism

b) conservation

c) water supply.

Think about the balance of the features that you have listed. What does this suggest is the main purpose of Kielder Water?

ICT

3 The internet can provide excellent images of geographical features such as Kielder Water. Some websites that provide images and detail about Kielder Water are listed below.

Official site for Kielder Water: www.northumberland.gov.uk/VG/kielder.html
Free picture download: www.freefoto.com/pictures/uknorthumbria/kielder_water/index.asp
Northumbrian Water online: www.nwl.co.uk/northum/default.asp

a) Which images might interest each of the following? Choose one photo from an internet site.

- a sportsperson
- a critic of Kielder Water
- a nature lover
- a water supply manager
- a Geography student in north-east England

b) Copy the selected images, using the right-hand mouse button, and paste them into a DTP program.

c) Write a caption for each photo to explain what it shows.

d) Write a message from Kielder. 'I saw this and thought of you because …'.

Solving the problem – managing demand?

In the past, attempts to solve water shortages in Britain have concentrated on increasing supply to meet rising demand. Now attention is turning to managing demand and, in particular, trying to reduce the wasteful misuse of water.

Water metering

Most householders in Britain pay to receive water from their local water company, but this is a fixed rate and is not related to how much water is used. Water metering means people pay for what they use. Trial schemes show that people *are* more careful when the amount of water they use is related to how much they have to pay. However, many feel that poorer people might suffer if such schemes were compulsory.

Education and information campaigns

Campaigns urging the public to use less water are quite effective in times of drought but less so when water appears plentiful. Education in careful use of water should begin at school and continue through life, but such education campaigns can be very expensive.

Increased prices

If water were made a lot more expensive, people would use less. However, all users would resist paying more. Businesses say it would harm their profits. Indeed the higher the price, the more the water companies might want to sell. They might not want to restrict demand. They might want to increase it!

Rationing

During water shortages rationing has been tried; for example, turning on the water for only two hours a day. However, this is not a possibility for everyday circumstances.

Water management in an LEDC

Case Study: Narmada River Project, India

So far you have investigated attempts to increase supply and manage demand in Britain. Many LEDCs suffer much more extreme water shortages than Britain and desperately need to increase their water supply in order to aid development.

The Narmada River in Gujarat flows through an area that often suffers low rainfall especially when the monsoon fails. A series of dams is being built to create reservoirs to provide water and electricity for the region (Map 12).

However, this project has aroused massive controversy. Arundhati Roy (author of the book *The God of Small Things*) spent a year publicising the plight of the people affected by the project. This was one of her encounters:

The last person I met in the valley was Bhaiji Bhai. He is a tribal [person] from Undava, one of the first villages where the government began to acquire land for the Wonder Canal and its 75,000 kilometre network. Bhaiji Bhai lost 17 of his 19 acres to the Wonder Canal. It crashes through his land, 700 feet wide including its walkways and steep, sloping embankments, like a velodrome for giant bicyclists.*

The canal network affects more than 200,000 families. People have lost wells and trees, and have had their houses separated from their farms by the canal, forcing them to walk two or three kilometres to the nearest bridge and then back along the other side. 23,000 families, let's say 100,000 people, will be, like Bhaiji Bhai, seriously affected. They don't count as 'Project-affected' and are not entitled to rehabilitation. But like Bhaiji Bhai they became paupers overnight.

* The dam system is linked by a system of irrigation canals for water distribution.

11 Bhaiji Bhai

From being a fight over the fate of a river valley it now raises doubts about an entire political system. What is at issue now is the very nature of our democracy. Who owns this land? Who owns its rivers? Its forests? Its fish? These are huge questions.

An extract from The Greater Common Good, an essay in *The Cost of Living* by Arundhati Roy, April 1999

Why are the dams necessary?

- India's population reached 1,000 million in 2001 and is rising by 15 million per year.
- 140 million people work in farming. Agricultural production must rise by nearly 4 per cent per year to support population growth.
- The recent improvements in agriculture have centred around high-yielding varieties (HYVs) of grain. However, HYVs need large inputs of water and traditional methods of irrigation cannot provide enough.

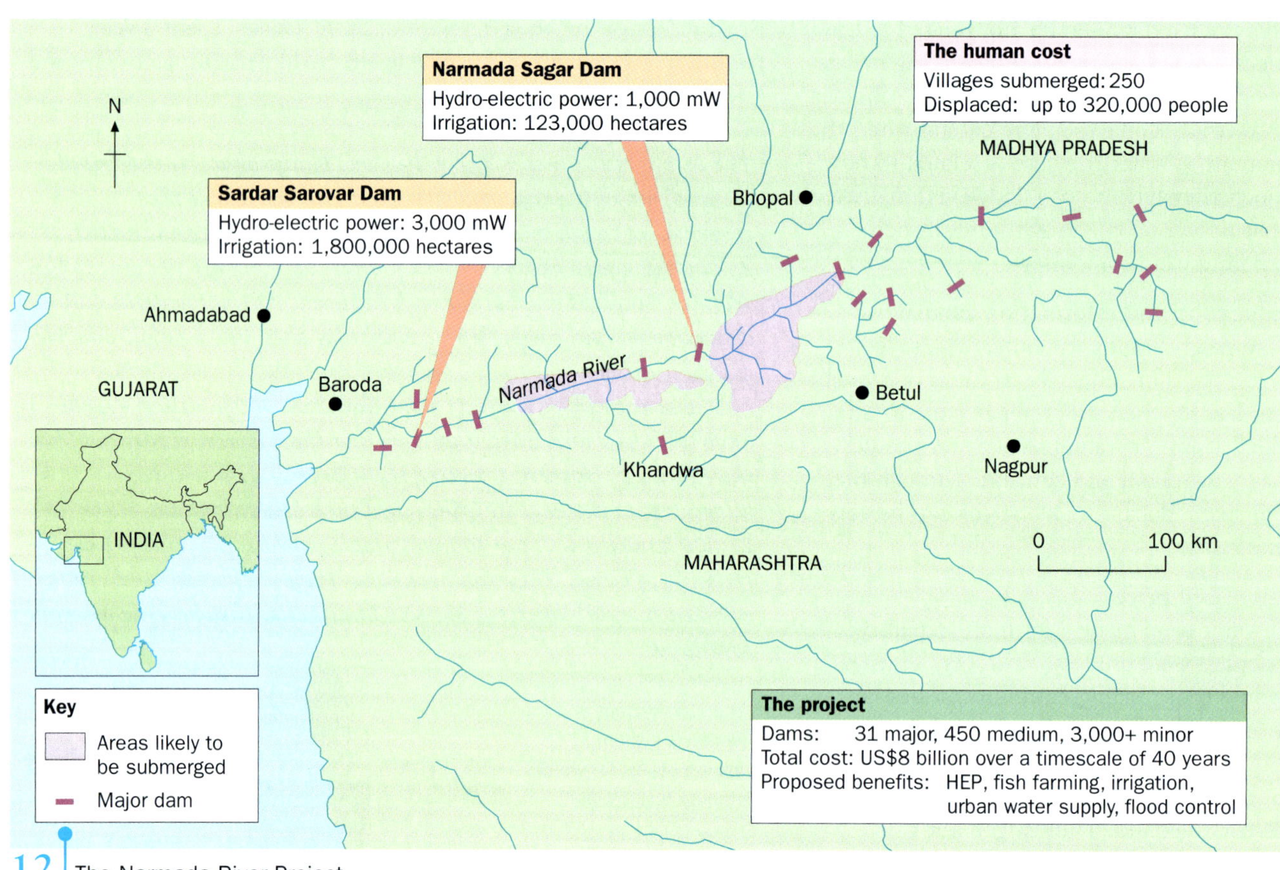

12 The Narmada River Project

What will be gained?

- The US$8 billion Narmada River Project will involve building 31 major, 450 medium-size and 3,000 smaller dams over about 40 years (see Map 12).
- The largest, the Sardar Sarovar dam, will:
  - provide 3,000 mW of much-needed electricity
  - irrigate 1,800,000 ha of land, in drought-prone areas
  - provide water for growing towns and cities
  - promote the fishing industry
  - attract tourists, providing jobs for local people.

What will be lost?

The project will dramatically affect the local population:

- Up to 320,000 people will be displaced as their villages are flooded.
- 100,000 ha of forest and agricultural land will be lost.
- Earthquakes might be triggered as the weight of water in the reservoir puts stress on the valley slopes.
- At best the project can hope to irrigate only 30 per cent of Gujarat despite having 80 per cent of the state's irrigation budget. Little money will be left for the remaining 70 per cent of the farmland.
- Unless soils are well drained they become waterlogged and salinisation occurs, as evaporation draws salts to the surface.

Focus tasks

Debate

1 A television debate programme has invited representatives from six groups to discuss the Narmada Project. At the end of the debate the audience will vote to decide which group presented the best arguments. The title of the debate is 'Should the Narmada Project go ahead?'

- **a)** Work in six groups. Each group elects one person as chair.
- **b)** Each group takes one of the roles from the panel on the right.
- **c)** Produce a group presentation lasting at most 5 minutes to outline your case for or against the scheme. Use the information on pages 119–20 and the information your teacher will give you. Your presentation could include interviews, maps and charts. A web search might also help you. Type the word 'Narmada' into any search engine, e.g. Ask.com or Google, to start your search. Save useful documents for use offline or bookmark interesting sites in your 'Favourites' file.
- **d)** After each presentation, questions can be asked by the other groups.
- **e)** After all presentations are complete, score the presentations using the mark-scheme below. Each person individually (and no longer in role) scores each presentation:

  +1 if you were convinced
  –1 if you were not convinced
  0 if you can't decide.

  You must abstain from scoring your own presentation.
- **f)** Total the scores and rank them to see who won the debate.

Debrief

2
- Why was the winning argument successful?
- Was it easiest to present? Why?
- Was it the most emotional or was it the most economically impressive? Why?
- Are all groups happy with the result? Why?

1 Narmada Action Group
Your group is convinced that prestige dam projects like the Narmada Project are flawed. These are huge, high-tech answers to problems that need small-scale solutions appropriate to the local community.

2 Indian Agriculture Committee
Your group is convinced that prestige dam projects like the Narmada Project are the only way to overcome a water supply problem that is likely to get worse.

3 Hydro-Power Ltd
You believe that projects like the Narmada Project are the only way to generate cheap electricity in a country with little oil, coal or gas

4 The World Bank
Your group invests in schemes like the Narmada Project if it can be proved that they will help a country to develop.

5 Environmental Concern Organisation
Your group is convinced that prestige dam projects like the Narmada Project pose an unacceptable threat to the environment.

6 Madhya Pradesh Villagers Co-operative
You feel that the Narmada Project will strip the local people of their homes and livelihood. The government is imposing a scheme that local people do not want

SAVE AS . . .

3 Write two paragraphs, using the example of the Narmada Project, to summarise the advantages and disadvantages of large dam projects to solve problems of water supply.

Case study: Drought and desertification in the Sahel

A drought is a period of abnormally dry weather that lasts long enough to produce a serious lack of water. Millions of people died as a result of drought and related famines in the 20th century. The severity of a drought depends largely upon the lack of water. However, the way that people manage the land often plays a major part in the problem. The Sahel experienced prolonged drought which resulted in famine for many people in 1984. It also led to increased desertification (Map 14).

Figure 13 shows that this region experienced below-average rainfall in the latter half of the 20th century. Rainfall variability was also a crucial factor. Rain that does not fall in a predictable pattern makes farming very difficult, because there are specific times of year when plants need water in order to grow.

How the drought increased desertification

- The cultivation of semi-arid, marginal lands meant that farmers relied upon limited rainfall arriving in time for the crops to grow.
- In the 1950s, heavier rainfall persuaded many people to move northwards as the desert margin retreated, creating more farmable land.
- Population in these areas expanded, then drought caused the water table to drop, leaving waterholes and rivers dry.
- Vegetation died, exposing the soil to wind and water erosion.
- Crops began to fail, leading to more and more marginal land being turned over to farming.
- In the desperate quest for food, no fallow (rest) periods could be allowed for the land to recover. By the early 1980s overcultivation was common, stripping the land of its nutrients.
- Cattle herds, a symbol of importance as well as a source of food, needed more land to support them as existing land was overgrazed.
- The soil on overgrazed and overcultivated land was easily blown away by winds and washed away when the rare rainfalls did occur. Without plants to aid infiltration the soil could not absorb water so easily, leading to excessive run-off and erosion.
- Famine caused many farmers to leave the newer cattle grazing areas to move back into areas they had left years before. Overcrowding spread desertification southwards.

Look back to Chapter 1.3 to see how people can combat problems caused by drought and desertification.

★ FACTFILE

Drought and desertification

- 30 per cent of the Earth is affected by drought.
- 70 per cent of the Earth's 'dryland' is affected by desertification.
- 1 billion people are affected by desertification.
- Desertification costs US$40 billion per year.
- 52,000 km^2 of land turns to desert every year.

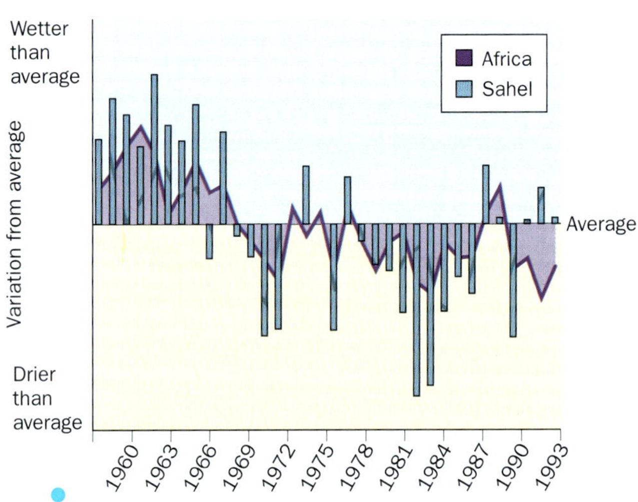

13 Rainfall variability in Africa and the Sahel

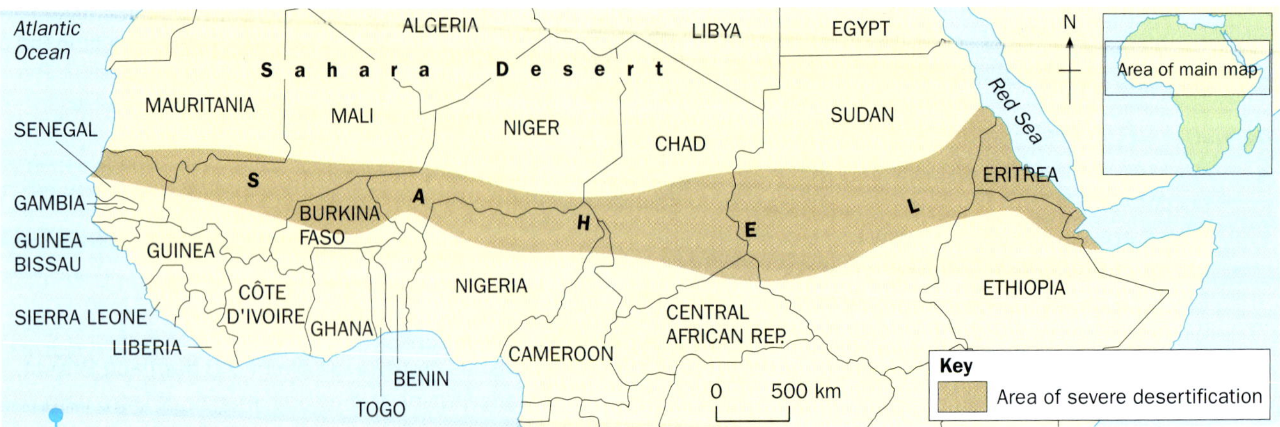

14 Desertification in the Sahel

Case study: Drought in the USA

In the USA, the extreme drought event that led to the great Midwest 'dustbowl' famine of the 1930s is remembered in literary classics like John Steinbeck's *The Grapes of Wrath*. Great advances have been made in farming since those days but, in spite of North America's immense wealth and technological expertise, the drought problem still exists today, even if the effects are less dramatic. Map 15 shows drought conditions in the USA for one day in September 2000 (note that this is specific to one day and is not the average), plus the financial cost in the form of heatwaves and fires. Some fires may be related to drought conditions.

The drought damage to farmlands east of the Rockies is critical, as this area comprises much of the American grain belt and large expanses of cattle grazing land. Farmers who find they have insufficient water to irrigate their crops are at the beginning of a long chain of suffering. Reduction of food supplies leads to increases in food prices, a loss of jobs in transport and retailing, and a rise in inflation that affects everyone. Global warming may be responsible for a spate of recent droughts. When temperatures rise, the interiors of large continents tend to dry out to a greater degree than coastal regions (this is described as CONTINENTALITY).

Tasks

1 Compare the droughts in the Sahel and the USA, under the headings:

- Duration
- Extent (the area affected)
- Effects.

2 Study Map 14, which shows the countries affected by desertification in the Sahel region of Africa.

a) What is the Sahel?

b) How large an area of land is threatened by desertification?

c) How many countries are affected?

d) This map needs to be regularly updated. What additional information might you need in order to update it? See what you can find at

www.cpc.noaa.gov/products/african_desk/

SAVE AS . . .

3 Using Figure 13 and the information on page 122 draw an illustrated timeline to describe the process of desertification in the Sahel. Start it in the 1950s.

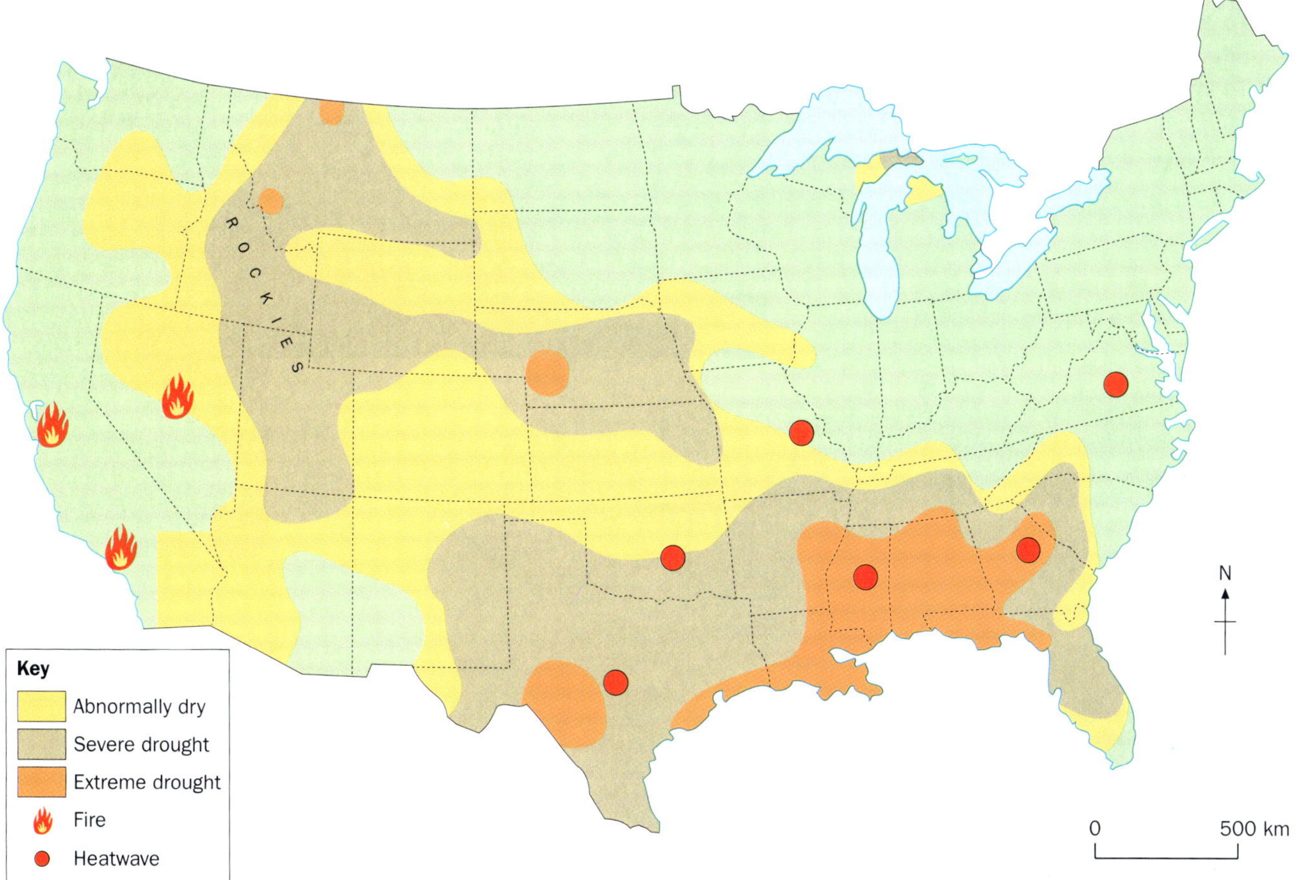

15 Drought in the USA, for one day in September 2000

3.4 How does glaciation affect us today?

DO YOU KNOW?

- During the last 1,000,000 years, Ice Ages occurred about every 50,000 years.
- The last Ice Age in Britain and Ireland finished about 10,000 years ago.
- During the last Ice Age the Earth was only 5 °C cooler than it is today.
- Alpine GLACIERS move quite slowly at 20–200 m per year, but in parts of the Himalayas and the USA they have moved at over 300 m a day!
- Glaciers are powerful agents of erosion and deposition, creating distinctive landscapes, and providing valuable deposits such as sand and gravel.

AIMS

- **To consider whether glaciers are advancing or retreating.**
- **To understand how ice shapes the land.**
- **To recognise how glacial deposition affects the landscape.**
- **To investigate how past glaciations affect our lives today.**

2 A mountain glacier

1 A walrus on a shrinking icefloe

July 27
We trekked for 2 hours over moraine which looked so weird that parts of 'The Empire Strikes Back' were filmed here! When we finally reached Hardanger glacier we were amazed to see that the snout of the glacier is filthy – covered in debris. From far away it looks so bright and clean!

When you peer into the crevasses you can see a clear blue translucent light shining from the pure ice deep inside the glacier.

I think the glacier must be retreating – the terminal moraine is covered in water, and it looks as though it's falling to bits!

Yuck – the whole area around the snout of the glacier is covered with the rotting remains of dead lemmings. What a disgusting sight!

July 28
We trekked to an adjoining valley where a smaller glacier was thought to have been in retreat about 15 years ago.

Well, the glacier had gone! Totally retreated. Only patchy snow and ice on the wrecked ground showed that this was a glacial valley only a few years before. But you could see plenty of evidence of the erosive power of the ice.

Our hypothesis that the glaciers were in retreat was right – the measurements of the Hardanger glacier proved it and so did the fact that the smaller glacier no longer existed.

3 Diary entry of a British sixth form student: Hardanger glacier, Frise, Norway

The last Ice Age?

The last glaciers retreated from around Britain and Ireland at the end of the last Ice Age (Map 4) 10,000 years ago. Since then, instead of the POLAR CLIMATE of the Ice Age, Britain and Ireland have had a temperate climate. But have glaciers gone from our lives for good? Some scientists believe that we are heading towards another Ice Age and that the warm cycle has come to an end. Others maintain that the effects of global warming will outweigh any natural cooling and that the next Ice Age is further away than we think.

To find evidence for one argument or the other we need to look at the behaviour of glaciers today. Are glaciers in retreat, or are they advancing? The evidence is not conclusive. Today many glaciers in the world are retreating (e.g. the Robert Scott glacier in Antarctica), but in other places (e.g. the Svaritsen glacier in Norway) there have been significant advances.

What causes an Ice Age?

Ice Ages begin as a result of global climatic changes. More snow falls in winter than can be melted in summer. The GLACIAL BUDGET is the balance between inputs of winter snow (ACCUMULATION) and loss through summer melting (ABLATION). If there is a continued imbalance over many years then glaciers will either grow or retreat. As more ice spreads over the globe, more solar radiation is reflected and so the cooling of the atmosphere is accelerated.

What is a glacier?

In high-latitude (near the poles) and high-altitude (mountainous) regions snow builds up and is compacted into ice. As the weight of the ice increases it begins to flow downhill by the force of gravity. This moving pack of ice is called a glacier. The vast weight of the ice has immense erosive power and transports material hundreds of kilometres from where it was eroded. You are now going to look in detail at how glaciers shape the land.

ASIA
EUROPE
AFRICA
Pacific Ocean
Atlantic Ocean
NORTH AMERICA

Key
- Areas now covered by ice
- Areas covered by ice during the last Ice Age

4 Northern hemisphere: glaciated areas

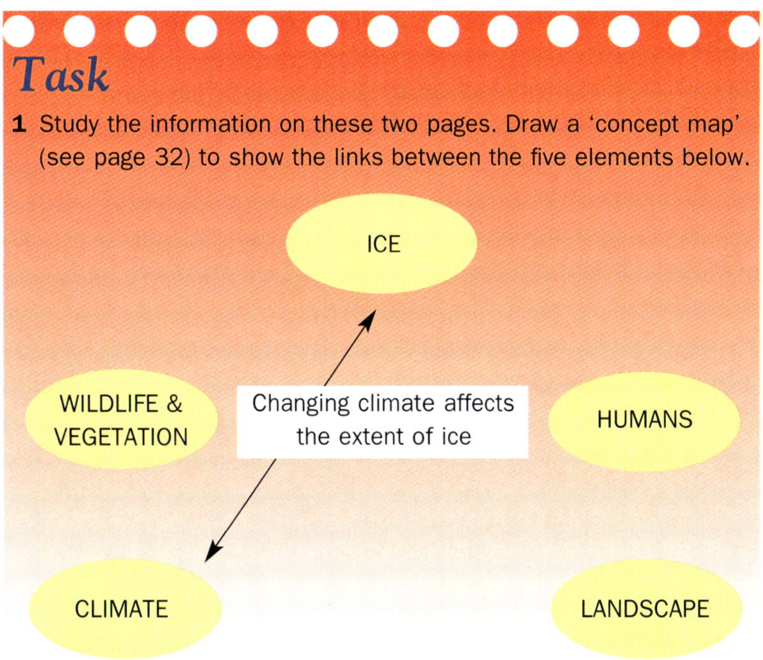

Task
1 Study the information on these two pages. Draw a 'concept map' (see page 32) to show the links between the five elements below.
ICE
WILDLIFE & VEGETATION
Changing climate affects the extent of ice
HUMANS
CLIMATE
LANDSCAPE

How does ice shape the land?

Although ice appears solid, it can behave like a fluid. When it pushes against an object it is solid enough to erode it by abrasion but fluid enough to move around it like water in slow motion. This process is helped by the fact that ice melts under pressure. Meltwater lubricates the glacier and allows it to slip. Meltwater can then refreeze (REGELATION) around solid rock, PLUCKING it away from its source and carrying it away (Figure 6).

The glacier in Photo 5 has the stripy surface common to many glaciers. The stripes are bands of rock fragments called MORAINE. These have fallen on to the glacier from the surrounding rock faces as a result of FREEZE–THAW weathering or frost-cracking. Water in cracks in the rocks expands by 9 per cent when it freezes. Continuous freezing at night and thawing by day causes fragments of rock to break off. The moraine may stay on the surface or be worked into the ice to act as an abrasive.

As a glacier retreats it leaves behind a distinctive set of landforms. These are weathered and changed over time. Glacial valleys may flood to become lakes, or in coastal regions, FJORDS. But they are still recognisable. Figures and Photos 5–14 on pages 126–129 show the main glacial landforms.

5 A moraine-striped glacier

Formation of a CORRIE

① 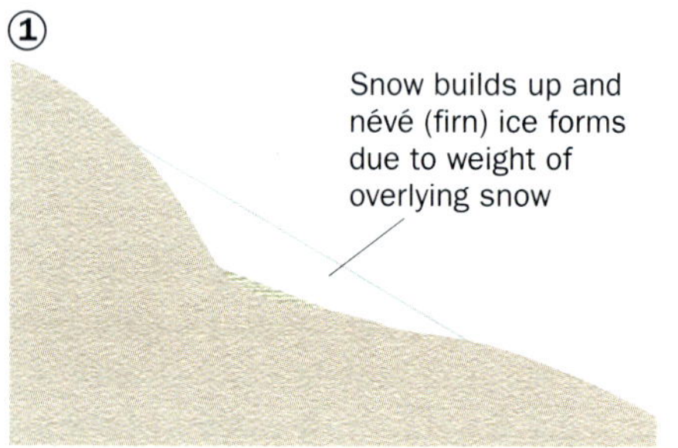
Snow builds up and névé (firn) ice forms due to weight of overlying snow

② 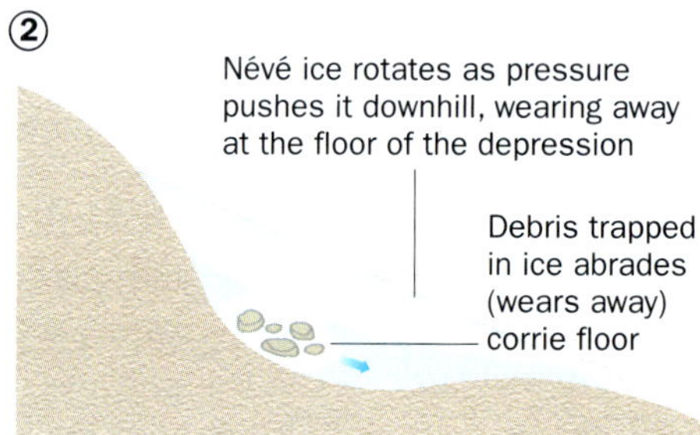
Névé ice rotates as pressure pushes it downhill, wearing away at the floor of the depression
Debris trapped in ice abrades (wears away) corrie floor

③ 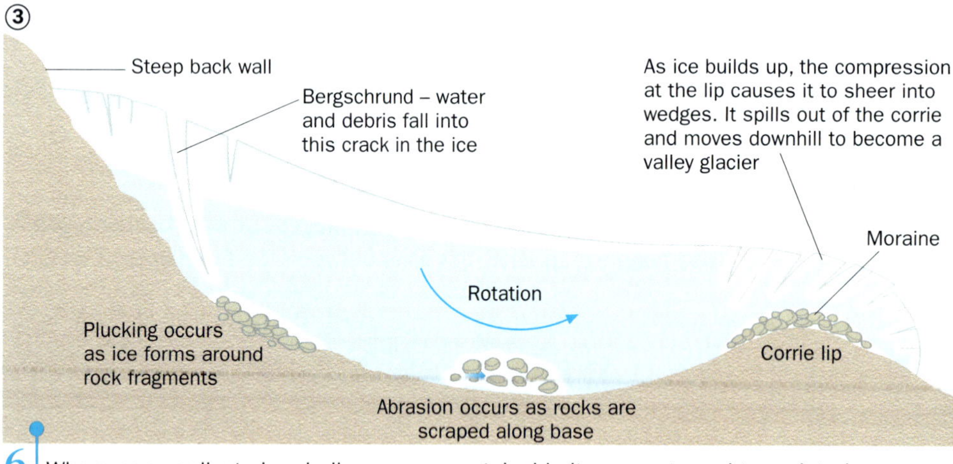
Steep back wall
Bergschrund – water and debris fall into this crack in the ice
As ice builds up, the compression at the lip causes it to sheer into wedges. It spills out of the corrie and moves downhill to become a valley glacier
Moraine
Rotation
Plucking occurs as ice forms around rock fragments
Corrie lip
Abrasion occurs as rocks are scraped along base

6 Where snow collects in a hollow on a mountainside it compacts and turns into ice. The ice erodes the hollow into a bowl shape called a corrie or cirque

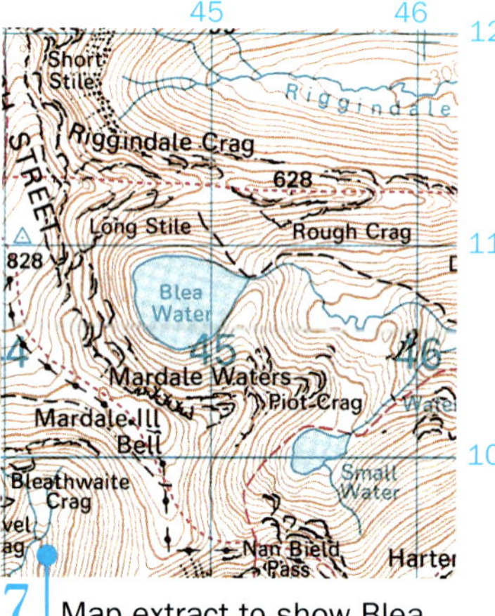
45
46
12
Short Stile
Riggindale
STREET
Riggindale Crag
628
Long Stile
Rough Crag
11
828
Blea Water
45
46
Mardale Waters
Piot Crag
Mardale Ill Bell
10
Bleathwaite Crag
Small Water
Nan Bield Pass

7 Map extract to show Blea Water in the Lake District. Reproduced from the 1997 1:50,000 Ordnance Survey map by permission of the Controller of HMSO © Crown Copyright

Tasks

SAVE AS . . .

1. On an outline copy of Photo 5, label five glacial features and explain how each one was formed.
2. Now do the same on an outline copy of Map 7.

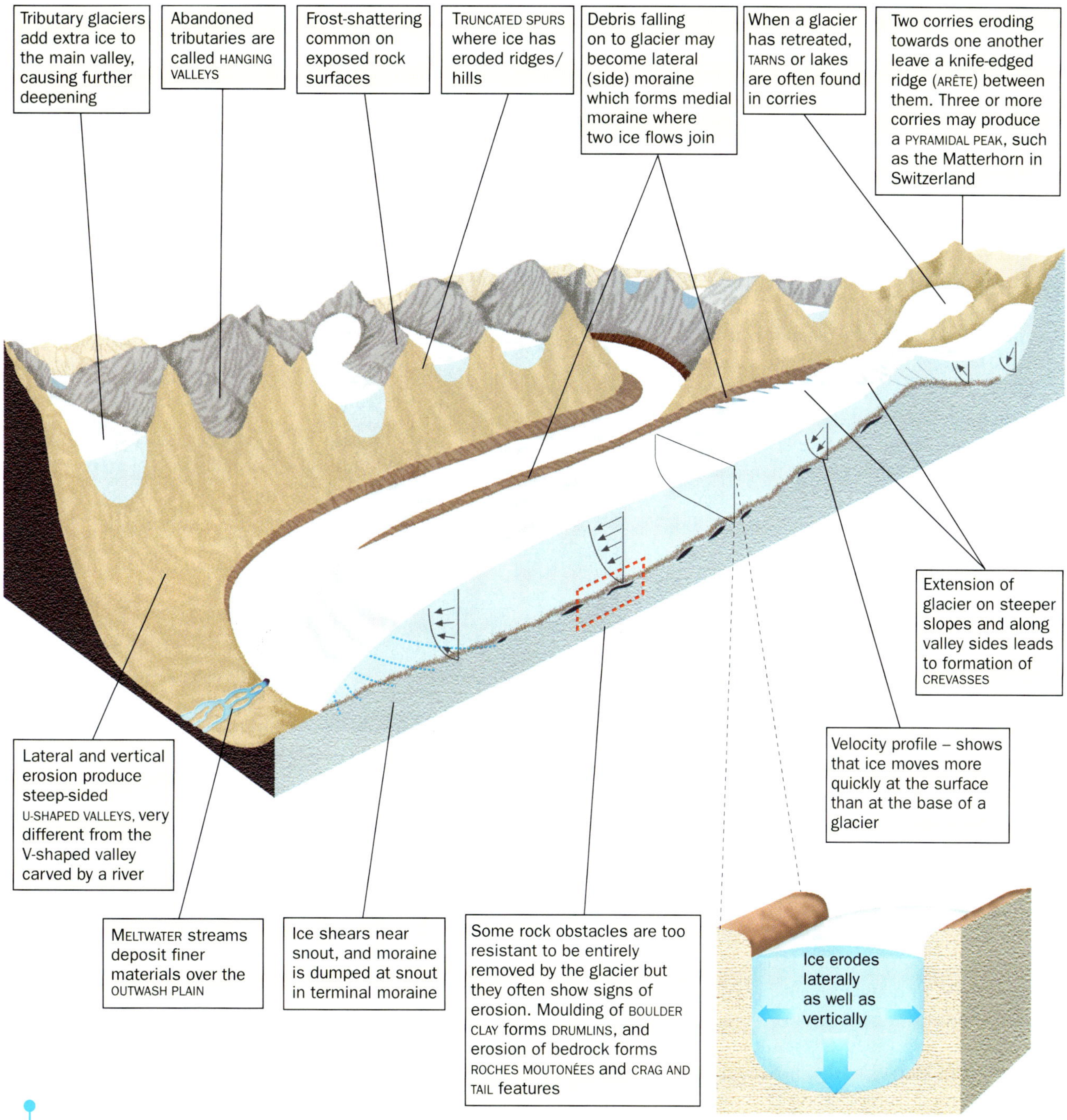

Tributary glaciers add extra ice to the main valley, causing further deepening
Abandoned tributaries are called HANGING VALLEYS
Frost-shattering common on exposed rock surfaces
TRUNCATED SPURS where ice has eroded ridges/hills
Debris falling on to glacier may become lateral (side) moraine which forms medial moraine where two ice flows join
When a glacier has retreated, TARNS or lakes are often found in corries
Two corries eroding towards one another leave a knife-edged ridge (ARÊTE) between them. Three or more corries may produce a PYRAMIDAL PEAK, such as the Matterhorn in Switzerland
Extension of glacier on steeper slopes and along valley sides leads to formation of CREVASSES
Velocity profile – shows that ice moves more quickly at the surface than at the base of a glacier
Lateral and vertical erosion produce steep-sided U-SHAPED VALLEYS, very different from the V-shaped valley carved by a river
MELTWATER streams deposit finer materials over the OUTWASH PLAIN
Ice shears near snout, and moraine is dumped at snout in terminal moraine
Some rock obstacles are too resistant to be entirely removed by the glacier but they often show signs of erosion. Moulding of BOULDER CLAY forms DRUMLINS, and erosion of bedrock forms ROCHES MOUTONÉES and CRAG AND TAIL features
Ice erodes laterally as well as vertically

8 Glacial landforms

Direction of ice flow indicated by scratches (striations)
Striated surface
Regelation and plucking on lee side
Roches moutonnées

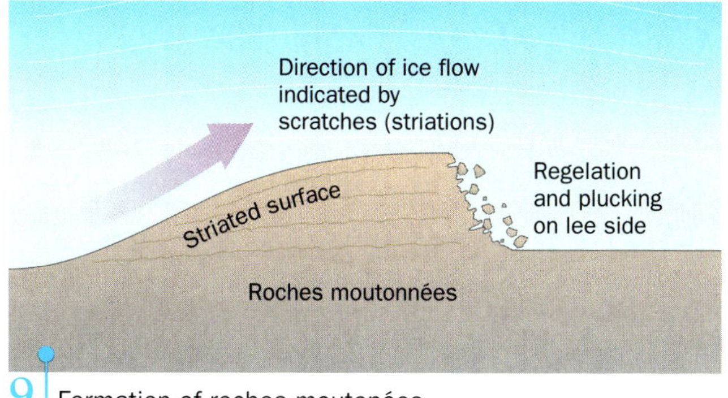

9 Formation of roches moutonées

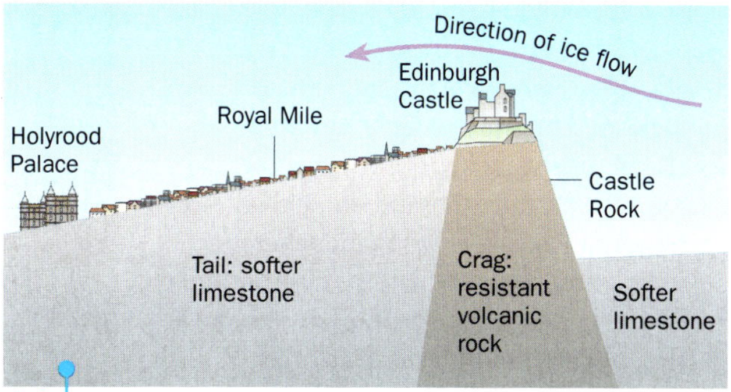

Direction of ice flow
Edinburgh Castle
Royal Mile
Holyrood Palace
Castle Rock
Tail: softer limestone
Crag: resistant volcanic rock
Softer limestone

10 Formation of the crag and tail features on which Edinburgh is built

Depositional glacial landforms

- ERRATICS – when ice sheets melt they often drop rocks which they have carried great distances (see Photo 11), e.g. Shap granite boulders eroded from Shap Fell are today left stranded as erratics on the limestone near Orton Scar, Cumbria.
- Moraine is a linear mass of weathered, angular rock fragments transported and then deposited by the glacier. *Lateral* (side) *moraine* builds up at the sides of glaciers where debris falling from steep valley sides accumulates. Where glaciers join, some lateral moraine is left in the middle of the combined glaciers and is called *medial moraine*. *Terminal moraine* is found at the glacier's snout. The size of the moraine is determined by the size of the glacier and type of sediment. Cromer Ridge in East Anglia (90 m high and 8 km long) and the 200 km-long central spine running down Long Island, New York are both moraines. Moraines can form significant landscape features. *Recessional moraine* can form behind terminal moraine, and marks places where the glacier stopped retreating for a period of years.
- Boulder clay consists of the abraded sediment and rock fragments crushed beneath the ice. This was deposited over much of Britain and Ireland (Map 13) after the last Ice Age. It is poorly SORTED, with fine and coarse fragments mixed together, and often forms an impermeable layer. It is also known as *ground moraine*.
- Drumlins are formed as a result of boulder clay being shaped by the moving ice into asymmetrical mounds (e.g. as found at Clew Bay, Co. Mayo, Ireland), with their long axes indicating the direction of ice flow (Photo 12).
- Meltwater landforms are formed when glacial streams redistribute material deposited by the glacier. They are characterised by better sorted, more rounded particles of sand and gravel. Long ridges called eskers are found where sediments, deposited by streams that once flowed beneath the glacier, have been exposed after the glacier retreated.
- KAMES are smaller areas of meltwater deposit.

11 An erratic

12 Drumlins (now partly covered by the sea) in the distance

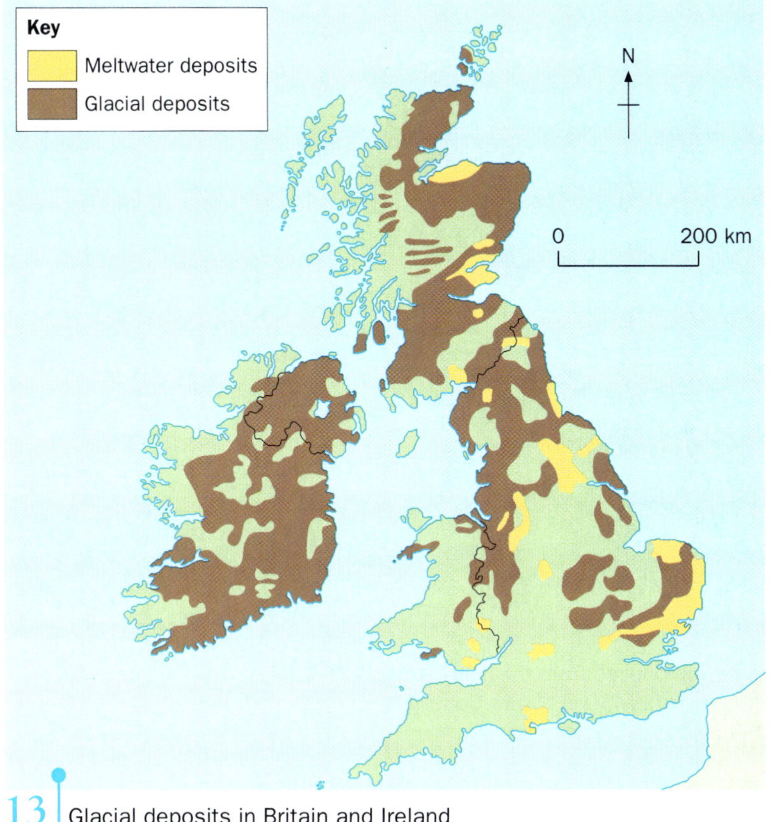

13 Glacial deposits in Britain and Ireland

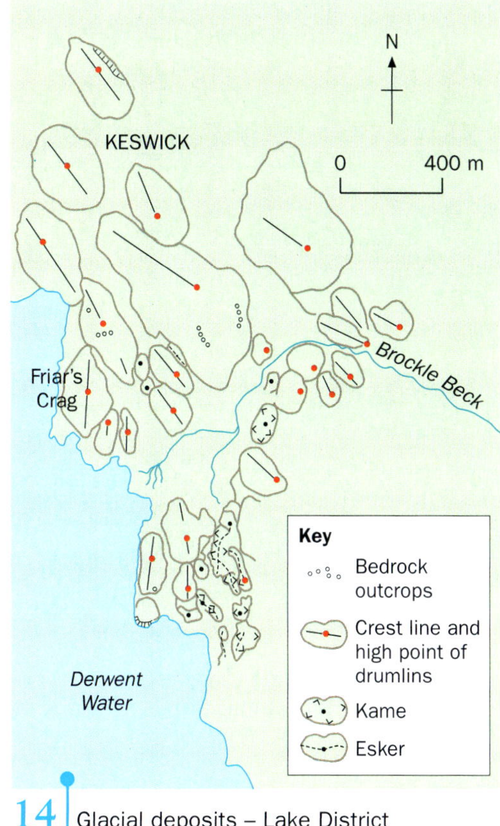

14 Glacial deposits – Lake District

Task

1 Work with a partner for this task.

a) Write each of these terms on a separate card

- corrie
- U-shaped valley
- freeze–thaw
- rock fragments
- hanging valley
- valley glacier
- abrasion
- plucking
- truncated spurs.

b) Study the information on pages 126–29.

c) Now, without looking at your book again, try to explain to your partner the work of a glacier from the first snowfall to the deposit of terminal moraine. Try to use all the terms on the cards.

d) Your partner should note all the terms you have successfully used, and then should repeat your description and try to add to and/or correct your version.

Discuss

e) Are there any glacial features that don't fit into the order of events that you have described?

f) What happens after 'terminal moraine'?

SAVE AS . . .

g) Make sure that you write up your corrected version of events.

2 Study Map 14. It is a map showing glacial deposition features near Keswick in the Lake District.

a) Using the compass arrow provided, work out the probable direction of ice flow.

b) Give reasons for your conclusions.

c) Use an atlas to suggest why the ice moved in this direction, and in which direction it would have moved once it had passed Keswick.

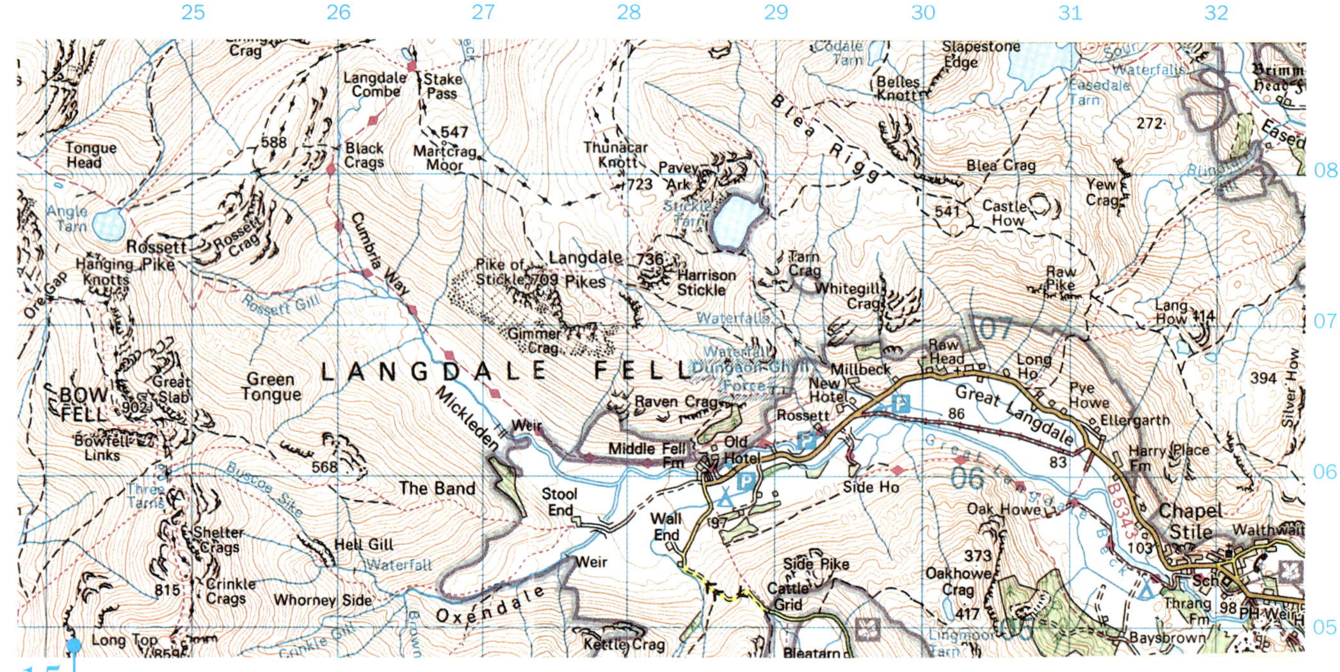

15 Map extract of the Langdales in the Lake District. Reproduced from the 1997 1:50,000 Ordnance Survey map by permission of the Controller of HMSO © Crown Copyright

Tasks

SAVE AS . . .

1 Study Map 15.

a) Identify the features and locations shown in the table below, and then complete a copy of the table. For the final column use the information on page 131.

b) Add at least three more glacial features yourself.

6-figure grid reference	Feature	Identifying characteristics	Land use and/or amenity value
	Stickle Tarn	*Contours show steep back wall – typical corrie with lip in south-east corner*	*Scenery and recreation*
261082		*High-altitude valley meets bigger U-shaped valley to south*	
303065	*North side of U-shaped valley*		

★ **THINK!**

Ordnance Survey (OS) maps are often included in exam questions and you need to be sure that you use them properly. It is important to show the examiner how you have used the map information. For example:

- If you are referring to a map feature, give a specific grid reference:
  - a four-figure reference for a large area, such as the U-shaped valley of Mickleden, 2606
  - a six-figure reference for a small feature, such as the piles of moraine at 263073 and 263071.
- Avoid repetition. Referring to the two piles of moraine (just mentioned) makes the point – you do not need to locate the rest of them.
- If you want to show links between features, be explicit, e.g. the New Hotel at 295065 is well placed to serve tourists climbing up to see the glaciated scenery of the Langdale Pikes (2707), but it is on flat land, just above the floor of the U-shaped valley.

Human uses of glaciated landscapes

You may think that Ice Ages are a thing of the past, and are of little importance today. It is not always obvious what impact previous glaciations have had on our lives, but think a bit more about it and you will find that they have affected our lives in a major way, as is shown below.

★ **FACTFILE**

Uses of a glaciated landscape

Highland uses	Comments
Recreation	Walkers, climbers and skiers enjoy rugged landscapes at different times of year. The inaccessible nature of glaciated uplands provides a refuge for rare plants and animals, attracting visitors, e.g. Glen Coe, Scotland.
Forestry	Coniferous forests (increasingly unwelcome in lowlands) planted on low-quality upland soils are processed for timber and pulp, e.g. Whinlatter Forest, Lake District.
HEP	Steep valleys and relief rainfall provide ideal conditions for hydro-electric power stations.
Water supply	Deep U-shaped valleys are easily dammed, and high rainfall ensures a regular water supply, e.g. Haweswater, Lake District.
Farming	In some areas deeper soils and sheltered sites on flat valley floors make good locations for hill farms, e.g. Langdale, Lake District.

Lowland uses	Comments
Recreation	Sandy deposits are used as golf courses. Fishing, watersports and nature reserves are found in flooded former gravel pits, e.g. Ashton Keynes Water Park, near Swindon.
Economic deposits	Well-sorted sand and gravel deposits are in great demand by the construction industry. Old gravel pits are needed as landfill sites.
Farming	Boulder clays provide rich farmland if adequately drained, e.g. Holderness.

Tasks

1 Ten statements are given below about the effects that past glaciations have on our lives today.

a) Working with a partner, each of you rank these statements in what you think is the correct order of importance – from least to most important – on a copy of the opinion line shown below.
Hint: Use a full landscape A4 page to do this.

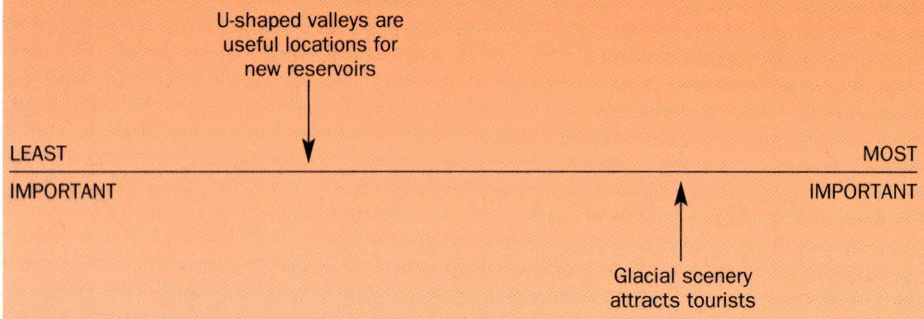

U-shaped valleys are useful locations for new reservoirs

LEAST IMPORTANT ——— MOST IMPORTANT

Glacial scenery attracts tourists

Statement bank

A Glacial sands and gravels are vital to the construction industry.
B Glacial scenery attracts tourists.
C U-shaped valleys are useful locations for new reservoirs.
D Glacial features tell us about past environments and help us to predict the effects of future ice ages.
E Glaciated regions are wildscapes, supporting habitats for rare flora and fauna.
F Glaciated regions often offer the best locations for HEP stations in Britain.
G Glaciated regions provide many recreational sites (golf courses in the lowlands, mountain climbing in the highlands).
H Fertile soils and sheltered deep valleys ensure continued existence for hill-farming communities.
I Valley gravel terraces provide dry sites for settlements.
J Glaciers/snowfields may contain historic atmospheric information trapped within the ice.

b) Compare your opinion line with your partner's. Are your rankings similar or very different? Suggest reasons for any differences.

c) Find out which statements are the three most important for the class as a whole. Do highland or lowland features appear to be considered most important? What is the explanation for this?

d) Do you think that glaciated landscapes have the greatest effects on our *economic*, *environmental* or *recreational* interests?

You are the glaciologist!

2 Your construction firm is planning to build a dam in a glaciated valley in the north of Britain. You have been sent information about a test borehole drilled in the centre of the U-shaped valley (Figure 16). This shows the type of sediment and rock found beneath the surface.

a) Explain how you can tell from this borehole that the valley was once glaciated. You must assume that this is typical of the whole valley.

b) Read the 'expert advice' on building a dam, and then write a letter to advise your construction firm whether you would recommend this site as a suitable place to build a dam.

Borehole data

- Quite well-sorted river sands and fine silts
- Well-sorted, partly rounded, fluvio-glacial sand
- Very fine layered clays and silts, well sorted – meltwater lake
- Mainly slate debris <25 cm – moraine
- Poorly-sorted mix of boulder clay <1.5 m
- Bedrock – impermeable slate, dipping at 45°

16 Sediments from a borehole taken from a U-shaped valley

Advice

When building a dam, remember that

- It must have foundations on solid rock.
- The bedrock should be impermeable, because the water will exert great pressure when the reservoir is full.
- The core of the dam will need a lot of 'fill' material, which can include any loose rock, sand or gravel.
- The greater the distance you need to import building materials, the more expensive it becomes.
- Permeable sediments can be made impermeable by grouting (pumping in cement).

Focus task

Glacial gravel extraction

Glacial deposits provide well-sorted sands and gravels which are used by the construction industry. The extraction of these deposits has a major impact on the environment. Workings can be landscaped afterwards, but some people believe that they should be left as part of our natural landscape and not be exploited.

3 United Aggregates (a major supplier of sand and gravel to the construction industry) has applied to extract a large amount of glacial sand and gravel from a site on the edge of the fictional town of Norton Keynes.

You are the town planner in charge of mineral extraction. You have to prepare a report for the chief planning officer, weighing up different views. You need to approve or reject the plan to extract gravel. Use the following format for your report:

- Outline the proposal and identify the possible benefits to the area.
- Identify the negative impacts on the environment and local population.
- Conclude your report with advice to accept or reject the proposal. You may decide to allow development on only some of sites A–D. If you accept it, the extent and location of the quarry should be indicated on a copy of Map 17.

Evidence given at the public meeting to discuss the proposal

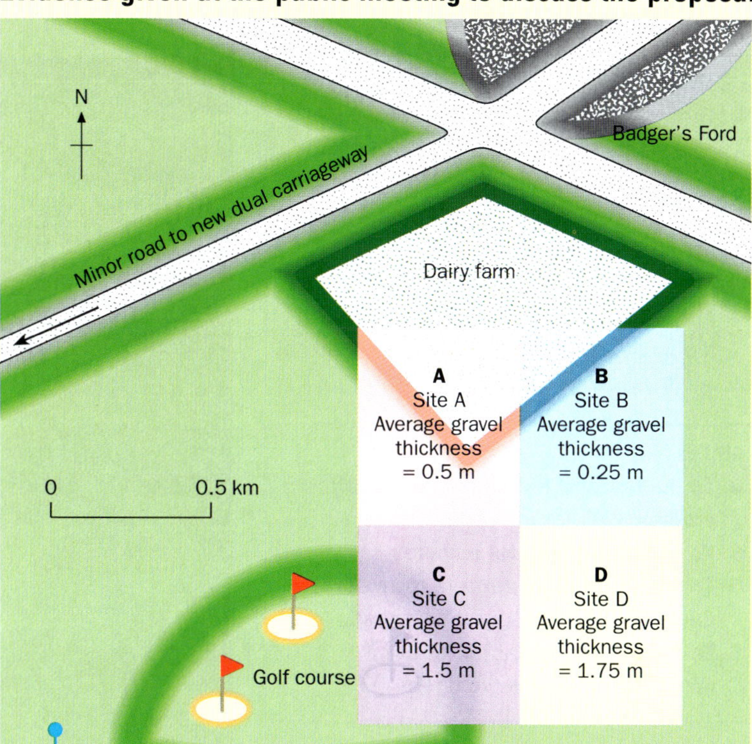

17 Proposed sites for development

United Aggregates (UA)

A 1 km^2 area of countryside has been identified as having significant deposits of post-glacial sediments beneath a thin layer of river deposits. United Aggregates would like to exploit all four sites (see Map 17) in time, but seeks to exploit Site A first, followed by Sites B, C and D over a five-year period. Each site covers 0.5 × 0.5 km of poor-quality (grade 3) agricultural land and would produce well-sorted gravel and sand from glacial outwash streams.

- Average volume of sand and gravel each site = 1,125,000 m^3
- £5,625,000 gross income @ £5/m^3
- £1,968,750 cost of extraction

The sediments will be processed for use in the building industry. After 5 years the extraction will be complete and the area will be landscaped to provide a nature reserve and country park. Estimates suggest that 30 jobs will be generated locally in addition to expected commercial spin-offs.

Highways Department

The cost of upgrading the minor access roads is estimated at £250,000 and would be borne by UA. The increase in heavy traffic is estimated at 400% and noise/dust pollution problems are anticipated. A minor rise in road traffic accidents in this area is likely.

Residents' Committee

The nearby edge-of-town Badger's Ford estate was sold as an up-market development, combining peace and quiet with easy access to the town. The residents are vehement in their opposition to this plan and see it as a threat to their quality of life and the resale value of their homes.

Municipal golf course

The site would spread about 75 m^2 on to the golf course, causing the re-siting of two fairways. The traffic, noise and dust pollution problems would reduce the quality of the local environment and its amenity value.

Co-ordinated Dairies

The proposal would affect a site on existing pastoral land supporting a dairy herd of 315 cows. The farm has three full-time dairy workers and an average annual turnover of £250,000.

Safemart Supermarkets

Safemart has earmarked the extraction site as the preferred location for its new out-of-town superstore. An estimated 150 jobs would be generated locally. If this site is refused it is unlikely that Safemart will accept another in Norton Keynes.

3.5 Should we defend our crumbling coastlines?

DO YOU KNOW?

- The coastline around Britain is being constantly eroded.
- On the Yorkshire coast at Holderness 2 metres of Britain disappears into the ocean every year.
- It is the job of the Environment Agency to maintain 1,400 km of sea defences around the 13,000 km of British coastline.

AIMS

- **To understand the causes of coastal erosion.**
- **To recognise coastal features.**
- **To understand the factors that affect decisions about coastal defences.**
- **To assess the impact of coastal defences on the environment.**

Task

1 Study Photo 1, which shows storm waves crashing on to Lyme Regis seafront in Dorset. Think about the effects that storm damage to the coast might have on the groups of people shown in the concept map below. Why would each group want to defend their coast from the sea? Redraw the map to show:

a) their reasons for defending the coast

b) the links between the groups. Some links have been started for you.

Why should we defend our crumbling coasts?

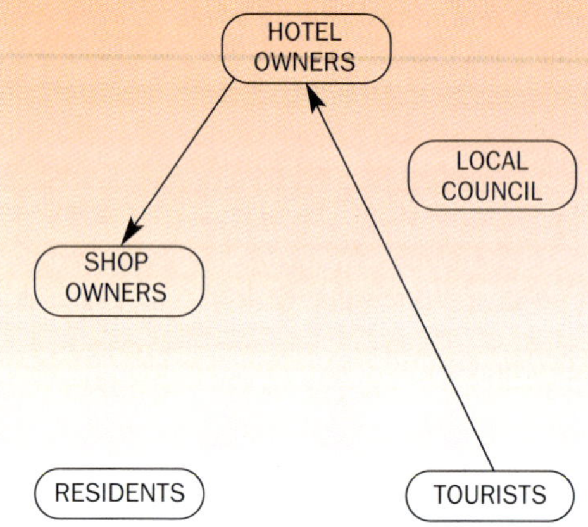

1 Storm waves battering Lyme Regis in Dorset

Well that's it! The Council has said that my farm isn't worth saving. It says here that had I been on the landward side of the main road I'd have been safe. But because I'm near the sea I'm on the wrong side of the planning 'green line' and 'no help can be offered at this time'. This is part of England being lost! I've worked for 50 years to build this place up. What are my grandchildren going to inherit now? I can hardly pay the mortgage as it is, and nobody will buy a farm that's condemned to fall into the sea. The Council hasn't got any idea what its stupid rules do to real people's lives!

2 Should we protect our coasts – or not?

Task

2 Read what Mrs York has to say in Figure 2.

a) Who does she blame for her problem?

b) Do you think she is justified in her opinion?

You will return to Mrs York later.

Ⓐ WHY DO COASTLINES CHANGE?

The energy of the sea can transport sediment and erode cliffs and beaches. Where the sea deposits its load, land may be created. Where the sea is eroding, land may be lost. Coasts absorb the energy of the seas and are constantly changing. The shape and form of coastlines depends on the geology (rock type and structure) of the area. This also determines how quickly the land erodes and what type of sediment is deposited.

Waves

Winds blowing over the surface of the sea are in frictional contact with the water. This transfers energy to the surface of the sea, rotating the water particles in a circular orbit to form waves (Figure 3). The stronger the wind the bigger and more powerful the waves. Wave direction is mainly determined by wind direction. As waves approach the coast, they are slowed down by friction with the sea-bed and they break when the water becomes too shallow. On gently shelving beaches the waves break early and with little force, but on steeper beaches they break later and with greater force. As the wave breaks, the SWASH or forward movement transfers energy up the beach, then the BACKWASH moves energy down the beach.

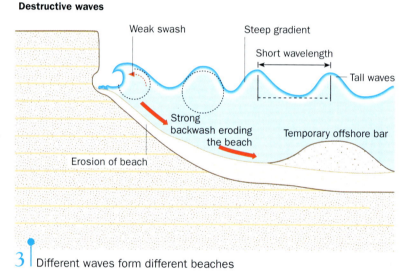

Storms produce frequent, tall, steep-sided waves that crash on to the beach with great erosional force (DESTRUCTIVE WAVES). SWELL WAVES that are created far out at sea have gentler profiles and push sediment up the beach (CONSTRUCTIVE WAVES) (Figure 3).

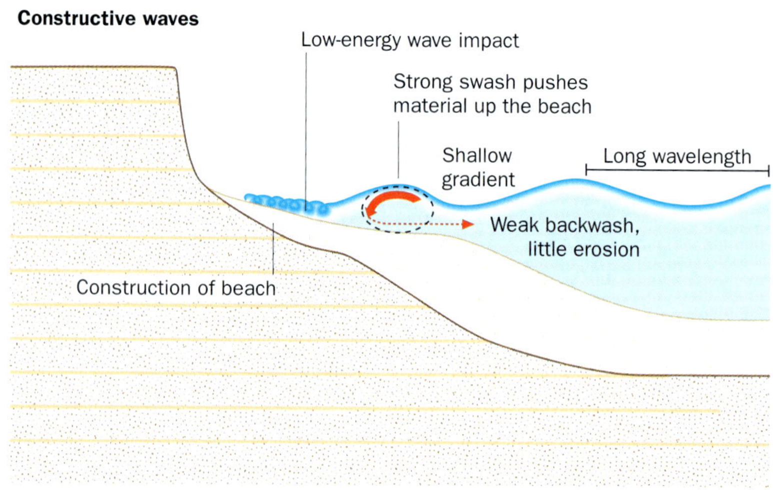

3 Different waves form different beaches

Processes of erosion

- Rocks may be hurled at the cliff, wearing it away by abrasion.
- The sheer pressure of the water forced into cracks in the rock may weaken it by hydraulic pressure (Figure 4).
- Eroded material will be further broken down as waves smash particles together (attrition).
- Some rock types, e.g. limestone, are dissolved by the sea (corrosion or solution).
- When cliffs have been undercut by the processes of erosion they may collapse.

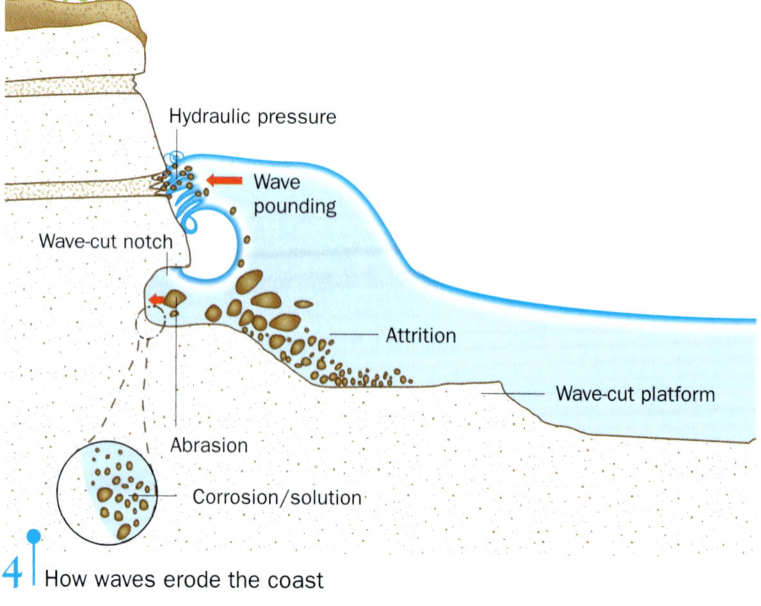

4 How waves erode the coast

Features of erosion

Where waves erode the base of a *cliff* they form a WAVE-CUT NOTCH (Figure 4). Over time the notch gets bigger and eventually the cliff above it collapses. As the cliffs are eroded further and retreat, the rock at the base of the cliff forms a WAVE-CUT PLATFORM (Figure 4). This eventually becomes so wide that it breaks up incoming waves, reducing their ability to erode (Figure 5).

Headlands and bays develop as a result of different rates of erosion in areas of hard and soft rocks. Soft rocks may erode quickly, forming bays, but more resistant rocks are left jutting out as headlands. Eventually a natural balance starts to take effect as the bays become sheltered from the main energy of the waves, leading to deposition of sediment, while the headlands become more exposed leading to quicker erosion (Figures 5 and 6).

Cracks in rocks such as limestone and chalk can be eroded by waves to form caves, arches, stacks and stumps (see Figure 6).

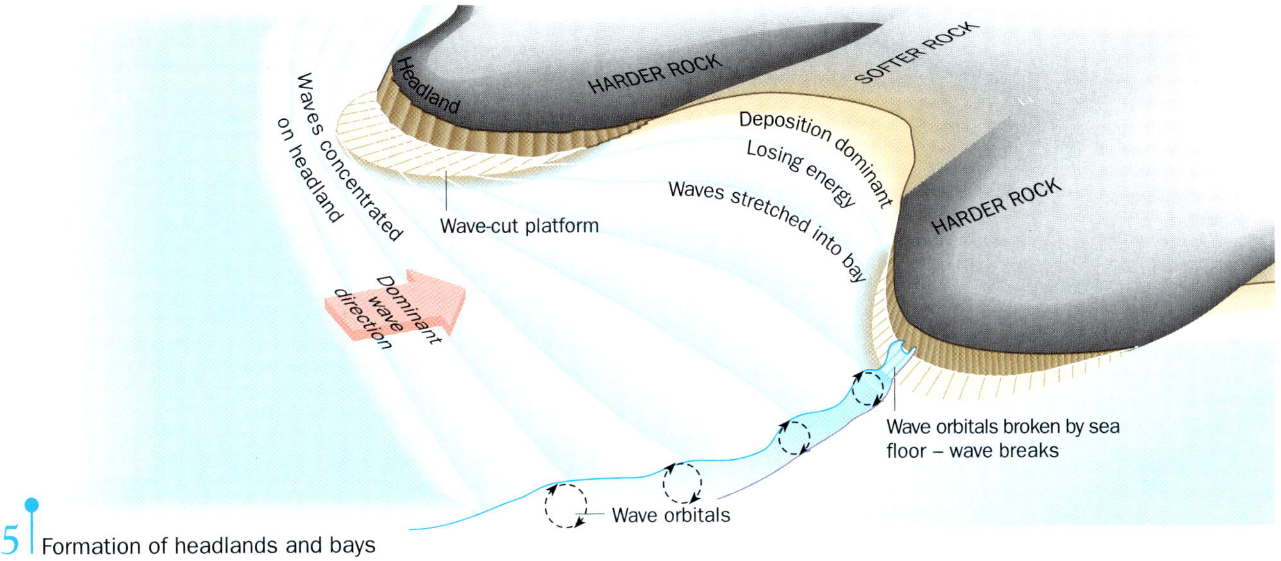

5 Formation of headlands and bays

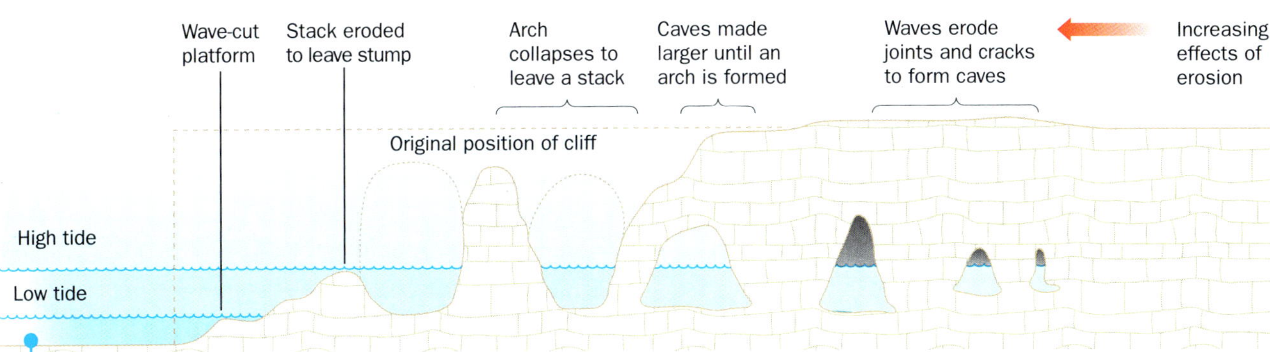

6 Formation of caves, arches, stacks and stumps in a headland

Tasks

1 a) You are going to play **Taboo**. You must each take turns to explain to your partner how waves can change the shape of the coast but you are *not* allowed to use any of the geographical terms in the box below. You must use other words instead.

abrasion erosion deposition swash backwash beach wind direction hydraulic pressure attrition cliff bay

b) Between you, decide which of the geographical terms in the box you couldn't do without (A words), and which ones you were able to avoid or improve upon (B words).

SAVE AS . . .

2 a) Write out a final version of your explanation in Task 1, using the geographical words. Put a box around each A word, and underline each B word.

b) Are there more A words than B words? Why are the A words and phrases more effective?

The geology of the coast around Flamborough Head, Humberside (Figure 7) has determined the form of the coastline there. The headland around Thornwick Nab is hard, resistant chalk and flint. It is eroding relatively slowly. The areas to the south and north of Flamborough, e.g. at Hunmanby Gap, are wearing away more quickly because they are made of boulder clay, which is a softer deposit.

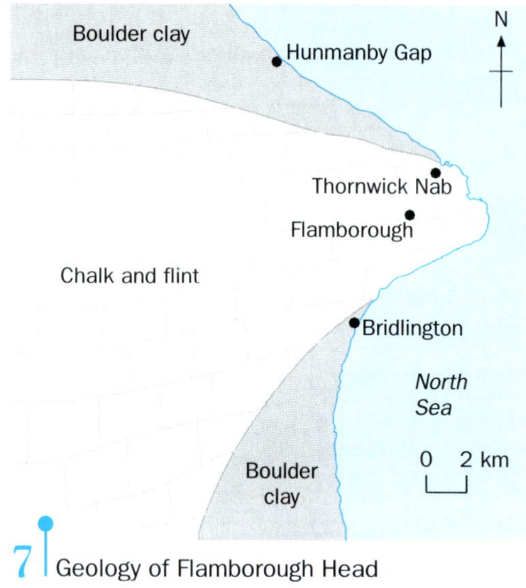

7 Geology of Flamborough Head

8 Map extract: Flamborough Head. Reproduced from the 1996 1:50,000 Ordnance Survey map by permission of the Controller of HMSO © Crown Copyright

9 Hunmanby Gap, north of Flamborough Head – boulder clay cliffs with broken jetty

10 North Landing, Thornwick Nab, Flamborough Head – chalk cliffs with boulder clay cap

Transportation and deposition

Deposition occurs when the sea moves so slowly that it can no longer carry any sediment. Some of the material is transported out to sea by the current but much is spread around the surrounding shoreline by the waves. *Beaches* are formed from deposited sediment. This is a positive effect of deposition. Beaches protect cliffs and attract tourists. Deposition may also have negative effects. It can silt up harbours and endanger shipping. Beaches can shrink and grow; at times more material is pushed on to the beach (constructive phase), whereas at other times storms erode material (destructive phase).

Larger particles (gravel, pebbles, cobbles) usually create steeper beaches than smaller particles (sand, silt, mud) which are easily moved by backwash.

Waves often hit beaches at an angle because their path is governed by the wind direction. As waves break, their swash pushes sediment up the beach, often at an angle to the shoreline. As the swash loses energy, the backwash returns down the beach under the influence of gravity and so travels at right-angles to the shoreline (Figure 11). The repetition of this procedure thousands of times a day gradually moves material along the beach in a process known as LONGSHORE DRIFT. In some areas this brings in sediment, but in other places it may remove beaches, so exposing cliffs to more erosion.

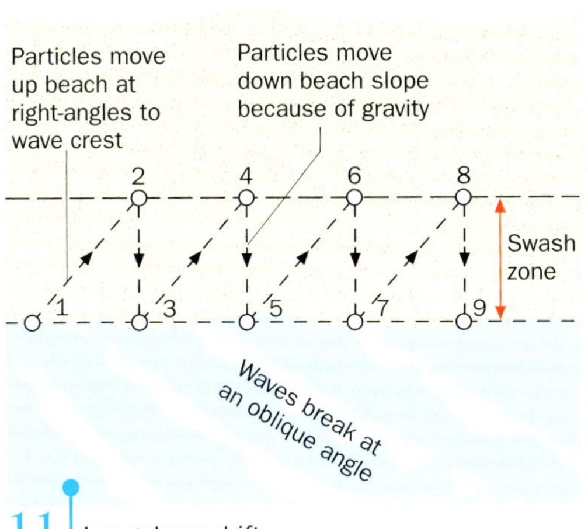

11 Longshore drift

Features of deposition

In areas of relatively shallow water, where there is a pronounced change in the direction of the coast, longshore drift can create elongated stretches of sand and shingle called spits (Figure 12). Sediment is transported along the coast until it stretches out across an estuary or bay. If it cuts right across a bay it is called a bar or barrier of sand. Spits rarely cut off estuaries, since the current of the river at its deepest point will remove sediment out to sea. Many spits are curved because winds blowing from other directions cause a hook-like structure to develop. In the calm waters behind the spit, finer sediments are deposited and *saltmarshes* may develop.

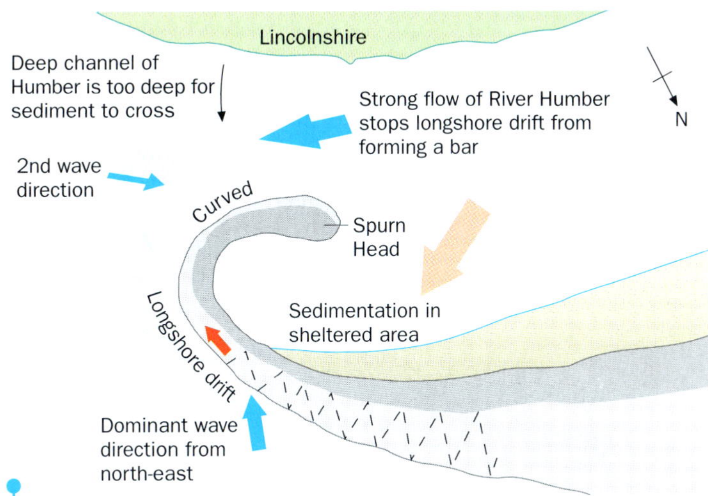

12 Formation of Spurn Head (a spit) in the Humber estuary (this is an enlargement of circled area of Figure 14 on page 141)

Tasks

1 a) Draw simple sketches of Photos 9 and 10.

b) Annotate your sketches to identify the key features. Map 8 will help you.

c) Study Figure 7. Explain how geology has affected the cliff form.

2 Study this diagram, which shows a new concrete harbour in a sandy bay.

a) Draw a copy of the diagram in the middle of your page to show how you think it will look after many years of longshore drift. Show the original positions by dashed lines.

b) Annotate your diagram with short paragraphs to explain how, and where, longshore drift might alter the distribution of sediment in the future.

3 Study Map 8.

Draw a table like the one below. Complete it by including at least five features of coastal erosion and deposition.

Feature	6-figure grid reference	Brief description	How formed	Mainly erosional or depositional

4 On a simple sketch of the coastline shown on Map 8, mark on the areas that you think will be in serious danger from erosion in 10 years, 100 years and 1,000 years if the coast is eroding at the rate of 1 m/year.

B HOW CAN WE PROTECT THE COASTLINE?

The coastlines of Britain and around the world are constantly changing. Coasts are used for a variety of very important activities, such as recreation, settlement and industry. There is a natural desire to protect coasts. Where the coastline is subject to constant erosion, three main responses have evolved:

1 **The structural approach** – this involves building defences against the erosive power of the waves, as shown in Figure 13. Sometimes this will mean 'hard engineering', such as building seawalls, but increasingly it includes 'soft engineering', for example 'feeding' beaches with extra sand to replace the eroded material. Many structural solutions are very costly, e.g. the Holderness coast in North Yorkshire has seawalls at places such as Withernsea. There are plans to support these 'hard' structures with 'softer' defences such as offshore reefs (Figure 14). This would provide a series of 'sacrificial bays' which should remain stable for up to 200 years.

However, reflecting wave energy or denying the sea sediment inputs tends to make the problem worse elsewhere. For example, at Holderness it is estimated that areas to the south of the defences would erode at over twice the normal rate.

In the USA, seawalls and groynes built at Cape May in New Jersey have caused increased erosion down the coast where it now averages 20 m/year and has created a bay 0.5 km deep and 1 km wide.

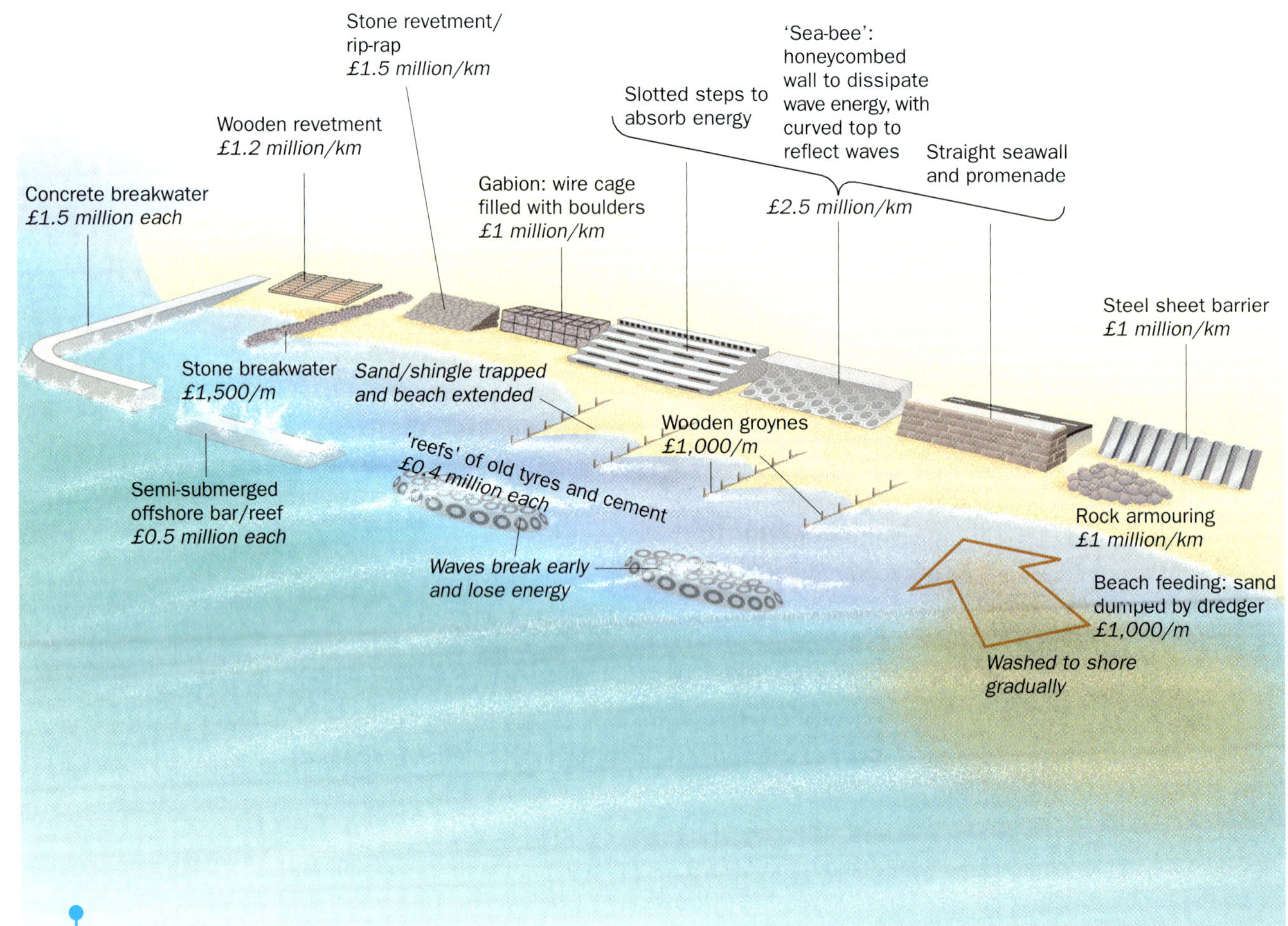

13 The variable costs of coastal defences

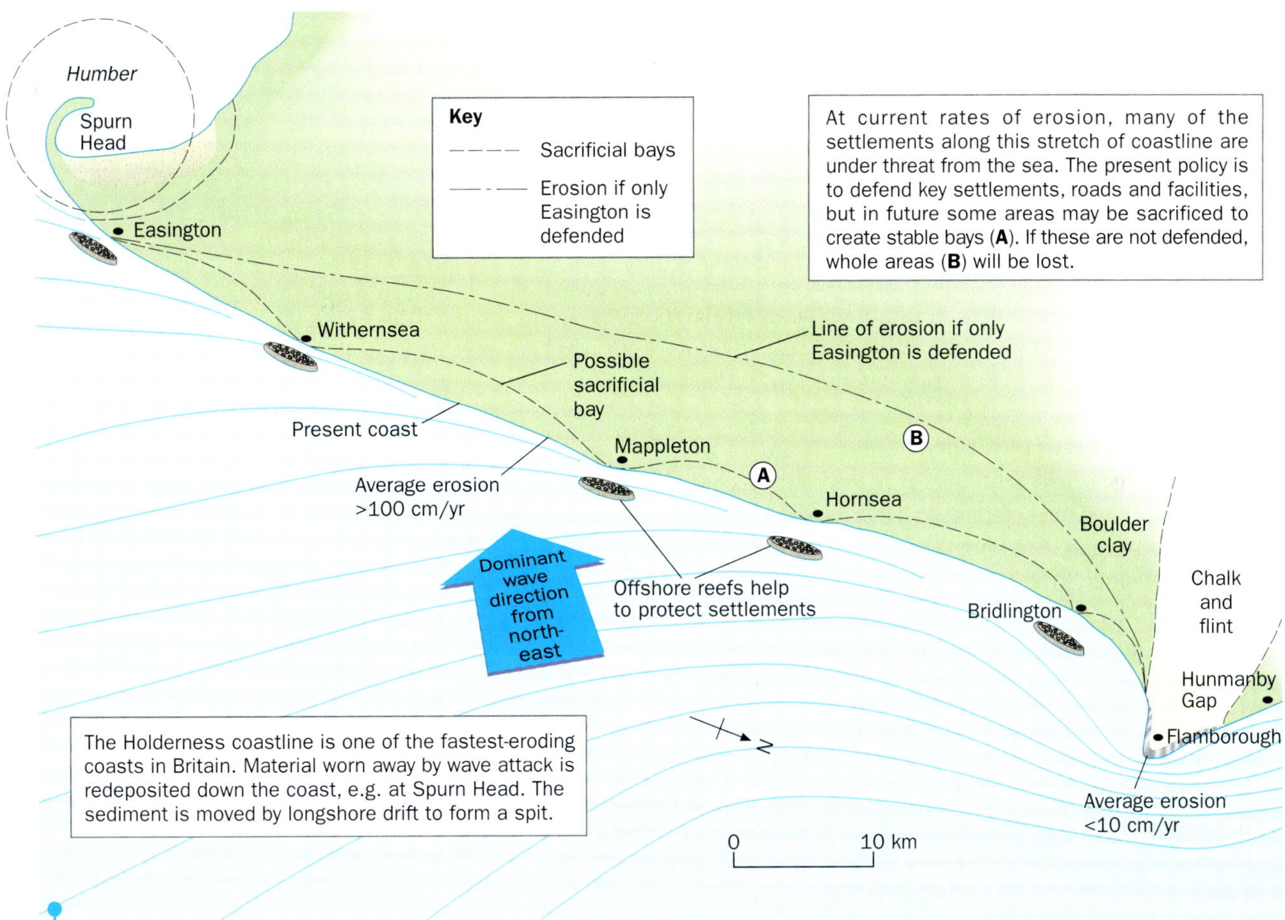
Humber
Spurn Head
Key
Sacrificial bays
Erosion if only Easington is defended
At current rates of erosion, many of the settlements along this stretch of coastline are under threat from the sea. The present policy is to defend key settlements, roads and facilities, but in future some areas may be sacrificed to create stable bays (**A**). If these are not defended, whole areas (**B**) will be lost.
Easington
Withernsea
Line of erosion if only Easington is defended
Possible sacrificial bay
Present coast
Mappleton
B
A
Hornsea
Average erosion >100 cm/yr
Boulder clay
Dominant wave direction from north-east
Offshore reefs help to protect settlements
Bridlington
Chalk and flint
Hunmanby Gap
Flamborough
The Holderness coastline is one of the fastest-eroding coasts in Britain. Material worn away by wave attack is redeposited down the coast, e.g. at Spurn Head. The sediment is moved by longshore drift to form a spit.
N
Average erosion <10 cm/yr
0
10 km

14 Coastal protection at Holderness

2 **The behavioural approach** – this tries to get around the problem by planning ahead to ensure that people are not greatly endangered or inconvenienced by erosion. Typically this might involve refusing planning permission in areas under threat from the sea. Parts of the east coast of England, for example Suffolk, have a 'green line' policy which states that any ordinary buildings seaward of a given line will not be defended. Only places of great importance, e.g. the gas terminals at Easington on the Holderness coast (Figure 14), would be protected under such a scheme.

3 **The 'do nothing' approach** – the supporters of this approach argue that coastal erosion is going to happen whatever is done by planners or engineers and therefore it is a waste of money to try to stop it. In fact, since reflecting the energy of waves only sends the problem elsewhere, it may be better to leave the sea to do its natural work. This approach has already been adopted in many areas where land is of low value or is lacking in important features.

Tasks

1 Look back to Mrs York on page 135. Have you changed your view towards her? Give reasons.

ICT

2 The Holderness coastline in Yorkshire is one of the fastest-eroding coastlines in Britain. Your task is to use the internet to identify three more fast-eroding coastlines where there is similar conflict between whether to defend the coastline or let nature take its course. You could start with these websites:

www.npm.ac.uk
www.lancing.org.uk/geography/geoindex.htm

For each area make notes under these headings:

- Causes
- Effects
- Management.

Case study: Defending Lyme Regis

On the next four pages you are going to look in detail at coastal problems in Lyme Regis in Dorset. But first we will summarise the key issues:

- Coastal defence is possible but it upsets the natural 'balance' of the coast (often creating other problems), and requires significant amounts of money to introduce and maintain.
- Decisions to protect or abandon areas threatened by the sea are always controversial. Even uninhabited land may support important species or have attractive landscape features, but it is unlikely to be protected because defences would probably cost more than the value of the land being defended.
- Around 20 per cent of the population live along the coastline of Britain. Large settlements and road and rail links are likely to be defended, but some smaller settlements and farms will not be protected because the long-term cost of protection will be greater than the value of the land. If one small village is saved then surely all should be protected. Difficult choices have to be made before taxpayers' money is spent.

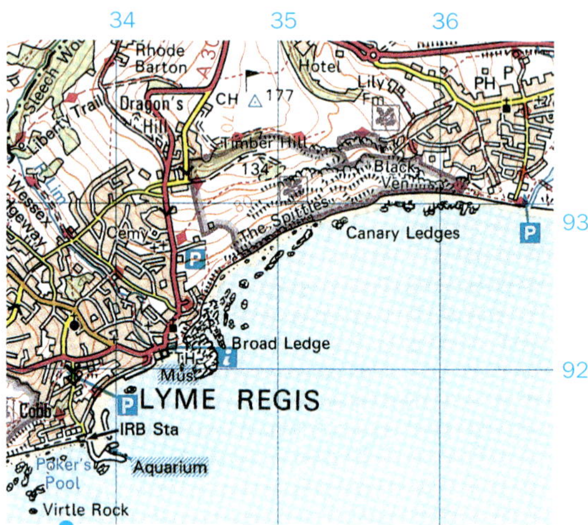

15 Map extract: Lyme Regis. Reproduced from the 1998 1:50,000 Ordnance Survey map by permission of the Controller of HMSO © Crown Copyright

The town of Lyme Regis in Dorset (population 3,500) is situated in a natural bay between high, crumbling cliffs to the west and east (Map 15 and Photo 16). It owes its prosperity to the sea. Tourists flock to visit the town, which has been immortalised in several novels, for example Jane Austen's *Persuasion* and John Fowles' *The French Lieutenant's Woman*. People enjoy strolling along the parade above the seawall, relaxing on the beaches behind the Cobb (the harbour), or hunting for fossils at the foot of nearby cliffs.

Lyme Regis has had to defend itself from the ravages of the sea since the 11th century. In the 1990s the repeatedly patched-up seawalls were on the verge of collapse. Winter storms like the one in Photo 1 (on pages 134–35) had removed beach material, allowing waves to undermine the base of the walls. But Lyme Regis was not going to be abandoned to the sea. Something had to be done and the solution had to be effective, affordable and environmentally suitable.

16 An aerial view of Lyme Regis

17 Cliffs to the east of Lyme Regis

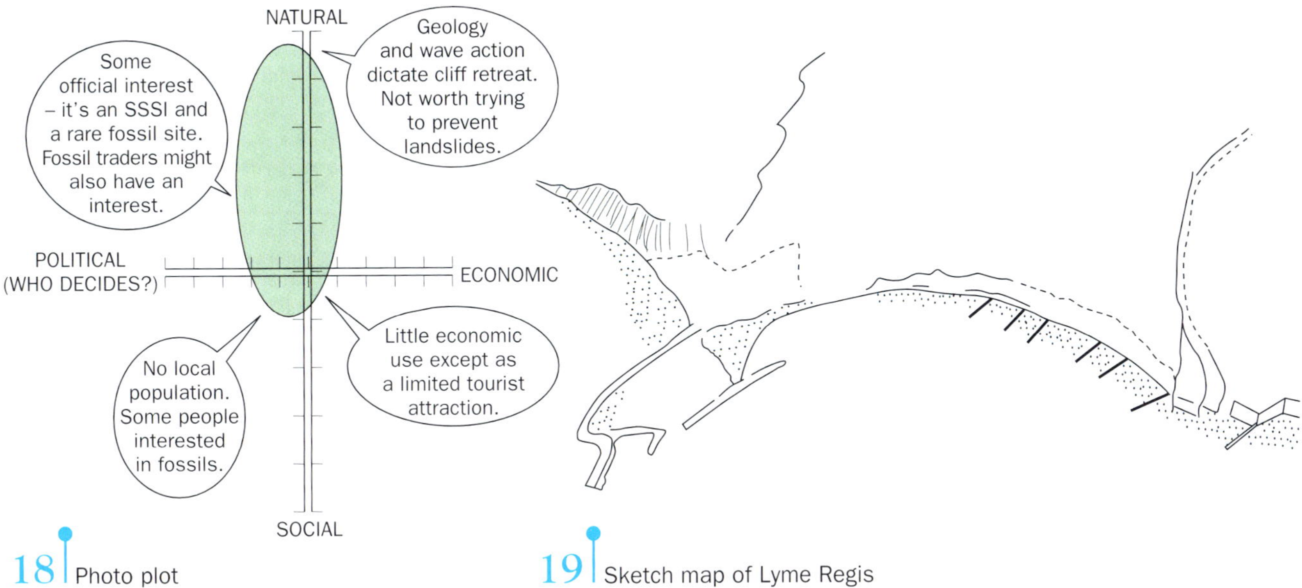

18 Photo plot

19 Sketch map of Lyme Regis

Tasks

1 Study Photo 17 which shows cliffs to the east of Lyme Regis. Different factors will affect the future shape of the coast here.

Figure 18 shows one student's interpretation of which of the following four factors he thinks will be most important:

Natural	Economic	Social	Political

The student thought that the future shape of the landscape would be most affected by natural forces, so the plot is stretched towards the top of the graph.

Study Photo 16, showing the harbour at Lyme Regis, and draw a photo plot (on a blank copy of the axes in Figure 18) to show how each factor will affect the future shape of the harbour. Annotate your diagram to explain it.

2 a) Use Map 15 and Photo 16 to annotate a copy of Figure 19. Draw it in the middle of your page. Colour-code any evidence of possible tourist attractions in blue, and any evidence of coastal defences in red.

b) Write a paragraph to explain how Lyme Regis' coastal features bring economic benefits as well as costs.

20 Local people in Lyme Regis were asked questions about the Council's proposals for new sea defences. These are their responses.

We must prevent further erosion. This town relies on tourists and they'll be driven away from an increasingly damaged sea-front. They come here because Lyme has a great image as a resort with something for everyone. Anything that damages our image will take years to recover from.

When I look at defending one location I have to think about what will happen down the coast. If the sea's energy isn't used up by eroding and transporting sediment in one place, then it will erode further along the coast instead. Even just trapping sand here will lead to sediment starvation, and more erosion, somewhere else.

I pay taxes to the local Council, and as far as I'm concerned it's their duty to protect my house. Any more damage to the town will mean that my house will become worthless. To me, the main cost should be met by those who live right on the sea-front. They stand to benefit most, so they should pay.

Town councillor

Regular visitor

Coastal defence engineer

Shop owner

Lyme resident

I love this beach. OK, it's a bit crowded in the summer, and the power-boaters make a lot of noise at times, but usually it's got a really unspoiled feel. I wouldn't want to see it changed too much. The promenade along the top of the wall is one of the best bits about the sea-front. They won't take that away, will they?

Any big scheme on the sea-front will spoil my trade. I've got this souvenir shop as well as a B&B down Marine Parade. I reckon we've got about three months each year to make money to live on for the rest of the year. We all know what it'll be like: roads blocked, trucks everywhere, dust and noise. If I lose my regulars and the shop trade falls off, I've had it – I'll have to pack up.

Focus task

Defending Lyme Regis

1 You have been commissioned by West Dorset County Council to draw up a plan to defend Lyme Regis, and in particular the area inside the present harbour to the east of the Cobb. The Council wants a scheme that will meet the economic, social and environmental needs of the town. The cost and effectiveness of the scheme will be key factors in deciding on the successful plan.

You will need to use all the information provided on Lyme Regis, and Figure 13 on page 140, showing different types of coastal defences.

a) Copy and complete the table below to evaluate all the defence options shown in Figure 13.

b) Consider local people's opinions about the Council's proposals for new sea defences, as shown in Figure 20.

c) Devise a plan to protect the town. You must cost each item in your plan.

d) Locate the defences on another copy of Figure 19 and annotate it to explain your choices and locations of defences.

e) Write up your plan as a report with simplified diagrams. Use your photo analysis from Task 1 on page 143 and your map of tourist attractions and coastal defences from Task 2 to help to justify your decisions.

f) Present your plan to the Council as a 5–10 minute talk. They will vote to decide which plan will go ahead.

Option	Advantages	Score 0 to +5	Disadvantages	Score 0 to −5	Rank

Focus tasks

Choosing the plan

2 The plan that the County Council finally decided on for Lyme Regis is shown in Figure 22 (page 146). Compare the Council's plan with the one you have devised. Think about:

- Scale (what you chose to protect)
- Location (where you put your protection)
- Funding (what your protection cost)
- Type of defence (what methods of protection you used).

Comment on any similarities or differences, and suggest reasons for the differences.

Environmental Impact Assessment

3 Photo 21 shows part of Phase 1 of the scheme under construction. Before any scheme of this nature can commence, the planning authority will ask for an Environmental Impact Assessment (EIA) to be done to predict how the scheme will affect the area concerned. The EIA examines the impact of the proposed scheme on the natural environment, and in this case must consider social and economic factors too. It assesses whether the impact will be positive or negative.

You are going to do a simplified EIA.

Criteria / **Score**	−3	−2	−1	0	+1	+2	+3
Physical							
Erosion							
Sedimentation							
Visual impact							
Wildlife							
Pollution							
Social and economic							
Traffic							
Amenity							
Business							
Local employment							
Property values							
Safety							

a) Draw a copy of the table below.

b) Referring to Photo 21, decide whether you think the building phase of the scheme will have a negative or positive effect on the criteria in the left-hand column. This will be quite hard. There are no right answers. The important thing is that you have good reasons for your judgement. Score each criterion from −3 to +3 by placing a small circle in the appropriate box.

c) When you have assigned a value to each criterion, join the circles to give a profile line.

d) Repeat parts b) and c) for what you imagine the impact might be ten years after completion. Look at Figure 22 to remind yourself of the scheme. Give each criterion a score. This time use squares for each score and join them with a different coloured line.

e) Which criteria show the greatest differences between impact during construction and impact after completion?

f) Compare the impact of your scheme with the actual plan. Would your scheme have had more or less overall impact?

21 Phase 1 of construction in Lyme Regis

Lyme Regis – what actually happened?

Making decisions on coastal defences is no easy matter. The final scheme proposed by Costain Engineering and accepted by West Dorset County Council is shown in Figure 22. Phase 1 (completed in 1995) provided the town with sea defences whilst improving Lyme's sewerage system. In the past, raw sewage had poured into the sea close to the beaches where thousands of summer visitors relaxed and swam. Costain Engineering built a combined sewage pumping station and stormwater storage tank inside the new seawall, which was protected by 21,400 tonnes of rock armour.

Originally the plan involved the siting of three offshore breakwaters (reefs or bars) to reduce the impact of waves upon the beach, but these were dropped because local opposition said they ruined the area's appearance.

The final scheme has to balance the need to protect existing land uses whilst preserving the attractive character of the harbour area. The immediate impact of 'hard engineering' solutions may outweigh the apparent advantages of 'soft engineering' options.

★ THINK!

In several instances in this book you are encouraged to think about the solution to a particular problem. By comparing your ideas with a model (often an ideal or average) or with what happened in reality, you may find similarities. However, if you find that your answers differ from what a model predicts or what really happened, it doesn't mean that they are wrong! By identifying differences you may be highlighting features unique to your solution. In Focus Task 1 for Lyme Regis you may have suggested a coastal defence scheme that would be effective, but possibly rejected only because of cost or appearance. By considering the options, you will have learned a lot more than by just memorising what was actually done.

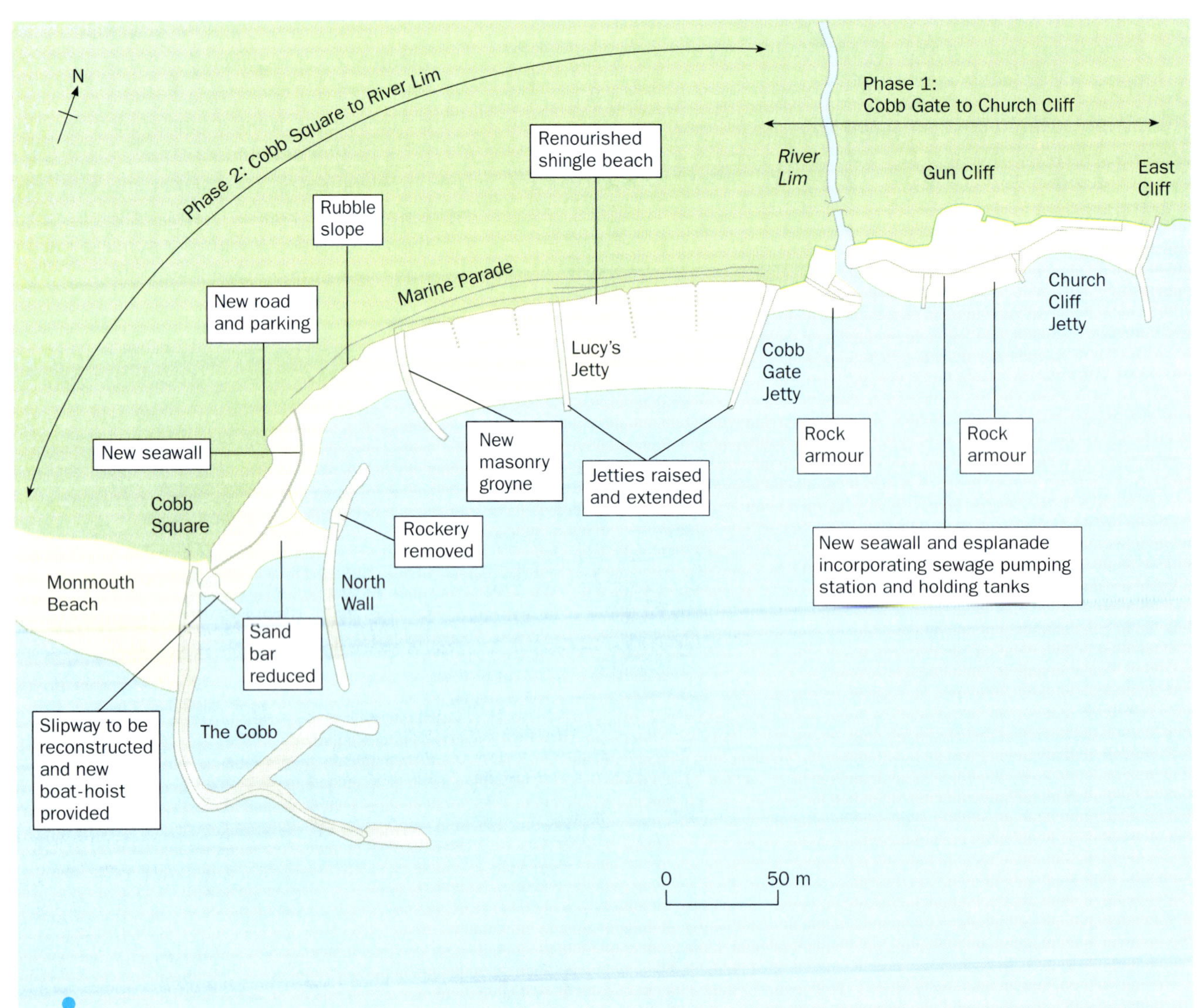

22 Plan of the proposed coastal defence scheme in Lyme Regis

4 Population

Why is population changing?

In this unit you will study the many factors that influence population change, and the many different consequences that flow from it. These nurses are describing some of them. On page 187 you will meet the nurses again and join in their conversation.

My name is Hope Matatu. I am from Zimbabwe. I am a trained community nurse. I do everything from pre-natal care, midwifery, child care, inoculations, right through to care of the elders.
My name is Blanca Garcia. I come from Seville in Spain. I am a midwife. In the last 30 years in Spain, because of the increased use of contraception, our birth rate has fallen rapidly. Our population is not replacing itself.
My name is Richard Moore. I live in Belfast. I am a nurse who specialises in geriatric care. Better healthcare means people can live longer, but many people become very dependent in their old age. Some of my patients are too old to look after themselves any more.
In Zimbabwe our birth rate is starting to fall and the death rate has started to come down too. But that might change soon. Southern Africa is suffering from an AIDS epidemic, and we cannot afford the education to stop the disease from spreading. Who knows what the long-term effects might be?
With an ageing population we soon won't be able to get enough skilled workers. We could recruit the workers we need from the young and growing population in North Africa.
But that would be exploiting the LEDCs again by taking away their best-trained and most adventurous workers. That would stop the countries from developing.

4.1 How does population change?

DO YOU KNOW?

- There are over 6 billion people in the world.
- The world's population is growing. In the late 20th century it took approximately 30 years to double.
- The fastest growth has been in LEDCs.
- Libya has the fastest-growing population in the world at 3.6% per year.
- In the MEDCs, population growth is slow. In some countries there was even a small fall in population at the end of the 20th century.
- Latvia (a Baltic Republic, formerly in the old USSR) has the fastest-shrinking population at –0.7% per year.

AIMS

- **To identify what is causing different rates of POPULATION CHANGE.**
- **To understand the meaning and the importance of POPULATION STRUCTURE.**
- **To understand the DEMOGRAPHIC TRANSITION MODEL.**
- **To examine the causes and possible consequences of Europe's AGEING POPULATION.**

1 Shanty towns and high rises in Mumbai, India

2 Harvesting wheat in England

3 City nightlife

4 Zairian refugees in Tanzania

6 Traffic congestion in Hong Kong

5 Trekking in the French Alps

Tasks

Look carefully at Photos 1–6 on these two pages.

1 Write one sentence about each of the photos. In each sentence make a comment about how crowded the area is and give reasons why it might be crowded or empty.
2 In which photo might you most expect to find:
   - people who really want to be there?
   - people who are forced to be there?
   - great extremes of wealth and poverty?
   - overcrowding getting worse?

Ⓐ POPULATION GROWTH

Shall we have a baby?

As geographers we study population. We look at numbers, distributions, density, structure, and so on. But don't forget that a country's population is made up of a lot of individual people. It is their decisions that lead to population change. Governments try to influence population, but it is not politicians who usually make the vital decisions about when, or whether, the people of their country will have babies. This decision is up to couples, and more especially women. When people are making the crucial decision about whether to have a baby they do not think: 'Shall we add another person to the world's total population?' They are more likely to think about:

How will having a baby affect my career?
How much will a baby cost?

When the baby grows up it will bring money into the family.
Will the baby love me?
How will a baby affect my lifestyle?

How will a baby affect our relationship?
I need a child to look after me when I get too old to work.
Is the baby likely to survive or die?
My religion says it is good to have children.

It would be nice for our parents to have grandchildren.
Are babies fashionable?
Will our friends laugh at us if they think we cannot have babies?

Tasks

1 Work out how much an average child costs its parents in the UK.

a) Estimate the costs of:

- buying a pram, cot, etc. needed for a baby
- decorating a child's room (or moving to a bigger house)
- food
- clothes
- leisure activities
- education

  (food, clothes, leisure activities, education: Estimate the cost per year, then multiply by 18.)
- loss of mother's earnings (include cost of possible lost promotion)
- childcare costs
- any other expenses you can think of.

b) Take off the total amount of money from all of the money gained from:

- child benefit, etc.
- any income the child brings in to the family. (This does not include money earned to spend on him or herself, but should include money that would be paid to support the parents when they are too old to work.)

2 In MEDCs such as the UK, do you think children are seen as an economic cost or an economic benefit to their parents? Give reasons.

Case study: Population in rural Kenya

Matooni village is in Kyuso, eastern Kenya (Map 7). The people in the village are mainly SUBSISTENCE FARMERS. They grow grain (mainly millet) and vegetables, and keep cattle and goats. The whole family has to contribute to running the farm and the household (Photo 8).

Mothers carry their newborn babies to the fields because they have to start work as soon as possible after the birth. When the babies become too heavy to carry, their older sisters and brothers look after them. From the age of about three, even toddlers have to work on the farm, doing jobs like scaring birds away from the crops.

Nowadays most children go to school, but often only part time (Photo 9). They still have to work when they are at home. The girls help their mothers to carry water, grind corn, work in the fields, and so on. Boys help their fathers to look after the herds. The children's work makes an important contribution to the family's survival and also means that both boys and girls are learning their adult roles in society. They will have to take over completely when their parents grow old. Adults know that there will be no pension to support them in old age. They will have to rely on their children for support when they are no longer fit to work.

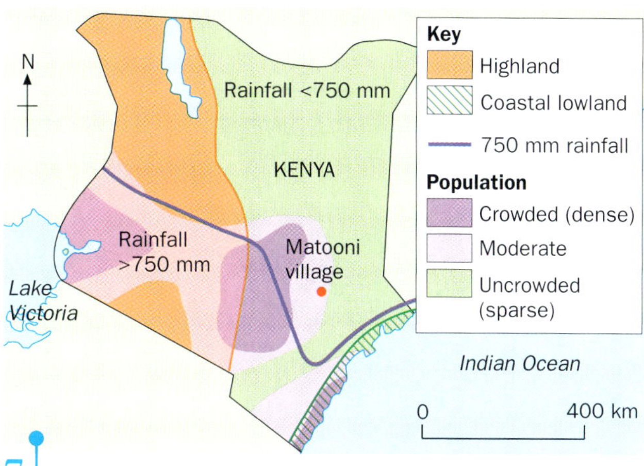

7 Location of Matooni village, Kenya

8 A boy tends the family's cattle

9 School pupils learn about farming

As in many other LEDCs, this creates great pressure to have large families. The INFANT MORTALITY RATE (see page 160) is high (although it now seems to be falling in many LEDCs, including Kenya). This means that parents need to have several sons in the hope that at least one survives.

By the time they are about 12, boys and girls are fully productive members of the society. However, at about 16, many young men leave the village, either to serve in the army or to look for temporary work in the city. They are expected to send money back to the family. These 'REMITTANCES' are the main source of cash in villages in this area. The money can be used to improve houses, to buy equipment, to pay taxes; it can be kept as a reserve to buy food during drought years; or it can be saved up to be put towards a girl's *dowry*.

Tasks

3 Look at the questions and statements on page 150. Which factor do you think would most affect a decision to have a baby in:
a) an MEDC such as the UK?
b) an LEDC such as Kenya?
Give reasons.

4 Make a list of pressures on parents:

- in the UK to have small families
- in rural parts of Kenya to have large families.

5 Do you think the need for large families in LEDCs will reduce in the 21st century? Explain your answer.

How is the world's population distributed and how is it growing?

- POPULATION DENSITY means how crowded a given area is. *Dense* population means a lot of people per square kilometre (km^2). *Sparse* population means few people per km^2.
- POPULATION DISTRIBUTION means the way people are spread over the Earth's surface. From looking at Map 10 you ought to be able to work out which places have dense population and which have sparse population.
- Rates of population growth vary around the world.

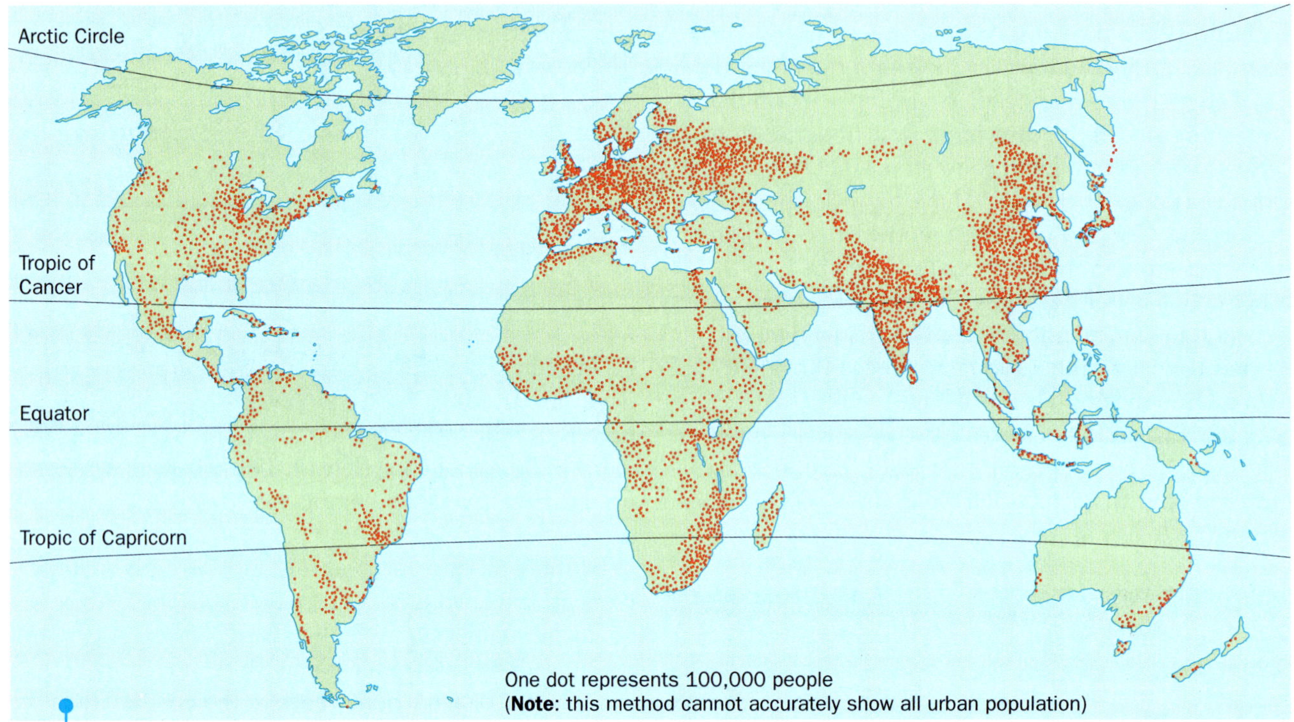

10 World population distribution in the late 1990s

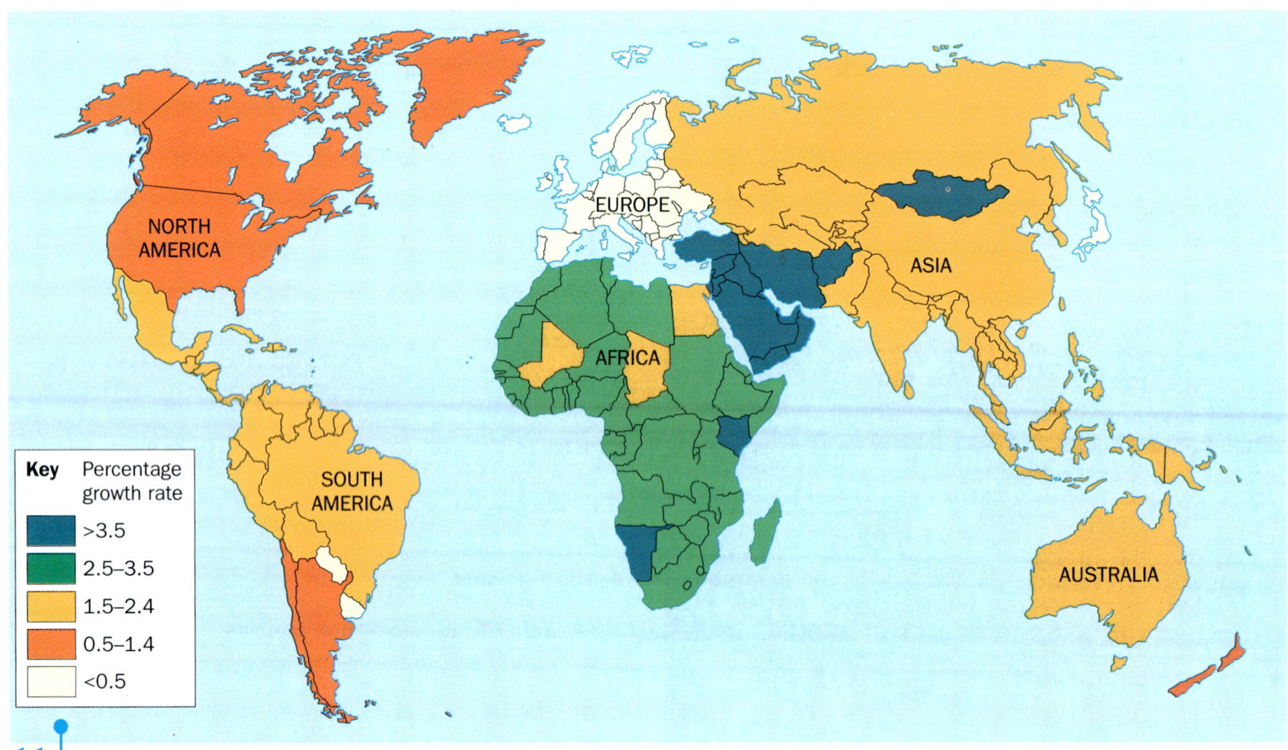

11 Average annual population growth rate in the late 1990s

Tasks

SAVE AS . . .

1 Using your atlas and remembering what you learned at KS3, you should be able to name examples of the following:

Sparse population in an area that:

A is very cold.
B is very steep and mountainous.
C is very hot and wet.
D is very hot and dry.
E is too far from the sea for trade.
F has exploited big deposits of oil.
G has good soil and climate but farming uses machinery more than people.

Dense population in an area that has:

H fertile alluvial soils and deltas.
I fertile volcanic soil.
J a river that irrigates a desert area.
K coalfields where industry grew in the 19th century.
L some excellent ports.
M climate and scenery that attract many tourists.
N a lot of industry but few raw materials.

Label them on a copy of Map 10 which your teacher will give you. Use a key to show whether they are examples of sparse or dense population. You may get some ideas from your work on page 149.

2 Copy and complete the following paragraph, choosing the correct words and phrases from the brackets:

The world distribution of population is (largely influenced/ completely determined) by the physical environment. People (only/mostly) settle in areas with an attractive environment. People will settle in areas with difficult conditions if there are some (resources/settlements) to attract them. Some other areas have sparse population (because of/in spite of) having quite good environmental conditions.

3 a) Compare Maps 10 and 11. With the aid of an atlas, locate three countries in each of the following categories:

- Dense population and rapid growth
- Moderate population and rapid growth
- Sparse population and rapid growth
- Dense population and slow growth
- Moderate population and slow growth
- Sparse population and slow growth

b) Name and shade these countries on a map of the world.
c) Complete a key, choosing the colours carefully.
d) Write a paragraph to describe the distribution pattern you have shown on your map.

4 Study the information in Table 12 which shows how long it takes for the population of some countries to double.

a) List the 18 countries in order of their doubling time or rate of shrinking.
b) What pattern can you see in your list?
c) Explain the pattern you can see.

5 Ask your teacher for details of the GNPs of the countries listed in the table.

a) Draw a scatter graph to see whether there is a link, or correlation, between GNP and population doubling time.
b) Explain what your graph shows.

6 A good way to check whether the relationship between the two sets of figures is significant is to use a test called the 'Spearman's rank correlation test'. This is difficult, but your teacher might give you a sheet which shows you how to do this.

Country	Doubling time (years)
Algeria	29
Australia	100
Brazil	48
China	67
Egypt	34
Germany	–*
Jamaica	41
Japan	289
Kenya	27
Malaysia	31
Russia	–**
South Africa	46
South Korea	75
Spain	1,386
Sweden	4,077
Thailand	63
UK	433
USA	116

* Germany's population is falling by 0.1% per year
** Russia's population is falling by 0.5% per year

12 Population doubling times at current rates

Can the world manage its growing population?

People have worried about population growth for many years. One of the most famous people who have looked at this issue was Thomas Malthus. In 1798 he predicted that population was likely to grow faster than food supply (Figure 13). He said that:

- Food supply could, at best, grow only at an arithmetic rate (i.e. 1 – 2 – 3 – 4 – 5, etc.) over a period of time.
- Population could grow at a geometric, or exponential, rate (i.e. 1 – 2 – 4 – 8 – 16, etc.) over the same period of time.
- This would lead to 'war, famine and disease' unless 'men curbed their sinful lusts' and so reduced the growth of population!

Fortunately Malthus's predictions have not come true in the UK. Some of the reasons for this are given below.

In the 1800s

- Large numbers of people migrated to 'new' colonies.
- The agricultural revolution (e.g. seed drill, four-field rotation) allowed food supply to increase more rapidly than the population.
- More land was farmed as a result of drainage and hedgerow removal.
- Prairies and other grassland areas were opened up for wheat growing in the USA, Australia, and elsewhere.
- Better methods were developed for food transport and distribution.
- New methods of food conservation were developed, e.g. canning, freezing and drying.
- Education was provided for all, including girls.

In the 1900s

- The BIRTH RATE declined. (You will find out more about the reasons for this on pages 160–61.)
- There was increasing use of contraception.
- Agriculture was mechanised.
- Scientific agriculture – use of pesticides, fertilisers and insecticides – made more food available.

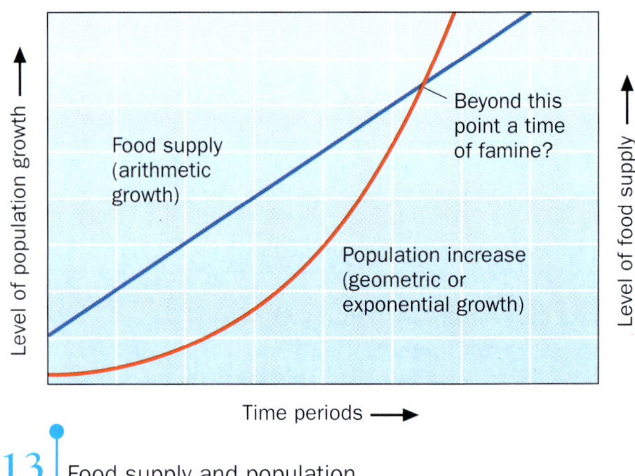

13 Food supply and population

14 The 20th-century world population explosion

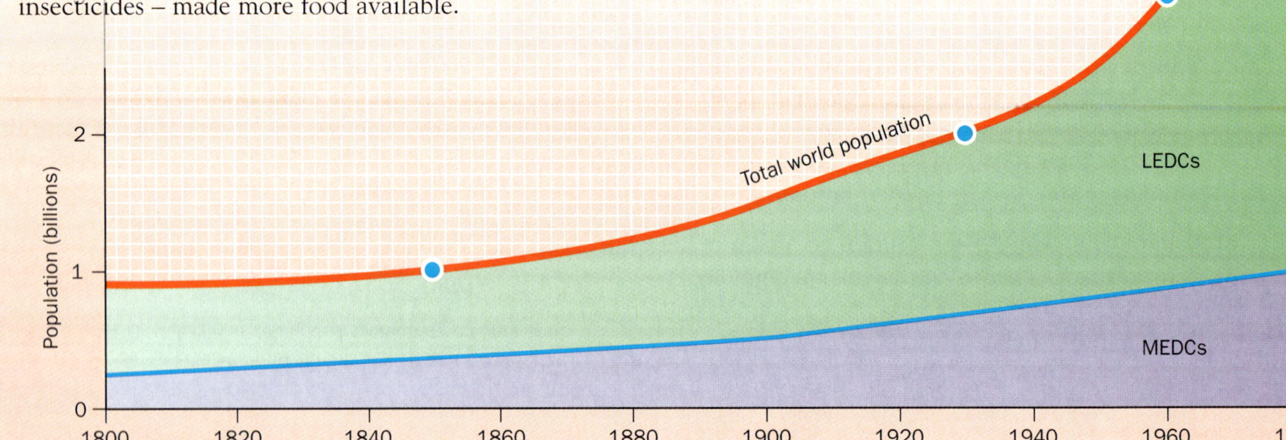

Should we be pessimistic or optimistic about population growth?

Pessimists	**V**	**Optimists**
Although Britain has not reached the point that Malthus predicted, some people think that the world as a whole is heading towards a disaster on a much bigger scale. These people are called Neo-Malthusians. They refer to the growth in the world's population over the last 50 years to support their views. The Neo-Malthusian model has been described as a *pessimistic scenario*.		In 1965, Ester Boserup, a Danish economist, put forward an alternative theory. She said that in the past an increase in population had always led to improvements in technology. As the demand for food increases, farming becomes more intensive to meet the demand. In other words, 'necessity is the mother of invention'. The Boserup model has been described as an *optimistic scenario*.

★ **FACTFILE**

The latest figures for world population suggest that:

- Between 1960 and 2000 the world population doubled from 3 to 6 billion.
- The world population will probably never double again.
- The birth rate is falling faster than expected in most LEDCs.
- In most MEDCs, the birth and death rates are about the same, although in many north and eastern European countries the birth rate is lower than the death rate.
- LIFE EXPECTANCY is increasing in most countries.
- Average world life expectancy is now 62 years and is increasing by one year every five years.
- The average age of the world population is now 26 (in MEDCs it is 37).
- By 2025 the average age will be 32 (in MEDCs it will be 43).
- The late 20th century was probably the last time when under-16s would outnumber over-60s.
- Women tend to have most of their children when they are aged between 18 and 30 – with so many people now in this age group, the population will almost certainly keep on growing well into the 21st century.

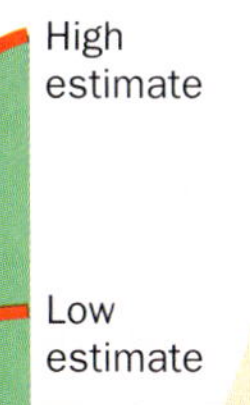

Figure 14 shows the rise in world population, sometimes described as a 'population explosion', that took place during the 20th century. The estimates show how the population might change between 2000 and 2025.

- The 'high estimate' shows what will happen if population goes on rising at the same rate as it was rising in the 1980s.
- The 'low estimate' shows what might happen if the rate of population growth slows down – especially in LEDCs. There is some evidence that this is happening.

Tasks

1 a) Read through the information given in the Factfile above. Sort the bullet points into three groups:

- Facts that seem to support the pessimistic views of the Neo-Malthusians.
- Facts that support the optimistic views of Boserup.
- Facts that do not seem to fall into either group.

b) Which scenario, pessimistic or optimistic, do you support? Give reasons.

You will return to this on page 157.

ICT

2 To help you to understand the rate of growth of the world's population, find a 'population clock' on the internet. This is a graphic and fun way of showing very serious information. Try:

www.worldgame.org/worldometers

which has a series of 'Worldometers' including population.

★ **FACTFILE**

- The world's total population passed the 6 billion mark in 2000.
- Over 4.8 billion people (80 per cent) live in LEDCs. The two countries with the largest populations are both LEDCs:
  - China, with 1.25 billion
  - India, with 1 billion.
- The USA is the MEDC with the largest population but at 270 million it has less than a quarter of China's population.
- On average the population of LEDCs is growing by 1.8 per cent per year (doubling every 38 years).
- Many people feel that this growth rate is not sustainable and that if it continues it will lead to disaster, as Malthus predicted.

Case study: Population control in China

In the England and Wales case study (see pages 160–61) you will learn that population growth started to fall because of a mixture of economic changes in society. Then, on pages 162–63, you will see how, in Thailand, government policy has affected birth rates. However, in China the government became even more deeply involved in developing population policy. The government had two reasons: they wanted to avoid the Malthus type of disaster, but they also realised that China could only enjoy a rising standard of living if its population growth was controlled. This section looks at how China cut its rate of population increase, and considers whether its methods should or could be tried elsewhere.

The 1950s

In the 1950s the philosophy of the Chinese government was 'a large population gives a strong nation'. The government encouraged people to have children for the good of the country.

Then, in 1959–61, there was a serious famine. Up to 20 million people died, including many children. The birth rate fell. More boys died than girls.

The 1960s

After the famine there was a population boom. Population increased by 55 million per year (equal to the total population of the UK). The government did nothing to try to reduce the birth rate.

The 1970s

From the mid-1970s onwards, however, policy changed. At first the government just urged people to reduce their families. They publicised the catch-phrase *Wan-xi-shao*, which means 'Later, longer, fewer', standing for:

- Later marriages.
- Longer gaps between children.
- Fewer children.

The 1980s

This did not work well. The population went on increasing, so in 1979 the government introduced the 'one child policy' which set very strict limits on who was allowed to have children, and when. Strong pressure was put on women to make sure that they used contraception. There were even people known as 'granny police' whose job it was to watch their neighbours to make sure that they were taking the pill. If they suspected that women were pregnant without permission they reported them to the authorities. Enforced abortions and sterilisation became common. The policy was successful in urban areas but not in rural areas where many couples disobeyed the ruling and had large families.

'Boys are better!'

One side-effect was the practice of 'female infanticide'. Couples wanted sons, so many baby girls were killed or aborted, so that people could try again for a son.

'Little emperors'

Some people say that there has been another undesirable side-effect of the policy. Children with no brothers or sisters get all the attention of their parents and grandparents. They can become very spoilt by all the attention, and they grow up not knowing how to share with other people. They are called 'little emperors'. Of course this view may well be an exaggeration. Many families have an 'only child' who grows up to be very sensible and well-balanced.

The 1990s

From 1989 the one child policy was relaxed slightly and this has speeded up the increase in the birth rate slightly (Figure 15). But in any event an increase in births in this period was unavoidable because the peak number of women born in the population boom time of the 1960s and 1970s are now of child-bearing age (Figure 16).

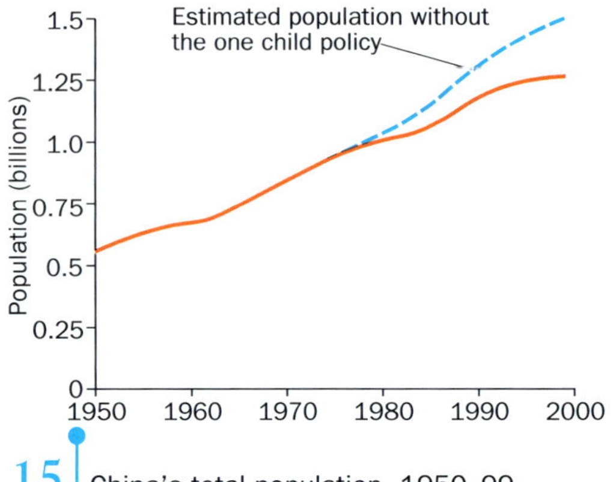

15 China's total population, 1950–99

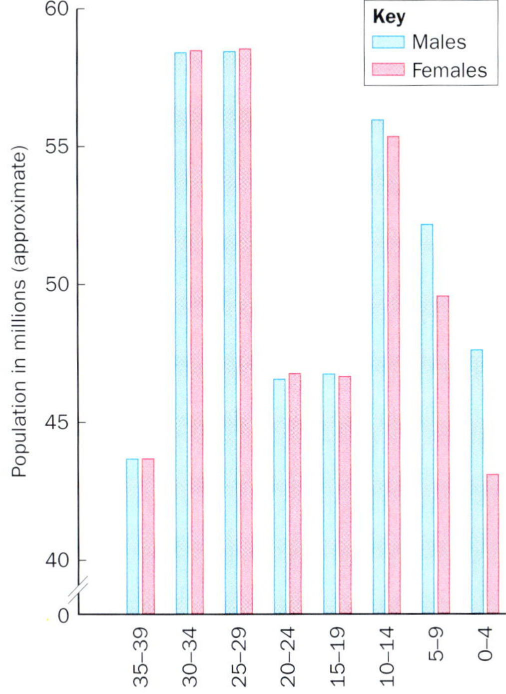

16 Population data for China, 1960–99

17 A one child family in China

Tasks

1 Look closely at the graph in Figure 15 and the text on these two pages. When might the following events have happened or be likely to happen?

a) Mrs Chang has an abortion because she already has one child.
b) Mrs Ho has her sixth child. She is very pleased to serve her country by having so many healthy children.
c) China's leading baby store lays off workers because revenue is down.
d) Mr Fung complains that all of his friends have found girlfriends but there are no girls left for him.

2 On your own copy of Figure 15 mark each of the government policies mentioned in the text.

3 What evidence can you find in Figure 16 of the following:
a) A population boom in the 1960s.
b) Parents choosing to have sons rather than daughters under the one child policy.
c) Success for the one child policy to start with.
d) A setback for the one child policy later on.

4 Discuss: do you think that the one child policy was
a) successful?
b) morally justifiable?
c) a model for other countries to follow?

5 Photo 17 is to be used as part of a campaign for the one child policy. Write some text to go with it explaining why the policy is necessary.

SAVE AS . . .

6 Using all that you have found out, write a paragraph describing the trends shown in Figure 15.

Focus task

7 On page 155 you compared two views of world population growth. Make a large copy of the table below. Find facts, figures and ideas as you work through this chapter to put in each column. You can also add any information that you found out in Task 1 on page 155.

The future of the world's population	
Reasons to be worried *The pessimistic scenario* 	**Reasons to be cheerful** *The optimistic scenario* 

Case study: Population growth in North Africa

North Africa (Map 18) provides some recent evidence that shows how population growth rates are slowing down in LEDCs even without drastic measures like China's one child policy. See Table 19.

What has caused the rapid changes in the birth rate in these countries? People who have studied the area say that there have been 'four revolutions' for women in these countries. All of these have tended to influence the birth rate.

	1983				1996			
	BR ‰	DR ‰	Increase in pop (%)	Fertility rate*	BR ‰	DR ‰	Increase in pop (%)	Fertility rate*
Algeria	46	14	3.2	7.0	31	7	2.4	4.4
Tunisia	35	10	2.5	5.6	26	6	2.0	3.3
Morocco	44	13	3.1	6.8	26	6	2.0	3.3
Egypt	43	12	3.1	6.0	29	8	2.1	3.6

* Fertility rate: the average number of children born per woman during her lifetime.

Key: BR = birth rate; DR = death rate. Birth rates and death rates are usually expressed in figures per thousand (‰), rather than per hundred (%).

19 Fertility in North African countries

18 North Africa

- **The urban revolution** Women who move to cities usually have fewer children than those who stay in rural areas. This is probably linked with the other revolutions described below.
- **The education revolution** There has been a slow but steady increase in education for women and girls in these countries. This is especially true for women in the cities. Once women have an education they are likely to:
  - marry later
  - become more aware of family planning
  - get paid work.

  All of these changes make them less likely to have large families.
- **The working revolution** Women in the countryside work on farms, as part of the family labour force. This does not stop them having children because farm work and looking after babies are shared amongst all members of the community. However, paid work in the cities is more difficult to combine with bringing up children. Working women don't want to lose their income.
- **The MIGRATION revolution** Many people from North Africa have migrated to Europe during the last 40 years. The migrants stay in contact with friends and relatives in their home countries. When they write or visit home they probably describe social conditions in their new countries and the attractions of a Western European lifestyle, with smaller families and more consumer goods. This can have a big effect on social attitudes of people at home.

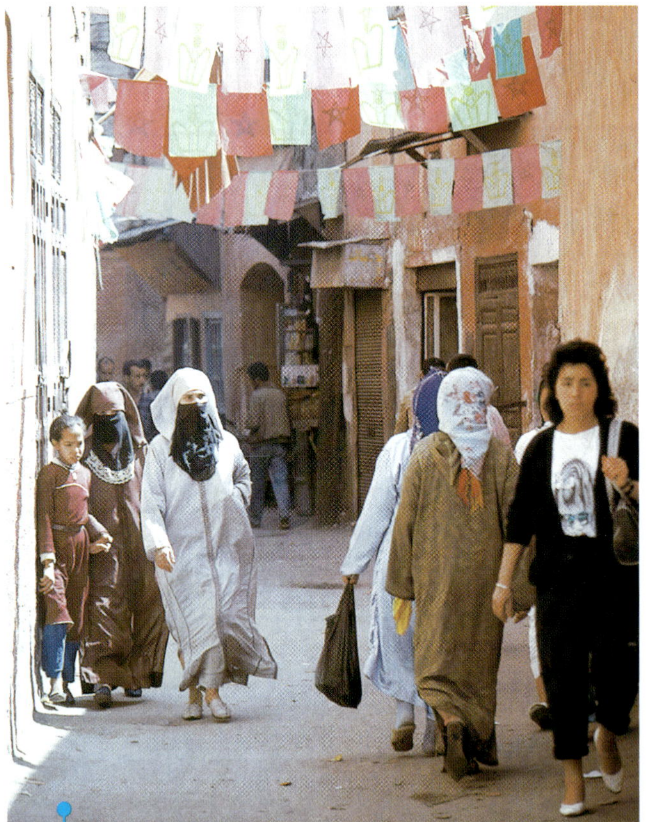

20 Moroccan women in the 1990s

Tasks

SAVE AS...

1 Write down three pieces of evidence which suggest that population increase in North Africa is slowing down.

2 a) List the 'four revolutions' affecting population change.

b) Next to each revolution write three key words or phrases to show why that revolution is causing a change in the birth rate.

How is population growth measured?

Imagine that you work for the manager of the flats shown in Figure 21. Your boss has asked you to collect some data.

Tasks

3 Count the population of the flats, using a data collection form like this:

Data collection form				
Age	**0–16**	**17–59**	**60 and over**	**Total**
Males				
Females				
Total				

4 You go back a year later to see how the population of the house has changed. This is what you find:

- Clare has had twin girls.
- Jack has died. Maud's sister has moved in to share the flat.
- Phil has moved out, and a female student has moved in.
- Alex has left to go to university. Denis and Eileen's son left university and has moved back home because he has a job in the area.

a) Count the population again, and complete another copy of the data collection form.

b) Compare your two forms, and then fill in a copy of the balance sheet below.

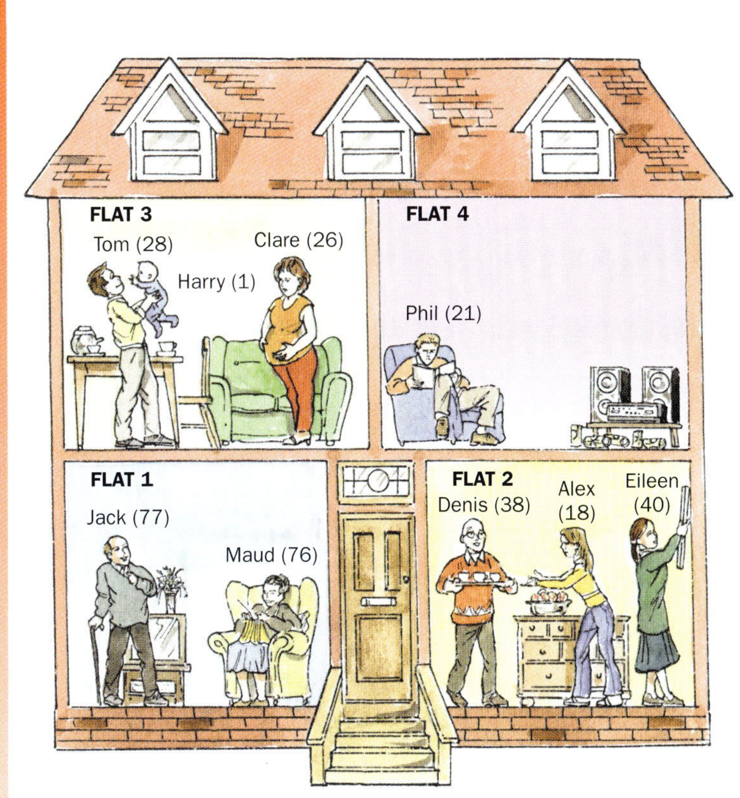

21 Population of a house divided into flats

Balance sheet	Increase in population	Decrease in population	Population total	% change
Population at start of period			______	
Births during period	+______			
Moved in during period	+______			
Deaths during period		–______		
Moved out during period		–______		
Population at end of period			______	
Total increase or decrease			______	
Percentage increase or decrease				______

Measuring population and population change in a country or a region is similar to the process described above. In the UK, government officials visit every household every ten years and collect the data. Every householder legally *must* fill out the CENSUS form.

The data must be precise. They are essential for government planning. The data are used to predict how the population is going to change in future.

B DEMOGRAPHIC TRANSITION

Case study: Population change in England and Wales 1700–1970

In England and Wales the first full census of the population was taken in 1801. Since then a census has been taken every 10 years, except in 1941 (why do you think that was?). Fairly good estimates of the population for 50 years before the first census can be made by studying other records, especially parish records of births and deaths.

Figure 23 shows population growth in England and Wales from 1700 to 2000. Notice that:

- there was very slow growth of population in the 18th century.
- it grew rapidly throughout the 19th century.
- it carried on growing, but at a slower rate, for most of the 20th century.
- the graph almost levelled out at the end of the 20th century.

Figure 22 shows how the birth rate and death rate changed over the same period. Some possible reasons for the changes are shown on page 161.

> ★ **FACTFILE**
>
> **Population data**
>
> - Birth rate = the number of live births, in a year, for every 1,000 of the population.
> - DEATH RATE = the number of deaths, in a year, for every 1,000 of the population (sometimes also called the mortality rate).
> - These are usually expressed as 'birth rate per 1,000' and 'death rate per 1,000', or '‰'.
> - NATURAL POPULATION CHANGE = original population + (births – deaths).
> - Population change = original population + (births and IMMIGRATION) – (deaths and EMIGRATION).
> - Population change is usually expressed as a percentage.
> - Infant mortality rate = the number of children, out of every 1,000 births, who die before they reach their fifth birthday.
> - Life expectancy = the average age of death in a country, at a particular time.

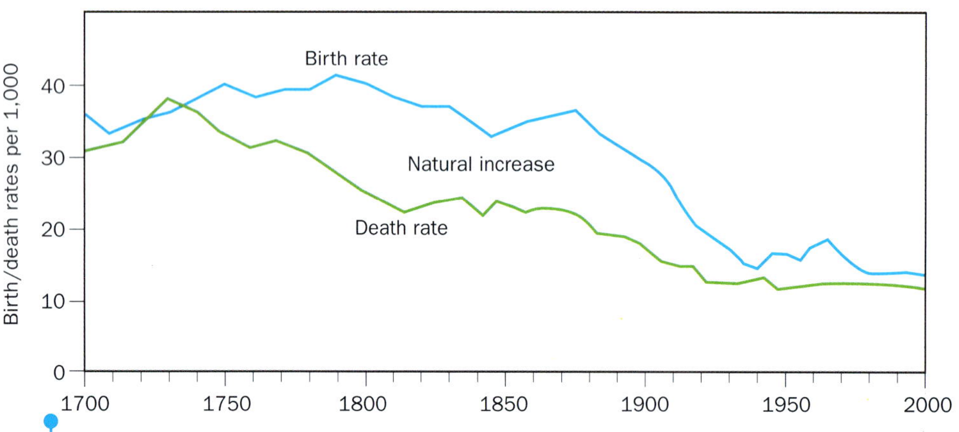

22 Births and deaths in England and Wales, 1700–2000

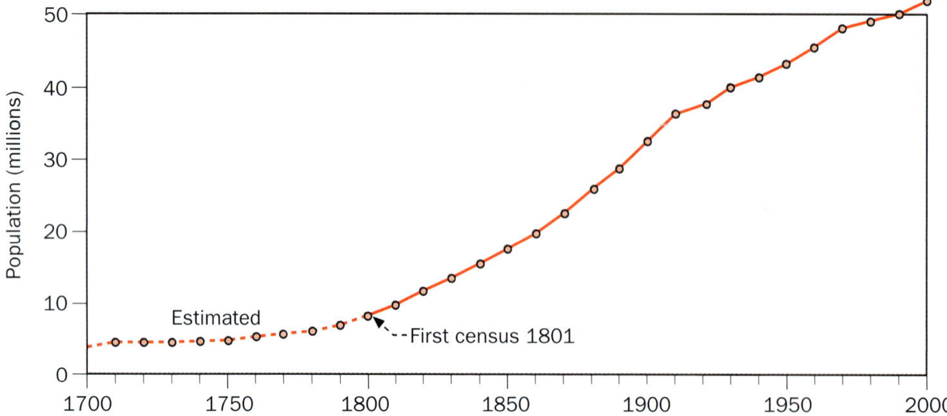

23 Population growth in England and Wales, 1700–2000

1751	Gin tax reduces drunkenness.
1770	Industrial Revolution causes migration to towns.
1780 onwards	Increasing employment of children in factories. Children's wages are essential for family survival.
1790 onwards	Cotton clothing gets cheaper. It is easier to wash than woollen clothing.
1796	First vaccination against the killer disease smallpox.
1812	Apprentices allowed to marry earlier.
1830s	Rapid growth of unplanned, dirty towns leads to infant mortality rates as high as 350‰.
1833	Factory Act makes it illegal for children under the age of 9 to work in factories.
1842	Health Report finds that poor sanitation is main cause of disease and death in Britain.
1846	Repeal of Corn Laws. Imported food becomes cheaper.
1848	Cholera epidemic leads to towns being cleaned up.
1850 onwards	Railways improve fresh food supply in towns.
1870	Annie Besant publishes first book on birth control.
1870s	Public Health Act improves clean water supply and sewage disposal.
1876	Education up to age of 13 made compulsory for all.
1911	First national old age pension.
1914–18	First World War: 1 million British men killed.
1942	Beveridge Report leads to better healthcare for all.
1960 onwards	The contraceptive pill becomes widely available. More women join the labour force.

24 Causes of changes in birth and death rates in England and Wales since 1750

Task

1 a) Look at the information in Figure 24.

b) Draw up a table like the one below to classify the factors.

	Factors affecting:	
	Birth rate	**Death rate**
Causing increase		
Causing decrease		

c) Choose at least three factors from Figure 24 to go in each box of your classification table. Some factors might appear as affecting both the birth rate *and* the death rate.

d) For each factor that you put in your classification table, write a brief explanation of why you put it there.

e) Copy Graphs 22 and 23. Use Figure 24 to add notes to your graph to show changes affecting people during that period. Use one colour for factors affecting birth rate, and another colour for factors affecting death rate.

★ THINK!

Figure 24 lists a lot of factors that have affected population change. They are all interesting and relevant to a study of changing population in England and Wales. However . . . you cannot be expected to learn *all* of them for an exam.

Completing your own table should help you to identify the *key factors*. Then learn your table. It should be quite easy to remember the structure and the headings for the table, because it follows a clear pattern.

Once you have learned the pattern it should be quite easy to learn the details that fill in the table.

It is always easier to learn clearly structured information rather than a random collection of facts and figures.

The demographic transition model

Tasks

1 Look again at Figure 23 on page 160. Try to divide it into four separate stages as follows:

Stage 1 Low population, increasing very slowly.
Stage 2 Population increasing at a faster rate.
Stage 3 Population still increasing, but rate of increase slowing down.
Stage 4 High population, almost stable again.

2 Now look again at Figure 22 on page 160. Think about what is happening to the death rate and the birth rate in each of the four stages in Task 1.

SAVE AS . . .

3 Copy and complete this table to record your findings.

	Stage 1	**Stage 2**	**Stage 3**	**Stage 4**
Death rate				
Birth rate				
Total population	Low, increasing very slowly	Increasing at a faster rate	Still increasing, rate of increase slowing down	High, almost stable again

In the 1960s, people noticed that many MEDCs had gone through the four stages of development described here in Task 1. They called this the demographic transition model.

They also noticed that LEDCs seemed to be in a situation very similar to Stage 2. Their death rates had fallen but their birth rates were still very high, leading to the population explosion shown in Figure 14 on page 154.

In Britain, Stage 2 had taken well over 100 years to complete because social, economic and technological changes were introduced gradually and the death rate fell slowly. In many LEDCs the death rate fell far more rapidly because these changes took place more quickly. As the death rate fell, the birth rate stayed high and so the population increased very quickly. This put pressure on resources. The Neo-Malthusians thought that disaster was approaching.

Optimists about world population hoped that the LEDCs would soon move into Stage 3 of the demographic transition. They felt that when people and governments realised a possible disaster was approaching, the birth rate would start to fall. In China the one child policy forced the country into Stage 3. In Malaysia and Thailand it has happened more slowly.

Case study: Population change in Malaysia and Thailand

25 South East Asia

Malaysia

Malaysia is a former British colony in South East Asia. It became independent in the 1960s. It had already entered Stage 2 of the demographic transition by 1960. Table 26 shows what has happened since.

	Birth rate ‰	**Death rate ‰**	**Total population ('000s)**	**% change per year**
1960	47	20	8,036	–
1970	44	16	10,396	2.8 ('60–'70)
1980	44	13	13,879	3.1 ('70–'80)
1990	37	11	18,065	2.6 ('80–'90)
1995	34	7	20,689	2.7 ('90–'95)
2000	27	5	23,264	2.3 ('95–'00)

26 Population change in Malaysia, 1960–2000

Tasks

4 Use the figures in Table 26 to draw a graph showing Malaysia's population change.

5 a) Why did the rate of population growth increase between 1960 and 1980?

b) Why did the rate start to fall between 1980 and 2000?

c) At what stage of the demographic transition was Malaysia in:

- 1960?
- 2000?

Explain your answers.

6 Study Graph 27.

a) When did Thailand move from:

- Stage 1 to Stage 2
- Stage 2 to Stage 3
- Stage 3 to Stage 4

of the demographic transition model? Explain your answers.

b) What possible reasons can you think of to explain why Thailand has moved through the demographic transition model more quickly than Malaysia?

ICT

7 a) Find the United Nations population website:

www.undp.org/popin

Similar information is available on the US Bureau of Censuses website:

www.census.gov

b) Select data related to *population growth* for a variety of countries. You could choose the countries in this chapter, but make sure that you choose at least three LEDCs and three MEDCs.

c) Insert the data into a spreadsheet, using a program such as MS Excel.

d) Use the program to produce both a pie chart *and* a bar chart to show the growth of population in your chosen countries from 1950 to the present.

e) Which do you think is the more effective way of showing the information? Explain your answer, saying why you chose one technique and why you rejected the other.

Thailand

Thailand's experience is similar to Malaysia's, but its birth rate has fallen even faster. The fall in Thailand's birth rate has been partly a result of the National Family Planning Programme, run by the Ministry of Health since 1970. The programme has included:

- public information programmes, to ensure that everyone knows about contraceptive methods
- advertising the benefits of the two child family
- establishing health centres throughout the country, to provide all common contraceptive methods, mainly free
- training paramedics ('barefoot doctors') and midwives; they are mainly from the local villages, so they are known and trusted in their communities. They provide healthcare, especially for mothers and babies, so that most babies survive to be healthy children.

27 Population change in Thailand, 1930–2000

28 A government poster to persuade people in Thailand to use contraception

© POPULATION STRUCTURE

29 In China nearly 50% of the population is under 20; in the UK the figure is around 35%. Conversely, in China around 7% of the population is over 65 whereas in the UK this figure is nearly 20%.

What is population structure?

To understand the population of a country we need to consider more than just 'crude numbers'. We need to understand the population structure. Population structure describes the proportion of the country's population in each age group.

Population structure is changing in many parts of the world. Some countries have a very youthful population. Other countries have an ageing population (see the photos in Source 29, for example). Changes in the population structure affect the need for jobs, services, goods and so on. Knowing the population structure of a country at the present time also allows governments to predict what will happen in the future and to plan for it. Today's babies will be having their own babies in 15–40 years' time, and so future birth rates can be predicted.

POPULATION PYRAMIDS are drawn to show population structure. A population pyramid is two bar graphs drawn back to back! The left-hand graph shows the number of males in each group (or *cohort*). The right-hand graph shows the number of females in each cohort. They often show 5-year cohorts, but other age bands can be used.

Pyramids A and B in Figure 30 show the population of England and Wales in 1801 and 1921. Population pyramids C and D in Figure 30 show the population of Malaysia in 1980 and 1998. They show that it took over a century for England and Wales to pass from Stage 2 to Stage 3 of the demographic transition (see page 162) but Malaysia made that transition in about 20 years.

Tasks

1 Look at Population Pyramid A for England and Wales 1801 on page 165. What percentage of the population is:
   a) Male aged 0–4, female aged 0–4, total 0–4?
   b) Male aged 0–14, female aged 0–14, total 0–14?
   c) Male 60+, female 60+, total 60+?

2 a) Describe the shape of Pyramid A.
   b) What does this suggest to you about:
   - birth rate?
   - death rate?
   - infant mortality?
   - life expectancy?

3 What stage of the demographic transition model does Pyramid A represent?

4 Compare Pyramid A for 1801 with B for 1921. How had the population structure changed by 1921?

5 Compare the two pyramids for England and Wales with those for Malaysia. Why do you think Malaysia moved much more quickly than England and Wales from Stage 2 to Stage 3?

(A) England and Wales 1801

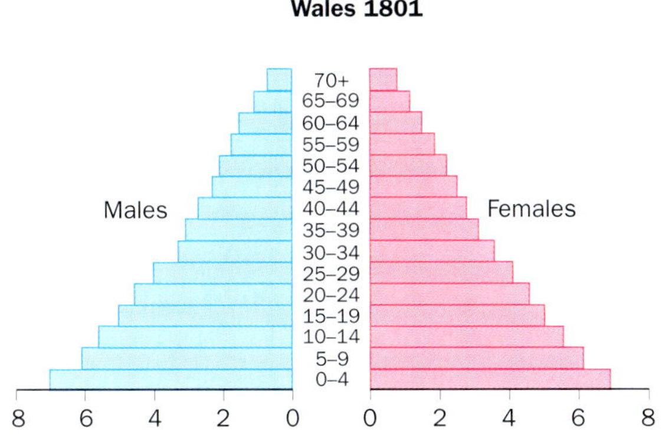

(B) England and Wales 1921

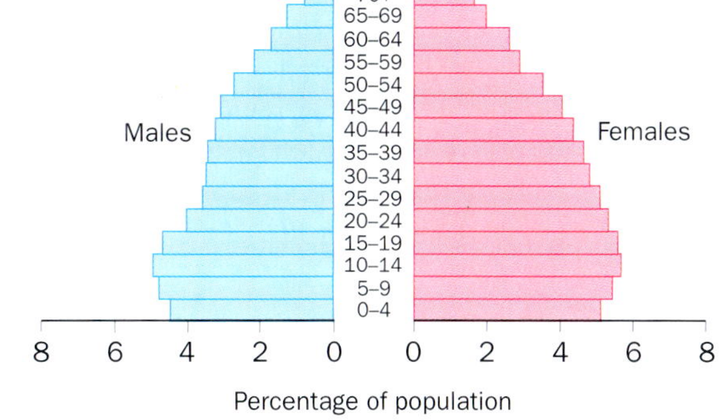

(C) Malaysia 1980

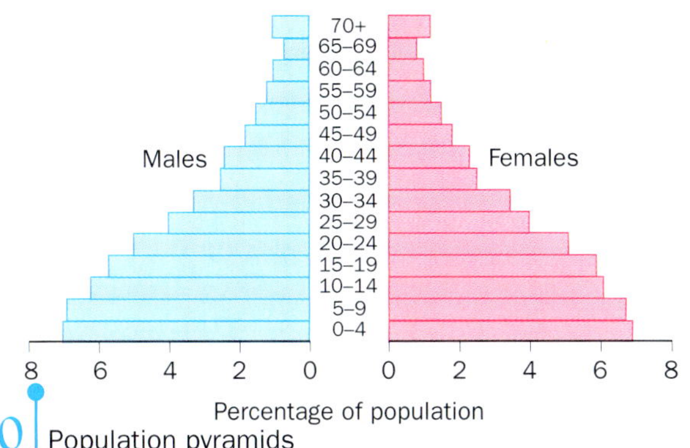

(D) Malaysia 1998

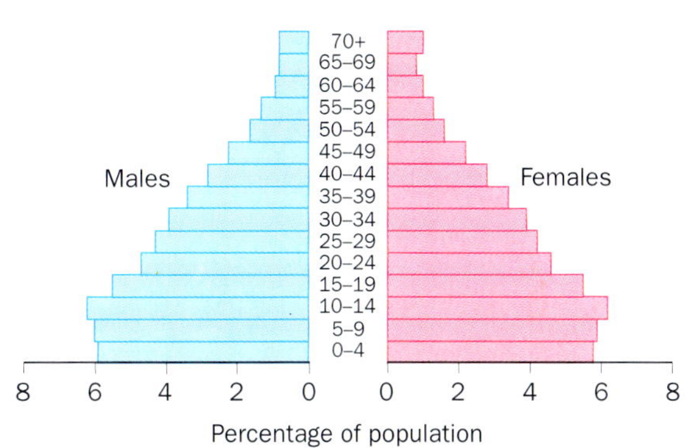

30 Population pyramids

SAVE AS . . .

6 On a copy of Pyramids A and B for England and Wales, add labels to either one to show evidence that:

- the birth rate has changed
- infant mortality has fallen
- the death rate has changed
- life expectancy has increased
- the First World War affected the cohort of males between 20 and 50
- the First World War affected the birth rate (both because soldiers were away fighting and because those at home felt insecure about the future and so did not want babies)
- the country has moved from early Stage 2 to late Stage 3 of the demographic transition model

7 Now add labels to a copy of Malaysia's pyramids. Find evidence of similar changes.

ICT

8 The US Census Bureau site at

www.census.gov

will enable you to find data on the population structures of most countries. It will also allow you to draw population pyramids for selected years.

a) Log on to the site.

b) Produce pyramids to predict Malaysia's population structure in 2025 and 2050.

c) Cut and paste the diagrams into a word-processing package such as MS Word. Then add your description of what is likely to happen to the population of Malaysia during the next 50 years.

d) Look at the United Nations website:

www.undp.org/popin/wdtrends/a99/a99cht.htm.

This also allows you to draw population pyramids for the world, MEDCs and LEDCs, with predictions of changes up to 2050. Which of the two sites is the more helpful to a Geography student in your opinion? Give reasons.

Sketch pyramids

Drawing accurate population pyramids is a long and rather difficult task. Sometimes geographers use sketch pyramids instead. The four sketches in Figure 31 represent the four stages in the demographic transition model. They have been annotated to show the key features of each stage.

★ THINK!
Learn to draw these four sketch pyramids quickly but accurately, and learn the key features. This will help you to remember important information for your examination.

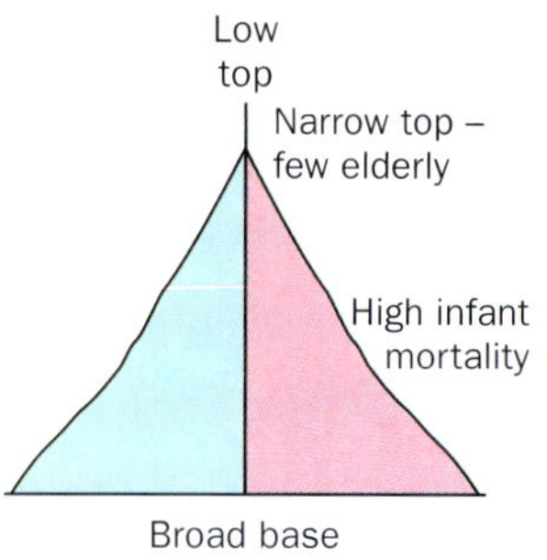

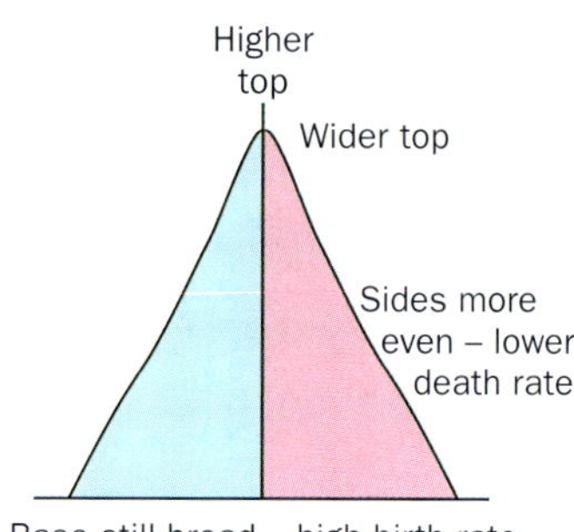

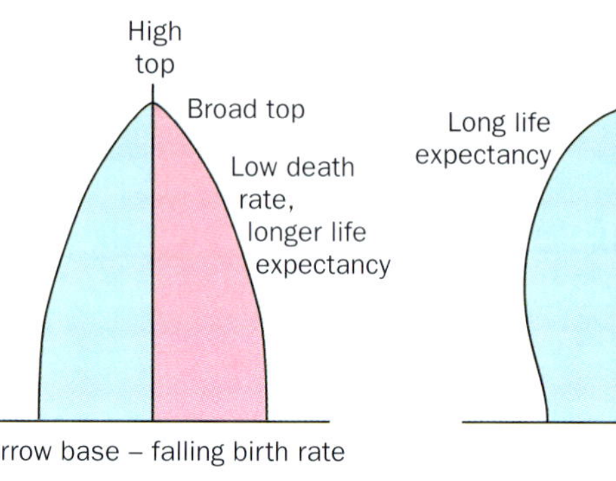

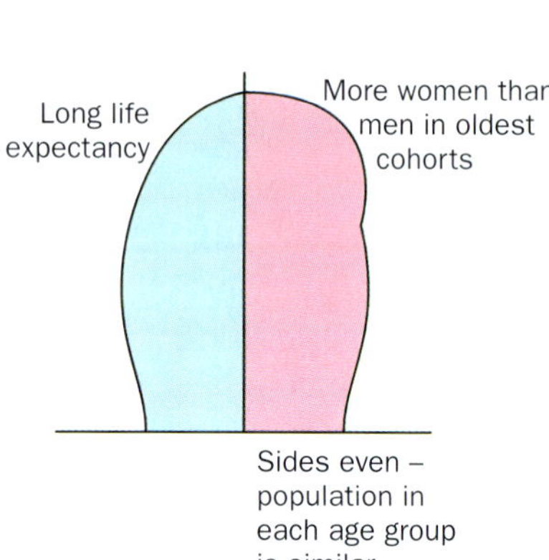

31 Sketch pyramids showing the four stages of the demographic transition model

Tasks

1 Make a copy of each of the sketch pyramids in Figure 32. Annotate each one to show its key features.

2 Give each of your sketches a title, to show which one represents:

- A retirement town on the south-east coast of England.
- A town that has grown rapidly, attracting many people to work in its ICT industry.
- An old industrial town where a steelworks closed ten years ago causing much unemployment.
- A new town that attracted many young couples about 15 years ago.

3 Give reasons why you chose each sketch for each town.

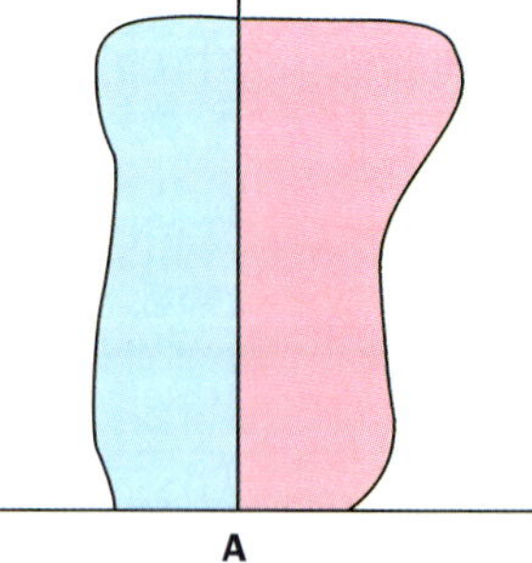

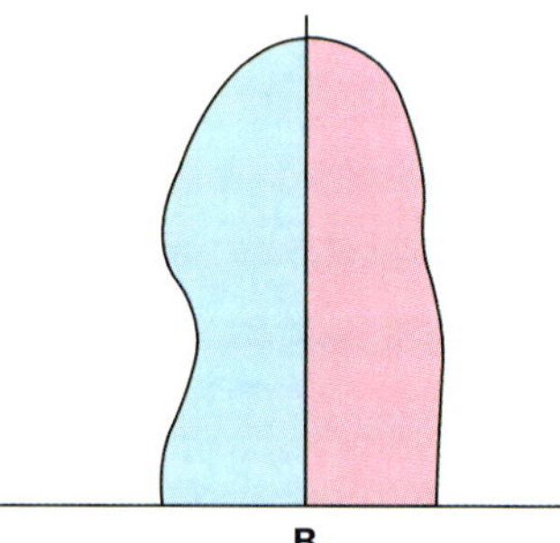

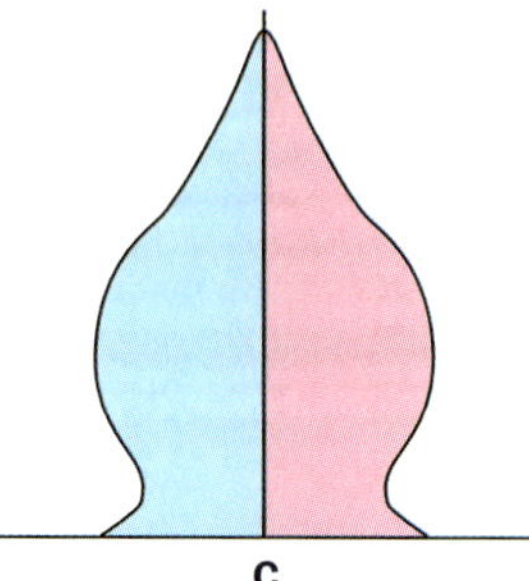

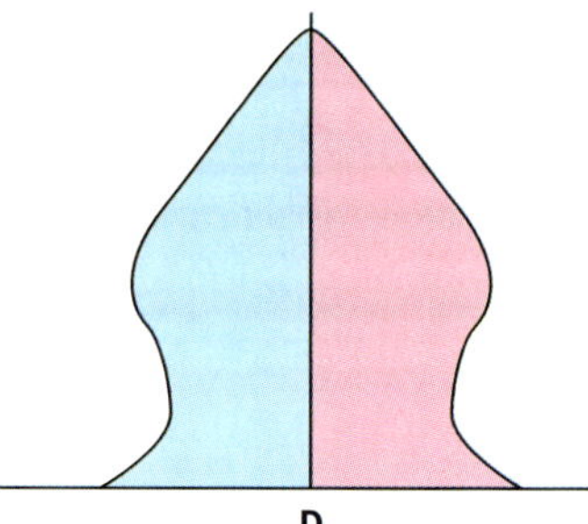

32 Sketch pyramids of towns in the UK

Dependency ratios

DEPENDENCY RATIOS are used to show how many non-economically active people are supported by each economically active member of a population. The dependent population includes children who are too young to work and adults who have reached retirement age.

A dependency ratio can be worked out from a population pyramid.

Assumptions

A dependency ratio is usually based on the following assumptions. (You may think that they are not completely true in all countries!)

- Children below the age of 16 do not work for a living. (In the European Union, however, children are regarded as dependent up to the age of 19.)
- People over the age of 60 do not work for a living. (Note that the 'official' retirement age in many countries is different for men and women. However, many people retire before the official age, while some continue to work after it.)
- People between 17 and 60 form the economically active population. They have to work to support the rest, who are known as the dependent population.

The ratio can be worked out using a simple formula:

$$\text{Dependency ratio} = \frac{(\text{16s and under}) + (\text{over-60s})}{(\text{17–60s})}$$

Your results from Task 4 on this page should show that the UK has an ageing population. An increasing number of people (particularly older people) are dependent on the economically active population. This will be one of the biggest issues facing the country in the next few decades.

Tasks

4 a) Use the following figures (showing population in thousands) to work out the dependency ratios for the UK for 1991 and 2021. The ratio for 1971 has been done for you.

	1971	1991	2021 (projected)
Over 60	7,300	8,700	14,000
17–60	31,600	37,300	32,000
16 and under	13,400	10,800	8,000

In 1971:

$$\textit{Dependency ratio} = \frac{13{,}400 + 7{,}300}{31{,}600}$$

$$= \frac{20{,}700}{31{,}600}$$

= *0.66 (which means that each worker had to support two-thirds of a dependant)*

b) How does the dependency ratio change between 1971 and 2021? How can you explain the changes?

5 Copy and complete the diagram below to show some of the possible results of an increasingly ageing population. How might it affect each of the categories?

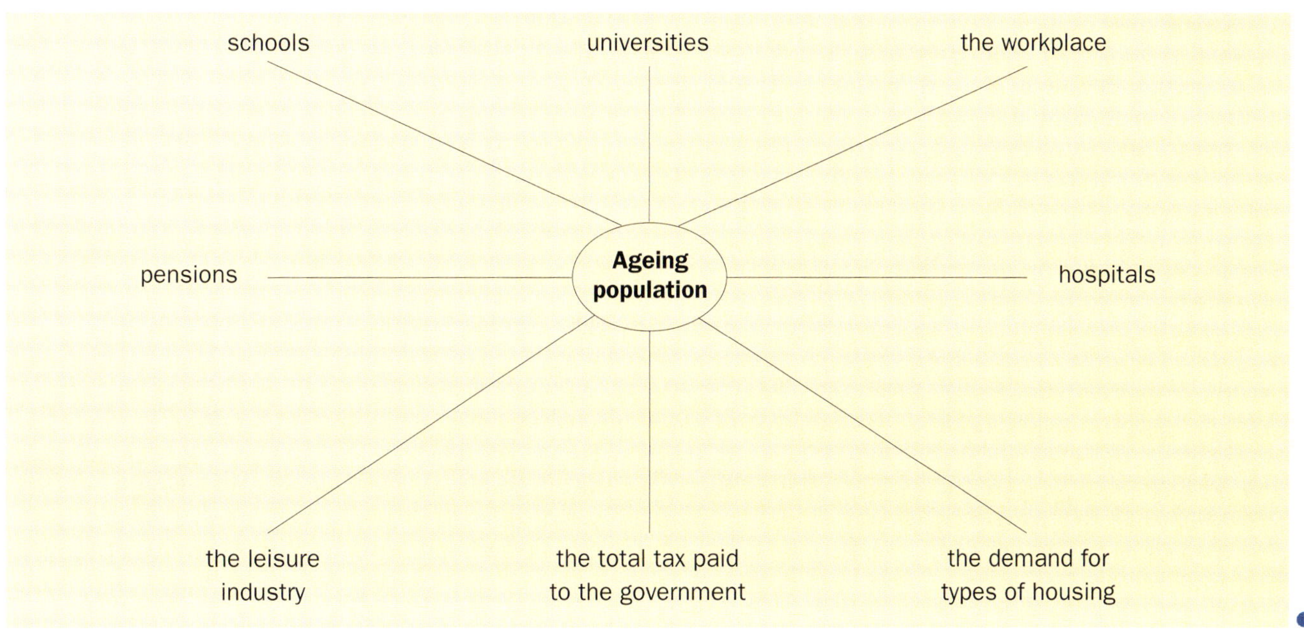

Case study: Ageing population in Europe

In March 1996 the European Union produced a report on predicted population changes in the current EU countries (Map 33) over the next 30 years. Their predictions included the following points:

- There will be an extra 37 million people aged 60 and over – an increase of nearly 50 per cent.
- The number of pensioners will rise to 111 million, or one-third of the EU population.
- The working population (aged 20–59) will shrink by 13 million.
- For the first time in recorded history, the over-60s will outnumber the under-20s.
- There will be three times as many over-80s as there are now.
- There will be 9 million fewer children and teenagers – a 10 per cent decline.

Note that the EU defines dependency as those aged under 19, and over 60. The EU totals are given in Figure 34.

Table 35 shows how the different cohorts of the population are expected to change in the EU, from 1995 to 2025.

There are two main causes of the changing patterns.

- **Falling birth rate** – a fall in the average number of children born to each woman. Note that in most of Europe the birth rate fell rapidly between 1945 and 1970. Now it seems to be settling down just below the rate that is needed to keep the total population stable. However, in Italy, Spain and Ireland the birth rate did not start falling until the late 1970s. The age cohorts born since the birth rate fell reach maturity from the late 1990s onwards. The birth rate fell later in Italy, Spain and Ireland than in the rest of Europe because:
  – they urbanised later than most other countries in western Europe. This meant that until recently there was an economic advantage for most people in having large families
  – they all had strong Roman Catholic traditions, and the Church was strongly opposed to artificial contraception.
- **Falling death rate** – life expectancy has increased in all countries of the EU (Table 36). In the past, increased life expectancy was chiefly due to a decline in infant mortality. Now the increase chiefly involves the over-60s. The main causes of the change have been:
  – advances in the treatment of cardiovascular (heart) diseases
  – advances in the treatment of cancer
  – changes in lifestyle, especially better diets, reduced smoking, more regular exercise and more frequent medical checks.

33 The countries of the EU

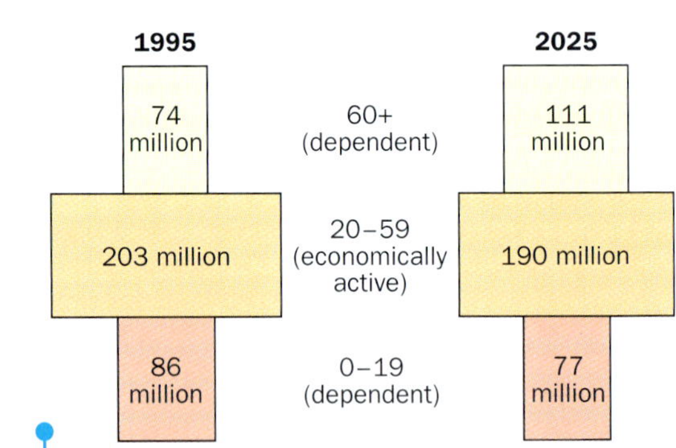

34 Dependent and economically active populations in the EU, 1995 and 2025

So far these changes have affected the better off more than the poor, because they have better access to medical treatment and also have a healthier lifestyle. If life expectancy of the poor is increased in the future then the death rate will fall still further.

Life expectancy for women is greater than for men in all countries of the EU. The average woman lives six years longer than the average man. Many reasons have been suggested to explain this, but the main cause is still not clear. What reasons can you suggest?

	0–19	20–59	60+
Austria	−11.4	−3.1	56.4
Belgium	−9.8	−6.1	46.1
Denmark	2.6	−3.0	41.5
Finland	−5.2	−4.1	66.6
France	−6.1	0.2	57.7
Germany	−12.1	−13.5	51.2
Greece	−2.1	0.5	41.4
Ireland	−25.1	2.7	67.7
Italy	−19.4	−15.2	40.7
Luxembourg	22.2	14.3	70.4
Netherlands	−1.2	−1.2	79.5
Portugal	−8.8	5.0	34.1
Spain	−17.2	−4.4	45.3
Sweden	1.2	3.7	38.1
United Kingdom	−8.2	−2.8	43.6

35 Predicted percentage change in population in the EU, from 1995 to 2025, by age cohort

Tasks

1 a) Use Figure 34 to work out the dependency ratio (DR) in Europe in 1995 and 2025. Use this formula:

$$DR = \frac{\text{(population 0–19) + (population over 60)}}{\text{population 20–59}}$$

b) What do the figures tell you?

SAVE AS . . .

2 Draw two maps to show the changes in the population shown in Table 35:

a) Draw two outline maps of Europe.

b) Complete one to show changes in the percentage of people aged 0–19. Add a key.

c) Complete the other to show changes in the percentage of people aged 60 and over. Add a key.

d) Look for patterns on your two maps. Do neighbouring countries show similar rates of change? Write a paragraph to describe the patterns on your maps.

3 What is likely to happen to the birth rate in Italy, Spain and Ireland from the late 1990s onwards? Explain your answer.

4 How can the life expectancy of poor people in EU countries be improved? Using your own experience make suggestions under the following headings:

- Access to healthcare
- Improved housing
- Changes in lifestyle (throughout life, not just in old age)
- Improved working conditions
- Better pensions

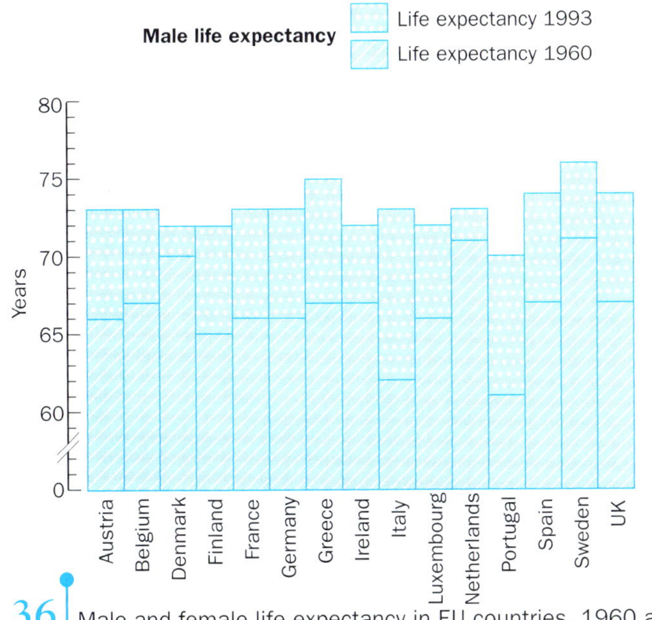

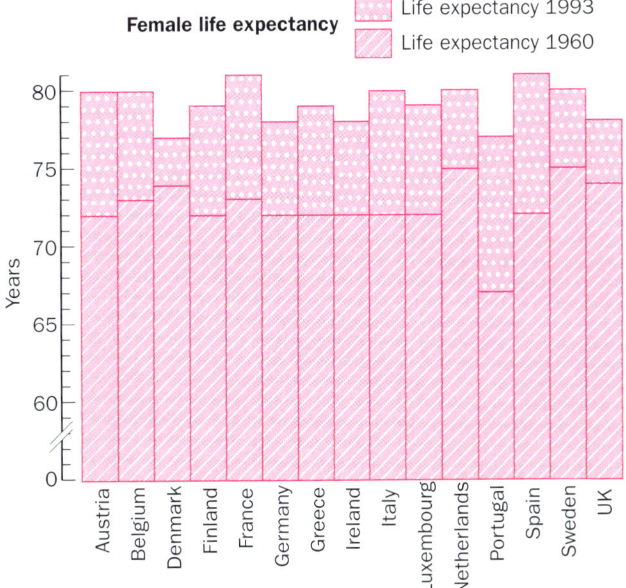

36 Male and female life expectancy in EU countries, 1960 and 1993

How can Europe deal with an ageing population?

Stage 1 **Retired but fully active and independent:**
Mr Thomas – a healthy 65 – has no significant health problems. He can still run a half-marathon and spend all day gardening if he wishes. He takes five holidays a year.

Stage 2 **Less active and less independent:**
Mrs Simpson, aged 75, has arthritis and she is unable to walk very far. She can care for herself in her own home – cook her own meals – but she needs help carrying anything heavy, getting around outside the home or going to the shops. She pays a home help to come and do her heavy housework for her for one hour per day and her son pops in every weekend to help out as well.

Stage 3 **Totally dependent:**
Mr Mann, 85, has had a stroke and is unable to walk or talk. He lives in a nursing home where he is cared for by professional carers who make sure he gets his medicines on time and that his quality of life is as good as it can be. He watches a lot of television although he finds most of it very boring and wishes he could be with his family.

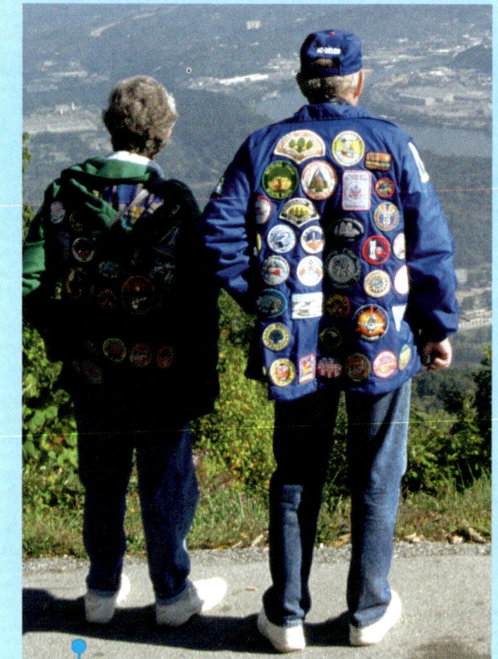

37 'Since our retirement we've been everywhere'

There are many different stages of ageing. At each stage of 'old age' people have different needs. These different needs present some opportunities and some problems for older people themselves, and for society as a whole. As the population of the EU becomes increasingly elderly, there are a number of issues that must be managed.

Governments and other policy makers have not worked out exactly how they are going to deal with many of the issues yet. So now it is your turn to have a go.

Focus task

1 Imagine that you are a delegate at a conference on the theme of Ageing Populations. Your task is to decide what issues are raised for an MEDC by an increase in an ageing population. Divide into four groups. Each group should discuss one of the Briefing Papers on pages 171–73. Questions to help you to structure your discussions are provided alongside each Briefing Paper. Each group will need to appoint a chairperson, someone to record decisions, and someone to present the group's Outcome Paper to the class.

Stage 1 brainstorm: simply brainstorm your ideas, recording them as a spider diagram.

Stage 2 choose some of the ideas from your brainstorming session and discuss them in detail.

Stage 3 write your Outcome Paper: select a maximum of three specific recommendations. Write a paragraph about each recommendation. Explain who needs to take action on it.

Your recommendations should be supported by evidence and examples. These can come from this book, from research, or from your own experiences. Here are some useful websites for research:

Age Concern: www.ace.org.uk
University of the Third Age: www.u3a.org.uk

38 The state provides a basic pension

Briefing Paper 1 – The 'Grey Pound'

Older people form a growing market for goods and services. Some older people have a great deal of disposable income. This comes from private pensions and possibly from selling a large family house and moving into a smaller one. On the other hand many older people have a fairly limited income.

Most old people have plenty of time. They are not tied by the traditional timing of the working day and holiday periods. Many people want to take up new interests or to develop new skills.

In the early stages of old age many people are still fit and active, although their fitness is likely to decline as they reach later old age. In the early stages of old age, travel and extended holidays are often a priority. Later, people need opportunities for leisure activities close to home. They may also need to adapt their homes to suit their decreased mobility. At each stage of their lives people are a market for goods and services which improve their lives, and which provide income and employment for others. People's needs do not cease in old age – they simply change.

Tasks

1 Older people provide a large market for leisure and other products. Consider how older people's spending power and their free time might affect the products and services offered by the following industries:

- the holiday industry at home and abroad
- education
- libraries
- sport and leisure clubs
- pharmaceutical and medical equipment industries
- transport services
- other.

2 Write up your three key recommendations.

Briefing Paper 2 – Pensions

Most countries have an official retirement age of between 60 and 65. Retired people are paid a pension by the state. The money for this comes from taxes paid by people who are still working. It usually provides only a very basic standard of living. Some people who receive only the state pension have to claim extra social security payments in order to have enough money to live on. However, some people do not claim this money, for all sorts of reasons.

Some people have an additional pension. This comes either from an 'occupational pension' or a 'private pension'. With occupational pensions the worker and employer both pay money into a pension fund. Later the pension is paid out of money in that fund. Private pensions come from money that the individual worker has saved with a bank or insurance company. However, people who have had low wages or irregular employment often have not been able to afford an additional pension. Women are often particularly badly off in this respect. New 'stakeholder' pensions should help people in those groups to save for old age.

Other benefits are available for older people. These include free prescriptions, help with public transport costs, help with fuel costs, free TV licences, and so on.

Tasks

1 Discuss the questions below. Make notes on the key points in your discussion.

- Why is it a problem if there are more pensioners and fewer people working and paying tax?
- Should the retirement age in the UK be set later?
- Should the retirement age become more flexible?
- Should the state pension be reduced to cut costs? How would this affect older people?
- Would this be acceptable? Remember that as the over-60s increase in number they have more voting power.
- If the state pension were cut in value, how could people be persuaded to have private pensions?
- How can people on low wages be given help to save for private pensions?
- Does it matter if the divide between the 'older rich' and the 'older poor' gets wider?

2 Write up your three key recommendations.

Briefing Paper 3 – Housing an ageing population

39 Sheltered housing provides a supportive community

As people grow older their housing needs change. In particular, many couples no longer need large family houses when their children leave home, so some people decide to sell up and move to a smaller house.

When people give up work they no longer need to live close enough to their work to travel there every day. This may mean that they can move to a more pleasant area. They might be attracted by:

- climate
- scenery
- open space
- more people of the same age/interests.

On average, women live six years longer than men. This means that there is a growing number of elderly widows, often living alone. However, some men outlive their partners, so there is a smaller, but growing number of single elderly men. As people are left alone, their housing needs may change again. As they become less independent they may need to move into sheltered accommodation (see below) or into a care home.

It is worthwhile supporting older people so that they can stay in their own houses for as long as possible. This is cheaper than their moving into an old people's home, both for the individuals concerned and for the state. Most people are also much happier staying in their own homes for as long as possible.

Briefing Paper 4 – Who cares for older people?

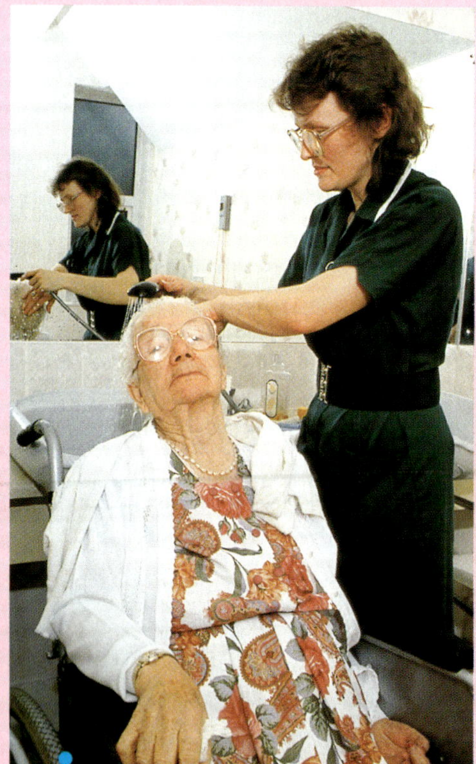

40 Carers provide a range of support services

Many people lead a full and active life well into their 70s. Inevitably, though, a time comes for many older people when they need to be cared for.

At first many can continue to live in their own homes, so long as they have support. Some people move into 'granny flats' in their children's houses and are cared for by their own families. Others move into sheltered accommodation where a warden keeps an eye on a group of flats or houses and where people can live more or less independently.

As people become older and less able to look after themselves, many move into a residential home. Some of these just offer general care. Others offer a high level of medical care from trained nurses.

Whether the elderly people live alone or with their family, they need a growing range of support services, which might include:

- health visitors
- meals on wheels
- home-help cleaners
- drivers for hospital visits

and so on

In addition, unpaid family carers need more support than they get at present (emotional support, training and money).

An ever growing number of people are needed in the caring professions. The need for trained carers in the EU will peak in 2020. While some carers are volunteers, many are employed full or part time. To attract enough carers, jobs must be well paid and rewarding.

Tasks

1 Discuss the questions below. Make notes on the key points in your discussion.

- As the population ages, should we expect a large-scale movement from some parts of the UK to others?
- Should we encourage this or discourage it?
- Might some older people move abroad? If so, who would be more likely to move?
- What new housing will be needed? For couples? For single people?
- How could large family houses be adapted and used?
- Where should old people's housing be built: close to the communities where people have lived all their lives, or elsewhere in quiet, secluded areas?

2 Write up your three key recommendations.

★ **THINK!**

Use your own knowledge

This task will work best if you use what you have learned from your everyday life. When you come into a Geography lesson, don't forget all the things that you already know about human behaviour and the world around you. You are surrounded by geography, so use your experience to illustrate geographical ideas. You'll be able to write more confidently about the case studies that you have found out about in your ordinary life than the ones you have learned about from a textbook or from the internet. However, in your exam, you must remember to explain your own case-studies by referring to the ideas and understanding that you have gained during your Geography course.

Tasks

1 Discuss the questions below. Make notes on the key points in your discussion.

- How do people's needs for care and support change as they get older?
- How can governments recruit enough carers when the population in the economically active 20–59 age range is falling?
- Carers need training. How can governments provide training?
- What will happen if there aren't enough professional carers?
- Would increased immigration solve this problem?
- Would immigration of working-age people badly affect their home countries?
- What would be the advantages and disadvantages of providing payments or other forms of support to people who look after their own elderly relatives?

2 Write up your three key recommendations.

SAVE AS . . .

1 When you have completed the work on the conference, collect copies of all four 'Conference Outcome Papers' and save them in your notebook.

2 Using these papers, write two newspaper articles under the headlines:

   a) 'More old people will cause problems, warns Government Minister'

   b) 'More old people will bring great benefits, says Age Concern'.

   The following websites could help you with data, pictures or diagrams to support your articles. For data on ageing populations, log on to:

   www.undp.org/popin/wdtrends/a99/ba99pwld.htm

   For maps and diagrams try:

   www.undp.org/popin/wdtrends/a99/a99cht.htm

4.2 What are the causes and consequences of migration?

DO YOU KNOW?

- 'Migration' is a movement of people from one place to another.
- When someone leaves a country it is called emigration. When a person enters a new country it is called immigration.
- It's easy for people to migrate within the EU, but most EU countries are introducing ever-stricter laws making immigration from non-EU countries more difficult.
- Most of the people now seeking entry into the UK come from eastern Europe.

AIMS

- **To consider the problems faced by people who live in MEDCs and who migrate mainly out of choice.**
- **To identify reasons for migration to Germany.**
- **To classify different types of migration.**
- **To understand the factors causing migrations in LEDCs.**

Sometimes psychologists try to list the events in a person's life that cause most stress. At the top of their list they always include:

- death of a loved one
- divorce
- being sacked or made redundant
- having a baby.

Almost equally stressful is:

- moving house.

Why is moving house stressful?

Many of you have probably moved house during your life. Perhaps some members of the class have moved several times. Here are some of the things that may have caused you stress. Perhaps you can add other things to the list:

- Having to pack up all your possessions, and then unpack them.
- Starting at a new school.
- Leaving your old friends and making new ones.
- Having to decorate your new room.
- Having to get your pets used to their new territory.
- Leaving behind places where you were happy.

These things that make moving difficult can be described as 'anchors' and 'barriers'.

- ***Anchors*** are things that attach people to their old home and make it difficult to leave the house and the area.
- ***Barriers*** are things that make it difficult to move to a new house and area. They are also the things that make it difficult to settle into the new area.

1 Anchors and barriers to moving

If moving house is so stressful then INTERNATIONAL MIGRATION – movement from one country to another – must be even more stressful. And the stress on people who are forced to migrate, as refugees, must be enormous.

2 Albanian refugees arrive in south-east Italy, hoping to gain entry.

Tasks

1 Make a list of six anchors and five barriers that make moving difficult. There are some clues on these two pages, but try to add some anchors and barriers of your own.

2 Draw a cartoon, similar to Figure 1, to illustrate a migration or move that you, or someone in your family, has experienced.

ICT

3 The UK has a tradition, going back several centuries, of giving protection to refugees who have been forced to flee their own countries. Many of these refugees have made valuable contributions to our economic and cultural development. However, the number of refugees varies from year to year. Sometimes MPs and other people become concerned about the number of refugees arriving in the country.

a) Imagine that you are working as a research assistant for an MP. You have been asked to find out the facts about the number of refugees seeking asylum in the UK. You should present these facts as a series of charts or graphs annotated with your own comments on the information.

b) Log on to the government website: www.statistics.gov.uk. Copy and paste relevant statistics on the number of refugees into a spreadsheet such as Excel. Use the data to produce a series of graphs showing changes over time. You could try to use stacked bar charts or line graphs.

Copy and paste your charts or graphs into a Word document and add your own comments to them.

There's more to migration than push and pull!

'Pushes' are bad things about an area that make people want to leave. 'Pulls' are the good things that attract them to a new area (see pages 184–86).

Once you think carefully about the stresses involved in moving, you will realise that the PUSH–PULL MODEL of migration tells only part of the story. The pushes and pulls have to be more than slight problems or attractions. They must be strong enough to overcome the anchors and barriers, which can be very powerful. For example, the refugees in Photo 2 are obviously in a very difficult situation which is causing them great suffering, yet they made the choice to migrate, knowing that they would face great problems. Think how serious the pushes must have been to make them choose to migrate.

Peggy's Story

A case study of migration in Germany

Peggy and Ella Schön are sisters who were born in Leipzig, East Germany, in 1965 and 1967 respectively. Their father and mother, Hans and Ilsa, worked in an engineering factory and lived in a flat in one of the large housing blocks built to replace houses destroyed by bombing during the Second World War. Both parents enjoyed listening to jazz music, and both girls were named after American jazz singers. Some of Ilsa's relatives lived in West Germany but the family were never able to visit them because of the strict border controls.

3 East German housing blocks

When the sisters went to school they both discovered they had special talents. Peggy was excellent at languages and Ella was a very talented gymnast. Ella was given a place at an elite training school in Berlin (see Map 6 on page 178) when she was 7. Here her talents could be developed for the benefit of the country, and herself. Later Peggy was offered a place at a special language school, also in Berlin. Both girls lived in boarding school during the week but usually travelled back to Leipzig at weekends. Their parents had been given a larger flat as a reward for the girls' talents.

Peggy did extremely well at school and went to the university in East Berlin to study English and Russian. Ella also did well, but it was realised that she would never make the national squad, so she transferred to the state school for circus skills. She used her gymnast's training to become an acrobat and contortionist. The East German government was keen to develop all talents! By the age of 16 she was given a job with the State Circus. She travelled all over East Germany and also performed in Poland and Czechoslovakia. She even travelled to perform in Russia twice. She was rewarded with a modern flat in Berlin.

German reunification in 1989 presented big opportunities for both Peggy and Ella. Peggy had completed her degree in East Berlin, so now she transferred to the university in West Berlin and did a course that linked languages with computer programming. Two years later she was very well equipped for a variety of work with transnational computer companies. She moved to Munich, attracted by the demand for people with skills like hers and by the beautiful climate and the closeness of the ski slopes in the Alps.

4

Ella developed a new stage act. She worked in theatres and nightclubs in the West, increasing her income enormously. However, in 1992, at the age of 25, she married a theatre producer, and settled in a modern suburb of West Berlin to start a family. She was too old for the life of an acrobat and contortionist.

Hans and Ilsa were disappointed by German reunification. The factory where they worked was soon taken over by a bigger firm and many workers, including Hans and Ilsa, were laid off. They were invited to go and live with Ella and her husband, to look after the grandchildren, but they found it difficult to settle in an area that was so different. They soon moved back to Leipzig, but that had changed too. Their state-owned flats had been privatised. They bought their own flat, but many of their younger neighbours sold their flats and moved west to look for work. The old close-knit community had broken up. After five years of doing odd jobs, Hans realised that his labour skills were not needed any more. He and Ilsa accepted pensions from the government. They live quite comfortably, but Hans wrote to Peggy: 'Leipzig has changed for the worse, but it is still home. We are too old and set in our ways to come and live in the West. Reunification was great for young people, but very sad for our generation.'

Peggy now works for a top German bank. She travels all over Eastern Europe arranging loans for companies that are trying to modernise and compete on the world market. She hopes to marry her boyfriend, Geert. He is a Dutchman who works for a transnational electronics firm based in Munich. Between them they speak seven different languages and they think of themselves as citizens of Europe. They like living in Munich but they do not feel tied to the place as Hans and Ilsa are tied to Leipzig. They expect that in the future they may move anywhere in Europe. Peggy finds that prospect really exciting after growing up under such strict control in East Germany.

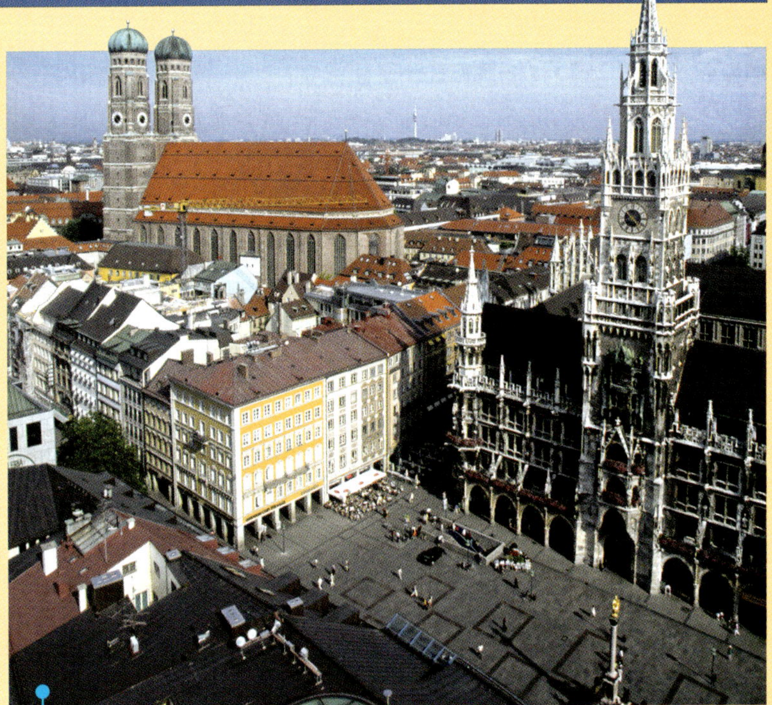

5 Munich

Tasks

SAVE AS . . .

1 Read through 'Peggy's Story' again. Look for pushes and pulls, anchors and barriers which affected the four moves, **A–D**. (If your teacher gives you a copy of the story, you can underline the different factors in different colours.)

A Peggy and Ella's move to school in Berlin.
B Peggy's move to university in West Berlin.
C Hans and Ilsa's move to live with Ella and her family.
D Their return to Leipzig.

Complete a diagram like this for each move, adding thoughts of your own to what you read.

Push factors	**Pull factors**
....	....
....	....
....	....
....	....
Anchors	**Barriers**
....	....
....	....
....	....

2 a) Geert and Peggy claim that there are very few barriers to stop them from moving anywhere in Europe. Explain why they say this.
b) Peggy may have children at some time in the future. Will this be a barrier or an anchor making migration more difficult? Explain your answer.
c) Until 1989 Peggy's mother, Ilsa, could not even visit her relations in West Germany. Why not? (In your answer, refer to anchors and/or barriers.)

How has Germany changed?

In order to understand and get the most out of Peggy's Story on pages 176–177, it is necessary to understand something about the changes in Germany since the end of the Second World War.

In 1945 Germany was in a desperate economic state. Much of its economy had been destroyed during the war, and many people had died. In addition, the country was split into two separate sections by the occupying armies. The border between the two sections cut roads and railways; divided families, villages and farms; and disrupted many aspects of German life. East Germany was occupied by the former Soviet Union, and a COMMUNIST GOVERNMENT came to power. West Germany was occupied by France, the UK and the USA and soon a DEMOCRATIC GOVERNMENT was elected. The two parts of the country developed in very different ways.

6 A divided Germany

West Germany

By the end of the 1950s, West Germany was enjoying what is often described as an 'economic miracle'. With the help of aid and support from the USA, and with hard work by the people, the economy was rapidly being rebuilt. Housing and the infrastructure of roads, railways, etc. had been repaired or replaced; the coal, steel, chemical and heavy engineering industries had been renovated; and modern industries such as electronics, motor cars and consumer goods were developing rapidly. In fact, the fresh start following the destruction helped West Germany to become a very successful, modern, industrial power. The standard of living of the people was improving rapidly too.

East Germany

East Germany followed a very different route. This sector of the country was in a much more difficult position at the end of the war. Its workforce was mainly employed in old, heavy industries, without the variety that was found in the West. The Soviet Union was not able to provide the aid that the USA gave to the West. In fact it even dismantled some German factories and took them to Russia to replace factories destroyed by the Germans in the war. As a result, the economy recovered more slowly than in the West. It was not modernised but continued to rely on heavy industry.

The standard of living did not rise as fast as in the West, but the government did try to make sure that all of the people were well looked after. No one was unemployed, although this meant that industry was over-staffed and less efficient than in the West. The government also made sure that everyone was housed and that basic pensions were paid to all old people – but it did not encourage individual enterprise and initiative like the market economy in the West did.

7 East German industrial area

8 Cars on a robotic welding line in West Germany

Why did West Germany encourage immigration?

The rapid economic growth in the 1960s in the West led to a problem of labour shortage. The government there encouraged East Germans (or 'Ossies') to come to the West, partly because it still wanted Germany to be one country, but also because it needed people for their labour and skills. This worried the East Germans, because it was usually their most skilled and dynamic citizens who left. So, in 1963, the authorities built a huge wall across Berlin, and a fortified fence all along the border between East and West Germany, to stop people from 'escaping' to the West. The only people who were allowed to leave were those who were too old or too ill to work.

The pull of West Germany's economy attracted workers from many other countries. Unemployed Britons went for temporary work as builders in the 1960s and 1970s, but the biggest source of workers was Turkey. Pushed by a poor economy and a growing population, over 2 million Turkish people went to work in West Germany. Unlike migrants from East Germany, they were not allowed citizenship and could not become permanent residents.

Very few migrants arrived from Germany's former colonies in south Asia and Africa. However, several thousand people migrated to the UK or France from their former colonies and then moved on to Germany.

The collapse of communism

As West Germany continued to grow and East Germany did not, the difference in their standards of living became wider. By the late 1980s the pushes and pulls had become enormous. East Germans were allowed to travel for holidays in other communist countries. Some of these countries, especially Czechoslovakia, had started to relax their border controls and allowed people to visit the West. Thousands of Ossies 'went on holiday' and then crossed into West Germany, migrating to find work. The old-fashioned Trabant cars, loaded with possessions, came to symbolise this flood of migrants.

9 Migrants, in a queue of Trabant cars, arrive in Bavaria

Back in the East, peaceful mass protests were taking place as people called for a change of government. Suddenly, in 1989, the communist government collapsed. The Berlin Wall was attacked and broken down, and soon after this the two parts of Germany were reunited.

10 The fall of the Berlin Wall

Inequalities

At first the Ossies were delighted by their new freedom, and many expected to become rich overnight. Many people moved West. Later, things became more difficult because the old East German factories could not compete against modern Western competitors. Many workers lost their jobs as factories were taken over and modernised, or else closed down. In fact many had to be closed because they were producing so much pollution.

Germany today is still 'a country divided'. Although massive development has taken place in some parts of the East, especially around Berlin, other parts still suffer from high unemployment rates and very poor housing.

However, despite its problems, Germany as a whole is seen as a very attractive place by many people from poorer countries to the south and east. Map 11 shows the flow of migrants to Germany from the rest of Europe and gives the reasons in broad terms. Germany has over four times more immigrants arriving each year than the UK. Every new economic problem or social conflict in eastern and southern Europe provides a push which brings a new wave of migrants to Germany. They continue to be pulled by the strong German economy, and will do whatever is necessary to overcome the barriers placed in their way.

Unfortunately, one of the barriers to immigration is the attitude of some Germans. There is a small but vocal Neo-Nazi movement in some areas. It is especially strong in poor areas of east Germany. These groups do their best to discourage immigrants. However, the government and police are trying to control the Neo-Nazi movement.

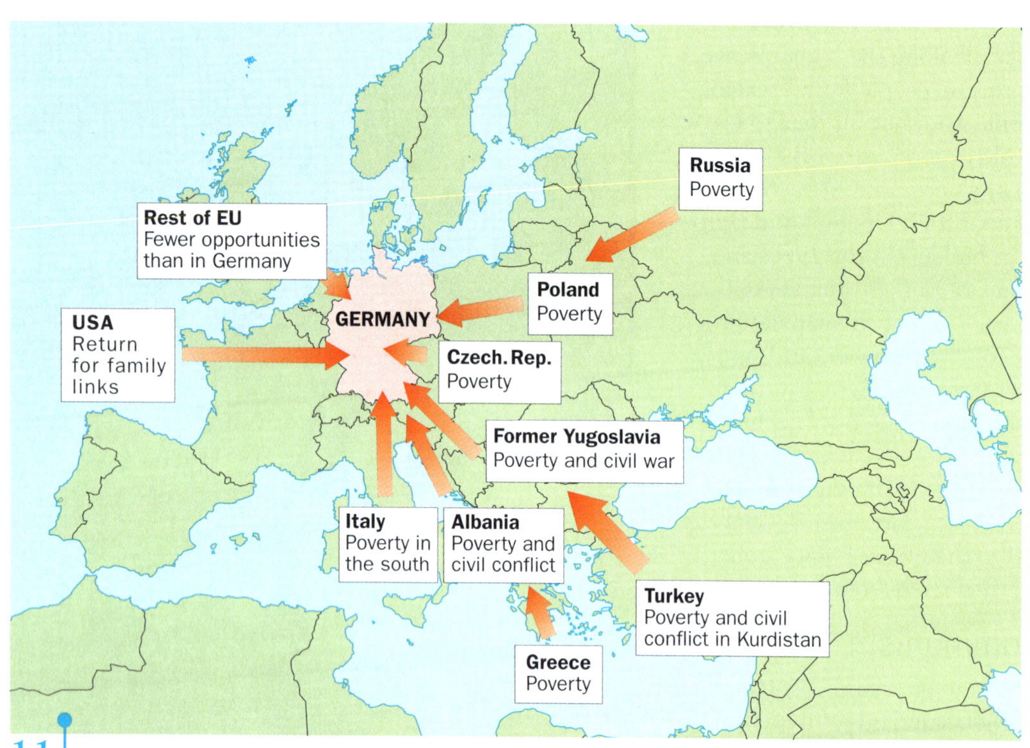

11 Migration into Germany, 1960–2000

12 A Turkish area of Berlin

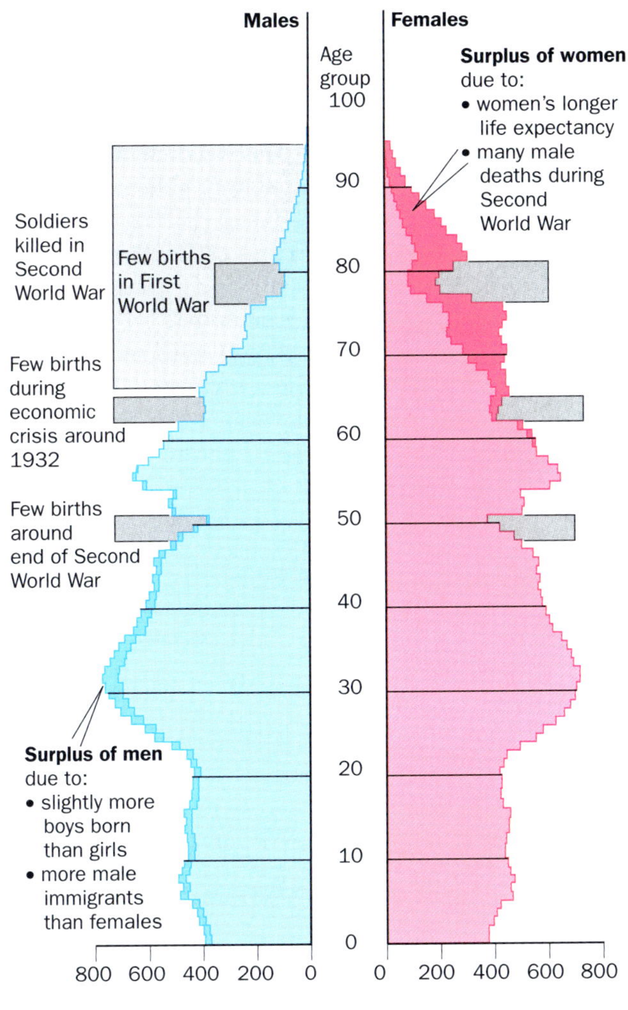

13 Population pyramid for Germany, 1995

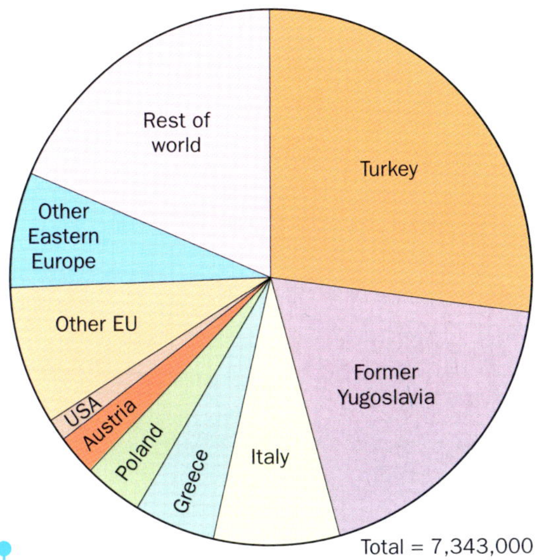

14 Foreign citizens living in Germany in 1995

Tasks

1 According to pages 178–80 what seem to be the main pulls attracting migrants to West Germany through the period since the Second World War?

2 Study Figure 13, and the facts below:

- Germany's birth rate in 1997 was 10‰. The death rate was 11‰.
- In 1997 the total population was 82 million. It is expected to be 81.7 million by 2010 and 76.2 million by 2025.

Explain how all of these figures might affect the government's attitude to immigration.

3 Figure 14 shows the nationalities of foreign citizens living in Germany in 1995.

Describe some of the pushes that might have made migrants move to Germany from Turkey, the former Yugoslavia, Italy and Poland.

Focus task

4 How has migration into Germany affected different groups?

Study all the text, diagrams and photographs on pages 178–81 and think carefully. Suggest how immigration into Germany has affected:

a) the migrants who have been successful in their new country

b) migrants who do not seem to have been successful

c) rich Germans who may have employed migrants in their businesses or as domestic workers

d) poor Germans who may have had to compete with the migrants for unskilled jobs

e) the people left behind in the migrants' country of origin, who may have had money sent home to them, but who miss their close relatives and friends.

Refer to cultural changes as well as to economic changes. Suggest some of the consequences for the different groups of people living in a multi-racial society. Re-read 'The migration revolution' on page 158 to remind yourself of one cultural change that has happened.

Classification: Is migration all one type?

What causes people to migrate, and what are the consequences? To be able to analyse and understand the causes and consequences of migrations, it helps if you *classify* the migrations. (You will find it is the same with other classifications – making them will help you to structure your answers in an exam.) You can use the questions below to help you to classify different types of migration.

? QUESTION 1 *Is the migration forced or voluntary?*

If the main reason for the movement is a push from the old home, the migration can probably be described as FORCED MIGRATION. If the main cause of the move is a pull from the new home, it is probably a VOLUNTARY MIGRATION.

For instance, in May and June 1999 many inhabitants of Kosovo fled from the country. Their homes were being burnt by Serb police and armed gangs. This was a case of forced migration. In 1997 people left the island of Montserrat because the eruption of Mount Soufrière was threatening their lives. This was also forced migration, but forced by a natural cause rather than a human cause.

15 Destruction from the eruption of Mt Soufrière in Montserrat

However, when a family called Barlow (see page 224) moved from Tottenham in North London out to Chelmsford in Essex, their move was entirely voluntary. They wanted a better quality of life in the countryside. Mr Barlow wanted to be able to travel more quickly to work and the two children wanted to be able to go out more on their bikes, away from traffic.

This migration can be shown on a *bipolar* diagram like the one below. A bipolar diagram is one with two 'poles' or ends. To use it, place a circle somewhere on the line to show whether you think the move was forced or voluntary, e.g. if you think the move was more forced the circle will be nearer that end of the line. If more voluntary, then it will be closer to the word 'Voluntary'.

Forced • • • • • • • • • **Voluntary**

The Barlows' move is shown like this:

Barlows' move

Forced • • • • • • • • (•) **Voluntary**

? QUESTION 2a *Is the migration temporary or permanent?*

Some migrants plan to stay in their new home for ever. This is PERMANENT MIGRATION. Others know they will live there for only a certain length of time. For instance, some of you will go to university in a few years' time. You may go to study in a town away from home, and then leave the university town when you graduate. This is TEMPORARY MIGRATION.

Other people migrate to earn very good wages for a few years. For example, many oil workers go to countries such as Saudi Arabia for a period of some years. They do not sell their house in the UK because they know that they will be coming back to their real home.

If the migration is temporary, you can ask another follow-up question:

? QUESTION 2b *Is the migration seasonal?*

SEASONAL MIGRATION used to be common in mountain regions of Europe, like the Alps and the Pyrenees, where there was a seasonal movement (called 'TRANSHUMANCE') of people and their animals to fresh pastures

Now this tradition has almost disappeared but a new type of transhumance has developed. In winter many people travel to Alpine ski resorts looking for work in hotels or on the ski slopes. They live in the ski resorts all through the winter then, when the snow disappears, they migrate to seaside resorts around the Mediterranean for the summer. Ski instructors may get work as lifeguards, and chalet attendants find work in hotels. These migrants follow a seasonal pattern which depends on climate, just as the herdsfolk used to do.

You can use a bipolar diagram to show how temporary or permanent a migration is. A worker in a ski resort is shown on this diagram:

Temporary • • • • • • • • • **Permanent**

Seasonal winter ski worker

QUESTION 3 *Is the migration international?*

International migrants move from one country to another. INTERNAL MIGRANTS move from one place to another in the same country. The main differences between international and internal migration are very important. The barriers to international migration are usually much greater. They can include:

- greater distance, and so greater cost
- differences in culture, language, customs, etc.
- the need to obtain visas, work permits, etc.
- insecurity; e.g. foreign migrants are often expelled if there is an economic slump, or deported for minor crimes
- difficulties travelling home to visit friends and relations
- possible racism or other forms of prejudice.

Many poor Mexicans cross the border into the USA to look for work. They often cross illegally. They may have to pay guides to show them the best points to cross. The border is patrolled by armed guards, with dogs, helicopters, mobile searchlights, etc. If the migrants are caught they may be imprisoned or deported. American employers can get away with paying the migrants very poor wages. If the migrants complain, the employers threaten to report them to the immigration authorities. The push of poverty and the pull of American jobs and wages must be very strong to make people face all these barriers.

The barriers for the Mexicans are especially great because they are moving to a country where they do not have much power or status. Some international migrants do have power and status in their new countries, so they can reduce the effect of the barriers. For instance, British doctors are encouraged to migrate to the USA.

This bipolar diagram shows the migration of Mexicans to the USA; it is international, but to a neighbouring country.

Internal • • • • • • • • • **International**

16 Mexican immigrants in the USA receiving payment for picking green chillies

Task

1 Read Peggy's Story on pages 176–77 again. Complete a 'migration profile' for each move that she made. For each move place a circle somewhere on each line between the two sets of words.

A migration profile

Forced • • • • • • • • • **Voluntary**

Temporary • • • • • • • • • **Permanent**

Internal • • • • • • • • • **International**

★ THINK!

There are three pairs of words in the migration profile: forced/voluntary, temporary/permanent, internal/international. It should be quite easy to learn and remember them. Once you have learned this structure, apply it to each migration that you study and you will find it easier to remember the details.

How is migration connected to urbanisation?

LEDCs are rapidly urbanising (see pages 200–202). The main reason for urban populations growing in the LEDCs is people migrating from the countryside into cities. This happens for many different reasons. Push factors are generally more important than pull factors. The main push which is found in many areas throughout the LEDCs is *poverty*. This affects huge numbers of people all the time and puts them under great pressure to move. In addition, *natural disasters* or *war* sometimes give an extra push.

17 In the Rishi Valley, India, the mechanisation that followed the Green Revolution cut the demand for farm workers

Push factor 1: Poverty

- **Population growth** means that the same area of land has to support more and more people. Over-farming the land can often lead to soil erosion (see page 24), as the soil loses its nutrients and is blown or washed away.
- **Plots of land are too small** In many parts of India, for example, a farmer's land has to be divided equally between his sons when he dies. If farmers have several sons, the land that each one inherits is not enough to support a family properly.
- **Patterns of land ownership** In much of South America, absent landlords own large estates (or *estancias*). Peasant farmers may be allowed to use some of this land in return for working in the owner's fields. However, the peasants know that they can be turned off the land at any time. It is not worth their while trying to improve the soil if they do not have secure rights to the land. They will not benefit from their hard work.
- **Debt** Money is often borrowed from moneylenders at very high rates of interest. In India some small farmers borrowed money to buy the seeds and fertilisers that were introduced by the Green Revolution (see page 62). If the price of their crops fell, or if the crops did not grow, often they could not pay back their loans. Their debt got worse until many were forced to sell their land.
- **Desertification**, or the spread of desert conditions, has occurred in the Sahel region of Africa and other regions. Areas which only just had enough rainfall to make farming possible became over-populated. Farmers tried to farm the land more intensively and this destroyed the vegetation and soil. The problem was probably made worse because rainfall totals were lower than usual during the 1970s and 1980s (see page 122).
- **Disease** is both a cause and a result of poverty. Poor water supplies, lack of proper medical services, malnutrition, lack of insecticides to destroy the carriers of germs, etc. all cause disease. Many people are weakened by the diseases. They cannot work efficiently, so they grow less food and become weaker. This is an example of a 'vicious circle of poverty'.
- **Use of land for CASH CROPS** In many LEDCs, land that was once used for growing food by subsistence farmers has been taken over to grow crops for export to MEDCs. This was often necessary because the LEDC was in debt to the MEDC and had to earn money to pay the interest on the debt (see page 47). However, the loss of subsistence farmland means there is less food for the local people. In Kenya, for example, land that used to grow maize and other food crops is now used for growing mangetout peas which have become popular in Europe.

18 Failed sunflower crop, owing to drought, in Kenya

19 Collecting water from a shallow pit in a dried-up river bed, Somalia

Task

1 Poverty can be a push factor but it can also be a barrier. Explain how.

Push factor 2: Natural disasters

Natural disasters occur only occasionally, but they provide a sudden, very strong push.

Country/Region	Disaster	No. killed	No. of refugees
China	Floods	4,150	180 million
North India/Nepal	Heavy monsoon	3,250	36 million
Bangladesh	Monsoon rains + tropical cyclones	1,300	31 million
Central America	Hurricane Mitch	11,000	7 million
Philippines	Nine typhoons	500	5 million
Vietnam	Typhoons	50	2 million
Indonesia	Drought + fires	not known	1 million
Caribbean	Hurricane Georges	4,000	600,000
Argentina	Floods	20	360,000
Sudan	Floods	1,400	338,000

20 Environmental disasters, 1998

21 Residents of El Progreso, in northern Honduras, evacuate the area amid torrential downpours caused by Hurricane Mitch

Table 20 lists the ten biggest natural disasters in 1998 and the number of people who were affected by them. Note that:

- most of the people affected lived in rural areas
- most left their homes when the disaster occurred and became refugees or migrants
- many of the refugees moved to cities
- many refugees later returned to their land when the crisis was over but a significant number did not. What is more, geographers have discovered that when people have moved once they are more likely to move again, permanently. It seems that the first move reduces the effect of the anchors that tie people to their homes.

The number of people affected by disasters in 1998 was higher than in any previous year. This is not just chance. There are two principal reasons for this:

1. The growing world population means that more and more people are being forced to live in areas that are prone to hazards. Safe, fertile land is just not freely available.
2. The number and severity of hazardous events is increasing. Global warming and deforestation are probably the main reasons for the increase.

Case studies of the following natural disasters are dealt with in other parts of this book:

- droughts (see page 22)
- tropical storms (see page 40)
- floods (see pages 48 and 62)
- earthquakes (see pages 65 and 70).
- volcanic eruptions (see page 75)

22 People flee the village of Bulhagarry in drought-stricken Ethiopia

Push factor 3: War and civil strife

War and civil strife have created countless refugees or migrants. In the late 1990s there were major conflicts in many regions of the world, including:

- Former Yugoslavia (Bosnia and Kosovo*)
- Rwanda*
- Turkey
- Iraq*
- Indonesia*
- Kashmir (between India* and Pakistan*)
- Afghanistan*
- Colombia
- The Horn of Africa (between Ethiopia* and Eritrea*)
- Sudan*.

All of these conflicts have created refugees. Sometimes towns are attacked and refugees flee to the countryside, but that is usually a temporary movement. When the threat is over they usually return to the city, or move to another city.

When fighting takes place in rural areas, many people are displaced. They often move to cities, or to refugee camps. When the fighting is over many people find it impossible to return to their land. Sometimes it has been taken by their enemies; sometimes the hatreds that caused the fighting still exist, so to return is dangerous; and often the land is dangerous because of land mines, unexploded bombs, or other forms of 'war pollution'. In fact each war creates more refugees, who often end up moving to other cities in order to survive.

23 In 1999 tens of thousands of refugees returned to Kosovo from Albania, ignoring warnings of mines and booby traps in their haste to go home

Pull factors

There are pulls that attract people from rural areas to cities. There is one pull that is far more important than all others: the opportunity to make a living. Opportunities for education and for enjoying 'the bright lights' of the city are also quite important in some cases – but survival is the first priority. You will return to this on page 200.

Tasks

SAVE AS . . .

1 a) Mark on a map of the world the countries with the ten biggest natural disasters in 1998 (see Figure 20). Then, using a different colour or symbol, mark on the same map the countries with major wars.

b) Describe the pattern shown by your map.

c) Explain why this pattern has developed.

2 Some countries in the list of wars are marked with an asterisk (*). They all have a very high rate of population growth – higher even than the average LEDCs. Suggest why areas with very high population growth are more likely to be involved in wars.

3 Suggest why all of the countries marked on your map have a very high rate of rural-to-urban migration.

Review: why is population changing?

Pages 187–88 bring together many of the ideas you have looked at in this unit, 'Population'.

You first read these characters' opinions on page 147. Now they are going to explain their views in more detail.

My name is Hope Matatu. I'm from Zimbabwe. I'm a trained community nurse.

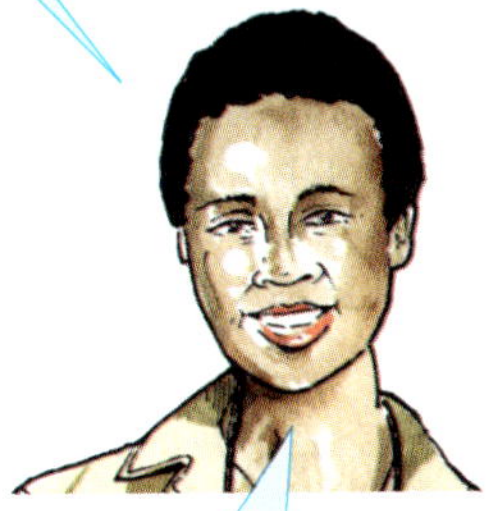

My name is Blanca Garcia. I come from Seville in Spain. I'm a midwife.

My name is Richard Moore. I live in Belfast. I'm a nurse, specialising in geriatric care.

1 It's very strange for me to hear you talk about your specialisations. In my work I'm just a nurse who does everything, from pre-natal care, midwifery, child care, inoculations right through to care of the elders. If there's no doctor available I even have to carry out minor operations. I'm responsible for all stages of life in five villages.

2 That sounds like real community caring. I'd love a job like that; not that I don't love my old people, but I would like your variety. Most of my people are too old to look after themselves any more. Some of them suffer from memory loss or even Alzheimer's disease so, lovely folk though they are, their conversation is not very stimulating. They show the success of the health service at extending people's lives, but they can do very little for themselves. They do need a lot of care. It's a pity they can't be looked after in their own homes, but that would need even more workers, because families aren't able to care for them in their present state.

3 In Spain most old people stay with their families, but it's starting to change even in our country. There was a social revolution in the last 30 years of the 20th century. Before that we had a strict dictatorship, and the role of women was very rigidly defined. My mother and grandmother had little education and were expected to stay in the house and care for their children, their husbands and their husbands' parents. That has changed during my lifetime. Now most girls get an education equal to the boys, and most of my friends have jobs. We are mainly still Catholic, but most of us use contraception, whether married or not.

The only trouble is – it means that there is less demand for midwives. Our birth rate has fallen to 9‰. Our population is not replacing itself. The proportion of old people has not reached the same level as in northern Europe, but it will start to do so in the mid-21st century.

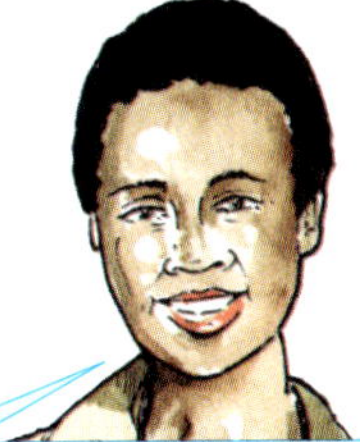

4 That has started to happen in Zimbabwe too. The birth rate is starting to fall, and the death rate has certainly come down. In particular the infant mortality rate has started falling – thanks to community nursing and increased female literacy. Women have learned to cope with simple childhood ailments, like diarrhoea, that used to be the big killers of babies. Inoculations mean that the other killers, like polio and smallpox, are being stamped out. Once couples know that their children are likely to survive, the birth rate starts to come down. Alas, that might all be changing now. Southern Africa is suffering an AIDS epidemic, and because we are such a poor area we cannot afford education to stop the disease spreading or drugs to ease the suffering. It's up to nurses and health workers to try to educate people, but this disease is damaging all our work in other areas of healthcare. Who knows what the long-term effects of the epidemic might be?

5 Maybe the drugs companies could subsidise AIDS treatment in LEDCs? In my work I have to worry more about the long-term problem of providing enough nursing care for the elderly. It will always be a labour-intensive area. We can't replace trained staff with drugs or hi-tech equipment. Unfortunately our birth rate has fallen, too. Our problem was made worse by young, trained people leaving during the Troubles. That served to speed up the ageing of the population. Some experts say that we'll soon have to encourage migration into Ireland. That will be ironic, considering all the people who have migrated out of Ireland over the last 200 years!

6 Spain doesn't need new migrants yet, but our economy and society are developing, since we joined the European Union. As living standards rise we're developing our service industries. The combination of an ageing population, more women with careers and an increased demand for services will bring big changes to our society. It's exciting being a young woman in Spain just now – but change is worrying for many people. Soon even my country might start to look across the Mediterranean to find a workforce from North Africa.

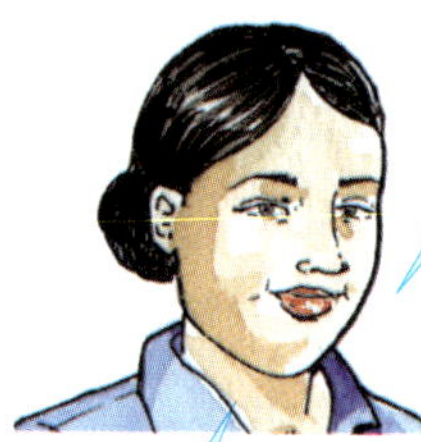

7 I do hope that Europe won't rely on LEDCs for their skilled workers. That could mean that we can't develop. My training wouldn't suit me to nursing in Europe, but I know many nurses from Zimbabwe who would like to try working over there. They would get far more pay, and better conditions. Who would blame them if they did move?

8 We might be exploiting LEDCs by taking away their best-trained workers.

9 At least when people travel to find work, they can send money back to help to support their home villages. Also, people who have travelled bring back new ideas. This helps to modernise village life. It might also help us to reduce the birth rate. Unfortunately, people migrating for work have also helped to cause the spread of AIDS.

10 Yes, there are many factors that influence health, lifestyles and population change. They're all linked together. It makes population change very difficult to understand and to manage and plan for.

Focus tasks

Read through the conversation between the three nurses again.

1 Working in a small group, pick out every factor that helps to explain why population is changing in each of the three areas that the nurses come from.

a) Write each factor in the centre of a piece of card. Write factors affecting the birth rate in *red*, the death rate in *blue* and migration in *black*.

b) In the top left corner of each card write the name of the region(s) that it affects.

In the top right corner write whether it is leading to growth or decline of the country's population (or write 'uncertain' if you need to).

Across the bottom of each card write whether you think that the factor will have just a short-term effect or if it will have a long-term effect.

2 Using your cards, write three paragraphs for each area, using these headings:

Paragraph 1
'How the population of . . . might change in the next 25 years'
Paragraph 2
'Why the population might change in that way'
Paragraph 3
'What impact will population change have on the area?'

★ THINK!

This conversation illustrates many ideas to do with population change, and brings together many of the topics covered in this unit. List the key factors you noted in each paragraph in Task 2, and learn them. In the exam be prepared to explain how the factors affect the specific places named in the conversation. (You do not need to learn the names of the nurses, or other 'background' details of the conversation.)

Settlement

How can we manage our cities?

Most people living in the UK live in urban areas. In this Unit you will investigate many issues facing people who live, or work, in cities around the world. You will explore:

- how settlements are changing
- what causes urbanisation
- how the structure of cities is changing
- how cities affect the way we live
- how cities can be managed.

5.1 *How do settlements develop?*

DO YOU KNOW?

- London was the first city in the world to reach a population of 1 million, in the mid-19th century.
- There are now more than 300 'million' cities – six of these have a population over 15 million.
- London is now only the 34th biggest city – it will fall further down the list in the next few years.
- By 2020 it is estimated that 77 per cent of Europe's population will be living in urban areas.

AIMS

- **To understand why settlements develop in particular places.**
- **To recognise problems caused by settlement sites.**
- **To identify different settlement functions.**
- **To consider the effects of ICT on the development of new settlements.**

Settlement, site and situation

The site of a SETTLEMENT is the place where it was first built. Its situation is how it links into its country or region. Many towns were first settled hundreds of years ago. They usually grew up at sites which were:

- **good for trade:** for example, where transport routes met, or at a bridge over a river; or
- **good for defence:** for example, on a hill where inhabitants could defend themselves from attack.

People also had to be sure that the site had:

- **clean water:** an easily available supply of fresh water
- **good drainage:** so that houses would not flood or get damp.

However, needs change. As settlements grow, the features that attracted people to the site in the first place often cause problems. For example, Paris in France was built on a very good defensive site on the Île de la Cité. The site made it difficult for enemies to attack. As trade grew, the need for defence was not as important as the need for easy access to the market place, and the original site proved difficult. Bridges were built to link the island to the mainland but they were expensive to maintain, and they easily became congested.

1 Île de la Cité, Paris

Edinburgh in Scotland was built on a volcanic hill (see page 127) which made the early settlement easy to defend. The original fortress, sited at the top of the hill, was built in the 7th century but by the 11th century Edinburgh was becoming an important market town. The steep slopes which offered such a strategic advantage became a problem, since they made the site difficult to build on and to travel round. However, they do help to ensure that today Edinburgh is a magnificent and beautiful city, which attracts many tourists.

2 Edinburgh

Settlement functions

The *functions* of a settlement are the main activities that go on there – the jobs people do, and the services provided. The functions of settlements change through time.

Phase 1: Pre-industrial settlements

Until the 19th century in Western Europe the most important function of most towns and cities was to provide services for the people in the settlement and the surrounding area. These services included:

- **trade:** markets and shops
- **administration**
- **defence** and law and order
- **education** and culture
- **religion:** centres of worship and organisation.

Towns and cities developed in the way that best served their function – as a market town, a commercial and administrative centre, a port, and so on.

There were often some small-scale industries in the towns. These were mainly craft industries, for example making clothing, weapons, tools and household equipment. Some crafts-workers specialised in the production of luxury goods for the important families in the town. Sometimes one town specialised in a particular craft.

However, before the industrial revolution much industry was scattered around the countryside. As cottage industries there was no need for these to be concentrated in towns, because they were often family businesses which did not employ many other workers.

Tasks

1 a) For both Paris and Edinburgh explain why each site was a good place to build a settlement.

b) Explain the problems that were later caused by that site.

SAVE AS . . .

2 Think about a town or city which you know well.

a) When was it first settled?

b) Why was the site chosen?

c) What are its functions?

i) How many of the functions named in the text can be found there today?

ii) Does the settlement have other functions not given here?

iii) List the functions it does have. Give specific details of where each function takes place within the town or city. For example, Guildford:

Education function – e.g. University of Surrey
Religious function – e.g. the cathedral
Administrative function – e.g. AQA examination board.

★ THINK!

All Geography specifications include the study of urban geography. You will need to know examples to illustrate the ideas that you learn. Every textbook will give you some examples. However, it is often easiest to learn and remember details about your own local town or city. Try to see how all of the ideas that you study can be applied to your own local example. Then, in the exam, you should be able to write answers that show a really good ‘sense of place’. You will be writing about somewhere that you know well, so you can show what makes that place special.

Phase 2: The development of industrial towns and cities

In Western Europe the development of factories and improvements in transport in the 19th century changed the way settlements grew. New settlements grew up around energy sources – particularly coalfields. Access became important. Industrial towns needed railways and canals so that coal and other raw materials could be transported easily. The new industries were labour-intensive so lots of workers were needed.

Many people left the countryside and moved to live as close as possible to the factories where they found work. Many small market and administrative settlements grew to become huge, sprawling industrial areas or CONURBATIONS (where several existing towns spread and join to form a single urban area).

This type of development is illustrated here in a case study of the Ruhr in Germany. This process took place in MEDCs in the 19th and 20th centuries. It is now taking place in some LEDCs, as manufacturing declines in MEDCs but grows in LEDCs.

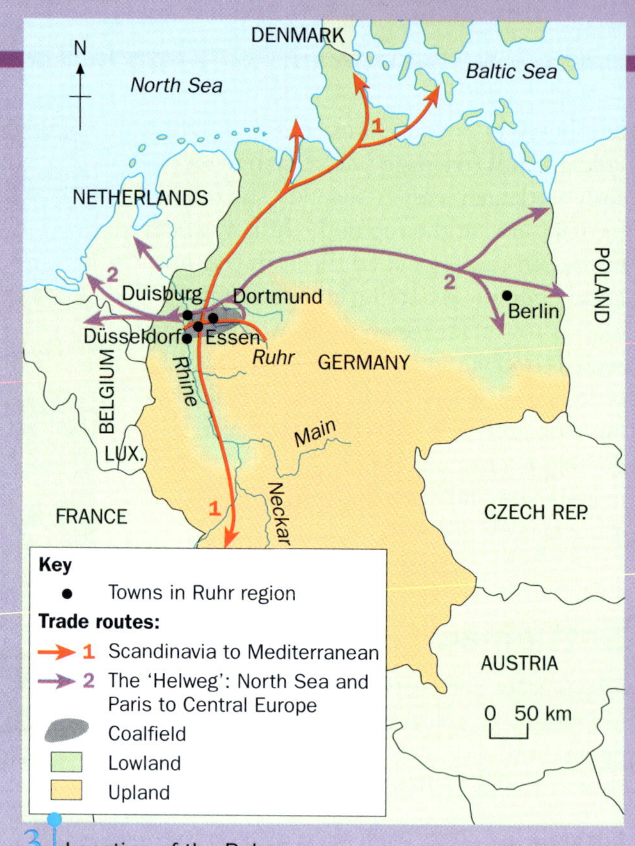

3 Location of the Ruhr

Case study: the Ruhr

The Ruhr region in Germany is one of the biggest conurbations in Europe. Several of its towns were established as market centres in the Middle Ages. Duisburg was the most important of these, due to its site features (see Map 3). Then, when coal was discovered beneath the surface, each of the towns grew as industrial towns and merged to form a massive mining and industrial conurbation, as you can see from Map 4. However, the conurbation is not a continuous built-up area. There are areas of woodland and parks between some of the towns. The authorities have planned carefully to protect these areas and stop the urban sprawl. Although its mines and industry declined in the late 20th century, the area's position in the heart of Europe will ensure its continued importance.

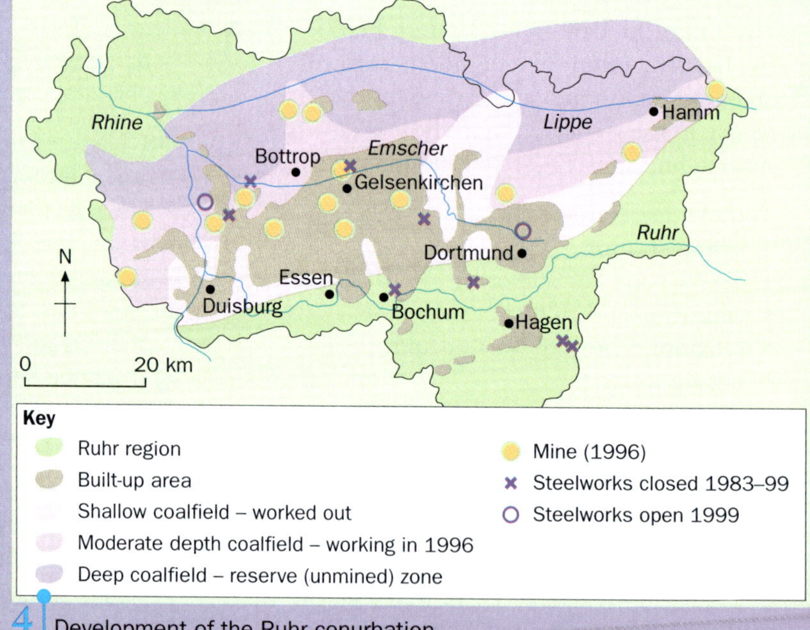

4 Development of the Ruhr conurbation

Task

1 a) Study Map 3. Why did the Ruhr towns become established in the Middle Ages? (Include the word 'situation' in your answer.)

b) Look again at Map 3. Why is the Ruhr region called a 'conurbation'?

c) Explain why the Ruhr conurbation developed in the 19th and 20th centuries. (Include the word 'function' in your answer.)

5 The Rhine at Duisburg

Phase 3: The decline of manufacturing and the growth of services

In the 19th and 20th centuries, many UK manufacturing towns grew and merged into sprawling conurbations just like the towns in the Ruhr had done.

These towns were often heavily dependent on a single manufacturing function; for example, Leeds and much of West Yorkshire was a centre for the wool industry, Newcastle upon Tyne ship building, Stoke-on-Trent pottery making (see Chapter 6.3). You probably know of many other examples yourself.

In the late-20th century, UK manufacturing industry declined. You will study the reasons on page 292 – here we are interested in the consequences. Many towns and cities, with the manufacturing function gone, either began a process of gradual and sometimes painful change as they developed new functions; or experienced a vicious spiral of decline (see Chapter 5.5).

Nowadays the strongest parts of the UK economy are the service industries, such as computing, shopping, entertainment, finance, advertising and information. In MEDCs generally the settlements that are growing are those that are geared to services. One of the fastest-growing service industries is tourism. Some settlements have become as dependent on tourism as others once were on manufacturing. You will examine three contrasting case studies of the impact of tourism on settlement on page 198. Another rapidly growing area of the service sector is Information and Communication Technology (ICT). Settlements specialising in ICT are growing quickly. ICT is also having other, more deep-reaching impacts on settlement patterns.

★ **FACTFILE**

A settlement hierarchy

'Settlement' is a broad term. It could mean anything from a small hamlet of three houses to a massive city of 20 million people. Geographers therefore talk about a settlement hierarchy, although you will quickly see that none of these divisions is watertight or rigid.

Type of settlement	Typical population	Typical service function
hamlet	about a dozen	none
village	up to 10,000	essential services such as a post office or village shop – mainly used by local people
town	tens of thousands	a wide range of shops and services used by many people from nearby villages
city	hundreds of thousands, or millions	comprehensive range of services used by people from all over a region or even from the whole country

Phase 4: How will the ICT revolution change the function of settlements?

Modern communication systems have changed many of the links between site and function that operated in the past. Developments in ICT now mean that many functions can take place almost anywhere in the world. Companies can locate their business wherever:

- the environment is most pleasant
- costs are lowest
- the government offers most support
- the labour supply is most flexible
- the modern communications infrastructure is best developed.

This is all part of the process known as GLOBALISATION.

These developments could change the appearance of cities in many ways. They also mean that many new settlements will appear in unexpected places, in fact wherever people have the enterprise and initiative to provide the needed communications infrastructure.

Over the next four pages there is a case study of one such place: Kuala Lumpur in Malaysia, home of Cybercity.

Tasks

Get out your crystal ball to predict the future.

2. Work with a partner to list ways in which you expect ICT to change life in the future. List as many as you can.
3. Now select all those that you think would have an effect on settlement, i.e. those which will affect where people live or work, or the way in which villages, towns or cities might grow.
4. Now use these to make your own concept map (see page 32). Write 'How ICT will affect settlement' in the middle of a large page. Write around it all the changes you have selected and alongside them explain why ICT will lead to each change.
5. Compare your conclusions with some actual developments in Kuala Lumpur, described on pages 194–96.

Case study: Kuala Lumpur, Malaysia

6 Kuala Lumpur

8 Chinatown in Kuala Lumpur

Dear Mr Smith

My 'gap year' before university is turning out to be brilliant. I'm in Kuala Lumpur (or KL, as everyone calls it), the capital of Malaysia. I'm writing this in an internet café on the edge of the central business district.

I remember reading about Malaysia being one of the rapidly developing countries of South East Asia. You called it one of the 'Tiger' economies, and said it was a Newly Industrialising Country or NIC.

Well, I never expected it to be like this. As we flew over the country we could see rainforests on the lower slopes of the hills, with steep, bare rock slopes above. Some parts were scarred with opencast mines where they extract tin.

Along the coast the forest had been cleared and plantations have been set up. You told us that rubber is grown in Malaysia, but we have been told that now all the plantations are palm trees, grown for their palm oil.

KL has one of the most modern airports I have ever seen. The road into town is a fast dual carriageway, just as modern as any road in the UK. We did not pass any shanty towns or SQUATTER SETTLEMENTS, although a man on the bus told us that there are quite a lot of these on land by the railway lines. The ones near the main road were cleared away before the tourists arrived for the Commonwealth Games in 1998. I did take some pictures of new housing developments being built along the road – I'll send some.

The area between the airport and KL is going through an incredible development. They call it 'Cyberjaya' or 'Cybercity'! If it works out as planned it will challenge many of the commercial and educational functions of old cities in the West. They have the advantage of building it all on a GREENFIELD SITE. I'm sending some details from a magazine. (See page 196.)

We've travelled around the central areas on the rapid transit system (RTS). It's a bit like the Metro system in Newcastle but it's raised up on stilts in the city centre. The carriages are all made of stainless steel, even the seats. This makes life difficult for Muslim women in traditional silk dresses. When the train brakes, they slide along the seat. We tried not to giggle – but they seem to think it's quite good fun! The RTS is one of the projects that they hurried to get ready for the Games, but it does seem well used now. The ticket-seller laughed when we asked if we would see any shanty towns from the trains. He said, 'There are none of those left in the redeveloped parts of the city!'

KL seems neat and well cared for – not like the stereotype of an LEDC city – but even without shanty towns I've seen incredible contrasts here.

7 The Petronas or 'Twin' Towers: the world's tallest building

We went shopping in China Town. (I did some Geography research: out of 17 million people in Peninsular Malaysia, 57% are Malays, 27% are Chinese and 9% are Indians, with about 7% of other races.) It was really exciting, with lots of shops and street stalls selling food, cheap clothes and electrical goods, religious ornaments, dragons, kites – everything. The Chinese 'shop-houses' are very long and narrow. They have a shop, restaurant or workshop on the ground floor and the family live above.

Our Malay friend, Azizah, took us to a new shopping mall built by Chinese businessmen. What a mixture! From the outside it's designed to look like an Egyptian pyramid. Inside it's a cross between a UK out-of-town shopping centre and China Town. The shops look very prosperous, and the main colour is red – symbolic of the Chinese New Year. Azizah says that lots of businesses are Chinese-owned. People from China have migrated all over South East Asia and taken their trading skills with them, but the government is encouraging Malays to be enterprising too.

At the weekend we saw another side of multicultural Malaysia when a Hindu friend took us out to the festival of Thaipusam in the Batu caves (dissolved from permeable limestone!). The people were making offerings to one of their gods. It was very colourful and exciting, but in the heat it all got too much for me, so we went back to the air-conditioned comfort of our hotel.

We also visited the Twin Towers. You probably know, it's the tallest building in the world. It looks so spectacular. It's entirely covered in stainless steel. To stand near the bottom and look up made me feel dizzy. I'll send you a picture – but it doesn't really show the enormous size of it. It has a shopping centre, but I bet it's not as lively as China Town.

The Prime Minister is very keen on massive status symbols like the Twin Towers and Cyberjaya. I'm not sure whether they're a good use of scarce resources in an LEDC. I shall visit Cyberjaya to decide whether it's right to build something like that in a country where low income earners find it difficult to afford housing and healthcare. I suppose that, if it's successful, it will generate the money to pay for houses, doctors and education.

I hope you can see from this e-mail that I'm doing what you always told us to do – living my Geography. Must dash now. Lots to see and lots to do.

Helen

9 A shopping mall in Kuala Lumpur

★ THINK!

Helen says that she is 'living her Geography'. You should try to do the same. Wherever you travel there is real geography around you. Try to get used to looking out for aspects of human and physical geography in the environment. It helps to make places more interesting, and it will help you to write answers that show the 'sense of place' referred to earlier.

Tasks

1 Describe each of Photos 6–10. Refer to:
- the physical geography (if you see much evidence) – look for details of the shape of the land, the weather and the vegetation
- the people
- the buildings
- transport
- what is happening now
- evidence of change
- issues in the area (for example, problems and opportunities).

2 Read Helen's e-mail again. What recent changes does it describe that have taken place in and around Kuala Lumpur?

3 Helen is still in KL. Write back to her with a list of questions that will help you to find out more about the geography of the city.

4 Would you like to live in KL? Give reasons.

10 Pots containing milk and honey – offerings to the Hindu god Lord Subramaniam – are carried on the head as part of the annual festival of Thaipusam.

Multimedia Super Corridor (MSC)

The MSC in Kuala Lumpur is an excellent example of globalisation. It is a zone 15 × 50 km, extending south from KL alongside the highway that links the capital to Malaysia's new international airport (see Map 11). On one side of the road is Putrajaya, the city that is being built for the country's government. To the west of the road is Cyberjaya (or Cybercity) – a perfect environment for companies making multimedia products and services. It has three key elements to attract companies from around the world:

- a high-capacity communications infrastructure, based on a digital optical fibre network, and a new international airport
- new laws to encourage electronic commerce and to help the development of multimedia applications – including laws to protect the ownership of material published electronically
- an attractive living environment with careful zoning plans (a planned settlement) integrating 'mega-projects' with green reserves – all houses will be linked to a system that will respond to alarm calls, connect households to workmen, find babysitters, and so on, all on the city website.

The government has targeted seven types of businesses to develop there. They are:

- electronic government systems
- telemedicine
- smart school teaching systems
- national, multi-purpose, smartcards
- research and development clusters
- manufacturing for worldwide export
- borderless marketing centres.

Malaysia's multicultural society, political stability and reliable support services provide the ideal platform for multimedia developers to launch export operations. Companies joining MSC are entitled to operate tax free for ten years, and they have access to other government incentives. BT, Microsoft, Nokia and Ericsson are already operating in the MSC, and many other companies are considering starting up there.

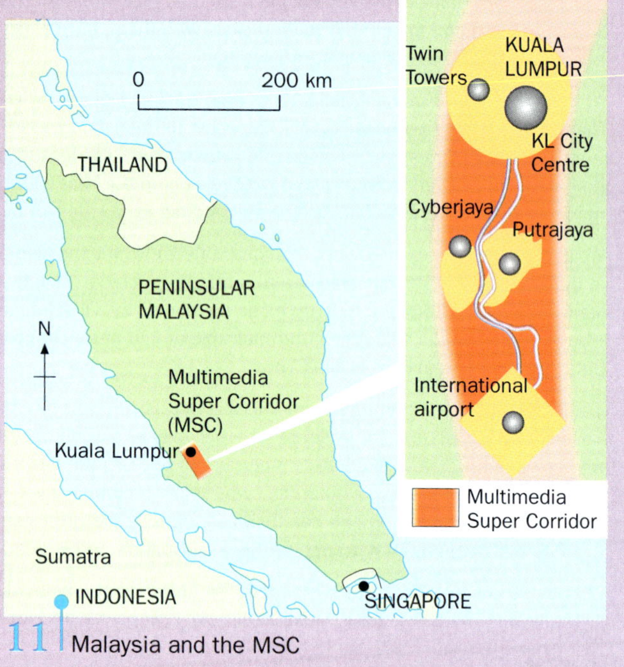

11 Malaysia and the MSC

★ **FACTFILE**

What the terms mean

Multimedia applications – linking computers to TV, virtual reality, etc.

Laws to protect electronic publishing – at present most countries have laws and systems that allow authors to collect fees when their works are copied, but most do not have similar laws which protect their work when it is copied electronically. Such laws would make people more willing to publish their work on the new media.

Electronic government systems – tax collection, census data, police records, and so on.

Telemedicine – like NHS Direct, where you ring a call centre and speak to a nurse to get a diagnosis of any health problems.

Smart school teaching systems – interactive, on-line lessons to be used in school or from home.

National smartcard – a combination of identity card, bank card, credit card, driving licence, etc.

Research and development clusters – especially for science and ICT-based research.

Borderless marketing centres – call centres linked up to the whole world to sell to businesses and private individuals.

Tasks

SAVE AS...

1 The ICT industry will become one of KL's main functions. Using information on pages 194–96, make a list of the needs of modern ICT-based industry. Next to each item state:

- *either* how KL meets these needs,
- *or* how KL is working to meet these needs in future.

2 Would you like to work in the Multimedia Super Corridor (MSC)? Give reasons.

★ THINK!

Connect your case studies

You have probably noticed that there is quite a focus on Malaysia in this book. For example, on page 28 you looked at the tropical rainforest ecosystem in Malaysia. On pages 302–303 you will examine the car industry in Malaysia. There are other examples as well on pages 97, 162 and 335.

This focus on one country is deliberate. We think it produces better work and better exam answers to know more detail about fewer places. It helps you to connect case studies together. It gives you a fuller picture of a country. The alternative – dotting around the globe from country to country – makes it hard to remember things.

So as you study this aspect of Malaysia – its role as a key player in the global ICT industry – keep in mind Badan, the traditional farmer in Sarawak (see page 8), and the other aspects of Malaysia you have found out about in the book. Together they give you a much fuller and more balanced picture of a country than they would on their own.

Malaysia websites

You would expect 'Cybercity' to have a good website. You can find it at:

www.cyberjaya-msc.com/

There are many other sites about Malaysia, for example:

www.asiatour.com/Malaysia

General information on modern Malaysia can be found at:

www.malaysiainfo.com/

Tourist information can be found at:

www.carcosa.com.my

Tasks

3 Choose three websites relating to modern Malaysia. Look carefully at them and then write a critical review of them. Use the following headings in your review:

- How informative was the site?
- What type of image of the country did the site give?
- Did you like the site? Give your reasons.
- Who was the site aimed at?
- Did the site suit its purpose?

4 Kuala Lumpur is well on the way to becoming a modern conurbation of the 21st century. One of the reasons why it is becoming a very attractive city for transnational corporations is the development of big prestige projects like Cyberjaya and Twin Towers. However, some Malaysians believe that these projects are a waste of money.

a) List the arguments for and against such prestige projects.

b) 'Was the government right to go ahead with big, expensive projects like these?' Write down your own conclusion to this question.

★ THINK!

What use is the internet?

Quite often in this book you will be asked to look up things on the internet. This can be very useful in Geography. Sometimes it is much more up-to-date than a textbook ever can be, and you can get a range of opinions. However, the internet can sometimes present inaccurate information. It can also be very slow. Sites often change. Sometimes they disappear altogether. So don't waste time – if the sites we suggest don't give you what you need, use the search skills you have developed to find better ones which are more up-to-date. And beware! Don't get side-tracked. You can waste an awful lot of time aimlessly surfing the net.

Finding sites is easy. Printing off the information contained in them is also easy, and provides you with a great mass of paper that is difficult to store and which you will probably never read again. So decide exactly what it is that you want to find out. Go through the information you have found, and *be selective*. Only keep what is relevant, and paste it into a new word-processing document, to then print out and keep with your notes.

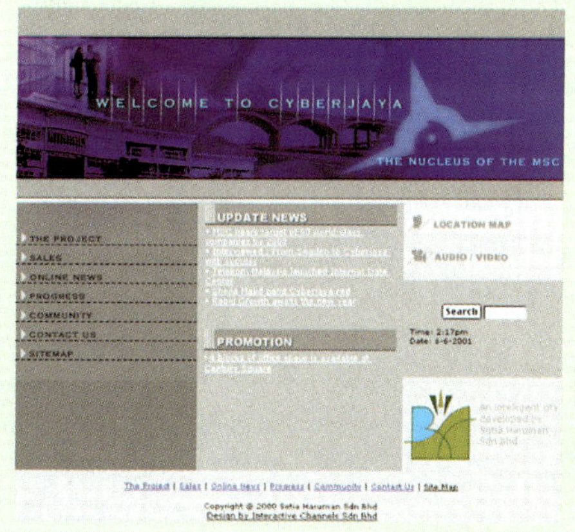

How does tourism change settlements?

Tourism is the fastest-growing industry in the world. Tourism can have a major impact on settlement patterns. It can change existing settlements or it can create totally new ones. Here are three contrasting examples.

Case study 1: Paris

Paris has gone through many stages of development (see page 190). For centuries it was the trade, finance and administrative centre of France. In the 19th century it became an industrial city. It grew quickly to become the biggest city in Europe – which it still is.

Nowadays the centre of Paris has a huge tourist function. Visitors are attracted by its famous art galleries and museums, and by its history, culture, and architecture. It has thousands of restaurants, a lively nightlife and a romantic reputation. It has one of the world's most famous landmarks, the Eiffel Tower. Visitors mostly use the efficient public transport system.

The biggest recent development is Disneyland Paris which has spread the tourist function out to the edge of the city. Five million tourists visit Paris each year. Tens of thousands of Parisian jobs depend on tourism. But in a city with 10 million people and many different functions, tourism does not completely dominate Paris.

Case study 3: Matalascañas

All along the Mediterranean coast – Europe's 'sun belt' – from Portugal in the west to Greece in the east, the massive growth in package holiday tourism since the 1960s has transformed settlement. Small fishing villages have grown into bustling resorts. Many totally new settlements have been created on previously unsettled coastline.

Matalascañas is on the south coast of Spain. It is a new tourist resort built in the 1970s. There was not even a village at this site before. Now it is a town which in peak season has a population of up to 80,000. It can cater for 40,000 tourists at a time.

The site was selected because of its climate, beaches, and the nearby national park. Unlike Ambleside, it doesn't suffer from a traffic problem. As a purpose-built resort it was made to cope with modern traffic. Most visitors come for a beach holiday and do not bring a car. Land is not a big issue either since this is not a protected, valuable or mountainous area. Indeed, one reason for choosing this site was that land was available for development.

Case study 2: Ambleside

Ambleside in the Lake District National Park was first built as a MARKET TOWN at the point where several roads met. Tourism to the Lake District increased rapidly in the 1960s following the building of the M6 motorway. With the rise in car ownership, millions of people a year now visit the Lake District, the vast majority coming by car. Ambleside is on the main road through the Lake District, so a large number of tourists visit the town or pass through at some point in their stay.

Tourism brings a lot of business to the area, but it has also changed Ambleside beyond recognition. Tourism has now almost completely replaced Ambleside's original function as a market town at the centre of a farming community. Almost all local employment is dependent on tourism. Shops are geared to tourism. There are plenty of hiking shops and gift shops but no major supermarket.

Tourism puts a strain on the settlement. Ambleside needs to change to adapt to its new function but is limited by many factors – particularly its site. There is little land available for new building. The National Park strictly controls what can be built, in order to preserve the beauty and character of the area. With a limited supply of housing, property prices have increased so much that people who work in the area often cannot afford to live there and have to drive in from Kendal, more than 20 kilometres away, where house prices are cheaper. Traffic congestion in the summer can be horrendous.

One solution to Ambleside's traffic problems would be a by-pass but this is resisted by environmentalists and has never been possible. Its site made Ambleside attractive as a tourist centre in the first place; now its site restricts its development.

12 Ambleside in the Lake District

However, the site was not perfect – one of the biggest problems was obtaining a fresh water supply, as you can see from Map 13. Now the Spanish Government is planning to build a second resort (Costa Doñana) nearby for another 35,000 tourists, which could lead to increased conflict over water supply.

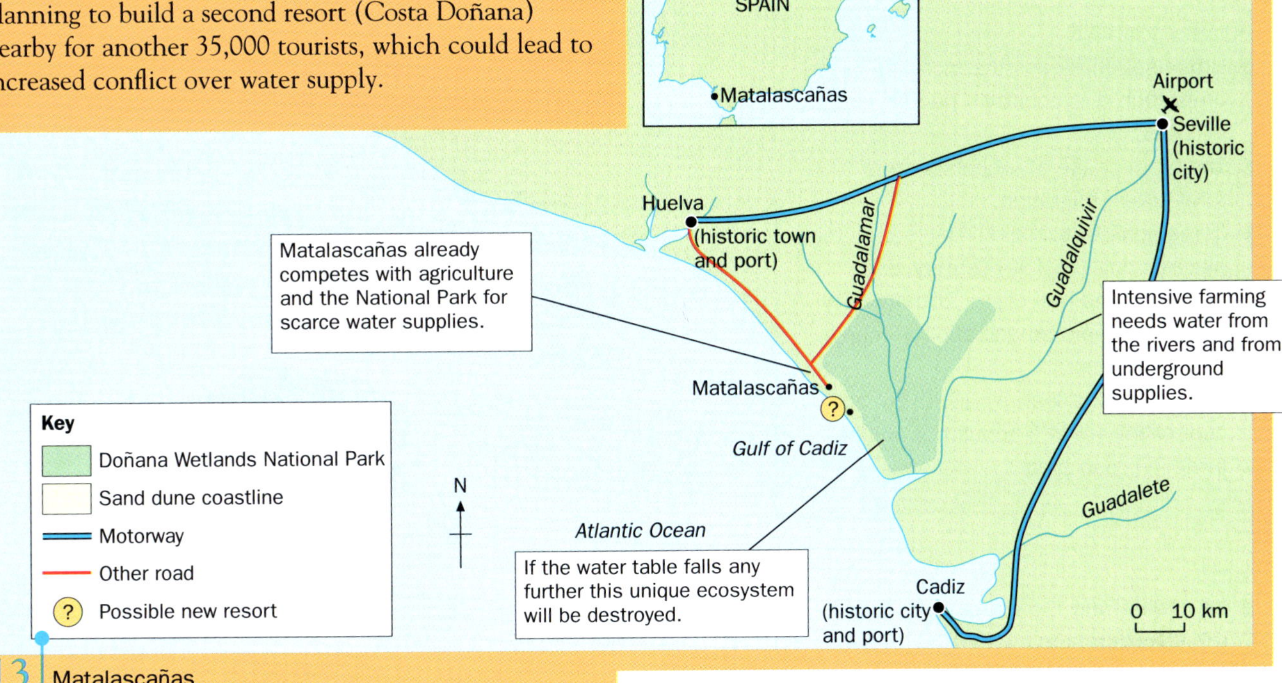

13 Matalascañas

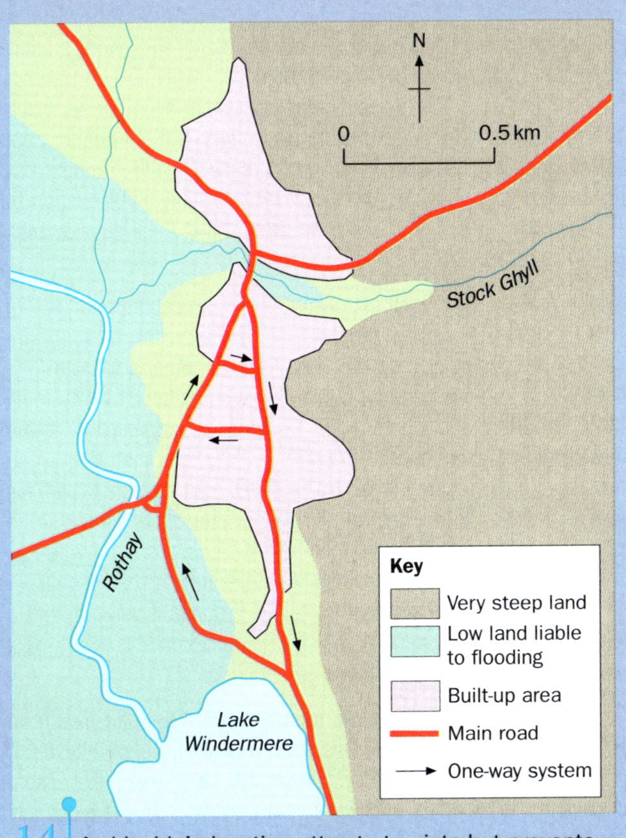

14 Ambleside's location attracts tourists but prevents expansion.

Tasks

1 Study the information about Ambleside.
   a) List the tourist attractions of Ambleside. Use the word 'site' in your answer.
   b) How has tourism changed Ambleside? Use the text to list at least five changes. Make sure you use the word 'function' in your answer.
   c) Explain why Ambleside's site both creates and limits its tourist function. Refer to the difficulties that would be faced in building a by-pass.

2 Study the information about Matalascañas.
   a) List the advantages and disadvantages of this site for building a purpose-built tourist resort.
   b) List the advantages and disdvantages to the local area of the existence of this resort.
   c) Who do you think might support the building of a second resort, and who might oppose it?

3 Compare the three case studies. Write at least one statement each about settlement, site and function which could fit into each sector of this Venn diagram. For example 'Tourism is its most important function' could describe both Ambleside and Matalascañas but not Paris.

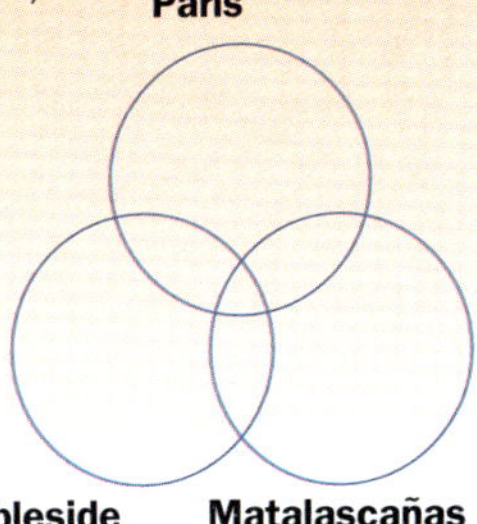

5.2 What causes urbanisation?

DO YOU KNOW?

- Urbanisation means the increase in the proportion of a country's people who live in cities.
- The main cause of urbanisation is rural–urban migration.
- In many MEDCs, rapid urbanisation took place during the 19th century.
- In the second half of the 20th century there was rapid urbanisation in many LEDCs.
- It is estimated that, by 2020, 57 per cent of the world's population will be living in urban areas.

AIMS

- **To understand the causes of urbanisation in many LEDCs.**
- **To examine stereotypes about squatter settlements in LEDC cities.**
- **To consider how squatter settlements can be improved.**

Carlos's story: A case study of rural–urban migration

Carlos Vega is 17 and lives in a *BARRIO* (see the Factfile, page 204) on the outskirts of Mexico City. He moved there three years ago from the village of Santa María, about 200 km to the north.

His father and grandfather are both farmers. They work a small plot of land which belongs to a big landowner. The landowner will not allow the peasants any more land, even though much of his land is only used for grazing his horses. Sixteen people from Carlos's family still live in Santa María, including his grandparents and cousins. The land produces only a little more than enough to feed the children and old people, so most young men leave to seek work in the cities.

Carlos's uncle, Emilio, built his own house in the squatter settlement of La Esperanza (Spanish for 'hope'). When he started, seven years ago, the settlement was on the very northern edge of Mexico City, about 12 km from the centre. Now there are some newer settlements built even further out. When he and his friends first moved on to the land, he slept on bare ground. Soon he built a rough shed from wood and sheets of corrugated steel. As he saved a bit of money from his work as a waiter in a restaurant, he paid a man in the settlement to make him some bricks. Emilio then built a single room. Now he has built a second one, but one wall is not yet finished. He has used all of the wood and corrugated steel for lean-to rooms, a veranda and a toilet. The toilet has a cesspit, which is emptied once a month by a tanker from the city council.

The men in the settlement dug a trench to put in a water-pipe from the mains, so there are five taps shared among the 65 houses in La Esperanza. Most houses have electricity. The men made illegal connections to the grid, but they are negotiating with the electricity company to legalise the connection. They are happy to pay for the electricity if the company will check that it is safe.

USA
Monterrey
MEXICO
Gulf of Mexico
Santa María
Guadalajara
Mexico City
BELIZE
GUATEMALA
Pacific Ocean
N
0 200 km

1 The village of Santa María and three of the largest cities in Mexico

The people in the settlement are working together to build a schoolroom too. They realise that education for the children is the only way out of the *barrio*. At the moment one of the residents of La Esperanza, Raquel Damore, runs a school for the children in the *barrio*. They meet in the shack that Raquel's husband built before he was killed in an accident at work. There are about 70 children aged between 5 and 11 in the settlement, and Raquel tries to teach all of them some basics. Each child can go for only about four half-days each week, but all the parents try to find the money to pay for some lessons. Quite a few of the adults even go to Raquel's evening classes to learn to write. When they have built the schoolroom they will try to get a properly qualified teacher who can teach more than just reading, writing and arithmetic. The authorities will pay her wages, and Raquel can concentrate on child-minding. She is not qualified to teach.

Carlos lives with his uncle Emilio, his aunt Ester and their two young children. Emilio and Ester have a bedroom which they usually share with the two children. Carlos and two other lodgers (two more of his cousins) put mattresses on the floor in the living-room when it is cold, but on warm nights they take their mattresses outside on the veranda. They have put up pictures of football stars in the living-room, next to Ester's pictures of the Pope and the Virgin Mary. It helps to make the place feel like home. All three boys pay rent to Emilio, but as they are family, he charges them only what they can afford.

None of the boys has a regular job in the 'formal' sector (see page 207) which would bring in a regular wage. Carlos gets some work as a trucker's mate. He does not get paid much, but he tries to get on trucks going north, so that he can go to his home village. He always tries to take some money to help out his family, and he brings back fruit and vegetables which his cousin Luis sells in one of the better housing districts near the city centre. Luis also looks after the fruit trees that Emilio has planted. They help to give shade to the veranda and they are starting to bear fruit that can be sold. Carlos is also becoming skilled as a bricklayer. He does labouring for builders in the *barrio*, and they show him how to lay bricks. There is so much building going on that there is always work for good brickies. All of these jobs are in the INFORMAL SECTOR (see page 207).

What of the future? Carlos says: 'Soon I plan to start to build my own house. Emilio says that I can use some of his spare land, but I would like to be part of a new 'invasion'. There's land that no one uses about a kilometre from here. I know some men who are planning an invasion. I have started to collect bricks and spare wood from my building jobs so that I will be ready to start building.

'If I had my own house I would bring Carolina, from my home village, down here. She could easily get work as a hairdresser. She could work in the house and visit women in their own homes. We would not ruin our chances by having children yet, not for several years. The clinics in the *barrios* can show you how to make sure! The priests turn a blind eye here, not like in the village. And I would carry on building work. My brother Juan wants to come down and live with me. He could do the labouring work for me and pay rent, just like I do now. He will have to sell vegetables too. The family back home have got used to the income it brings in for them.'

2 This photo shows a roughly built shelter in another *barrio* outside Mexico City. The homes of Emilio and his friends in the early stages of La Esperanza would have looked something like this.

Tasks

SAVE AS . . .

1 List the *pushes* and *pulls* that made Carlos move to the city. Also list the *anchors* and *barriers* that made the move difficult. (Look back to pages 174–75 to check what these words mean.)

2 Describe the work done by Carlos, Luis and Raquel. For each of them describe where they work, what their qualifications are, and what their plans for the future might be.

3 People in the city still have close ties with people in the countryside. Explain how both sets of people benefit from these ties.

4 Explain how the people in La Esperanza are improving:
   a) their housing
   b) electricity supplies
   c) sewerage and water systems
   d) education opportunities.

5 Suggest why the birth rate in the city might be less than the birth rate in the countryside.

6 Carlos plans to build a house even farther out of town than Emilio's. Why?

Urbanisation – MEDCs and LEDCs compared

There are many similarities between 19th-century urbanisation in MEDCs such as the UK, and 20th-century urbanisation in LEDCs.

★ **FACTFILE**

In the UK in the 19th century:

- People were pushed by the mechanisation of agriculture and by population pressure in the countryside.
- They were pulled by the hope of work in the factories set up during the Industrial Revolution.
- When they arrived in the towns they found conditions were often very squalid, with poor housing and much poverty.
- Conditions gradually improved towards the end of the century.

In LEDCs in the 20th and 21st centuries:

- People are being pushed by the mechanisation of agriculture and by population pressure. Natural disasters and war are also major pushes in many places (see pages 185–86).
- People are pulled by the hope of finding work in factories and service industries. However, too many people are competing for too few jobs, and so many find work in the informal sector (see page 207).
- When they arrive in towns, conditions are often very squalid, with only squatter settlements available.
- Conditions are gradually improving, in some areas at least.

Rural–urban migration

Region	1950	1970	1990	2020 (est.)
World	29.2	37.1	45.2	57.4
Europe	56.3	66.7	73.0	76.7
North America	63.9	73.8	74.3	78.9
Oceania	61.3	70.8	71.3	75.1
Latin America	41.0	57.4	75.1	83.0
Asia	16.4	24.1	28.2	49.3
Africa	15.7	22.5	33.9	52.2

3 The percentage of world population living in urban areas, 1950–2020

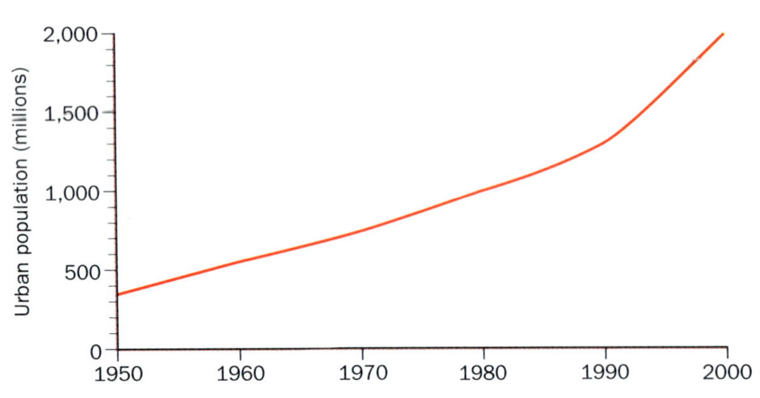

4 Total urban population in LEDCs, 1950–2000

Tasks

SAVE AS...

1 Use the figures in Table 3 to draw a line graph showing how the percentage of people living in urban areas has changed. Use a different-coloured line for each region.

2 Using evidence from your graph, divide the world regions into two groups:

a) Group 1 – rapidly growing urban population

b) Group 2 – slowly growing urban population.

3 Write two sentences explaining:

a) what all the regions in Group 1 have in common, apart from their rates of urban growth

b) how the regions in Group 2 are different.

What happens when people move to cities?

Poor people from the countryside who move to cities have two main priorities: to find shelter and work. Unfortunately most rural–urban migrants also have two big problems:

- They have no money (or very little). Poverty is usually the main push which forces them to migrate.
- They have few skills (or few that are needed in the city).

These problems seem enormous, but most people do manage to survive. The migrants have a number of advantages, including:

- Many have friends or relatives already living in the city, who help them when they first arrive.
- Most of the LEDCs are in tropical or subtropical countries, so people do not need protection from the cold as much as in the UK.
- There is a large informal sector (see page 207) which relies on casual labour. In fact, many companies from abroad have set up branches in LEDCs specifically to take advantage of cheap labour – wage rates are low compared with the UK, but they do allow people to survive.
- Many newcomers keep in touch with people back home, who provide them with food and other goods to sell in the city (as you saw in Carlos's story).
- Many governments realise that they have to allow people to live somewhere – the police often turn a blind eye to people who are squatting or working illegally.
- The people who migrate to the cities are often the most dynamic people from the countryside – their determination and adaptability make them great survivors.

5B A worker in a computer factory in Bangalore, India

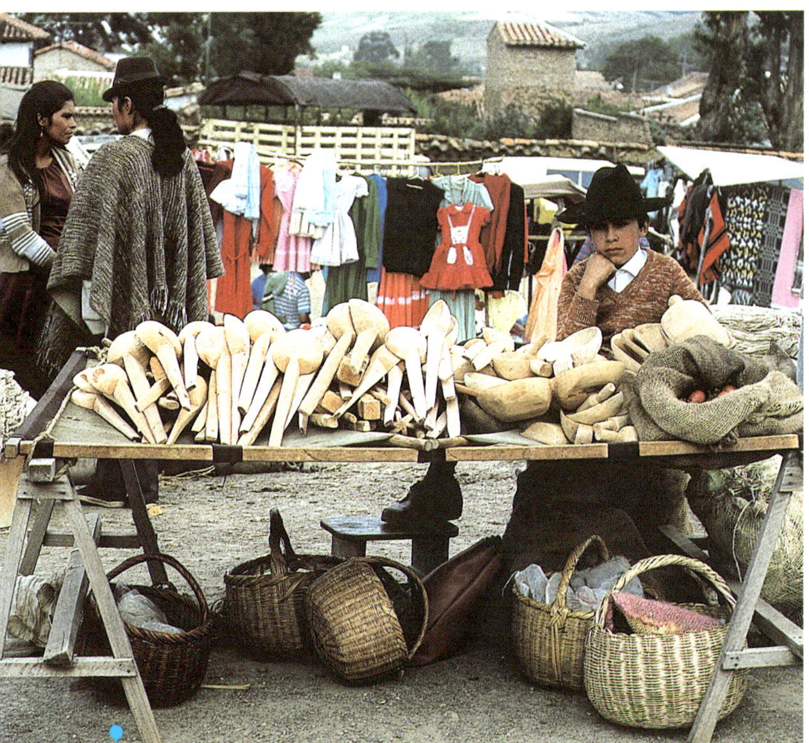

5A People who move into the cities work hard to succeed. This picture shows a street trader in Colombia.

5C A market in Indonesia

Finding shelter

Most newcomers to the city would like to rent a proper house but they have no money and even if they could afford the rent there are not usually enough houses to go round. Instead:

- many move in with friends or relatives, as Carlos did in La Esperanza
- some sleep on the streets (see Photo 6). In some cities thousands of people live rough. Most railway stations in India have a well-established resident population. People know who 'owns' each patch of pavement, and the street people protect each other from outsiders. They include families in which three generations live together, and also gangs of orphaned or abandoned children
- many squat – i.e. they build a makeshift house on some unused land.

6 Living on the street. This photo shows people living outside the railway station in Phnom Penh, Cambodia.

How do squatter settlements develop?

People squat on three main types of land:

- land that is not really suitable for building, because it is too steep, too marshy or too polluted
- land close to the city centre that has not been built on because no-one knows who owns it, or else the owner has left it empty, hoping it will go up in value
- land on the edge of the city that was once farmland but was abandoned as the city spread.

You might expect the local authorities to object to people doing this but it is far more common for the city authorities in LEDCs to help people in squatter settlements by providing them with water and electricity supplies, while leaving them to do most of the building work themselves.

★ **FACTFILE**

Barrios or bustees

In Spanish, *barrio* literally means a district of a town. In Spain it could describe a run-down inner suburb or a modern outer suburb. In most of Latin America, however, *barrio* is used to describe a poor district, often a squatter settlement on the edge of the built-up area.

In Brazil, where the language is Portuguese, they use the word *favela* to describe the same thing.

In India, squatter settlements are called *zopadpattis* in Bombay and *bustees* in Calcutta.

People in MEDCs often refer to these settlements by the cover-all term 'shanty towns'.

Task

When people in the UK build a house they usually complete the tasks in the order shown in the flow diagram below:

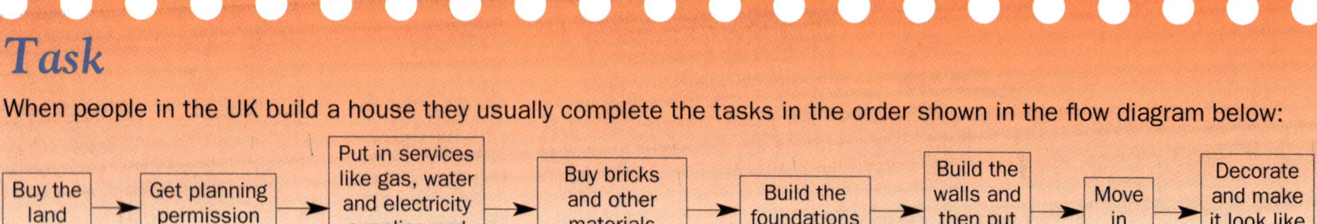
Buy the land
Get planning permission
Put in services like gas, water and electricity supplies and sewage pipes
Buy bricks and other materials
Build the foundations
Build the walls and then put a roof on
Move in
Decorate and make it look like 'home'

1 Complete a flow diagram, like the one above, to show the development of a home in a squatter settlement like La Esperanza (described on pages 200–201).

Hope or despair?

Squatter settlements are often rather rough and ready places. Most squatter settlements look very primitive to someone who is used to housing in the UK. However, we can often develop stereotypes of what conditions are like in such places. We may remember the worst points about some of the worst settlements and imagine that all settlements are as bad as these – we assume that they are all the same. *This is not true*.

Some squatter settlements can indeed be described as 'slums of despair'. Here physical, economic and social conditions are bad, and are steadily getting worse. Others, however, can be described as 'slums of hope'.

Tasks

2 Read Carlos's story again (pages 200–201) and the lists below. Decide whether La Esperanza is a slum of despair or a slum of hope. Draw up a table and list the features of the settlement that suggest 'despair' in one column and 'hope' those that suggest in another.

3 Many people in MEDCs have stereotyped views of squatter settlements (this means that they think that they are all the same). Ask your teacher for a list of the common stereotypes of squatter settlements.

Check through them to see whether they match what you have read about La Esperanza in Carlos's story.

4 Look back at page 194. What did the Malaysian government do with KL's squatter settlements? Why?

Slums of despair	Slums of hope
High unemployment and underemployment.	Some formal employment and much informal employment.
Weak family and friendship structures.	Strong family and friendship structures.
Poorly built housing and little improvement going on.	Housing improvement through individual and group action.
Poor water supply and sewerage system.	Water supply being improved with help from authorities. Sewage usually stored in septic tanks and removed by tanker.
Easy spread of infections and disease.	Infections and disease under control.
Illegal hook-ups to the electricity mains, or no supply at all.	Illegal hook-ups to the electricity mains, which are often being replaced by legal connections.
Widespread crime, prostitution, drug dealing and other social problems due to poverty and lack of police control.	Crime etc. not widespread because of strong social structures and co-operation between community and police.
Settlement appears untidy and poorly organised. Much litter, rubbish and piles of junk around houses.	Settlement appears tidy and fairly organised. Piles of stuff around houses are the raw materials of earning a living. Much evidence of informal economy.
City authorities opposed to settlement. Threats to bulldoze houses to remove threat of crime and disease and to clear the 'eyesore' of the slum. These threats make squatters insecure and cause settlement to deteriorate.	Co-operation between the settlement and the authorities, who realise that people have to live somewhere, and so do their best to provide an infrastructure of roads, bus services, education, healthcare, electricity, water, etc.

NB In reality, most settlements have some features of both hope and despair.

Case study: Improving the *bustees* in Calcutta

People living in squatter settlements in LEDCs can do a lot to improve conditions themselves, working either on their own or in community groups. However, the authorities also need to help if squatter settlements are to be improved. One example of what is being done can be seen in Calcutta in north-east India.

Calcutta lies in the Ganges delta, at the centre of an area that has a very dense, overcrowded rural population. The delta's soils are very fertile, but the area suffers many natural disasters. It is often flooded by monsoon rains or by cyclones (see pages 62–63). The area also suffered several wars and civil conflicts in the late-20th century. Each new war or flood brings refugees flocking to Calcutta. Map 7 shows the distribution of squatter settlements in Calcutta.

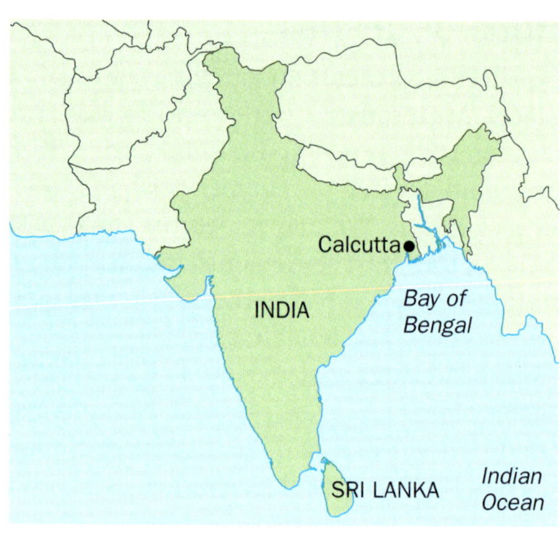

Because the land is so low-lying, many of the settlements, known locally as *bustees*, flood very easily. The floods not only destroy many homes but also bring disease, as the drains cannot deal with all of the polluted floodwater.

Until recently Calcutta had a reputation for having some of the worst slums in the world. It was here that Mother Theresa cared for thousands of dying street people. However, while Mother Theresa was caring for the dying, the Calcutta Metropolitan Development Authority (CMDA) was responsible for improving conditions for the living. The Authority has tried to improve the infrastructure. Since 1960 the CMDA has:

- reinforced the banks of the River Hooghly, and tried to stop people from squatting on the lowest-lying land near the river
- improved sewage disposal – in the 1960s there were 1,000 deaths a year from cholera, but in recent years there have been none
- improved the water supply – there is now at least one tap for every 25 *bustee* houses.
- made concrete roads to replace mud tracks between the shacks
- installed street lighting in many *bustees*, to improve safety and to give some light to people with no electricity in their homes
- widened roads and improved public transport from the *bustees* into the city centre

The CMDA does not work on the *bustee* houses. The occupiers must improve their homes themselves.

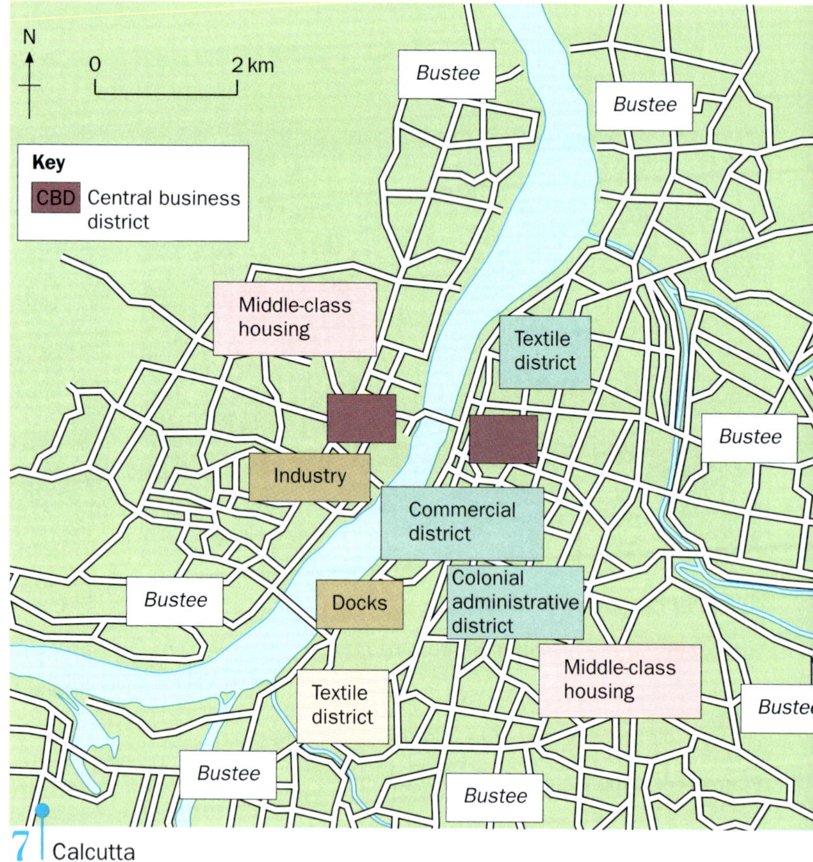

7 Calcutta

Task

SAVE AS...

1 a) List five problems faced by *bustee* dwellers in Calcutta.

b) For each problem you have listed, explain how the geographical site and/or situation of the city has created or affected that problem.

c) Explain how each measure taken by the Calcutta Metropolitan Development Authority has attempted to manage that problem, or, if it has not been managed, how the CMDA might manage it in future.

Finding work

Many newcomers to LEDC cities dream of finding well-paid work in a factory. The ideal job would have a contract, regular hours, weekly wages, pleasant working conditions, the chance to join a union, training, and the prospect of promotion. Jobs like this are in the FORMAL SECTOR of the economy. Other formal sector jobs include civil servant, train driver and shop assistant.

Such dream jobs are rare for migrants. Most newcomers find work in the informal sector. These jobs have at least some of the following features:

8 In Delhi's main railway station porters wait for the train to stop. They will then carry bags for passengers. This is a labour-intensive service.

- Irregular hours – people work when they are needed, if they are needed.
- Few regulations – e.g. health and safety regulations.
- No set workplace – many informal workers work on the street, selling or making goods or providing services.
- Irregular wages – people get what they can, and if they work for someone else they will be paid only if there is a profit.
- No taxes to pay – often the work is not registered with the authorities and payments are made in cash; but if people pay no taxes they do not get any benefits.
- Any training is given 'on the job'.
- Supplies come from friends or family, often from the home village.
- Transport is by bike, rickshaw, or on people's backs.
- Informal sector businesses are small and often run by family or friends.
- Such businesses often exist to provide for the needs of other poor people living in the same areas, and so they have to change quickly to meet the changing needs of a shifting population.

9 Ram Lal uses his one-room building as a house and a workplace. His informal manufacturing business makes the 'barrels' for ballpoint pens. Note the Hindu symbols on the wall and painted on his machine.

10 Part time, informal primary employment in a shanty town. This girl is selling recycled bottles.

The informal sector traditionally includes primary, secondary and tertiary work, but now it includes quaternary jobs too (see page 306).

- PRIMARY SECTOR jobs involve obtaining raw materials, in industries like farming, fishing, mining and quarrying.
- SECONDARY SECTOR jobs involve manufacturing – turning raw materials into finished products.
- TERTIARY SECTOR jobs provide services for people. Tertiary workers do not produce any finished products.
- QUATERNARY SECTOR jobs involve work in the ICT sector of the economy, including research.

Tasks

SAVE AS...

1 Various types of work are listed on the cards below. All of these take place in some, or all, squatter settlements. Make your own set of cards showing the name of the job. Add details to each card to show:

a) what training is needed

b) what equipment is needed.

2 Classify the jobs by sorting them into 4 groups. Then list them all in a table like this:.

Informal sector jobs			
Primary	**Secondary**	**Tertiary**	**Quaternary**

3 In small groups, discuss which tasks you are:

a) most surprised

b) least surprised

to see taking place in squatter settlements. Give reasons.

Focus task

'Acción Communal' role play

4 Imagine that you live in a squatter settlement like La Esperanza, but bigger. There are about 150 houses and about 800 inhabitants. You have been elected on to the committee of a group called 'Acción Communal' (Community Action) which works to improve conditions in the *barrio*. You work in three ways:

- raising money by persuading all inhabitants to pay into the Community Development Fund
- persuading people to do voluntary work building new facilities for the community
- working with the politicians to try to get them to provide services and facilities for the people.

The following projects have been proposed to the committee of 'Acción Communal'.

A
Build a schoolroom, and then persuade the city authorities to provide a teacher.
There are 68 children between the ages of 5 and 11 and there is no school in the *barrio*. Some go to a small private school in a nearby *barrio*, but this is expensive and the mothers do not like their children crossing the busy main roads to get there.
Cost: $1,000 for materials; 200 hours of work.
Time: Possible to open next September if the city backs us.

B
Set up a Credit Union (CU).
People save money with the CU. Once enough has been saved, the CU makes small loans to people who want to set up their own small businesses. They will have to pay interest, which will go back to the original savers. At first, all of the loans will go to businesses that will improve conditions for everyone in the *barrio*.
Cost: Acción Communal would need to provide a loan of $500 to start up, which might be paid back.
Time: An organiser would need to work 4 hours a week at first.

C
Build a police station, and then persuade the police authorities to pay a full-time policeman to work in the *barrio*.
He will keep an eye on the young hooligans who sometimes cause disturbances, and will stop some of the crime caused by homeless people who wander into the *barrio*. The police will have to be persuaded that the people here, although they are squatting, would like to buy the land if only they could find out who owns it.
Cost: $1,000 for materials; 200 hours of work.
Time: Possible to open next September if the police authority backs us.

D
Clear the polluted waste ground on the edge of the *barrio*.
This land was polluted by a factory that used to dump chemicals. Flatten the land and make it into a football pitch for the youngsters who have nowhere else that is safe to play. Also create some market stalls along one side. People could sell fruit, vegetables and meat and be shaded from the hot sun.
Cost: $300 for materials; $300 to pay for proper waste disposal; 200 hours' work.
Time: Could be finished in 4 weeks.

E
Save up and buy a small van for the *barrio*, and build a garage for it.
We could use it as a taxi to take people to and from work in the city centre 13 km away. People could also hire it to go out into the countryside to visit their home villages, and bring back fruit and vegetables.
Cost: $5,000 for van; $200 for materials for garage. Hire of van would pay running costs and help to pay for eventual replacement.
Time: We could go ahead as soon as the money is raised.

F
Build concrete platforms around the communal taps.
This would mean that people were not standing in mud while they filled their buckets. They could also rest the buckets on a firm surface, so they would be less likely to spill over. Also build a channel so that spilt water would drain away and so flies would be less likely to breed in stagnant puddles.
Cost: $50 for materials; 30 hours' work.
Time: We could complete the job in a week.

G
Raise money to send one of the women from the *barrio* to train as a local health worker and midwife.
These courses are run by a Catholic charity, so the cost is subsidised. The health workers specialise in maternity health and child health, especially giving vaccinations and treating tummy problems. They are also trained in good-hygiene practices for poor homes.
Cost: $1,000 for the training course; $500 for equipment to set up basic clinic. (Note that this is only 25% of the full cost. The charity pays the rest.)
Time: The course lasts 3 months.

The committee is going to meet a local politician to persuade him to help the people of the *barrio*.

a) Working in groups, discuss each of the possible schemes.

b) List the good points and bad points about each scheme.

c) Decide which is your favourite scheme. Also decide on at least one reserve scheme.

5 For both your favourite and your reserve schemes from Task 4, write a paragraph explaining how each will help your *barrio*.

5.3 How is city structure changing?

DO YOU KNOW?

- Shopping and service functions used to be focused in the CENTRAL BUSINESS DISTRICT of a city – the CBD.
- This is no longer the case in modern cities which have a more decentralised structure.
- The areas of most development in the early-21st century are right at the edges of cities at the urban–rural fringe.

AIMS

- **To understand the structure of today's cities.**
- **To consider how well your own local town or city matches the model of the decentralised city.**
- **To consider how traffic problems can be solved.**
- **To consider how pressures on the urban–rural fringe can be managed.**

A THE DECENTRALISED CITY

This chapter is about cities and the changing structure of cities. Studying one family's travel patterns within a city in a typical week will help you to understand the structure of a modern UK city.

The Sheridan family live in Jesmond (see Map 2 on page 212). Sally Sheridan is studying GCSE Geography at a secondary school in Newcastle upon Tyne. Her teacher asked her to take part in some research for the town planning department at the university. They were looking at the way people move around the city in their daily lives. The project was called 'Movement in the Decentralised City!' Ten people from the class were asked to keep a travel diary for their family, for a week. Sally agreed to take part. Figure 1 shows a part of her family's diary. Maps 2 and 3 (page 212) show all the places she refers to.

The Sheridan family

Sally travels to school in Gosforth. This is not her nearest school but it was chosen because of its good reputation as a language college.

She has a weekend job at the Silverlink multiplex cinema.

Will, Sally's twin, goes to a private school near the city centre, mainly because of its reputation for rugby.

He has no job, because weekends are spent playing rugby.

Jenny (mother) is a teacher at a school in Whickham, a suburb of Gateshead in the south of the Tyneside conurbation.

Steve (father) is the manager of a call centre, built two years ago on a GREENFIELD SITE just north of Newcastle.

	Journey	Reason	Approx. distance	Means of transport
SALLY SHERIDAN				
Fri	Home–Gosforth	School	4 km	Metro
	School–High St–School	Lunch and shops	1 km	Walk
	Gosforth–Home		4 km	Metro
	Home–Silverlink	Work	6 km	Car (Dad)
	Silverlink–Home		6 km	Taxi
Sat	Home–Newcastle centre	Shopping	2 km	Car (Mum)
	Newcastle–Tesco (Kingston Park)	Food shopping	5 km	Car (Mum)
	Kingston Park–Home		4 km	Car (Mum)
	Home–Silverlink	Work	6 km	Car (Dad)
	Silverlink–Home		6 km	Taxi
Sun	Home–Leisure centre	Play tennis	1 km	Bike
	Leisure centre–High Heaton	Visit friend	1 km	Bike
	High Heaton–Home		4 km	Bike
WILL SHERIDAN				
Fri	Home–Jesmond	School	2 km	Metro
	Jesmond–Kingston Park	Rugby club meeting	5 km	Car (Dad)
	Kingston Park–Home		4 km	Car (Dad)
Sat	Home–Durham	School rugby match	30 km	Metro/Coach
	Durham–Jesmond		28 km	Coach
	Jesmond–Newcastle	Looking round	1 km	Walk
	Newcastle–Jesmond	Visit friends	2 km	Metro
	Jesmond–Home		2 km	Car (friend)
Sun	Home–Gateshead	Watch Rugby League match	8 km	Metro
	Gateshead–Home		8 km	Metro
JENNY SHERIDAN				
Fri	Home–Whickham	Work	15 km	Car
	Whickham–Home		15 km	Car
	Home–Newcastle	Visit opera	3 km	Metro
	City centre–Quayside	Meal	1 km	Car
	Quayside–Home		4 km	Car
Sat	Home–Newcastle centre	Shopping	2 km	Car
	Newcastle–Tesco (Kingston Park)	Food shopping	5 km	Car
	Kingston Park–Home		4 km	Car
	Home–Heaton	Visit mother	4 km	Car
	Heaton–Local shops	Take mother shopping	1 km	Car
	Heaton–Home		4 km	Car
Sun	Home–Dunstanburgh	Plan Y9 field trip	60 km	Car
	Dunstanburgh–Home		60 km	Car
STEVE SHERIDAN				
Fri	Home–Longbenton	Work	5 km	Car
	Longbenton–Newcastle	Meeting/Lunch	7 km	Car
	Newcastle–Longbenton	Work	7 km	Car
	Longbenton–Home		5 km	Car
	Home–Kingston Park–Silverlink	Children's taxi service	14 km	Car
	Silverlink–Newcastle	Opera	6 km	Car
	Newcastle–Quayside	Meal	1 km	Car
	Quayside–Home	Take Jenny home	4 km	Car
	Home–Kingston Park–Home	Collect Will	8 km	Car
Sat	Home–Jesmond Dene–Home	Jogging	10 km	Foot
	Home–Silverlink–Home	Take Sally to work	12 km	Car
Sun	Home–Gateshead	Rugby League match	8 km	Metro
	Gateshead–Home		8 km	Metro

'journey string' – a series of journeys by a single form of transport to a series of destinations without returning to a base

1 The Sheridan family diary

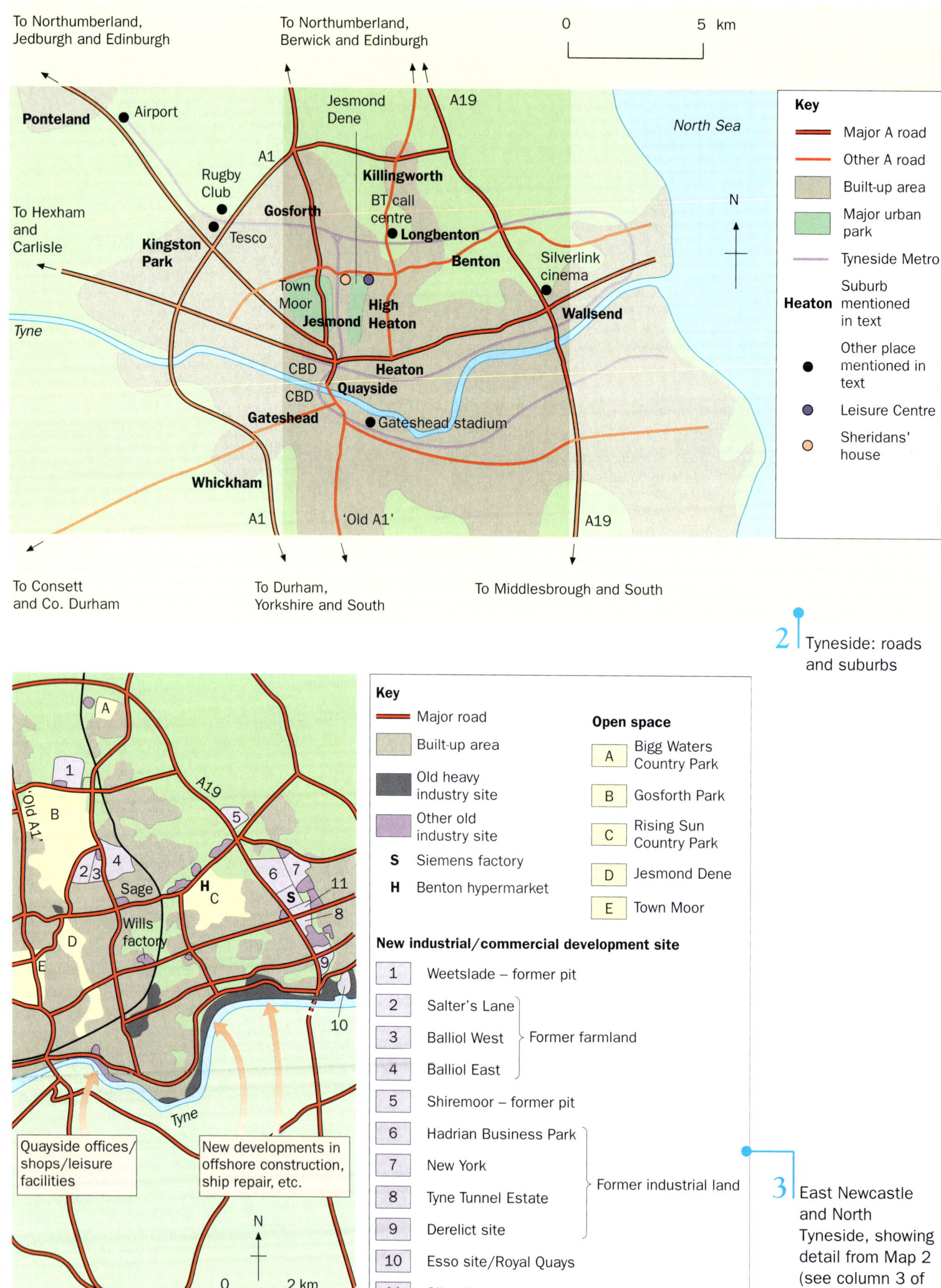
To Northumberland, Jedburgh and Edinburgh
To Northumberland, Berwick and Edinburgh
0 5 km
Ponteland
Airport
Jesmond Dene
A19
North Sea
A1
Rugby Club
Killingworth
Gosforth
BT call centre
To Hexham and Carlisle
Tesco
Longbenton
Kingston Park
Benton
Silverlink cinema
N
Town Moor
High Heaton
Jesmond
Wallsend
Tyne
CBD
Heaton
Quayside
CBD
Gateshead
Gateshead stadium
Whickham
A1
'Old A1'
A19
To Consett and Co. Durham
To Durham, Yorkshire and South
To Middlesbrough and South
Key
Major A road
Other A road
Built-up area
Major urban park
Tyneside Metro
Heaton Suburb mentioned in text
Other place mentioned in text
Leisure Centre
Sheridans' house
A
1
A19
'Old A1'
B
5
2 3
4
6 7
Sage
H
C
11
S
8
Wills factory
D
E
9
10
Tyne
Quayside offices/shops/leisure facilities
New developments in offshore construction, ship repair, etc.
N
0 2 km
Key
Major road
Built-up area
Old heavy industry site
Other old industry site
S Siemens factory
H Benton hypermarket
Open space
A Bigg Waters Country Park
B Gosforth Park
C Rising Sun Country Park
D Jesmond Dene
E Town Moor
New industrial/commercial development site
1 Weetslade – former pit
2 Salter's Lane
3 Balliol West
4 Balliol East
Former farmland
5 Shiremoor – former pit
6 Hadrian Business Park
7 New York
8 Tyne Tunnel Estate
9 Derelict site
Former industrial land
10 Esso site/Royal Quays
11 Silverlink

2 Tyneside: roads and suburbs

3 East Newcastle and North Tyneside, showing detail from Map 2 (see column 3 of the table on page 215)

Tasks

1 On a copy of Map 2, mark each of the Sheridans' journeys which are listed on page 211. Use a different colour for each person, and use a different style of line for each mode of transport (car, bike, Metro, etc.).

2 a) Complete a copy of the table below to show the different types of journey made by each person. In column 1 mark a tick for every journey from suburb to the city centre, or back. In column 2 mark ticks for journeys that start in one suburb of Tyneside and end in another. In column 3 put ticks for journeys that start and finish in the same suburb.

	1 suburb ↔ city centre			**2** suburb ↔ suburb			**3** within a suburb		
	car	**public transport**	**foot**	**car**	**public transport**	**foot**	**car**	**public transport**	**foot**
Sally									
Will									
Jenny									
Steve									

b) From your own experience explain why cars, rather than public transport, are often used for 'journey strings'.

3 The university's research was trying to show that movement patterns in cities have changed. In the past, most journeys were from the suburbs into the city centre, or stayed within one suburb. Their hypothesis was that 'in the modern city nowadays most journeys start in one suburb and end in another'. Does the travel pattern of the Sheridan family support or contradict this hypothesis?

ICT

4 The diary the Sheridan family kept provides lots of data on travelling around the city which will be easiest to analyse using ICT. For example, a spreadsheet will help you to work out:

What is the average length of journeys by each mode of transport?

Are there any links between the types of transport and the distances travelled?

a) Enter the data into a spreadsheet. You could lay it out, and start to fill it in, like this:

Sally	**Distance walked**	**Distance Metro**	**Distance car**	**Distance taxi**	**Distance bike**
	1	4	6	6	3
		4	2	6	1 etc

b) Use your spreadsheet to calculate:

i) the distance travelled by each person

ii) the average length of journey by each mode of transport.

c) Produce graphs to present the information visually.

5 Using what you know about the Sheridans' travel patterns, explain the problems faced by transport planners providing public transport for journeys from suburb to suburb.

6 A researcher visited the Sheridans a few days after they completed their diary to ask them some follow-up questions about their reasons for using certain types of transport for certain types of journey. Your teacher will give you a sheet with the questions and their replies.

a) Before you read their answers think about and list:

- reasons why people choose to use the car rather than public transport
- reasons why people use public transport rather than cars.

b) Compare your list with the Sheridans' replies to the researcher.

SAVE AS . . .

7 Suggest at least five ways in which the increasing use of cars is changing the structure of cities. You could refer to examples of change in Newcastle, or to another city that you know personally.

Understanding the structure of cities

Geographers sometimes use MODELS to help them to explain how cities have grown and developed. These help to explain why different activities are located in particular parts of cities, and the reasons for land use patterns. In these models cities have zones and each zone has a different function.

The earliest model was drawn up by Burgess in the 1920s (Map 4). He saw that cities had usually spread outwards from a central area. Most of the routeways met in the centre, so this was the most accessible place, and this was where the central business district (CBD) developed. Shopping and service functions were focused in the CBD. Around the CBD was a zone of mixed industry and commerce. Housing had developed in rings around these central zones. The oldest (and poorest) housing was nearest the centre. The further from the centre, the newer, bigger and, usually, better the housing. In this model the URBAN ZONES formed concentric rings:

- Each new wave of settlement created another ring of suburbs around the city.
- The wealthier you were the further out of the city you lived, so the movement of population was towards the outer suburbs – a process called SUBURBANISATION.

The Hoyt model (Map 5) was developed in the 1930s and added extra ideas to the Burgess model. It still showed the CBD, and areas of housing with the oldest near the centre and newest furthest from the centre. However, this model showed that industry developed in sectors along railways or canals which were used to transport the bulky raw materials needed by industry. Some industrial areas also grew alongside main roads into the town centres. In addition the model showed poor-quality housing developing close to the industrial areas to house industrial workers who could not afford to travel far to work.

Tyneside in the 1960s fitted the Hoyt model very well – see Map 6.

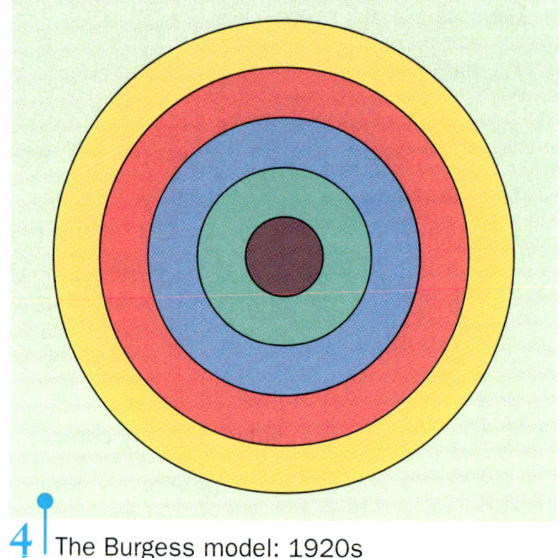

4 The Burgess model: 1920s

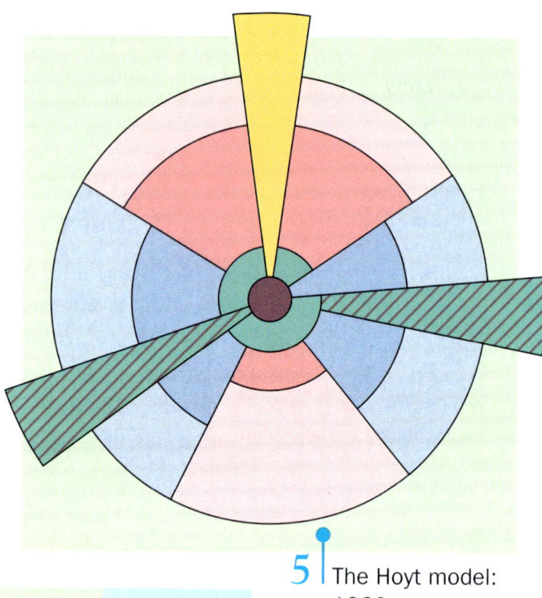

5 The Hoyt model: 1930s

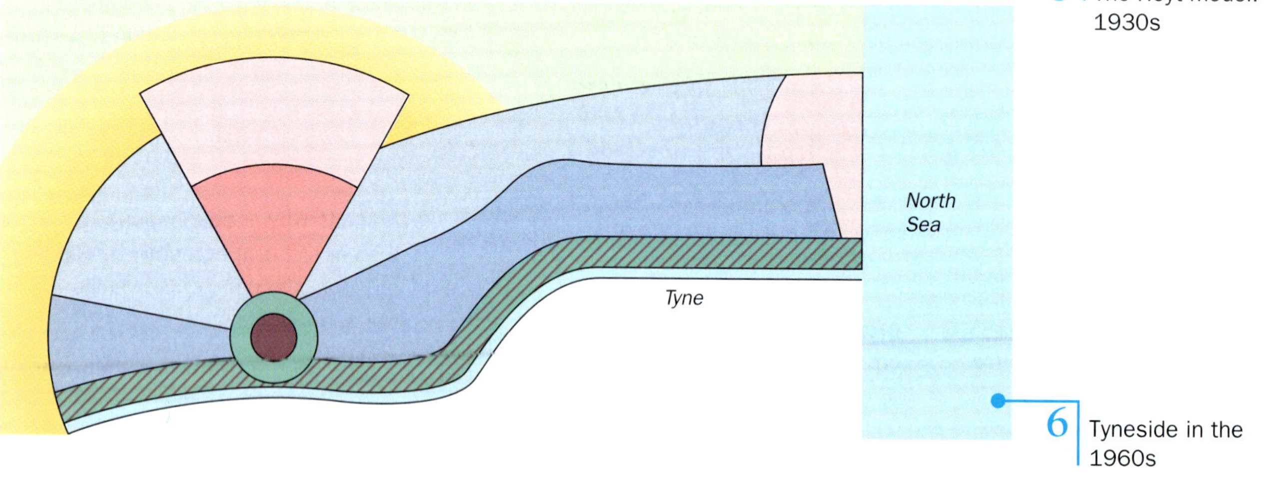

6 Tyneside in the 1960s

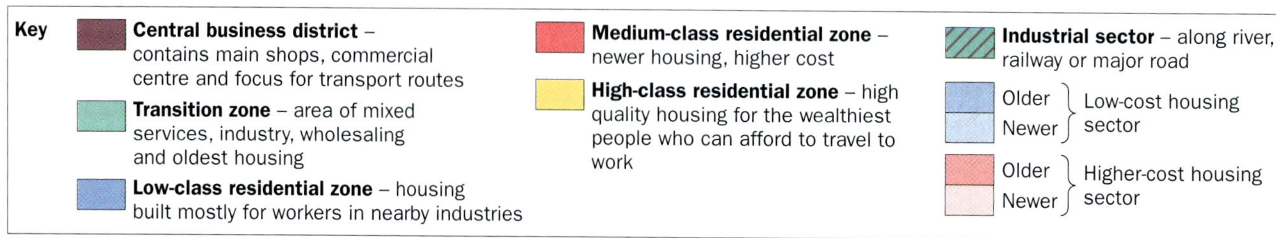

From the early 1960s onwards, the structure of most towns and cities in the UK changed rapidly. Some of the main changes are described in Table 7. This also shows the way these changes affected Tyneside.

Task

SAVE AS . . .

1 On a copy of Table 7 add a column for your own examples. Working as a class or group, fill in as many examples as possible from your own town or city.

★ THINK!

There are many changes shown in Table 7. You cannot possibly remember them all for an exam. The important thing is to remember the *patterns*. When you come to do your revision, pick out one pattern and one example for each category. Better still, use Task 1 to help you to revise examples for a city that you know. You will probably not need more than four or five for the exam.

Category	Pattern	Tyneside examples (see Maps 2, 3)
1 Inner city decline	Old heavy industry gradually declined.	Shipbuilding at Wallsend reduced. Engineering in Heaton reduced.
	Old working-class housing, close to industry, redeveloped.	Terraced housing in Byker demolished. New housing (e.g. Byker Wall) built.
	Many people from inner cities moved to out-of-town estates.	New estates built at Longbenton and Killingworth New Town.
2 Transport networks improved	New roads built to by-pass the city, reduce congestion in the centre and provide access to the edge of the city.	A189 built to by-pass the city to the east, through the Tyne Tunnel. A1 western by-pass built, linking to the national motorway system.
	New public transport systems set up to reduce cars in city centres and to improve access from suburbs.	Metro light railway system.
3 Out-of-town developments	New hypermarkets and out-of-town shopping replaced local shops and city centre shops.	Benton hypermarket built in 1980s. Silverlink multiplex cinema in 1990s.
	New industry attracted to GREENFIELD SITES on edges of cities, especially if close to good road connections.	New industrial developments at Silverlink and Salter's Lane
4 Former industrial sites redeveloped	Some BROWNFIELD SITES redeveloped for industry, but must have good road infrastructure.	Old colliery reclaimed at Weetslade; old oil refinery reclaimed at Esso site.
	Other brownfield sites landscaped and site used for nature conservation/leisure.	Old collieries reclaimed at Rising Sun site to make Country Park.
	Some old industrial buildings redeveloped for new uses.	Wills cigarette factory converted into luxury apartments. Quayside redevelopment of former dock area.
5 Inward investment	Government attempts to attract INWARD INVESTMENT from abroad, but . . .	Siemens factory: semi-conductors for computer industry . . .
	. . . in times of financial crisis these may be the first places closed by foreign firms.	. . . Siemens factory closed before it was open due to fall in world price of semi-conductors; loss of £40 million.
	Area tries to attract high-tech industry because of better long-term prospects than old, heavy industry.	e.g. Sage (computer programs for accountants).
	Area tries to attract LABOUR-INTENSIVE INDUSTRIES because of high unemployment.	e.g. BT call centre.
6 Inner city renewal	City centres suffer decline as shops and services move to out-of-town sites.	Grainger Town, empty office space in old buildings.
	Government and local authorities form partnerships in an attempt to bring activity back to old city centres.	Grainger Town Project to revive inner areas: leisure, tourism and housing in renovated buildings.

7 Changes in the structure of UK cities

The development of the decentralised city in the UK

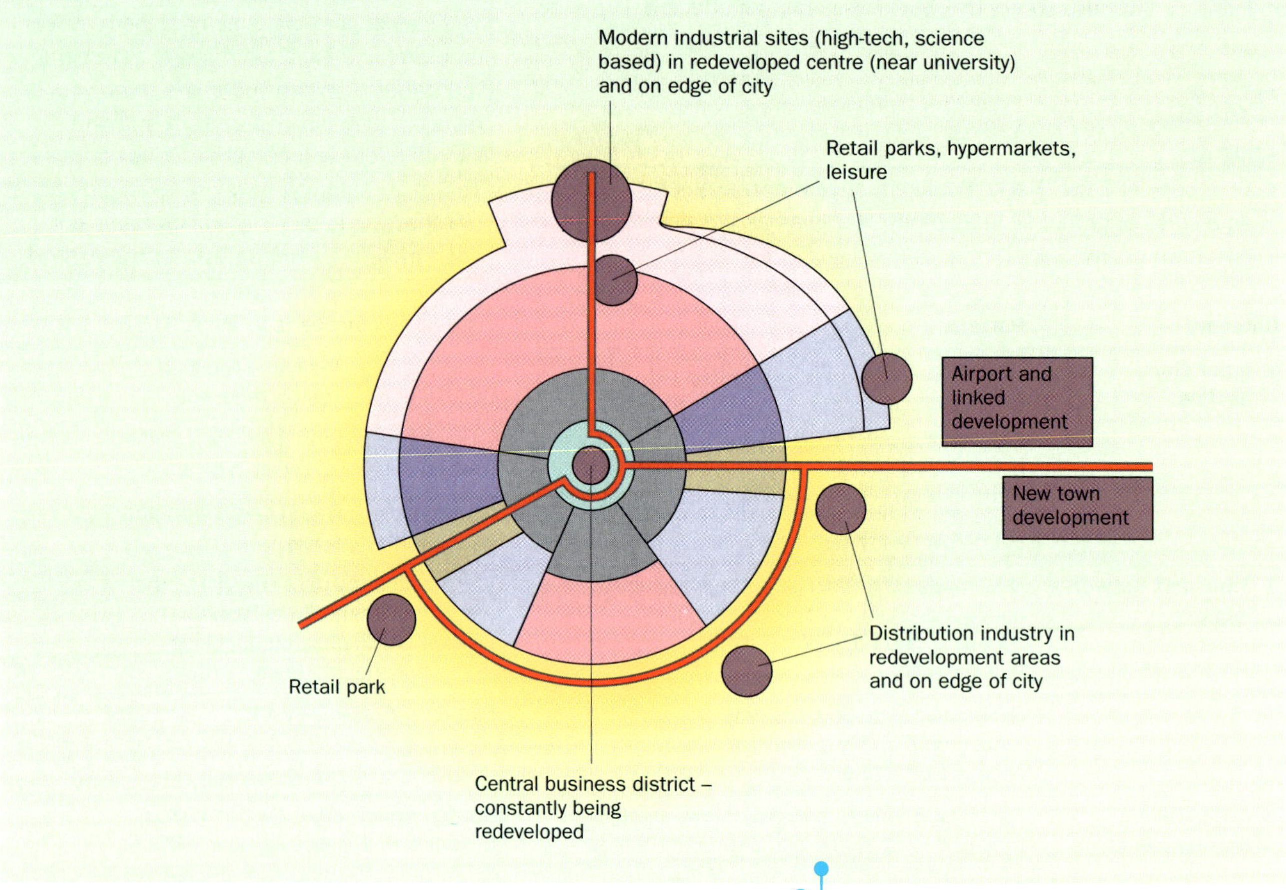

8 The decentralised city

Changes in industry and transport have together helped to create 'the decentralised city'.

In the 19th and early-20th centuries:

- most city dwellers worked in factories or in shops and offices in the CBD
- many people had to travel from the suburbs into the city centre, where they also did most of their shopping
- most travelled into the city by public transport

Mass-ownership of cars grew in the late-20th century, which meant that:

- people could move freely around the city, and did not have to rely on public transport
- work and services did not have to be centralised
- businesses began to rely on road transport rather than rail and canal
- motorways and ring roads were built to give easy access to sites on the edge of the city
- land at sites on the edge of cities was cheaper than in the centre, allowing cheap car parking as well as cheaper building.

As a result, the 'decentralised city' has many of its functions spread around the suburbs and the urban–rural fringe.

These changes have caused geographers to develop new urban models, adapting the ideas of Burgess and Hoyt. The new models may seem more complicated, but they are more up-to-date and more realistic (Figure 8).

Task

SAVE AS . . .

1 Draw your own concept map (see page 32) to summarise the links between increased use of road transport and changes in the structure of cities. Put 'transport' in the middle; the changes mentioned on pages 214–16 around the outside; then draw links and label them to explain each link.

Case study: Los Angeles – the ultimate decentralised city

Los Angeles, on the west coast of the USA, is very much a decentralised city because it is a fairly recent urban development. It did not grow on the remains of a pre-existing concentric or sector-based pattern. It has no real CBD or city centre. Instead it consists of separate suburbs linked by massive, multi-lane freeways. The city could grow in this way only because land was very cheap, there was plenty of room to expand and car travel was cheap.

Los Angeles is now so big that everyone has to travel by car, and so roads and car parks take up enormous areas of land. The car allowed Los Angeles to develop, but many people think that it is harming the city now.

Division

The car has led to a massive division between the rich and the poor. The rich are able to live further and further away from the city centre, to travel to shops in out-of-town malls, and to travel to all of the area's wonderful leisure facilities. Meanwhile the poor are trapped in run-down areas like Watts and West Hollywood. In these areas minority ethnic groups and recent immigrants – especially illegal immigrants from South America and Asia – make up a large part of the population.

Pollution

Los Angeles is one of the world's most polluted cities, because the unusual atmospheric conditions often trap car exhaust gases in the air over the city (see page 90).

In addition, the USA has some of the world's lowest fuel tax rates, so car use and pollution both continue to increase.

Task

ICT

2 Los Angeles is well represented on the internet. You can find Landsat images of LA on the US NASA education site at:

http://landsat.gsfc.nasa.gov

Choose one picture, copy it and print off three copies. Then, using an atlas to help you, annotate the pictures to show:

- Picture 1 – the main features of the city
- Picture 2 – how the physical geography encourages the growth of the city in some directions and stops it from growing in others
- Picture 3 – how LA compares to the decentralised city model in Figure 8.

9 A shopping mall in Los Angeles

10 Freeways in Los Angeles

B ISSUES IN THE DECENTRALISED CITY

How can we solve urban traffic problems?

Let's return to the Sheridans (page 210). They make many journeys between different suburbs of Newcastle upon Tyne. Many other people living in decentralised cities do the same. However:

- they do still travel into the city centre for shopping, leisure and other activities
- they do still travel into the centre by car
- Newcastle has excellent public transport (the Metro), but even in Newcastle traffic congestion in the city centre is a very serious problem
- URBAN REDEVELOPMENT and renewal schemes, aimed at bringing more people and services back into city centres, will make it worse.

Note Not all groups in society have equal access to cars. This can have a serious affect on their mobility. Which groups have best access?

- men or women?
- rich or poor?
- young, middle aged or the elderly?

How does this affect their lifestyles?

Traffic congestion is now a major problem in many city centres in the UK and around the world. Many different methods are being tried or considered to help to reduce congestion (see Table 11). They work in two main ways:

- 'carrots' to encourage people to use public transport, to walk or to cycle.
- 'sticks' to discourage people from using cars.

Description	Example	Advantages
Mass transit systems Vehicles run on tracks – separate from cars and often underground in the most congested city centre.	Tyneside Metro, Sheffield Supertram, Manchester Light Rapid Transit, London Underground	• Separate from cars • Not held up by traffic jams • Reduced air pollution • Can be quick and efficient
Bus priority lanes One or more lanes of main roads can be reserved for buses at all times or at peak times only.		• Speeds up buses • Should make buses more reliable • Should encourage bus use • Slows down cars so discourages their use
Road pricing or road rationing Cars can be automatically registered when they enter city centre areas, and then charged for using roads. Cars allowed into centres only on alternate days.		• May cut down traffic • Provides money for transport improvements
Pedestrianising city centres Many CBD shopping areas have closed to all road traffic, except for emergency vehicles, disabled people's cars and delivery vehicles at non-peak periods.		• Safer for pedestrians • Reduces air and noise pollution • May increase shopping in city centre shops
Increased car parking prices Land in city centres is expensive. Building car parks is also very expensive. Costs of car parking can be made very high. Penalties for illegal car parking can also be made very high.		• Discourages car use • Increases public transport use • Profits can be invested in public transport systems
Cycle lanes and other cycle-friendly policies Separate cycle lanes can be designated on roads, or separate cycle paths are built. Traffic lights and other junctions can be adapted to protect cyclists and give them priority over cars.		• Encourages bike use and discourages car use • Health benefits for cyclist • Reduced pollution

11 Solutions to traffic congestion

Tasks

1 Study Table 11. Classify the measures in Table 11 as 'carrots' or 'sticks' for the motorist.

2 Discuss: Are there any groups in society who you think should be exempt from car-restriction measures and allowed to use their cars as much as they need? Give reasons.

3 Use the internet and/or towns or cities that you know well to complete the 'example' column of your own copy of Table 11.

4 Using a town or city that you used in Task 3:

- **a)** describe an area where traffic congestion is (or was) common
- **b)** describe one or more schemes that have been used to ease traffic congestion in that area
- **c)** outline how you might assess how successful these schemes have been.

5 Choose one or more of the schemes to reduce congestion listed in Table 11. *Either* draw a cartoon *or* write a radio advertisement to try to convince motorists that your chosen scheme(s) will be the most effective.

Disadvantages
• High cost of investment • Limited network cannot serve whole conurbation
• Annoys motorists • Illegal use by cars • Accidents can block the bus lanes
• Unproven technology • Expensive to run • The less well-off will be discouraged but it won't affect the rich who will be able to afford to pay
• Some shopkeepers fear reduced trade • Less convenient for motorists • Increases congestion outside the pedestrianised area
• Drives customers away from city centre shops towards out-of-town developments
• British weather! • May annoy car users • Quite expensive given the present low use of cycles • Difficult to police • Cycling can be dangerous • Cycling accidents are particularly likely where cycle lanes rejoin ordinary roads

12 Sheffield Supertram built in the 1990s to link the city centre with the poorer northern and eastern parts of the city. It does not travel out to the richer southern suburbs where most residents still travel by car.

13 The London cycle network – 2,000 miles of marked cycle routes – is intended to make London a cycling city. However, in 1990 cycle journeys into central London were still less than one per cent of all journeys.

Conflicts on the rural–urban fringe

The rural–urban fringe is the area on the edge of a city. This area is not totally part of the city, nor is it totally part of the countryside. It is where the two meet and has features of both. It has the attractions of being near open countryside without being too far from the conveniences of the city. For many economic activities it is the dream location because there is space to build; it is accessible by car because it is usually well served by motorways or good main roads; and it is within reach of the large population concentrations of the city and its suburbs.

So, as cities become more decentralised, many different businesses have their eyes on the rural–urban fringe:

- House-builders want to build housing estates there because people want to live there.
- Developers want to build shopping centres there because land is available and parking is easy.
- The leisure industries want to build leisure clubs, golf courses or hotels there.
- Businesses want to move their offices out there because it is a more pleasant environment for their workers than the city, and costs are lower.

The downside of this is that developments on the rural–urban fringe can have many unintended consequences on the physical and human geography of the area:

- Some of the best farmland is in the rural–urban fringe and this can be lost.
- Wildlife habitats can be lost to new developments.
- Urban road congestion can spread into rural areas.
- Over-development puts strains upon water and sewerage systems that were not designed to cope with large populations.
- The character of the rural areas is changed by activity moving out from the city.
- In some areas the flood risks are increased by the extra building taking place on floodplains in the rural–urban fringe.

14 Maidenbower, a commuter village in West Sussex

At the same time, the more the urban fringe is developed the more the life is drained out of the inner city. Competition from the rural–urban fringe has been a contributor to the decline of city centres over recent decades.

For all of these different reasons and because of these pressures, the rural–urban fringes around cities in the UK have usually been specially protected over the past few decades. Since the 1950s UK policy has been to prevent urban areas from getting ever bigger as in Los Angeles. So there is an area called a GREEN BELT that surrounds many UK cities. All new building in this green belt is in theory forbidden – although some developments do get round the restrictions. The green belt helps to preserve the countryside closest to the town but it has another, less desirable, effect; out-of-town development can leap-frog the green belt and spread further out into the countryside. Villages far from the edge of the city become the focus for new development – particularly for housing developments. Some villages even 50 or 60 km from the edge of London have effectively become suburbs of London, with the vast majority of residents commuting by train or car into London to work.

Task

SAVE AS...

1 Look at Map 3 on page 212.

a) List the new developments that have taken place in recent years along the A19 to the east of Newcastle. Suggest how these have affected:

- farmland
- wildlife
- the air quality
- sewerage
- drainage
- public open space
- traffic on the roads
- water supplies
- local village communities
- the likelihood of flooding.

b) Why will the development of industry and commerce in this area have affected the value of housing in villages to the north of Newcastle that have easy access to the A19?

c) The development of this area is causing an upward spiral of economic growth. Draw a diagram to illustrate this. (To help you with this, look at the 'downward spiral of decline' on page 240. Your spiral should go the opposite way.)

d) 'Changes that bring economic development can cause social and environmental problems.' Do you agree with this statement? Explain your answer.

Some people resent us moving in here. They say that people from the town are ruining the character of country villages. But is it their village? The people who complain the most are the people who arrived five years before us! The long-term locals are more likely to say we are bringing life back to the village.

We worked hard to get here. We earned enough to buy our old cottage and to extend it. We employed a local builder who was glad of the work.

And why shouldn't our children have a large garden? Why shouldn't they breathe fresh air? Why shouldn't they learn how to ride a horse? Everyone would want this life if they could afford it.

Farmers have benefited. One leased the land he wasn't using to be made into the golf course we belong to. We're all winners!

One other complaint is that we don't use the village shop. Well, we're not the only ones. A lot of the locals go to the superstore in their muddy Land Rovers. It's only the older people who don't have cars who buy all their shopping in the village. The prices are higher than in the superstore on the edge of town, and I don't think the food is as fresh.

At least the landlord of the Belben Arms understands us. He realises that it's people like us who bring his profits. He is improving it to make it a real country pub, with lots of shiny horse brasses and old photographs of people cutting corn with scythes.

When I was a boy, my dad used to grumble about all the old High Street stores being taken over by big chains. 'Every town in the country will soon look the same,' he used to say. I'm glad he's not here to see what's happening on the ring road on the edge of town.

There it is – cheaply built, boxy buildings round a huge car park. A computer store, sports goods store, fashion store, DIY store, fast food outlet, multiplex cinema and ten-pin bowling. Every town has one just the same. Every city has dozens, just the same. What a mess!

And the traffic – all weekend cars clogging the roads in and out, exhausts chugging. All this where my dad used to walk and ride his bike. The final insult is that they have even built a 'garden superstore' right where he used to have his allotment.

Focus tasks

2 Read the two speech bubbles. They give extreme and stereotyped responses to developments on the urban–rural fringe. Many people have strong opinions on urban developments. They find it hard to be objective or neutral because most people are affected by these changes in one way or another. In a moment you are going to try to be an objective geographer. But, first, try being biased.

- **a)** Working with a partner, role play a conversation in which one of you supports development on the urban–rural fringe, the other is completely against it.
- **b)** Swap roles and continue.
- **c)** Discuss which role you found easier.

3 Now you are going to be a geographer. List the geographical issues that lie at the heart of this debate.

15 A retail park in Crawley

Sustainable urban living: how should we deal with household waste?

The average person in the UK produces 350 kilograms of household waste per year. This may not sound much of a problem to you but that is partly because waste disposal in the UK is so efficient that we are hardly aware of the waste we produce. However, being efficient at disposing of waste and being good at managing waste are very different things. You can see from Figure 16 that the UK is lagging behind other countries in recycling and reusing household waste. Many people feel that this is unsustainable and that we need to change our attitude to household waste – to see it as a resource to be managed rather than a nuisance to be disposed of.

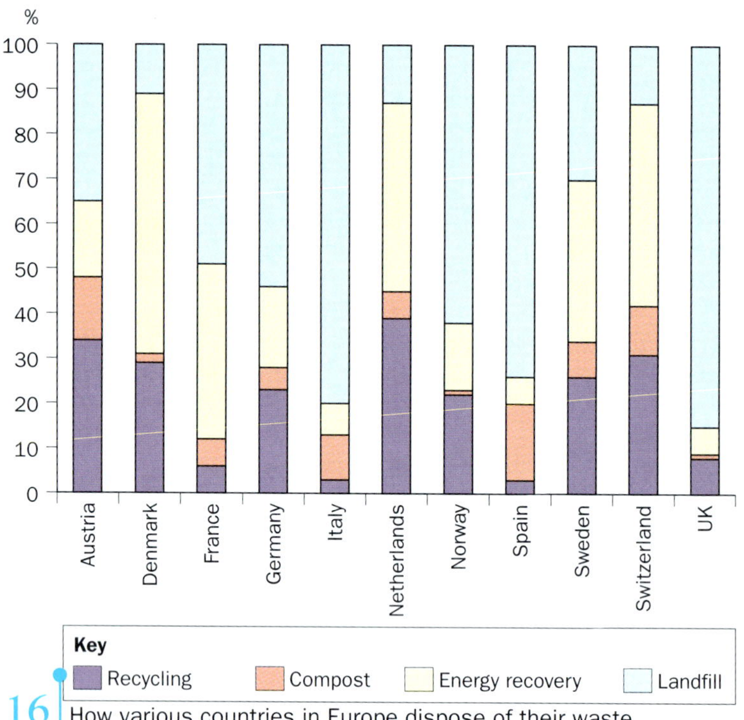

16 How various countries in Europe dispose of their waste

Learning from LEDCs

In LEDCs recycling is often an important part of the everyday economy. It is common for people to collect and make use of other people's rubbish. Waste tips are scavenged for any recoverable or recyclable materials. For instance, in Kenya:

- old car tyres are cut up and used to make cheap sandals
- tin cans and old oil drums are used to make charcoal stoves, lamps, buckets and metal tips for ploughs
- glass bottles are collected and returned to stores for refilling
- food waste is collected and fed to animals or composted for vegetable plots.

Waste is seen as an economic opportunity. In the informal sector (see page 207), where people do not have reliable paid employment, it is often the easiest and sometimes the only way of making money.

Landfill
Waste is dumped in old quarries or hollows. This is convenient and cheap in the UK (although not in countries like the Netherlands – can you think why?). However, it is unsightly and is a serious threat to groundwater and river quality, as toxic chemicals can leach out into the water. Decaying matter at landfill sites also produces methane gas which is not only explosive but also a very potent greenhouse gas. To discourage use of landfill sites a landfill tax can be imposed.

Energy recovery
This is any method of waste management that takes waste material and converts it into energy. The main one is *incineration*. In the past many councils burned their waste. Nowadays this has gained a bad reputation because it adds to CO_2 emissions and releases pollutants into the atmosphere. Many old and polluting incinerators have been closed down. However, some modern incinerators generate electricity, or power neighbourhood heating schemes, so these are considered to be a sustainable option.

Composting
On a small scale, organic waste (kitchen scraps and garden waste) can make compost which can be used to fertilise gardens or farmland. On a much bigger scale, 'anaerobic' digestion is an advanced form of composting in an enclosed reactor. Biological treatment of organic waste speeds up the breakdown process. It is then possible to use the biogases produced to provide an energy supply, and the solid residue can be used as a soil conditioner. Germany, Denmark, and Italy all have such plants but they are expensive to set up.

Recycling
Waste products such as paper, glass, metal cans, plastics and clothes can be reused if they can be collected economically. However, the initial start-up costs of recycling schemes may be high, and the market value of the material may be low. Householders may also be unwilling to sort recyclable waste from their other household waste.

17 What can we do with waste?

Tasks

1. Study Figure 16. Write a sentence to describe how the UK's approach to waste disposal compares with that of other European countries.
2. Study Figure 17. Draw up a table to show the advantages and disadvantages of each method of disposing of household waste.
3. Discuss how the examples of recycling from Kenya, an LEDC, are relevant or not relevant to people in an MEDC?

Learning from MEDCs

You can see from Figure 16 that the Netherlands has almost the best record of recycling in Europe. This is for various reasons: it is a very densely populated country with very little waste land; it is flat and does not have quarries or natural indentations to use for landfill. The government has therefore taken active measures to discourage waste. All neighbourhoods have recycling points. Waste is collected free from these points but households have to pay for all other waste to be taken away. Each bin is weighed and people are charged by the kilo.

All figures are %	Aluminium	Steel	Paper	Glass
Energy use reduced by	90–97	47–74	23–74	4–32
Air pollution reduced by	95	85	74	20
Water pollution reduced by	97	76	35	0
Mining waste reduced by	10	97	0	80
Water use reduced by	0	40	58	50

18 Benefit of using recycled materials instead of new raw materials

Case study: How is waste managed in Newcastle upon Tyne?

For many years Newcastle upon Tyne has based most of its waste disposal at the Byker Incineration Plant.

- Waste is collected in lorries and taken to the separation plant. Giant magnets take out all the waste that contains iron or steel. This is sent for recycling.
- What remains is broken up and put into giant drums called centrifuges. These whirl round the waste to separate out material of different densities. Valuable metals like copper, lead and aluminium are then also sent for recycling.
- The residue is burnt in the incinerator.
- The heat from this incinerator heats water which produces steam which is used for central heating in the flats in the Byker Wall development and to heat the nearby swimming pool.
- The small amount of ash residue is then either taken by barge and dumped at sea, or taken to be buried in landfill sites.

This sounds very efficient. However, there are some environmental problems with this system. Much of the material burned in the incinerator is plastic. If the incinerator is not hot enough it can produce fumes which can cause health problems. Also, ash dumped at sea has damaged some fishing grounds, because it makes the seabed sterile. Finally, this system ignores glass, paper or compost. These are being incinerated instead of being recycled. The city is now taking action, as described in Figure 19.

HOUSEHOLD WASTE IN NEWCASTLE

Newcastle residents produce almost one tonne of waste per household per year. Doing any of the following will help you reduce the environmental burden on your city.

1. REDUCE YOUR WASTE

- ✔ Reduce the junk mail you have to dispose of by getting your name removed from mailing lists.
- ✔ Reduce the unnecessary packaging you bring into your house. (An estimated one eighth of household waste is packaging.) Buying goods loosely or in bulk may help you to do this. Reuse old carrier bags at the supermarket.

2. RECOVER A PORTION OF YOUR WASTE FOR RECYCLING

- ✔ Newspapers, glass bottles, cans, plastic bottles, textiles, books, telephone books, aluminium cans, waste motor oil, cardboard can all be recycled at the city's recycling sites (6 shopping centre sites; 28 school sites; 14 other sites).
- ✔ Recycle small amounts frequently. Combine your recycling with your weekly shop – avoid making special trips to recycle. Deposit your recyclable material in the appropriate bank. Do not leave it at the base of the bank.

3. COMPOST

Composting in your back garden is the simplest way to improve the environment. The Council's Compost Project provides free training to novice composters. It also sells 220-litre plastic compost bins to City residents at £6 each, a saving of £6. You save money, help improve the environment and contribute to organic gardening.

4. BUY RECYCLED

Breaking into established markets is difficult for producers of recycled products. You can help to 'close the recycling loop' by actively buying material with recycled content.

19 Advice to Newcastle residents adapted from the City Council website, www.newcastle.gov.uk

SAVE AS . . .

4 a) Draw a diagram to show how household waste is dealt with in Newcastle at the Byker plant and through the recycling schemes.

b) Do you think Newcastle's is a good scheme? Write a paragraph to explain the strengths and the weaknesses of the scheme.

5 Figure 19 comes from the Newcastle City Council website. Add a feature which might help to persuade people in Newcastle to adopt these recycling measures, using any of the information found on pages 222–23.

5.4 How do people decide where to live?

DO YOU KNOW?

- Quality of life means how good your life is.
- There is a growing shortage of housing in the UK. The number of people wanting to buy homes is growing faster than the number of homes.
- There are over 400,000 homeless people in the UK – nearly 1 per cent of the population.

AIMS

- **To consider the factors that influence quality of life in urban areas.**
- **To understand how access to services varies within an urban area.**
- **To make decisions about where to live and who should be housed.**
- **To understand how access to housing varies for different groups of people.**

A WHICH IS THE BEST HOUSE FOR THE BARLOWS?

On average, people in the UK move home once every seven years. Where people decide to move to will depend on various factors, including jobs, education and their changing family circumstances. Most people move in the hope that their quality of life will also improve. (You will find out more about quality of life in Chapter 7.1.) They will consider the house itself – its size, its amenities, the quality of the building – and the area in which it is located. These factors influence the price of the house, and so where people move to will also depend on where they can afford to live.

The Barlow family – Gordon, Louise, Chloe and Jack – live in London, but they want to move to Chelmsford for a better quality of life. They visit some of the local estate agents in Chelmsford. There are hundreds of homes for sale, but the family have narrowed it down to four that they really like (Figure 2). Now they have to decide which house to move to.

I'll have a lot more space to play football once we move out of London.
I'll be able to commute to my job in London in just over half an hour.
Chelmsford has all the attractions of a town but it's much closer to the countryside than London.
I'm just about to start secondary school, so it's a good time for me to move.

1 Chelmsford

Chelmsford is one of the fastest-growing towns in the UK. It is in Essex, just 50 km north-east of London, and it attracts many people moving out from the capital. Some of them continue to commute into the city for work, but Chelmsford also has many large employers of its own, particularly in the growing information services sector.

The town faces huge pressure to provide homes for the growing number of people in the South East. Its population has more than doubled since 1951, to 160,000 in 2001. Most of the recent growth has been to the north-east of the town. Development to the south-west is restricted by the green belt, the land surrounding an urban area where new building is limited to protect the countryside.

Lower Anchor Street, Chelmsford

£120,000

Three-bedroom terraced cottage built during the early 1900s in the fashionable Moulsham area. This property benefits from many original features from the period, including high ceilings, open fireplaces and sash windows. Chelmsford town centre with its mainline railway station is within easy walking distance of less than 1 kilometre.

Lounge • Dining room • Kitchen • Three bedrooms • Modern bathroom Gas central heating • Stripped timber floorboards and doors 25 m rear garden • Parking permit available

Fourth Avenue, Chelmsford

£170,950

Extended four-bedroom semi-detached house built during the 1930s, the property has been modernised and improved in recent years. Fourth Avenue is one of the most sought-after cul-de-sac locations in Chelmsford and offers excellent access to Chelmsford town centre with its mainline railway station within 1 kilometre.

Entrance hall • Cloakroom • Lounge • Dining room • Kitchen/breakfast room • Four bedrooms • Bathroom with shower • Gas central heating • Double-glazed windows • 20 m south-facing garden • Garage and driveway

Chelmer Village, Chelmsford £161,995

Built in the 1980s and situated on the popular Chelmer Village development, a three-bedroom detached house with a pleasant south-facing garden. The property is still new and in immaculate condition throughout. Chelmer Village is to the north-east of Chelmsford town centre and has its own superstore, school and retail park.

Three bedrooms, one with en-suite shower • Fitted kitchen with wooden units • Lounge • Dining area • Entrance porch • Gas central heating • Garage and driveway • 16 m garden

Springfield, Chelmsford

£227,995

Luxury newly-built mock Georgian town house on the very edge of Chelmsford overlooking open countryside. This new development offers all the convenience of urban living in an exclusive setting. Springfield is north of Chelmsford, about 3 kilometres from the town centre, with easy access to the A12 trunk road to London.

Four bedrooms, one with en-suite bathroom • Large fitted bathroom • Lounge • State-of-the-art fitted kitchen • Dining room • Downstairs cloakroom • Double garage • Modern conservatory overlooking garden

2 Some of the properties for sale in Chelmsford

Tasks

1 Read the four estate agents' descriptions in Figure 2.

a) List all the factors that would have a positive influence on quality of life.

b) You would not expect these descriptions to mention factors that would have a negative influence on quality of life, but can you think of any? Make another list of those factors.

c) Rank the houses in your own order of preference for the Barlow family.

d) Prepare a list of questions that you would like to ask about each house if you were going to view them before you decided which one to buy.

ICT

2 You can use the internet to find out about houses in Chelmsford or other towns and cities in the UK. Businesses are now using the web to attract customers to view property. One site – www.homes-on-line.com – has thousands of properties listed. You need to specify the county (Essex) and the town (Chelmsford) when you search the site. Property details can be printed out and compared with those in Figure 2.

a) Suggest the advantages and disadvantages of using the web to search for property.

b) Suggest the advantages and disadvantages of estate agent's descriptions for studying the geography of a town or city.

Note that the prices quoted in the adverts used in this book are year 2000 prices. These will probably have gone up – but may also have gone down – when you do this research.

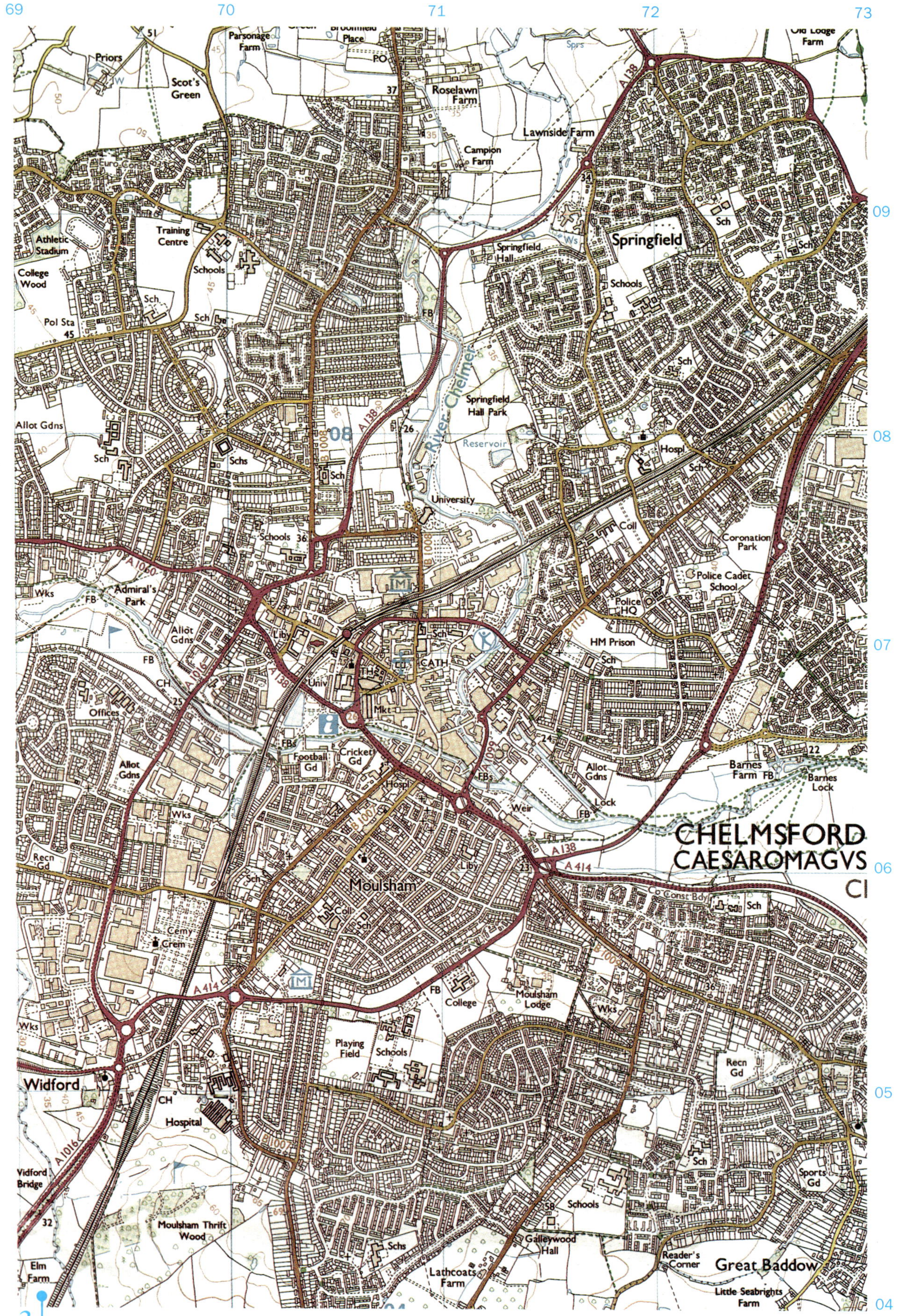

3 Map extract: Chelmsford. Reproduced from the 1998 1:25,000 Ordnance Survey map by permission of the Controller of HMSO © Crown Copyright

Where can the Barlows afford to live?

4 Chelmsford town centre

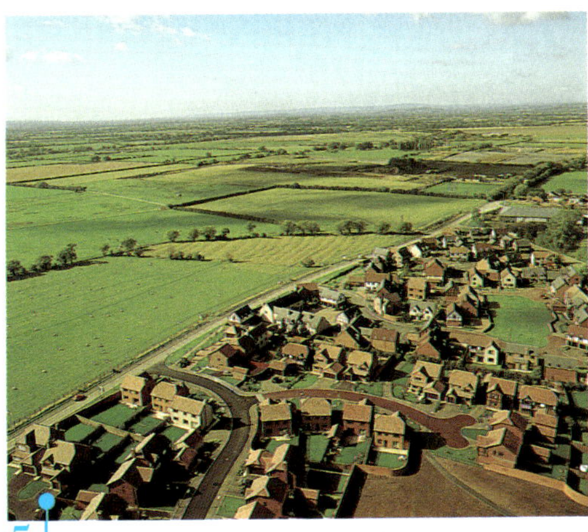
5 A typical new housing development

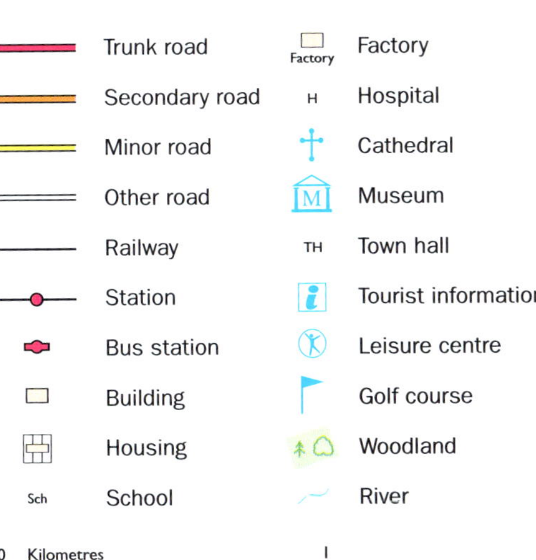

Most people buying a house have to borrow money from a bank or building society, especially if they are buying their first home, or buying a larger home in a more expensive area. Banks and building societies can offer a *mortgage* – a long-term loan that has to be paid back over a period of 20 or 30 years. Usually, homebuyers need to find at least 10 per cent of the value of the property themselves.

Before the lender (the bank or building society) agrees to give a mortgage, it needs to know the household income of the prospective buyers. This is to ensure that the people borrowing the money can afford to pay it back. Most banks and building societies have a similar formula to calculate the maximum amount of money they are willing to lend. For example, for a typical household, where two people are working, the maximum that a bank or building society will lend is 3 times the highest annual income plus 1 times the lower annual income.

Tasks

1 Look at Map 3 and Photos 4 and 5.

a) Match each of the houses in Figure 2 with one of the following grid references: 706084, 729075, 710062, 729092 (one of them is a newly developed area which has been built since this map was made).

b) Using information from the map and photos, describe the advantages and/or disadvantages of each location as an area to live in. This is called an environmental quality assessment. Consider access to the town centre, services such as shops, open space, and transport routes. Complete a table to show the advantages and disadvantages of each house.

c) Look back at the rank order in which you placed the houses (Task 1c on page 225). Now that you have more information, how has your order of preference changed?

2 Work out which of the four houses the Barlow family could afford to buy.

a) Gordon earns £36,000 a year from his job. Louise can earn about £12,000 a year from her job. Work out how much a bank or building society would be prepared to lend them.

b) They also expect to gain £60,000 when they sell their house in London. They will be able to put this towards the cost of a new house, as well as the money that they can borrow. Work out which houses they could afford to buy.

Where is the best area for the Barlows to live?

The Barlows go to Chelmsford to view each of the houses they have seen in the estate agents' descriptions. Not only do they want to see each house, they also want to find out what the ENVIRONMENTAL QUALITY of its area is like. An old friend has given them a simple piece of advice: 'It's more important to get the right area than the right house. You can change the house but you can't change the area!'

Like many families these days, the Barlows are trying to reduce the number of car journeys they make. This is partly because they want to be more environmentally friendly, but also because cars are expensive to run, and they don't want to become a two-car family. It is important that access to services like shops and schools is good, so that they don't have to make too many unnecessary journeys.

I'd like to be near to local shops. It's handy to be able to pop out to the shops when you've forgotten something, rather than make a special journey in the car. I often need to get something photocopied at the local newsagent as well. In the morning the children walk to school. I think it's important that they get some exercise and besides there's so much congestion these days outside the school gate.

I hate to waste time travelling. In London I can spend up to an hour each way getting to work and back. When we move to Chelmsford I hope to be able to do the journey to the City in less time, mainly because the train service is much quicker. So it's important that we live where I can reach the station easily in the morning. Apart from that, I'd like to live somewhere greener where I can breathe a bit of fresh air!

Gordon

In London it's not safe to go out on our bikes because the roads are so busy. When we move to Chelmsford there will be more parks and fewer busy roads, so we'll be able to go out more on our own.

Jack

6 Taken at grid reference 710062

7 Taken at grid reference 729075

N

Prison

Sewage works

0 1 km

Key

Main road

Industry

School/playing field

Shopping

Public open space

Housing

8 Land-use map of Chelmsford

Tasks

1 Look at Photos 6, 7, 9 and 10.
   a) Using the grid references in the captions, locate them on Map 3.
   b) What do you like or dislike about the environment in each photo?

2 Make two lists showing things that have:
   a) a positive
   b) a negative
   impact on the quality of life in urban areas.

3 Compare the environmental quality in the four photos. *Either* devise a survey sheet to assess environmental quality, *or* use one that your teacher may give you. (You have done an environmental quality assessment before, on page 227.)

4 Look at Maps 8 and 11.
   a) Use these maps to identify areas of Chelmsford that are service-rich (with lots of services) and those that are service-poor (with few services).
   b) Draw similar maps to show areas that are more/less than 0.5 km from the nearest:
   i) shops **ii)** school.
   (Your teacher may give you base maps of Chelmsford to use.)
   c) Complete another map of Chelmsford to combine the information on Map 11 and your own maps. Highlight areas that are service-rich and service-poor.

Focus task

5 Look again at the location of the houses on Map 3. Locate them on Map 8 as well. Now think about the environmental quality and access to services of each location.
   a) *As a group*, and using all the information on pages 224–29, make the final decision about which house you would choose for the Barlows to buy. Is this the same as your earlier choices? Why?/Why not?
   b) *Individually*, imagine you are Chloe or Jack. Write a letter to your grandma to tell her about the new house, the area it is in, and why you have decided to move there. Also mention the houses you rejected, and explain why.

9 Taken at grid reference 706084

10 Taken at grid reference 729092

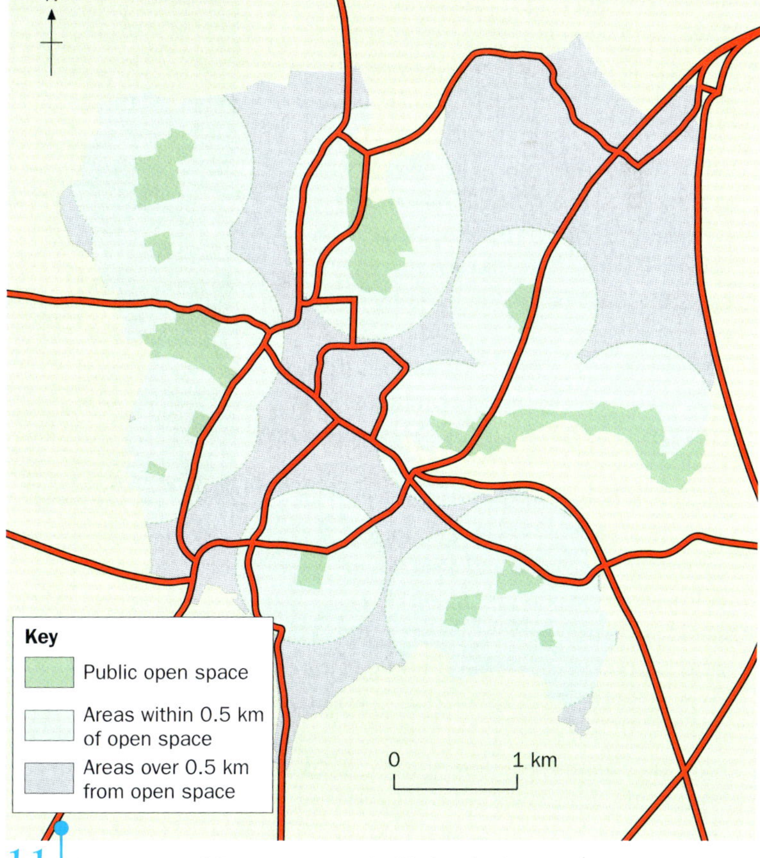

11 Access to public open space in Chelmsford

B VARIATIONS IN THE QUALITY OF LIFE IN URBAN AREAS

Quality of life varies from one area to another in towns and cities. These variations produce patterns of inequality that can be found in all urban areas. How do these patterns develop, and how similar are they from one urban area to another?

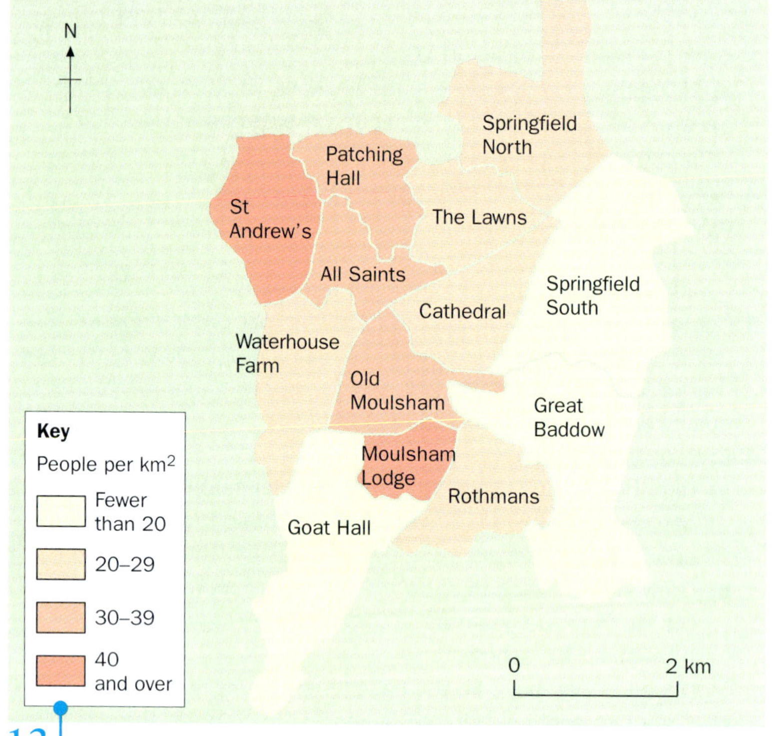

12 Population density in Chelmsford's local government wards

- Patterns of inequality develop as a result of the decisions taken by many individual families in the housing market. People with more money tend to move to areas with a better quality of life, while people who are less well off live in areas with a poorer quality of life. Typically, richer people move to the less crowded outer suburbs, leaving behind poorer people in the inner city. This pattern, which has been repeated in many cities, led geographers to propose the concentric ring model of cities (see page 214). In recent years, inner city redevelopment and suburbanisation have begun to change the established patterns. Some inner city areas have become more desirable while some suburban estates have deteriorated. While most of the high-quality housing in Chelmsford is further from the centre, areas like Moulsham, close to the centre, have also become more popular.
- Physical features such as rivers, relief and natural vegetation can affect the quality of life, as well as the location of major land uses such as transport, industry and commerce. Most of the industry in Chelmsford follows the line of major road and rail routes. This in turn has an effect on the quality of life for those living nearby.
- Government planning regulations in the UK have made an impact on urban patterns. The government designated land to the south-west of Chelmsford as part of the green belt. This has restricted urban growth in that area but encouraged growth to the north and east, where green belt restrictions do not apply.
- Local government has played an important role in housing people who cannot afford to buy a house. This can alter the pattern created by the housing market. Nationally, until 1979, almost a third of all homes were provided by local authorities for people to rent. In Chelmsford most of these local authority homes were built to the north-west of the town centre. Although many of these homes have been sold off since 1979, most of Chelmsford's local authority housing is still in this area.

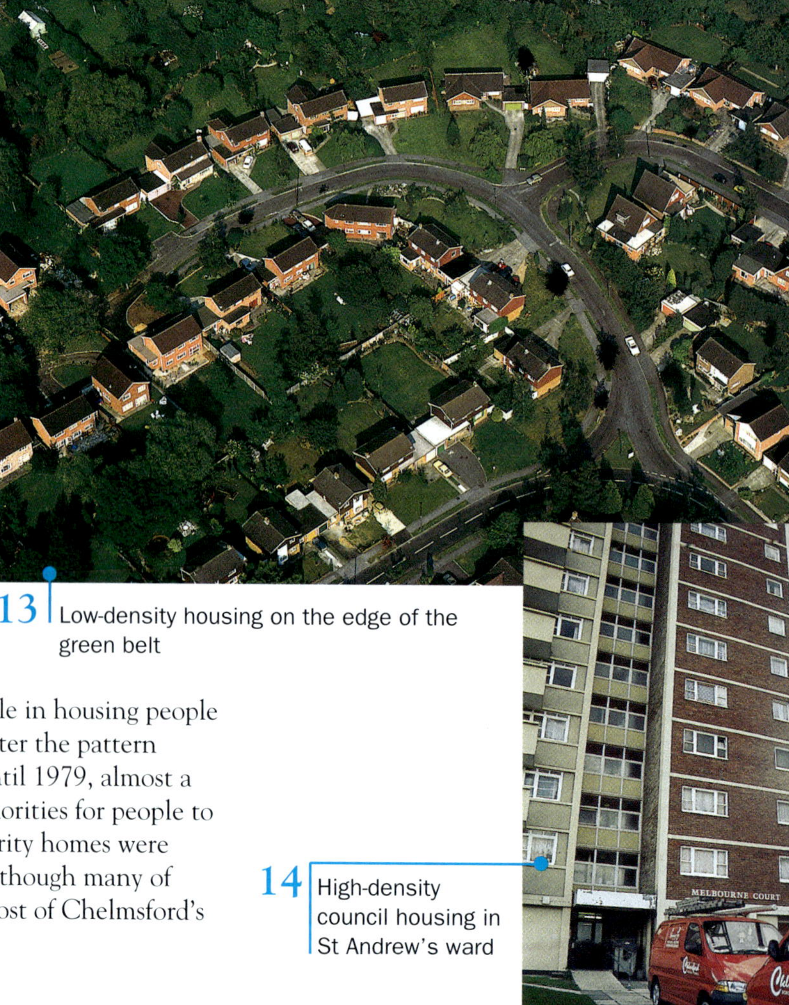

13 Low-density housing on the edge of the green belt

14 High-density council housing in St Andrew's ward

Ward	Population change (%)	Percentage of total population who: are under 16	are over 65	have long-term illness	own their home	rent their home from council	have no car	%age of women in employment
All Saints	0.8	24.1	17.8	11.7	44.8	46.3	41.1	56.7
Great Baddow	−7.4	17.5	20.7	8.2	69.6	21.6	26.9	69.2
Cathedral	−4.2	18.5	19.7	11.6	68.6	17.7	31.2	67.6
Goat Hall	−4.9	24.0	10.1	6.3	91.7	3.1	13.2	67.2
Moulsham Lodge	−12.6	19.9	10.4	6.2	97.6	0.1	14.3	71.6
Old Moulsham	3.1	15.5	17.9	10.0	81.0	2.9	28.7	72.0
Patching Hall	5.6	19.1	20.0	9.7	79.7	14.5	23.4	70.0
Rothmans	−10.2	18.1	18.0	10.1	67.4	19.1	25.0	65.8
St Andrew's	12.4	21.6	22.6	11.0	59.2	37.6	30.4	63.1
Springfield North	68.7	25.2	6.7	5.5	88.3	6.3	12.0	73.4
Springfield South	197.0	24.4	6.9	4.8	85.4	8.9	11.1	71.6
The Lawns	−12.7	18.0	14.1	7.6	94.8	1.7	13.1	70.2
Waterhouse Farm	−9.6	16.8	20.6	11.1	63.6	30.1	29.5	67.0

15 Population data for wards in Chelmsford

Tasks

1 Look at Map 12.

a) Describe the pattern of population density in Chelmsford.

b) Is this the sort of pattern you would normally expect to find in a town? How could you explain the pattern here?

2 Work in a group of four. Study the data in Table 15.

a) Discuss how each of the indicators might influence the quality of life. Choose four that you think would be the most important.

b) On a copy of the ward map of Chelmsford (Map 12), draw a *choropleth* map to illustrate one column of data in the table. Each person in the group should draw one map. See page 328 for an example of how to draw a choropleth map.

c) Compare the maps you have drawn. Can you see any similar patterns? How do you explain these?

d) Explain how these patterns may reflect quality of life within Chelmsford.

Focus task

3 Your task is to carry out an investigation into quality of life in your own town or city (or nearest urban area). Think of a question about quality of life that you could investigate, for example:

Which part of the town has the best quality of life?

Alternatively, think of a hypothesis that you could test, for example:

Quality of life improves as you move further from the city centre.

You could do this investigation as part of a group, or as a whole class. You could divide the town or city into smaller areas for each group to investigate. In a large city you will need to choose one section of the urban area to investigate. You will need to obtain both PRIMARY DATA (which you find yourself by doing fieldwork) and SECONDARY DATA (which you get from other sources).

Use a map of the urban area divided into ward areas. This will help you to choose the areas that you will investigate.

a) Obtain secondary data about each ward from the local authority's planning department. Your teacher may give you the name of a person to contact, or provide this data for you.

b) Carry out fieldwork in each of the areas where you have chosen to obtain primary data. You could take photographs, do environmental quality assessments (see pages 227 and 229) and interview people who live in the area. Never work alone.

c) When you return to the classroom, share information with other groups. Draw maps, graphs and tables to show your findings. Describe the patterns that you can see.

d) Write up your investigation. When you come to your conclusion, remember to focus on your original question or hypothesis.

What happens if you can't afford to buy a home?

Not every household in the UK owns a home. People who cannot afford to buy a home usually rent, while others prefer to rent rather than buy. Many of those who cannot afford to buy a home fall into the safety net of social housing – housing provided by local authorities and, increasingly, by housing associations. Until 1979 nearly half of all homes in the UK were rented, either from local authorities and housing associations, or from private landlords (privately-rented housing). In 1979 a new Conservative government was elected that wanted to create a nation of home-owners. To achieve this it gave local authority tenants the option to buy the homes that they were renting. Since 1979 many people have accepted this offer, so the number of owner-occupied homes has steadily increased while the number of rented homes has fallen. But it is unlikely that the goal of everyone owning their own home will ever be achieved, not least because house prices have risen considerably since 1979, and continue to do so. People on a low income simply cannot afford to buy their own home.

Local authorities still play an important role in housing people, even though they own fewer homes than they used to. They have the responsibility to allocate social housing in their area to people who cannot afford to buy. They also have a responsibility to find accommodation for homeless families. Each local authority has to keep a housing register – or waiting list – of households that need accommodation. Each household is awarded points according to its need. The greater the need, the more points the household gets and the faster it moves up the waiting list.

Total number of homes:	**1979** 17.6 million	**1999** 20.8 million
Tenure:		
Owner-occupied	56%	68%
Privately-rented	13%	11%
Local authority	29%	16%
Housing association	2%	5%
New house building:		
Private enterprise	118,000	121,000
Housing association	16,000	20,000
Local authority	75,000	300
Total	209,000	141,300
Private housing market:		
Average house price	£20,000	£101,000
Purchases	715,000	360,000
Of which first-time buyers	323,000	164,000

17 Housing in the UK – a comparison between 1979 and 1999

16 Blocks of local authority high-rise flats built during the 1960s to provide cheap rented accommodation. Local authorities now build very few homes and many high-rise blocks are being demolished because most people don't want to live in them.

Insecurity
- ☐ You have very few rights to stay in your home.
- ☐ You are threatened with homelessness within 28 days.
- ☐ You have no fixed address or are sleeping rough.

Harassment
- ☐ You are the victim of harassment from a neighbour.

Social factors
- ☐ Your present home has a harmful effect on your health.
- ☐ Your present home is unfit for human habitation.
- ☐ All your family cannot live together because of your housing.
- ☐ You are pregnant (or a member of your family is).

Lack of facilities
- ☐ Your home does not have a kitchen.
- ☐ Your home does not have a bath or shower.
- ☐ Your toilet is outside your home.
- ☐ You do not have a piped hot-water supply.

Too few rooms
- ☐ You lack one bedspace.
- ☐ You lack two bedspaces, or more.
- ☐ Children over 8 years old of the opposite sex have to share a bedroom.
- ☐ A child under 8 has to share a bedroom with parents.
- ☐ A child over 8 has to share a bedroom with parents.

18 Criteria used by local authorities to award points for households on the housing register

Trudie O'Dowd is a single mother. She lives with her two children, aged 4 and 1, in a small terraced house that she rents from a housing association. She suffers racial abuse from her neighbours because she is white and her boyfriend is black. He does not live at the house. She has now found a new part-time job on the other side of town and needs to move to be closer to work.

Samantha Day and her partner, Danny, share a two-bedroom local authority flat with Samantha's parents. Samantha is pregnant and they want to move into their own home. Danny has a low-paid job in a supermarket. They could not afford the high rents in the private market, let alone think about buying a home. There are lots of arguments between Danny and Samantha's dad. She is worried that they will have a fight.

Osman and Zulika Kuzio are refugees from the war in Kosovo. Their application for asylum has now been approved by the government. They have two children – a boy aged 10 and a girl of 6. They have two bedrooms in a house that they share with another refugee family. The local authority pays their rent but they are badly treated by the landlord who does not like refugees.

Harry and Eileen Tanner are both in their 60s. Harry used to work on a building site and was disabled there. He has not been able to do much work around the house since then. The house, which they own, has fallen into disrepair and is unfit to live in. If they tried to sell it they would not get enough money to buy another place. The damp is giving Eileen breathing problems, especially as they don't get out very often.

19 Case histories of households on a local authority housing register

Tasks

SAVE AS . . .

1 Look at Table 17.

a) Describe the housing changes that have happened between 1979 and 1999. (Updated figures can be obtained from the government website at: www.housing.detr.gov.uk.)

b) How does the information in the table explain the current housing shortage in the UK?

2 Work in a small group. You are going to devise a points system for people on the housing register of a local authority, to help in the allocation of housing.

a) First rank the criteria in Figure 18. Give each one a score from 1 to 10. Give 1 point to those that you think are least important, and 10 points for those you think are most important.

b) Compare your points system with the system used by a local authority (your teacher will be able to give you a copy of this). How do they differ? Do you want to change yours, or not?

3 Use the points system that you have devised in Task 2 to decide how housing should be allocated among the families in Figure 19.

a) Read the case histories and work out how many points each household would score under your system.

b) The local authority has the following properties available:

- a three-bedroom house
- a two-bedroom flat in a high-rise block
- a one-bedroom ground-floor flat.

Decide which people you should house and where you would house them. Explain your decisions.

c) What will happen to the people you were unable to house?

Should anyone be without a home?

In 1999, 160,000 households in the UK were officially recognised as being *homeless* by the government. This represents about 400,000 people but, according to the homeless charity Shelter, this is only the tip of the iceberg.

When you hear the word 'homeless', what image comes to your mind? Most people think of someone sleeping rough in a doorway on the streets of London. But it is not just London that has homeless people. Every night, while there are about 600 people sleeping rough in London, Shelter estimates that there are 2,000 more people on the streets in other towns and cities around the UK. However, this is just the most visible part of the problem. For every homeless person who sleeps rough, there are thousands of others whom we don't see, who are also in need of a proper home. These people may be staying with relatives, sleeping on a friend's floor, living in a hostel, or squatting (living in a home that has been left empty).

There are many reasons why people find themselves homeless. These are the most common:

- breakdown of a relationship with a parent or partner
- being asked to leave by friends or relatives who can no longer provide accommodation
- domestic violence or abuse
- loss of job or income, which leads to people falling behind with their rent or mortgage
- asylum seekers or refugees who have been allowed to stay in the UK.
- mental illness.

However, the reasons for homelessness go much deeper than people's personal circumstances. The housing shortage in the UK is a result of political decisions that have been taken – or avoided – over many years.

- The amount of new social housing is falling. Local authorities have almost stopped building new homes due to lack of central government funding.
- There are around 750,000 empty homes in England alone, most of them privately owned.
- Most empty houses are in inner city areas, especially in Scotland, Wales and northern England. There are a few empty homes in the south east of England.
- Around 50,000 households in the UK own a second home, the majority of which are empty for most of the year.
- Housing is not always affordable. Between 1990 and 1999, house prices and rents in the private sector doubled, while incomes increased at a slower rate.

You are 16. You've just left school. Your parents' marriage has been difficult for a while now, and your dad often hits you. One particularly bad night, you decide that from now on you're going to make a go of it on your own. You bundle some clothes, a sleeping-bag and some money into a bag. **What are you going to do?**

- Go to a friend's house?
- Go to the council, to ask them to find you somewhere to live?
- Try and get a flat or a place in a B&B?
- Go to a squat you know of in town?
- Try sleeping rough?

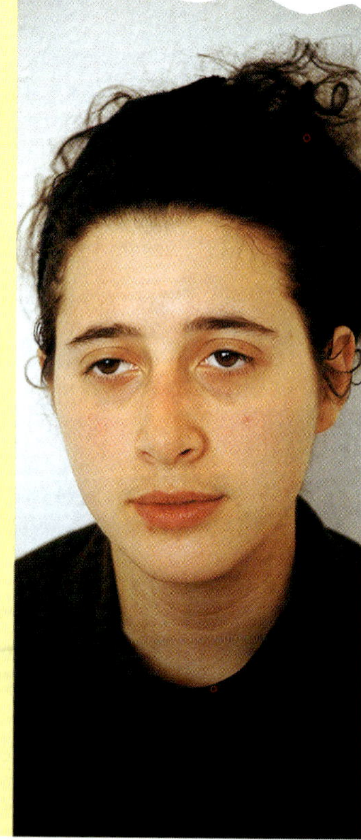

Your mum tells you that you can't go to school today and that you must stay at home to do the housework for her. Your mum is an alcoholic and has been drinking. Later that night you have to put your mum to bed – she's so drunk she can't do it herself. You're 15. **What are you going to do?**

- Phone Childline?
- Talk to a family friend?
- Leave home that night?
- Phone Social Services in the morning?
- Put up with it?

20 Homeless young people feature in the game 'Virtual Homelessness' on the website of the national youth homeless charity, Centrepoint. If you visit their website at www.centrepoint.org.uk you can put yourself in the position of homeless people and make the sort of decisions that they have to make every day.

★ **FACTFILE**

Dealing with homelessness in the London Borough of Newham

Local authorities have a duty to house homeless people who are in *priority need*. This includes: families with children, pregnant women, people who are considered vulnerable because of their age (under 18 or over 60) and people with physical or mental disability. It does not include the majority of single homeless people or couples without children.

Newham has a population of around 210,000. It has one of the largest numbers of homeless people in London. The Homeless Person's Unit of the housing department deals with over 3,000 applications for housing from homeless people each year. The local authority accepts responsibility for housing about half of these people.

Applicants have to pass a number of tests.

- Are they homeless, or threatened with homelessness within 28 days?
- Are they in priority need?
- Have they been living in the borough or do they have any local connection?

If accepted, the applicant is given temporary accommodation. At any one time the borough has about 1,900 people in temporary accommodation who are awaiting re-housing – 300 in B&B, 210 in one-bedroom local authority flats, 60 in hostels and 1,360 in property rented from private landlords.

Eventually, people in temporary accommodation can be re-housed in permanent housing. Average waiting times are 18 months for a one-bedroom property, two years for a three-bedroom property, and six years for a four-bedroom property.

Tasks

1 Study the information on pages 232–35.

a) Who do you think is mainly responsible for causing the problem of homelessness – the government, homeless people, local authorities or other people? Discuss this question with a partner.

b) What could each of these groups do to reduce the problem of homelessness? List all your ideas.

2 Look at the Factfile above. Complete a flow diagram, as below, to show the stages that homeless people have to go through to be housed by a local authority. Write the relevant details in each box.

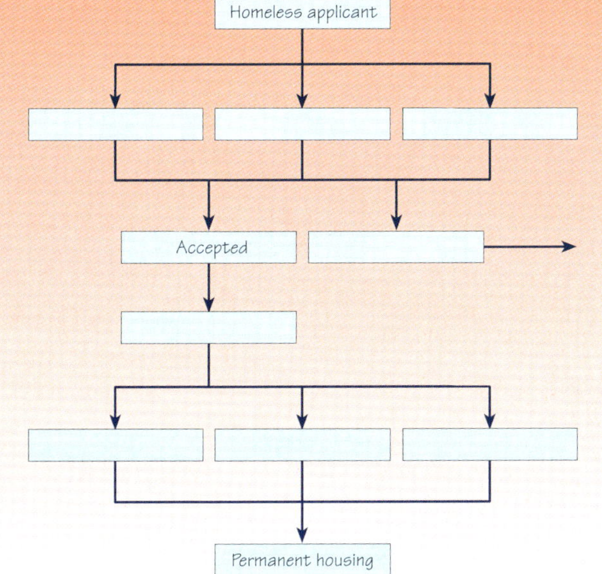

3 Look at Figure 20. If you were in the same situation as these young people, what would you do? You could choose any of the suggestions listed, or think of your own alternatives. Give reasons for your decisions.

ICT

4 When local authorities cannot help, homeless people can also turn to charities. Find out what help some of these charities offer by contacting their websites. You can try:

www.shelter.org.uk www.crisis.org.uk

What could *you* do to help the problem of homelessness?

Focus task

5 *'In the UK, access to housing for different groups of people is unfair.'*

Debate this statement with your class.

From all the work that you have done in this chapter, do you agree with the statement or not? Collect evidence to help you make up your mind. Think about how people become homeless and what this has got to do with the way that the housing market works.

Write a short speech, either in favour of the statement or against it. Hold a class debate where everyone has an opportunity to give their opinion. Listen to what other people have to say. At the end of the debate hold a vote to find out what the majority think.

5.5 *Improving the urban environment*

DO YOU KNOW?

- 89 per cent of the UK population live in urban areas but the population of most large cities is falling.
- The number of households in the UK is growing at a faster rate than the increase in total population.
- It is more environmentally sustainable for people to live in cities than to live in the countryside.

AIMS

- **To understand the reasons for the housing shortage in the countryside and why people are leaving cities.**
- **To consider alternative strategies to reduce the housing shortage in the UK.**
- **To assess the success or failure of attempts to improve the urban environment.**
- **To consider the needs of different groups of people in planning the urban environment.**

A A VILLAGE MYSTERY

Why do Greg and Diane have to leave Steeple Bumpstead?

You are going to solve a mystery.

Greg Nash and Diane Smith are planning to get married. Both of them were born in the small village of Steeple Bumpstead in Essex (see Map 2) and have lived there all their lives. They have known each other since primary school. They hoped to get a house in the village when they married so that they could still be near their parents. They could think of nowhere better to bring up their own children than the village where they were happy when they were young. But that will not be possible. The question is why?

And here is another mystery! Why does this chapter on the urban environment start in an Essex village, far from the nearest city?

1 Greg and Diane

2 Location of Steeple Bumpstead

3 Steeple Bumpstead

Focus tasks

1 Work with a partner.

First, think about the mystery you have to solve. Do you have an idea, or a theory, about why Greg and Diane have to leave Steeple Bumpstead? And what has it got to do with the urban environment?

2 a) Look at the information in Figure 4. (Your teacher may give you a copy of this on a separate sheet. If so, it would help to cut out the statements and lay them out on your desk.)

b) Read the statements aloud. As you do so, locate any places mentioned on Map 2. Find links between any of the statements and sort them into groups. For example:

> 12 There is no train station in Steeple Bumpstead and the bus service runs only twice a day.

could be linked to statement 9:

> 9 Greg learned to drive when he was 17. It's difficult to live in Steeple Bumpstead without a car.

c) You should find that each statement can be linked with at least one other, and sometimes more than one. Eventually you will be able to create a map linking all the statements. In case the lesson ends before you have finished, use the numbers beside the statements to draw your map so that you can easily continue next time. This will help you to remember the links that you have made.

3 Together with your partner, solve the mystery. Try to use all the information in Figure 4. Look back at your original theory. Does this prove it?

Write the answer to the mystery in your workbook or file.

mystery statements

1 The population of Steeple Bumpstead is 2,000. It has grown slightly in the past 20 years.

2 Barratt's are building one- or two-bedroom 'starter homes' in Haverhill which cost from £80,000 upwards.

3 Cambridge is one of the fastest-growing towns in the south-east of England. It has a lot of well-paid jobs in the computer industry.

4 There are few local authority homes or other rented accommodation in Steeple Bumpstead. Families with children get priority on the waiting list for local authority housing.

5 Greg is a newly qualified aircraft engineer at Stansted Airport where he earns £500 a week.

6 Diane's parents have divorced. Her mum still lives in the village and her dad has bought a flat in Cambridge.

7 Residents in the village want it to keep its character. There were protests over plans to convert the old vicarage into flats.

8 Since the M11 was built in the 1980s, the journey-time from London to Cambridge has been reduced to less than an hour.

9 Greg learned to drive when he was 17. It's difficult to live in Steeple Bumpstead without a car.

10 Most of the land around the village is high-quality farmland. It is hard to get planning permission to build.

11 Diane works in a hairdressing salon in Haverhill. She earns £180 a week.

12 There is no train station in Steeple Bumpstead and the bus service runs only twice a day.

13 In the past ten years house prices in Steeple Bumpstead have more than doubled. It is hard to find anywhere for less than £120,000.

14 Few people ever move away from Steeple Bumpstead. It is a quiet, friendly place to live.

15 More city people want to live in the countryside for a better quality of life.

16 Greg's parents have run the village pub for 30 years. The pub conversation used to be all about farming – now it is about share deals.

17 Stansted Airport is growing. It now employs over 3,000 people.

18 House prices in London and the South East have gone up faster than in the rest of the country.

19 Car ownership in rural areas is higher than in urban areas. Most families in Steeple Bumpstead own two cars.

20 Greg and Diane don't want to live with either of their parents, even though they'd like to be nearby. Like most young people, they want to be independent if they can.

4 The mystery of Steeple Bumpstead

★ THINK!

There is no simple solution to this mystery. It is rather like a jigsaw with many pieces to fit together. You should try to use as much of the information you are given as you can. It would not be wrong to say that Greg and Diane have to leave Steeple Bumpstead because they cannot afford to buy a house. But there is much more to it than that. The activity will show you that much wider issues are the real cause of their problem. As geographers we have to look beyond the obvious answers to find deeper reasons. At GCSE level this could be the difference between getting an average grade and a good grade.

Can we all escape from the city?

People are moving out of UK cities to find a better quality of life in the countryside. An estimated 300 people a day move from urban to rural areas – about 100,000 every year. This trend is called COUNTER-URBANISATION.

City dwellers often want to escape from overcrowded housing, traffic congestion, poor schools and the high level of crime. Many of them are attracted by the prospect of living in a quiet little village. The problem is that the more people who move into the countryside, the fewer quiet little villages there will be! What's more, the demand for housing in rural areas pushes the price of a home beyond the reach of many ordinary people who already live there – people like Greg and Diane.

The situation is made worse by the growth in the number of households in the UK. Although the total population is growing only slowly, the number of people living on their own in a single-person household is growing much faster. The proportion of single-person households in the UK grew from 18 per cent in 1971 to 29 per cent in 1996. This is due to older people living longer; more couples getting divorced, forming two households where previously there was one; and more adults choosing to live independently.

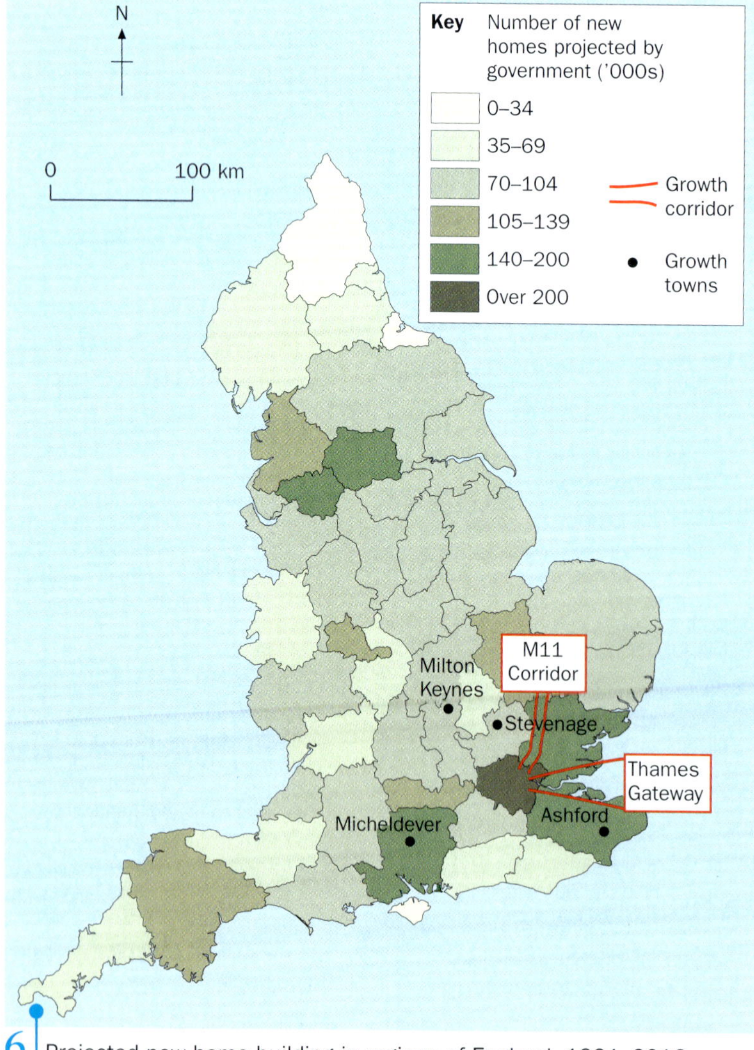

6 Projected new home building in regions of England, 1991–2016

5 Building on a greenfield site

New settlements are planned in south-east England to meet the demand for homes in the countryside. Micheldever (Map 6) will be a new town in Hampshire with 5,000 new homes. It will have its own public transport system to discourage people from depending on cars.

Growth towns are existing towns which will be expanded by building on greenfield sites. They may take up valuable green belt land but this may cause less overall damage to the countryside than new settlements being built outside the green belt. At West Stevenage in Hertfordshire, 10,000 new homes are planned.

7 Strategies to provide new homes in urban and rural areas

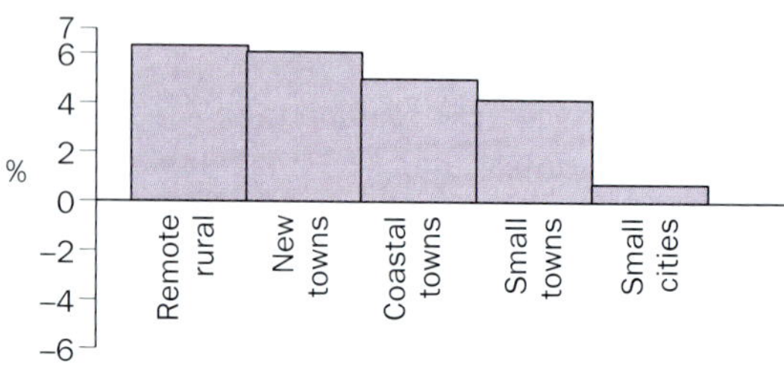

8 How population of different types of settlement increased or decreased in 1981–1991

9 A brownfield site in a city

To meet the growing housing demand in England the government has set a target of 5.5 million new homes to be built between 1991 and 2016. Map 6 shows in which parts of the country these will be built. But there is a huge dilemma about where to build. More homes could be built on so-called greenfield sites – undeveloped land in the countryside – to meet the demand. This would make housing in the countryside more affordable but could attract even more people out from the cities. Such a policy would not be sustainable since it would lead to the loss of land from the countryside and to people making more car journeys to get to work and to amenities such as shops.

Alternatively, more homes could be built in cities on brownfield sites. This means using derelict land and bringing empty buildings back into use as housing. This policy could actually help to improve the quality of life in cities and make them more attractive places to live. Eventually, if cities become better places to live, it might also reduce the demand for homes in the countryside.

Empty buildings in urban areas could be brought back into use. There are an estimated 750,000 empty homes in cities, with a further 800,000 potential homes in unused office blocks, warehouses and factories.

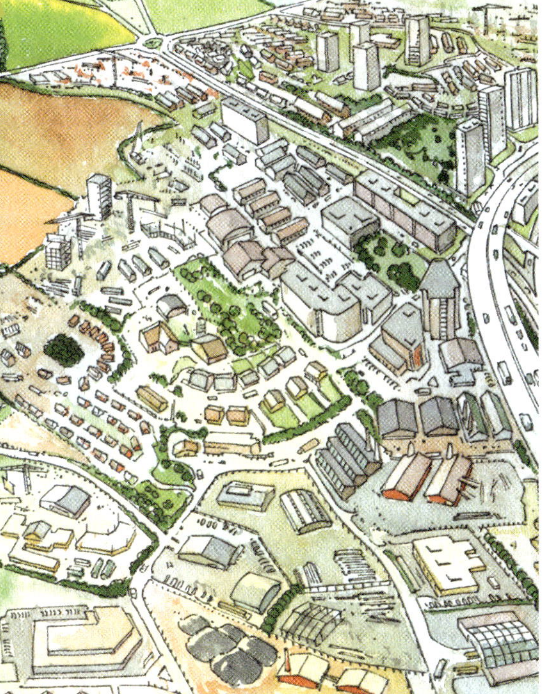

Derelict land in cities could provide brownfield sites for new building. Local authorities have powers to make builders use derelict land before they turn to greenfield sites. Up to 40 per cent of derelict land is currently used for car parking. If the number of cars in cities could be reduced there would be more space for people to live.

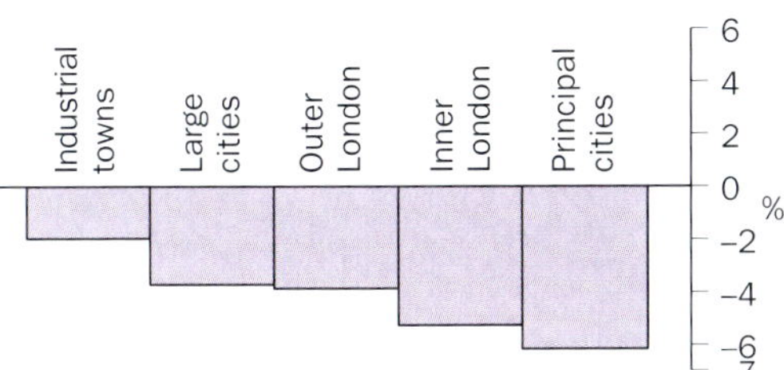

Tasks

1 Look at Map 6. Describe and explain the pattern shown on the map.

SAVE AS...

2 Study all the information in Figures 5–9. Consider each strategy to provide new homes in Figure 7. Draw a large table like the one below to compare the strategies. Give at least one advantage and one disadvantage for each strategy.

		Advantages	**Disadvantages**
Greenfield	New settlements		
	Growth towns		
Brownfield	Empty buildings		
	Derelict land		

3 Work with a small group. The government is seeking advice from different interest groups before it decides on the best strategy to meet the demand for more homes. Your group should represent *one* of these interest groups:

- building companies
- environmental organisations
- urban dwellers
- rural dwellers.

Which strategy in Task 2 would your group prefer? Prepare the arguments to support your case. Say why you do not like the other strategies. Present your advice to the whole class as a speech, from either one member or all the members of your group.

B INNER CITY DECLINE AND RENEWAL

What went wrong with cities?

To see how to make cities better we need to understand how they went wrong.

When cities in the UK were growing during the 19th century and first half of the 20th century, they offered people new jobs and homes, and a wide range of services that were not available in the countryside. People migrated to cities in their millions. But in the latter half of the 20th century many of the pull factors that drew people to the cities disappeared. Jobs in industry were lost, the quality of housing deteriorated and many services moved from city centres to out-of-town locations. This led to the decline of INNER CITIES.

The decline of inner-city neighbourhoods is a complex process. A number of factors link to each other in a 'spiral of decline' which is summarised in Figure 10. This depicts an extreme scenario. Not all inner cities experienced all these problems, or suffered them all as badly as this, but this tells you the main issues.

Tasks

1 Make your own copy of Figure 10 then, using the text on these two pages, add notes to explain how the closure of industry caused decline in the inner city.

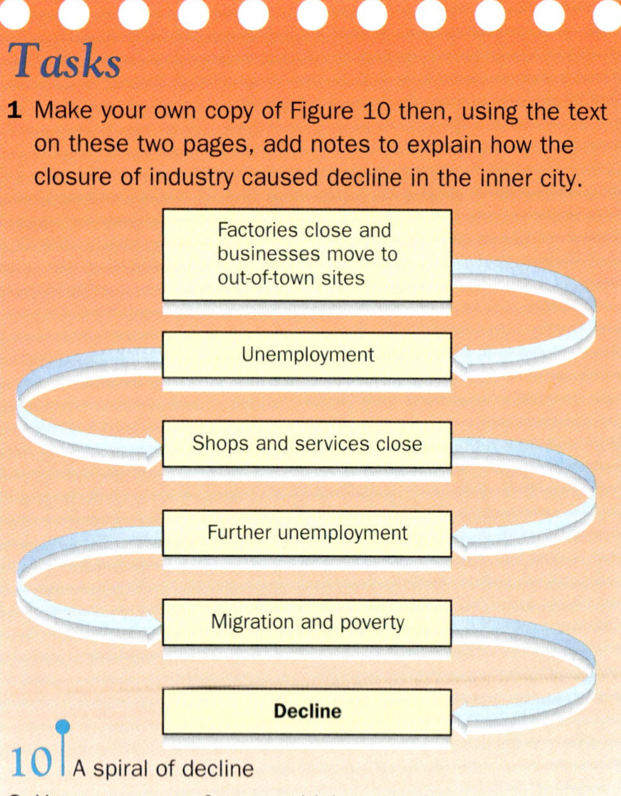

10 A spiral of decline

2 Here are some factors which caused industry to move from city centres:

- Greenfield sites
- Old buildings
- Purpose-built factories
- Poor, congested roads
- Poor environment
- Near motorways and A roads

Group them into push factors and pull factors.

The living city

1 Unemployment

From the 1950s traditional employment gradually disappeared from inner cities. Some factories made things which were no longer needed, so they simply closed down. Others moved out of town where they had space to grow. All industries mechanised, so needed fewer workers. Unemployment spiralled. Some cities depended on a single industry (even a single employer). When it collapsed, the resulting unemployment devastated whole communities.

11 The closure of industry leads to unemployment …

2 Outward migration

With few jobs available in the cities, the youngest and best qualified people left to find jobs elsewhere, leaving behind mainly those who, through lack of skills, opportunity, their age or health, could not work.

The dying city

Early-20th century inner cities were a thriving mix – bustling docks, railways, warehouses, factories and offices. The jobs were mostly skilled and semi-skilled and there was lots of work to be had. Hundreds of thousands of people lived right in the heart of the city. Conditions were often cramped, and cities were smoky, polluted places, but the cities were alive.

3 Loss of shops and services

Local services gradually closed down. Public transport reduced. With mainly poor people in the area there were not enough customers for services to survive. Shops, banks, cinemas, cafés and pubs had a dwindling number of customers with less to spend. Businesses could not attract enough people in from the wealthier suburbs to keep going. Some businesses struggled on but without income could not invest in making their service better. The most enterprising businesses moved out of town.

12 ... which means shops and services close down ...

4 Poor housing

Alongside these other problems, and making them all worse, was a housing problem.

Many houses in the inner cities had been built in the 19th and early 20th centuries. Houses built in the 19th century were cramped, small and basic. They had outside toilets. They had no gardens.

In the 1960s many councils decided to get rid of much of this old housing. They began slum-clearance programmes. In place of the houses they usually built high-rise flats. Some were well built. Many were not. People hated living in them. They lacked community feel; they were poorly ventilated and suffered from damp; they were expensive to heat; the open spaces designed to develop a sense of community spirit actually belonged to no one and so no one cared for them and they were soon vandalised. Poor design led to many 'hidden' places where hooliganism and criminal activity took place.

Many inner cities were therefore saddled with a nightmare combination of run-down old housing and unpopular new housing.

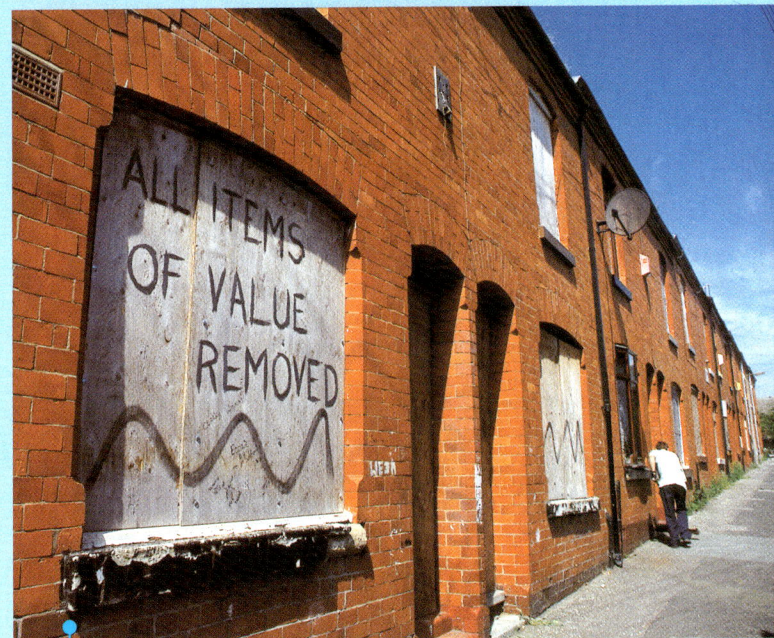

13 ... and people leave the city.

By the 1970s and 1980s many inner cities had become off-putting places.

- The Central Business District (CBD) lacked investment.
- The decaying environment was depressing and even dangerous.
- Inner-city housing areas were poverty-stricken and crime-ridden.

It was not the kind of place that people wanted to live if they could avoid it. Nor was it the kind of place where people from the suburbs wanted to come to shop or to relax.

So what did people do about it...? See over...

How are inner cities being renewed?

In the 1980s the government attempted to reverse the process of decline through urban redevelopment. Initially it set up Urban Development Corporations (UDCs), such as the London Docklands Development Corporation, to transform many old industrial and dock areas.

UDCs were set up in four phases:

- London Docklands; Merseyside (1981)
- Trafford Park in Manchester; Teesside; West Midlands; Tyne and Wear (1986)
- Bristol; Leeds; Central Manchester (1987)
- Lower Don Valley, Sheffield (1988).

Each UDC took over planning responsibility from local councils. It was felt that they could cut 'red tape' and make decisions more quickly. UDCs invested government money to reclaim derelict land and to improve the infrastructure (roads, etc.). Through the UDCs the government also offered companies reduced taxes if they set up in the area.

In ten years the London Docklands Development Corporation (LDDC):

- reclaimed 600 hectares of derelict land
- built 90 km of new roads, the Docklands Light Railway and London City Airport
- put in new infrastructure for gas, electricity, water supply and drainage services
- attracted 41,000 jobs to 2.5 million m^2 of industrial and office space
- built 15,200 new homes and refurbished 5,300 more
- built three new shopping centres
- built a new Technology College
- planted 100,000 trees and set up 17 conservation areas.

Local residents claim that the development helped to speed up the destruction of the traditional 'East End' of London. Most of the housing was too expensive for local people. Most of the jobs required a workforce skilled in technology and so were unsuitable for local people.

In general, there were success stories but the UDCs did not reverse the long-term decline in population of the major cities. The UDCs in England have now been replaced by partnerships between local authorities and private industry, managed by the governmental organisation, English Partnerships. They recognise that URBAN REGENERATION needs to be on a much wider scale if people are to be brought back into cities. Just as decline has many elements, so has renewal.

1 Attracting employers to locate in the city

- Make the inner city more attractive to industry by:
  - improving road access
  - improving image by cleaning up and planting greenery
  - offering tax breaks and grants
  - training the labour force.
- Make out-of-town locations more difficult to use by:
  - putting a tax on greenfield development
  - refusing planning permission.

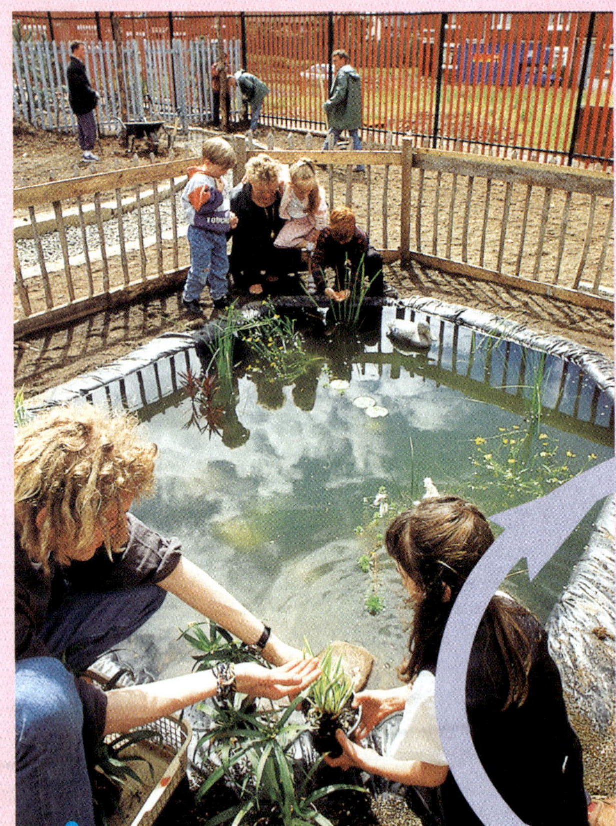

14 'Greening' the environment

The dying city

The renewed city

2 Attracting shops and services to the city

- Make the inner city more attractive for shops and services by:
  - improving the environment
  - building shopping centres
  - improving car access and car parking
  - improving public transport.
- Refuse planning permission for out-of-town shopping centres.

3 Attracting people to live in the city

- Stop the outward movement of traditional inner city residents by improving housing, the environment and job opportunities in the inner city.
- Attract new, wealthy residents to the inner cities by building new, high-cost housing in inner city redevelopment areas.

4 Improving housing in the city

- Provide investment through housing associations to:
  - build kitchens and bathrooms, install damp proofing, add new roofs, provide ownership of space around houses
  - carry out 'crime hardening' to protect houses from burglary
  - improve the local environment.
- Encourage GENTRIFICATION, that is, encourage private individuals, usually wealthy young people to, buy old houses in run-down neighbourhoods and 'do them up'.

Tasks

1 The main aim of the LDDC was to attract jobs into the area. Explain how each of the following made the area more attractive to firms thinking about locating there:
- reclaiming derelict land
- new transport systems
- new infrastructure for gas, electricity, etc.
- new Technology College
- planting trees and starting conservation areas.

2 Many of the companies that moved into the area were involved in newspaper publishing or finance. Why do you think that the LDDC was keen to attract these types of business rather than heavy industry?

3 Choose one of the following aspects of urban renewal:
a) attracting industry
b) attracting shops and services
c) improving housing
and write a paragraph or draw a diagram to explain how it might help to set off an upward spiral of development.

4 If you know of a local example to illustrate this, write a second paragraph to describe it. If not, save your answer to Question 1, then use the example of Birmingham described on pages 244–50 to write your paragraph.

5 Explain why people who live in villages on the urban–rural fringe (see pages 200–21) might be pleased to know that the spiral of decline in the inner city is being slowed down or stopped.

15 A conservation area in Toxteth, Liverpool which is being regenerated

C CASE STUDY: WHAT CAN WE LEARN FROM BIRMINGHAM'S EXPERIENCE?

Over the next seven pages you are going to investigate Birmingham, the UK's second largest city, to find out what has been done in the past to improve the urban environment and what is being done now. Does this have any lessons for the future, and how will it help to persuade people to live in Birmingham?

There is so much urban redevelopment going on in the UK that there is even talk of an 'urban renaissance' as people are drawn back into cities by an improvement in the environment, jobs, homes and services. But people have talked like this before and there have been 'false dawns'. The redevelopment of Birmingham city centre in the early 1960s was hailed as the way forward for cities. Forty years later it is seen as an urban disaster.

The Bull Ring: 1964 – hopes are high

A woman's world with no worries about the traffic or weather

The new Bull Ring Centre that opens in Birmingham this week is the largest shopping centre ever built. Its developers say that 'it is the most advanced shopping centre of its kind in the world', and will make Birmingham the envy of other cities in Britain.

Gone are all the things that make shopping in the city such a nightmare – busy streets to cross, long queues for the bus, and traffic jams. Now everybody can come to Birmingham in the comfort of their own car, park in a new multi-storey car park just a short walk from the shops, and all without even putting up their umbrella! The modern new urban ring road – a dual carriageway right around the city centre – will do away with traffic problems. Pedestrian subways and bridges will take shoppers to their destinations after they have parked the car. There will even be automatic lifts for the use of invalids or people with prams. For the average woman shopper it will provide a real fillip – the kind of spree that, until now, only a day in London could afford her.

● **The newly opened Bull Ring in 1964**

 Extract adapted from the *Birmingham Post*, 1964

The Bull Ring: 1999 – a different reality

Bulldozers enter bleak Bull Ring

Watch out Paris. A grand plan was unveiled in Britain's second city yesterday which aims to put Birmingham on the map of glamorous international shopping destinations.

The bulldozers are finally moving in on the eyesore Bull Ring. The shopping centre that came to symbolise the dreary brutalism of 1960s urban redevelopment and undermined the reputation of an entire city is at last to be torn down. Today the shopping centre is a maze of grimy underpasses, walls covered in graffiti, and boarded-up shops – the Bull Ring is a national joke.

Now the earthmovers are arriving and the developers plan to restore Birmingham's reputation. They reckon there is money to be made in the shape of 7 million shoppers living less than an hour from the city. The west side has already been regenerated. The canalside Brindleyplace development, the Birmingham Rep and the International Convention Centre have already done the seemingly impossible and pulled people back into downtown Brum for both business and pleasure.

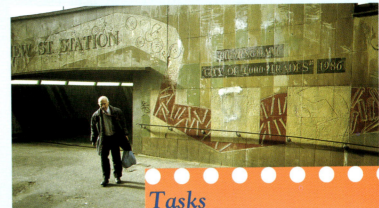

17 Extract adapted from the *Guardian*, 26 February 1999

Tasks

1 Read Figures 16 and 17.

a) Compare the expected benefits of the Bull Ring in Figure 16 with the reality in Figure 17. Why do you think this happened? (Map 18 might give you a clue.)

b) Describe the similarities between the two extracts. Are the latest plans for the city centre likely to be any more successful, do you think? Give reasons.

Look at this question again when you have studied all the information on pages 246–50.

ICT

2 Carry out an internet search for Birmingham past, present and future. You can focus your search on the Bull Ring. There is an official website for the Bull Ring:

www.bullring.co.uk

Or try the City Council website:

www.birmingham.gov.uk

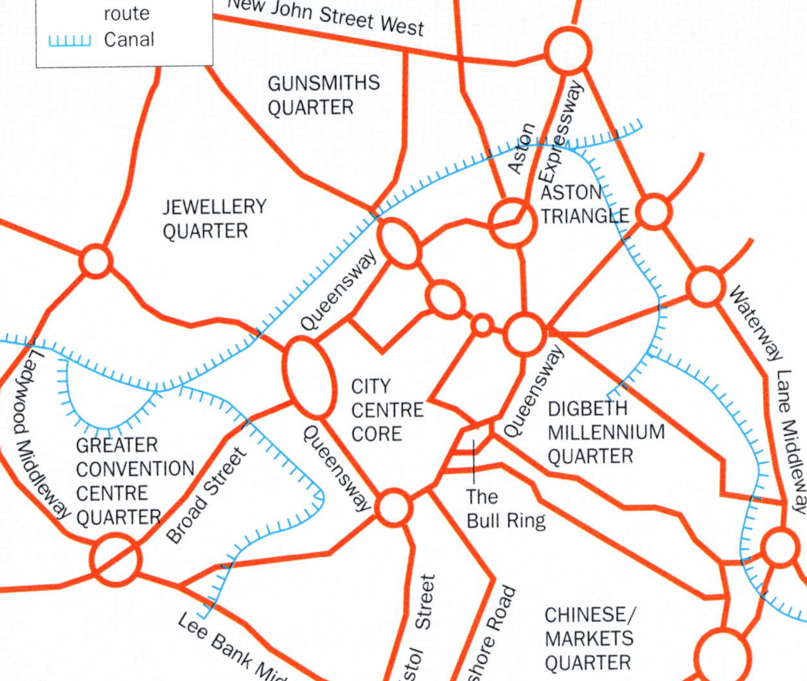

18 Birmingham city centre divided into seven areas for regeneration

New developments in Birmingham

19 The International Convention Centre in Birmingham

20 Shopping in Brindleyplace

Brindleyplace, just to the west of Birmingham city centre (in the Greater Convention Centre Quarter – see Map 18) was redeveloped during the 1990s. It has become a model for redevelopment of other parts of the city. The development cost £250 million, covering an area of 6.1 ha alongside some of Birmingham's old canals. At the heart of the redevelopment is the International Convention Centre (ICC), built in 1991 and the largest such centre in the UK. Around it are 79,000 m^2 of offices, 11,500 m^2 of shops, 140 luxury flats, the National Indoor Arena, a sea-life centre and the new regional headquarters for BT. The main features of the development are shown in Map 21.

The Brindleyplace redevelopment has made a dramatic improvement to the physical appearance of Birmingham. It has attracted many more people to the city for both business and pleasure. However, there are mixed feelings among Birmingham residents about the development. Although many new jobs have been created, fewer than half of these have gone to Birmingham residents. Many of the jobs are low-paid, short-term jobs in the retail and tourism sectors, and unlike many of the better-paid skilled jobs that Birmingham has lost in recent years. The development has provided no affordable housing for people living in the inner city. Indeed, it has forced up house prices in the surrounding area, making affordable housing harder to find for many people.

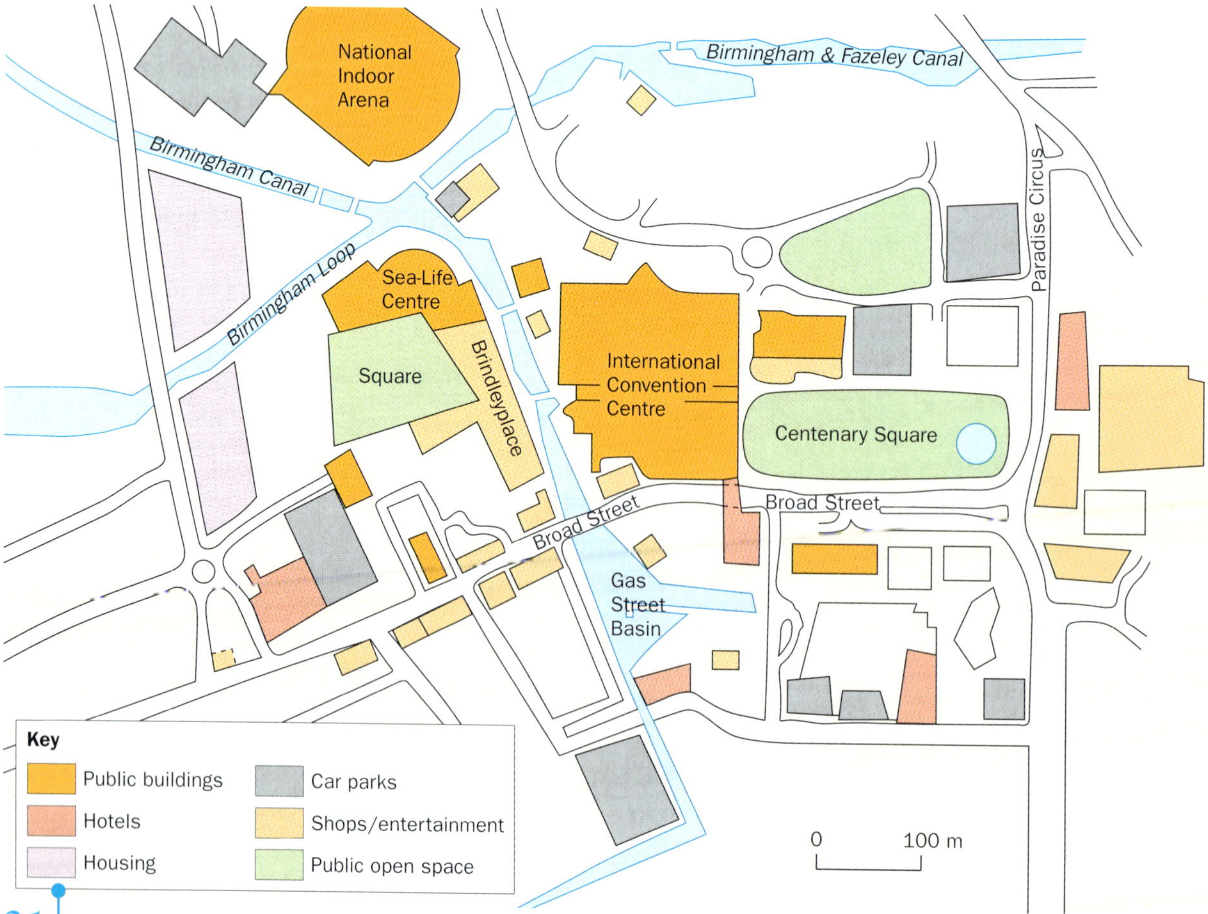

National Indoor Arena
Birmingham & Fazeley Canal
N
Birmingham Canal
Birmingham Loop
Sea-Life Centre
Square
Brindleyplace
International Convention Centre
Centenary Square
Paradise Circus
Broad Street
Broad Street
Gas Street Basin

Key

- Public buildings
- Car parks
- Hotels
- Shops/entertainment
- Housing
- Public open space

0 100 m

21 Main features of the redevelopment of Brindleyplace

A Business person

B Tourist

bus station pubs hospital information office
residents' parking cycle lanes cafés/restaurants clinics
retail markets shopping malls affordable housing
wholesale markets car parks clubs luxury housing
parks railway station cinemas schools
public squares theatres factories trees one-way streets
concert halls warehouses river/canals
motorways sports stadium offices speed humps
convention centre university CCTV bus routes
libraries police stations pedestrianised shopping area
sports halls museums ice rink supermarkets
art galleries law court department stores swimming pools
churches metro/light railway specialist shops
community centres mosques hotels
convenience shops youth clubs synagogues
town hall banks/building societies temples

C Inner-city resident

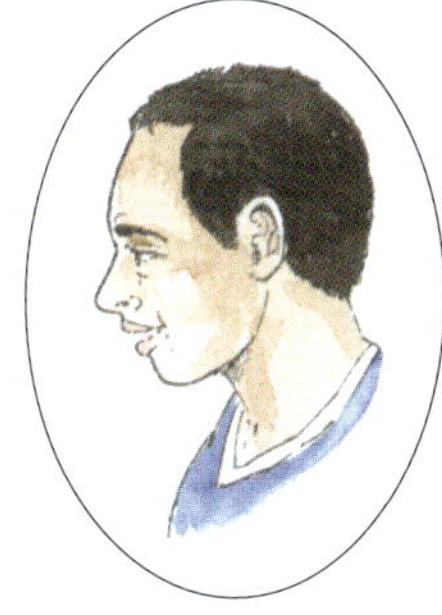
D Suburban resident

22 What features do people need in a city?

Tasks

1 Study all of the information on the opposite page.
   a) Describe the Brindleyplace development. List at least 10 important features.
   b) Compare the Brindleyplace development with the Bull Ring development of the 1960s. What are the differences? Is it likely to be any more successful? Give at least three reasons for your answer.

2 Look at Figure 22.
   a) Write the list of features on a sheet of paper. (Your teacher may provide you with the list on a sheet). Cut out each feature.
   b) Draw a large copy of the Venn diagram shown here. (Your teacher may give you a copy of this too.)
   c) Think about which of the features each of the four people would need in the city centre. Place each feature in the most appropriate space on your diagram. For example, you would put *hotels* in the space 'AB' because they are likely to be needed by business people and tourists. Omit any features that you think no one would need. Stick the features on to your diagram.
   d) Use the diagram you have made to evaluate the Brindleyplace development. How would the development meet the needs of each of the four groups of people? Should any group or groups get more consideration than the others? If so, explain why.

3 Think about your own town or city. Which of the features listed in Figure 22 does it have and which does it lack? Explain how you would improve it. Which groups of people would this benefit? Would anyone lose out?

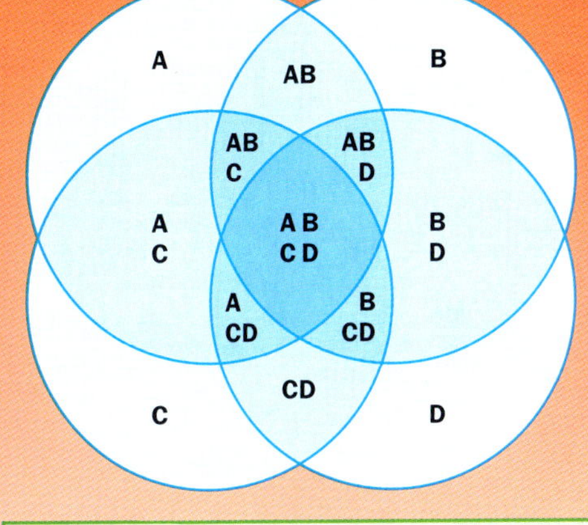

★ THINK!

You might feel that cutting out bits of paper and sticking them in circles is an insult to your intelligence. It's a pity that some city planners in the 1960s and 1970s didn't try it, because this sort of activity really helps you to think! And it's not as easy as it appears. As you think about which features each group of people needs, you will realise that there are difficult decisions to make, and you may have to compromise. Today, planners try to consult everyone who will be affected by urban redevelopment, so let's hope that cities in future will meet more people's needs.

Planning for the future

One of the key problems that Birmingham faces is its transport network and people's ability to move around. In the 1960s planners believed that more roads would provide easier access to the city and its services. This hasn't happened. The more that people depend on cars, the more traffic and congestion they create. Fewer people are willing to use public transport and so it has declined. People without cars – the young, the elderly and the disabled – have poorer access to the city than other groups of people (see pages 218–19). The inner urban ring road that was built in the 1960s to speed the flow of traffic has become a 'concrete collar' (see Photo 23). It prevents shops and businesses expanding outwards from the city centre and restricts free movement of pedestrians and cyclists who want to get into the centre.

Tasks

1 Read the statements in Figure 25.

a) Sort them into *facts* and *opinions*. (Facts are statements that can be proved, but opinions cannot be proved.) You will use the facts in Task 2.

b) Suggest which of these groups of people is most likely to hold each of the opinions:

- business people
- inner-city residents
- tourists
- suburban residents.

2 Look at Photo 23 and Figure 24.

a) Draw a sketch map of the area shown in the photo. Label the main land uses on your map, including roads, car parking, commercial (shops and offices), industrial and open space.

b) Annotate your map with some of the facts that you sorted in Task 1, to describe the problems of Birmingham city centre.

c) Do you think that the plans shown in Figure 24 will help to solve any of the problems? Give your reasons.

24 Plans for Birmingham city centre include the replacement of the ring road with pedestrianised tree-lined boulevards to create a more attractive environment.

23 Birmingham city centre from the air. The large road in the centre of the photo is Queensway – part of Birmingham's 'concrete collar'.

80% of the traffic on the ring road is not going to the city centre.

100,000 people come to Birmingham city centre every day, three-quarters of them from outside the city.

It is more pleasant to shop out of town rather than go to the city centre.

More people move out of Birmingham each year than move in.

It is easy to find a place to stay in Birmingham. Prices range from £350 a night in the best hotel to £30 in a B&B.

A city centre apartment costs about £200,000. Only the richest people can afford to live there.

Unemployment is concentrated in the inner city. It is over twice the rate found in the suburbs.

There should be more affordable housing for ordinary people to live near the city centre.

The city centre is an ideal place for single people to live, near to their work and entertainment.

People are unlikely to come to the city centre unless there is convenient access by car.

Only 40% of jobs at the ICC and NIA went to people living in the city.

Car parks take up too much land in the city. This could be better used for housing or other purposes.

The CBD in Birmingham is being choked by the 'concrete collar' all around it. Without it the business area would expand.

The worst problem in Birmingham is traffic. The city needs a better public transport system.

25 Facts and opinions about Birmingham

A new approach to URBAN PLANNING is needed in the 21st century if cities are to become places where people will want to live. In Birmingham the east side of the city centre – the Digbeth Millennium Quarter, see Map 18 – is soon to be redeveloped. It offers another opportunity to bring people back to the city.

Digbeth lies just outside the city centre at the bottom of the area that you can see in Photo 23. It is bounded on all sides by major roads and is also bisected by canals which these days are hardly used (see Photo 26). Traditionally Digbeth was an industrial area with many small engineering and metal-working companies. There are still some long-established companies but the old buildings are no longer ideal for modern activities. Other companies have moved out, leaving behind empty and derelict buildings. It is hard to attract new business into an area that is cramped and congested. However, 4,000 people are still employed in Digbeth.

One new development in the area has already begun. Millennium Point opened in 2001 (see photo 27) and will be a major new visitor attraction based on Birmingham's industrial heritage. In addition, it is expected to be a world-class centre for technology and learning, providing a focus for developing the technological base of the region. It is likely to bring more people to Digbeth and could be a catalyst to regenerate the area.

26 The canal at Digbeth

27 Millennium Point

The inner urban ring road is congested and the traffic causes local air pollution.
How could traffic be kept out of the city centre?
Millennium Point
The old canals in Birmingham are no longer used by industry. They attract vandals and are covered in graffiti.
How could the canals and towpaths be used now?
There is a bus station and rail routes to nearby New Street Station.
How could more people be persuaded to use public transport?
Industry and business provide jobs but also take up most of the land. Many buildings are left empty.
What is the best way to use the land?
Car parking is a major problem. Visitors and residents park on the streets, making them congested.
How can the problem of car parking be solved?
N
Jennens Road
Ring Road (A4540)
Millennium Point
Curzon Street
Digbeth Branch Canal
Masshouse Circus
Curzon Street Station
Moor Street Queensway
Moor Street Station
Digbeth Basin
Warwick Bar
New Street Station
Bull Ring
River Rea
Grand Union Canal
Digbeth Deritend
Bordesley Station
Camp Hill
0
0.5 km

28 How should the Digbeth Millennium Quarter in Birmingham be redeveloped?

Focus task

1 You are going to be city planners. You will produce your own plan for the redevelopment of the Digbeth Millennium Quarter in Birmingham.

Look at Map 28. It highlights five key questions that you must consider as you produce your plan. (Your teacher may give you a large outline map on which you can draw your land use and transport proposals for the area.) It will also help if you consider each of the following:

- the facts and opinions in Figure 25
- the needs of different groups of people (see page 247)
- previous experience of redevelopment in Birmingham (pages 244–50)
- solutions to urban transport problems in other parts of the UK (see pages 218–219).

Produce a plan that includes your map together with a written report. In the report, explain how the decisions you have taken will improve the quality of life in Birmingham and persuade people to live in the city. Use the five key questions on Map 28 as paragraph headings in your report.

6 Economy and the environment

How do our economic activities interact with the environment?

- *Is our use of the environment sustainable?*
- *Can we learn to use the world's resources in a sustainable way?*
- *How is our work changing?*

6.1 What is the future for farming?

A WHY IS THE UK FARMING INDUSTRY IN CRISIS?

DO YOU KNOW?

- Farmland covers more than 80 per cent of the land in the UK.
- Farm incomes fell by more than half between 1995 and 2000.
- Farmers in the UK produce 60 per cent of the country's food, but farming employs only 2.4 per cent of the workforce.
- As well as producing food, farmers look after the landscape.

AIMS

- **To consider why the UK farming industry is in crisis.**
- **To understand that there are alternatives to INTENSIVE FARMING.**
- **To look at ways of increasing agricultural output.**
- **To assess public opinion on the controversial issue of GENETIC MODIFICATION of crops.**

1

FARM INCOMES FALL BY 60% IN FIVE YEARS

The UK's farmers are in crisis! Many farmers are being forced to sell the land which their families have farmed for generations, as farm incomes tumble. Farmers can do little to improve matters. They are victims of forces outside their control. These include:

- Cuts in subsidies as over-production in the EUROPEAN UNION is tackled.
- Cheap imports from North America. Agricultural imports used to be taxed when they were brought into the EU. This protected European farmers from competition, but these protective tariffs have been outlawed by the World Trade Organization.
- Cheap imports of pig meat from eastern Europe. Farmers there do not have to follow the strict animal welfare and hygiene rules that UK farmers have to meet, so their costs are lower.
- Falling meat consumption in the UK due to health scares about meat.

Farmers have felt proud of their efficiency, and have met all the calls for increased production and cheap food since the end of the Second World War. Many now feel betrayed and feel like giving up the struggle to survive in the changed economic world. Those who remain used to look forward to handing on a well-run business to the next generation. Now they encourage their sons and daughters to plan for a future away from the farm. They see no security, and no prospects, in farming. They just see hard work, long hours, low rewards and large debts. Those who are determined to continue farming know that they will have to adapt to changing economic, political and environmental pressures.

2

WHY WAS BSE A DISASTER FOR THE UK'S LIVESTOCK FARMERS?

In the late 1980s farmers became aware that a disease called BSE (or 'mad cow disease') was affecting their cattle. The obvious sign of this disease was that cows lost control of their legs. When examined, the brains of these cows were found to have spaces in them, like sponges, thus the name BSE which stands for 'bovine spongiform encephalopathy'.

This was very similar to a well-known disease called 'scrapie' that affected sheep. BSE was probably caused by a new development in the feeding of cattle. Cattle normally eat grass. However, to cut costs and produce cheaper beef, feed manufacturers had been adding meat to their feed, including the remains of sheep with scrapie. The disease probably spread from one species to another.

Enter CJD! A small number of humans were found to be dying from a disease called CJD (Creutzfeldt–Jakob disease). Their brains had been damaged in a way that was similar to the way that cows' brains were damaged by BSE. Had the disease made another leap to another species: sheep to cow; cow to human?

The government did not want to cause panic before it was sure that BSE and CJD were linked. It did not want to damage UK farming. Little action was taken when the problem was first suspected in the early 1990s. Meat from affected cows was still allowed to go on sale. Even the diseased parts of the cow might be used for making beefburgers, or recycled as animal feed to feed more cows.

However, increasing attention on the topic by the media eventually led to a public outcry. The government eventually admitted a possible link between BSE and CJD. Confidence in beef crashed. The EU banned exports of UK beef. In the UK many people stopped buying beef altogether. The meat-processing industry and the transport industry lost work. Even farmers who had never used manufactured feed and had fed their cattle entirely on grass found that they could not sell their cattle. At last action was taken.

- Using the remains of diseased animals in feed was banned.
- Many infected or potentially infected cattle were slaughtered and their bodies burnt.
- 'Tracking' of cattle was introduced – a computerised record of the life history of each animal – so that animals can be certified as not having had contact with BSE.
- Strict controls on abbatoirs (slaughter houses) led to many small, local ones being closed.

Slowly, these measures restored the reputation of UK beef. The export ban was lifted by many countries. However, the whole BSE crisis cost the farming industry, and the taxpayer, hundreds of millions of pounds in lost sales and compensation payments – and many people and animals have suffered. BSE has now also appeared in cattle in France, Spain and Germany.

Tasks

1 Read Article 2. The government did not publicise the possible links between BSE and CJD until it was certain that the link might exist. Was this right? Give at least one argument in favour of the delay, and one argument against it.

2 **a)** Draw a flow diagram to show the stages by which people think that scrapie in sheep became CJD in humans.

b) Below your diagram write a sentence to summarise what could be done to stop diseases from being transferred from one species to another in future.

3 Read Articles 1–4, then:

a) Work in small groups to compile a list of problems facing farmers in the UK.

b) Discuss which of these you think is the most serious problem. Give your reasons.

4 Some people think that many of these problems are caused by public demand for cheap food. See if you agree.

a) Write 'Demand for cheap food' in the centre of a sheet of paper. Around the outside write the problems that you have identified in Task 3. Draw any links that you can see and label each link to explain it.

b) If you cannot connect a problem to the 'demand for cheap food', think about what might have caused that problem and add the cause to the centre of your diagram.

5 Discuss whether people should be made to pay more for food:

a) so that they can be sure it is safe

b) so that animals on farms are not treated cruelly

c) so that farmers get a fair wage.

SAVE AS . . .

6 Imagine that farmers and supermarket owners have agreed that the price of meat in shops is to rise by 10 per cent so that they can be sure it is produced safely, profitably and without cruelty. *Either* write an article for a magazine (to fill about a page) *or* design a poster to display in shops, to explain why the prices are rising, and why this is a good thing.

3

FARM CHEMICALS POLLUTE WATER AND KILL BIRD LIFE

In their drive to grow more and more food at low prices, farmers have been damaging the environment on which we all depend, according to environmental campaign groups. Spurred on by the European Common Agricultural Policy (CAP), farming in most of eastern England has become increasingly intensive. Fertilisers, pesticides (to kill insect and soil pests) and herbicides (to kill weeds) have been poured on to the fields in ever-increasing amounts.

When fertiliser is spread on fields, some is absorbed into the soil and some is taken up by the crops. If too much is added, the excess can be dissolved by rainwater and LEACHED downwards as the rainwater percolates through the soil. The fertiliser is washed down to the water table, where it can then pollute supplies used for drinking water. Fertiliser can also be washed into streams. Here it encourages the growth of green plants, which may grow so fast that they use up all the natural oxygen in the water. This means that fish and other water life forms cannot breathe, so they die. This process is called EUTROPHICATION.

Pesticides and herbicides wreak havoc with the food chain. Killing the insects can poison or starve the birds and small mammals that eat them. In turn, predators such as owls and foxes that live off the small creatures also lose their food source. Pesticide residues can build up in the higher-order predators, and make them infertile. Their eggs may not hatch, and the species can be threatened. Pesticide residues in food can also threaten human health.

This policy is not sustainable. It might even be too late to reverse some of the damage that has been done.

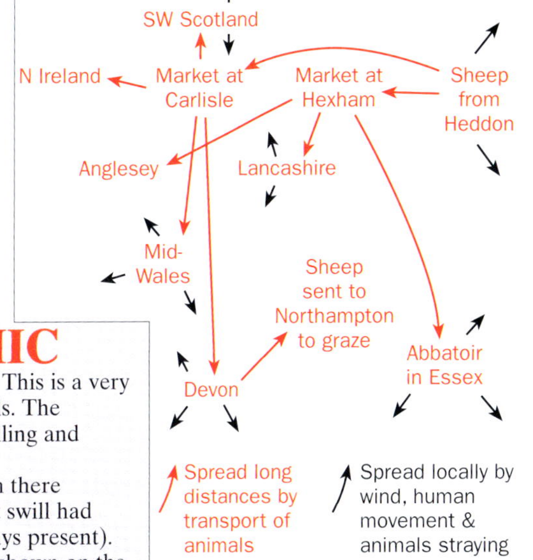

4

FOOT AND MOUTH EPIDEMIC

In late February 2001 another crisis struck UK farms – foot and mouth disease. This is a very infectious disease that affects sheep, cows, pigs and other cloven-hooved animals. The country's last outbreak was in 1967 and it was brought under control only by killing and burning over 100,000 animals.

In 2001 the outbreak started in Heddon, Northumberland. Animals on a farm there probably caught the disease from infected swill (pig feed). Infected meat in that swill had probably been imported from countries where foot and mouth is endemic (always present). The disease spread very quickly. Within ten days it had been found in all areas shown on the diagram. Within 18 days over 130 farms had been affected.

Causes

Causes of the rapid spread were:

- Animals were transported round the country for sale and slaughter. Closure of local abbatoirs (see Article 2) increased the amount of movement.
- Infection can be carried by the wind.
- Infection can be carried on people's feet, clothes, car tyres.

Remedies

To try to stop the disease:

- The government banned all movement of animals.
- Farmers did not leave their farms.
- Millions of animals were killed and burned.
- Rural footpaths were closed.
- Race meetings and other sporting events were cancelled.

The crisis was so severe that the government took the virtually unprecedented step of postponing the general election. By the time of the election on 7 June 2001 the crisis had peaked but by then over 3 million animals had been slaughtered. Not only was farming grievously damaged but the cost to tourism was estimated to be as much as £5 billion, with many jobs lost.

What can bird life tell us?

According to a government booklet, *Sustainability Counts*, the state of the country's bird population is an indicator of ENVIRONMENTAL HEALTH. The document states:

We value wildlife for its own sake and because it is an integral part of our surroundings and our quality of life. Birds are good indicators of the health of the wider environment.

If so, then there is indeed cause for concern. Between 1972 and 1996 there was a disastrous fall in numbers of many species of birds on farmland in the UK. Modern farming techniques threaten birds in many ways, most notably by destroying their habitat. Here are some examples:

Species	% decline
Turtle dove	85
Grey partridge	82
Spotted flycatcher	78
Corn bunting	74
Bullfinch	62
Linnet	52
Yellow wagtail	25
Kestrel	24

● SONG THRUSH

Needs: hedges to nest in and plenty of worms, insects and snails to eat.
Problems: modern chemicals can kill snails, and song thrushes seem to depend on them for part of the year. Many hedgerows have been removed.
Decline on farmland in last 25 years by 73%.

● REED BUNTING

Needs: boggy or marshy areas, farm ponds, rivers or wet ditches. Reed buntings rarely nest far from water, and need a good supply of weed-seeds and insects to feed their chicks.
Problems: cereal crops grow better in well-drained soils. Many ponds and marshy patches have been drained.
Decline on farmland in last 25 years by 61%.

● SKYLARK

Needs: short grassland or crops to nest in; large fields so that it can keep clear of predators. If crops grow too tall it won't nest. Skylarks also avoid hedges, tall trees and large buildings.
Problems: the move away from grazing animals on lowland farms towards growing autumn-sown cereals means that there are fewer suitable fields for the skylark.
Decline on farmland in last 25 years by 58%.

1 Decline in bird species on UK farmland, 1972–1996

Tasks

1 Copy and complete this table.

Bird	Needs	Problems	Solutions
Song thrush			
Reed bunting			
Skylark			

In column 4 suggest how farmers could become more 'bird friendly'.

2 a) Suggest why some people think that bird life is an indicator of 'environmental health'.

b) Do you agree? Give reasons.

SAVE AS . . .

3 In your exam you may be asked about the effect of agriculture on wildlife.

a) Write a paragraph to summarise the impact of farming change on one species of bird.

b) Draw a series of three diagrams to illustrate the process of eutrophication – see Article 3 on page 253.

ICT

4 Design a poster to appeal to landowners to preserve bird habitats. You will need to think hard about why a landowner might disagree with you that this is either possible or necessary.

You could use images and information from the website of the Royal Society for the Protection of Birds:

www.rspb.org.uk/

★ THINK!

The information about different bird species is important in helping us to understand the crisis in farming, but you will not need to learn it all for an exam. Learn details about one bird species, and the causes of eutrophication. That will be plenty to illustrate any question you are asked on the effect of agricultural chemicals on wildlife.

B WAS THE COMMON AGRICULTURAL POLICY A GOOD THING?

One problem for UK farming is over-production. To understand this problem we need to look back to the Second World War, and to consider changes in Europe as a whole.

Before the war, Europe imported a lot of food from North America, Argentina, Australia and New Zealand. It was produced much more cheaply there and European farmers could not compete.

During the war, however, many ships bringing food to the UK (and Germany) were attacked and sunk. Imports were drastically cut, while food production in the UK increased.

Even so, by the end of the war, people in many parts of Europe were starving. Food rationing continued well into the 1950s in the UK. When the European Economic Community (EEC) was set up in 1957, it developed the Common Agricultural Policy (CAP) to boost food production. This identified three problems which stopped European farmers from producing enough food, and offered three solutions.

Problems	Solutions
• **Unreliable prices** When the weather and other conditions were good, farmers produced high yields – but then market prices fell because there was more food than was needed. But in poor years the prices paid to farmers did not rise much, because cheap imports made up for the shortages. • **Low investment** Farmers could not be sure of getting a good price so it was not worth modernising their farms. Taking out loans would lead to serious debt problems. • **Wasted land** Much farmland had poor, badly drained soils. Land was also taken up with many hedges around small fields.	• **Tariffs** (or import taxes) were put on imported farm produce. This made imports more expensive than home-produced food. • **Guaranteed prices** were paid for the main farm products, such as milk, butter, meat or wine. If farmers could not get a good price, the CAP promised to buy up all that they produced at a guaranteed price. Surplus production in good years could be stored for use in poor years. • **Subsidies** were paid to farmers to encourage them to drain marshland, to add manure to poor sandy soils, and to rip up hedges to make fields bigger and more efficient for machinery.

2 CAP solutions to problems in food production in Europe after the war

The policy was very successful. By the 1970s the EEC was self-sufficient in most types of food that could be grown in Europe's climate. At this stage the CAP should have been changed. The policy had in fact become too successful. Huge surpluses started to build up. You may have heard of wine lakes and butter mountains!

However, farmers had great political power in many EEC countries. Politicians would not risk upsetting them, because they might lose votes. So the CAP continued to pay farmers to produce food that no one wanted. The EEC spent more on the CAP than on anything else.

Major Major's father was a sober God-fearing man . . . a rugged individualist who held that federal aid to anyone but farmers was creeping socialism. His speciality was alfalfa, and he made a good thing out of not growing any. The government paid him well for every bushel of alfalfa he did not grow. The more alfalfa he did not grow, the more money the government gave him, and he spent every penny he didn't earn on new land to increase the amount of alfalfa he did not produce. Major Major's father worked without rest at not growing alfalfa. . . . He invested in land wisely and soon was not growing more alfalfa than any other man in the county. Neighbors sought him out for advice on all subjects, for he had made much money and was therefore wise. 'As ye sow, so shall ye reap,' he counselled one and all, and everyone said, 'Amen.'

Major Major's father was an outspoken champion of economy in government, provided it did not interfere with the sacred duty of government to pay farmers as much as they could get for all the alfalfa they produced that no one else wanted or for not producing any alfalfa at all. . . . He was a devout man whose pulpit was everywhere.

3 Extract from *Catch-22*, by Joseph Heller. In the USA in the 1930s, over-production led to falling prices. Some farmers were paid *not* to grow crops. (Alfalfa, or 'lucerne' as it is called in the UK, is grown as cattle fodder)

SAVE AS . . .

5 What was the main success of the CAP?
6 Write a paragraph to explain what went wrong with the CAP. Use the terms 'guaranteed prices' and 'over-production'.
7 Write three sentences to summarise how each of the CAP's solutions helped to overcome farmers' problems.

Note: In Table 2, problems and solutions do not match directly – you will have to think about the connections.

8 Sensible policies can become silly if they are not adapted to changing circumstances. Read Figure 3 which comes from an American novel.
a) What was the original problem?
b) What was the government's solution?
c) What lessons might the planners of the CAP learn from this problem and this solution?

How did changes to the Common Agricultural Policy affect UK farmers?

In the 1990s, changes were made to bring the CAP under control. For example:

- **Milk quotas** Farmers were allowed to produce only a set amount of milk. If they produced more than this they were fined. This cut over-production of milk.
- SET-ASIDE Subsidies were paid to farmers who took land out of production. This was aimed at cutting the wheat mountains, and also at leaving some land untouched so that wild flowers, insects, birds and small mammals would find homes there.
- DIVERSIFICATION GRANTS Money was provided so that farmers could use land in other profitable ways, e.g:
  - setting up motorbike scrambling courses
  - making ponds and lakes and stocking them with fish
  - converting buildings for bed and breakfast use
  - starting pony-trekking and horse-riding centres
  - setting up rare breed survival schemes.
- **Support for conservation schemes** Environmentally Sensitive Areas (ESAs; see page 12) are areas that have a delicate ecosystem. In the UK they include wetlands, moorland, limestone uplands and chalk downlands. In these areas farmers were paid to maintain the traditional ecosystem. The 'Countryside Stewardship Scheme' later extended this type of payment to areas which have attractive countryside, even if they don't have a delicate ecosystem.

By the late 1990s less money was paid as subsidies to farmers. This had a severe effect on small family farms. Many farmers saw their incomes cut by over 50 per cent between 1995 and 2000. This cut forced many of them to sell their farms.

Often these farms were bought, at a very low price, by large firms which already owned many thousands of hectares of land. In difficult times they have enough resources to survive, unlike small family farms.

Task

SAVE AS...

1 This graph shows how production from farms in the EU has increased since the CAP was introduced. It also shows how demand for food has increased over the same period.

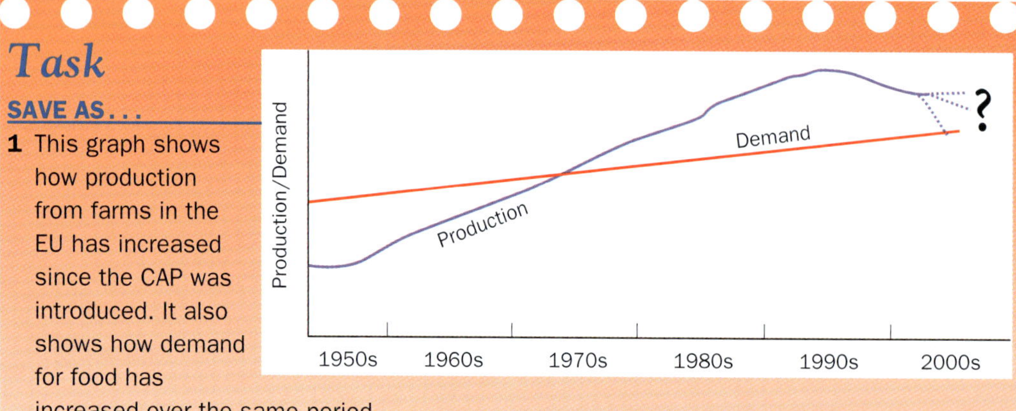

a) Make your own copy of the graph. Add three labels to show the relationship between production and demand:

- before the 1970s (for example: 'Demand exceeded production')
- in the 1970s
- in the 1980s and early 1990s
- in the late 1990s.

b) Using information from pages 255–57, make a list below your graph of three factors that led to the changes in production between the 1950s and the 1990s.

c) Add three more labels to your graph to show how you think the relationship between production and demand might change in the 21st century.

d) Make another list below your graph to show three factors that might cause these changes.

Business efficiency versus stewardship

- Farming employed over 1.5 million people in the UK in 1950 (more than 5% of the country's workforce).
- In 1998 farming employed only 600,000 people (2.4 per cent of the country's total workforce).
- Over the same period agricultural production more than doubled.

Efficiency has increased massively. Farmers produce more food, more cheaply, using less people, than ever before in history. But farmers in the UK would say that there is more to their role than this. They also act as stewards of the countryside.

Tasks

2 a) How did agricultural output per worker change between 1950 and 1998?

b) Suggest why this change happened.

3 Using Figure 5:

a) How did milk and wheat production change between 1973 and 1997?

b) Suggest why this change happened.

c) Suggest why the change was not steady from year to year.

4 Look at Figure 6. The 'agricultural trade deficit' is the difference between the amount of farm goods exported and imported.

a) How has the trade deficit changed over the period shown on the graph?

b) Suggest reasons for this change.

c) Why will the UK never be able to grow all its own food?

d) Suggest how the BSE crisis might have affected the agricultural trade deficit.

Focus task

5 UK farmers are efficient but is efficiency a good thing? As you read through this chapter, look for evidence for or against. Record evidence in a table like this:

Efficiency is good	
Arguments for	**Arguments against**

Then write a concluding paragraph which gives your view.

Agriculture in the UK produces an annual turnover of £17 billion and employs 2.4% of the nation's workforce. However, these measures underestimate the unique role that agriculture plays in the UK economy.

First, agriculture produces 60% of the nation's food. This affects inflation, the balance of trade and the quality of the food that we eat. Second, in many rural areas farming is the main industry in terms of employment, and so it is vital to the rural way of life and the environment. The NFU's goal is to ensure that farming can continue to provide the bulk of the nation's food whilst maintaining the rural economy and protecting our countryside.

4 Extract adapted from a National Farmers' Union leaflet

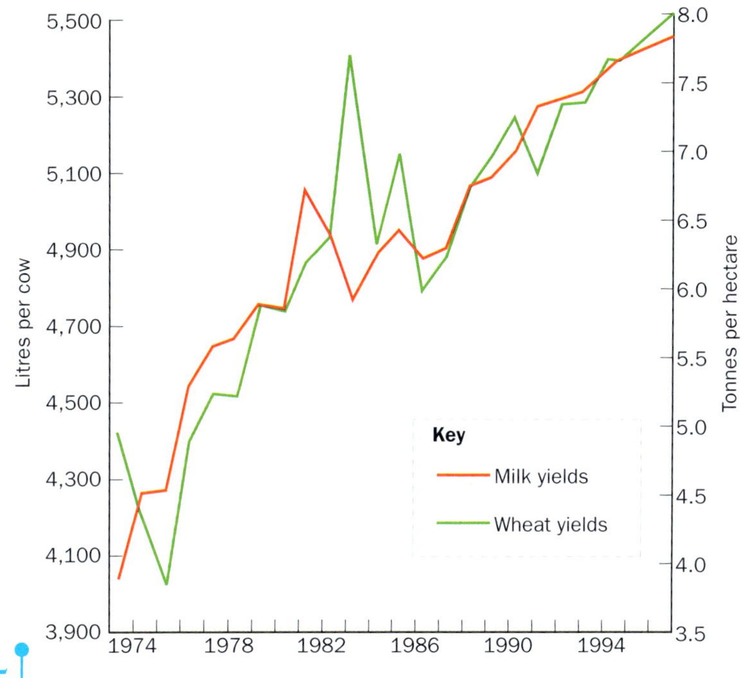

5 Milk and wheat yields in the UK, 1975–1997

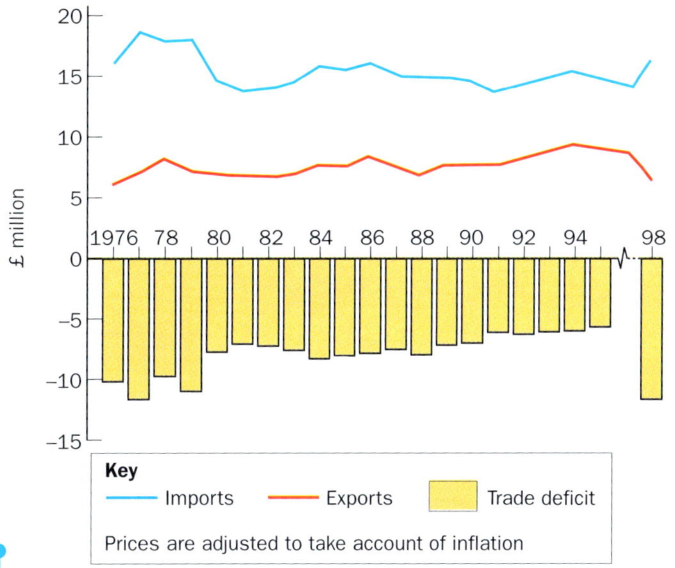

6 UK imports and exports of food, 1976–1998

© MUST FARMING BECOME MORE INTENSIVE TO SURVIVE?

- Intensive farming is farming on a small area with many inputs; for example, growing lettuces in greenhouses near to London.
- Extensive farming is farming over a wider area using few inputs; for example, grazing sheep on moorland in the Pennines.

Intensive farming means increasing inputs to grow more crops, or rear more animals on the available land. Farmers cannot usually increase the amount of land they farm, but they can increase the amount of crops the land produces by intensive farming. It is the growth of intensive farming that has made UK farming so efficient. Many farmers are therefore under economic pressure to be more intensive because it is one of the ways to increase farm income at a time when income is falling. However, intensive farming adds to the problem of over-production, and can damage the environment. Quite a dilemma! The three case studies on pages 258–65 look at how different farmers are dealing with the dilemma.

★ FARM FACTFILE

- The farm lies at a height of 600–700 metres and covers 300 ha.
- The area receives about 2,000 mm (70 inches) of precipitation a year on average, much of this falling as snow during the winter.
- The farm has thin soils on the steeper slopes and deep, acidic, peaty soil on flatter land in the valley bottom and on the top of the moors.
- 70 ha of the farm are enclosed in fields surrounded by drystone walls. About half of this enclosed land is flat and dry enough to be used for hay and/or silage.
- The rest of the farm lies on open moorland.
- There is a flock of 300 Swaledale ewes. They lamb each year. About 90 'hoggs' (female lambs) are kept each year to replace old ewes. The rest are sold for meat.
- 30 cows are also kept, most of which are pedigree Aberdeen Angus. They produce calves which are fed on their mothers' milk. No milk is sold because the costs of collection and transport in this remote area would be too high. The calves are kept for two years and then fattened up for sale.

Case study 1: Farming in an Environmentally Sensitive Area

Richard Betton is a hill farmer in County Durham. He runs Waters Meeting Farm in the Pennines. William, his son, is studying A-levels (including Geography) at the local comprehensive school in Middleton-in-Teesdale, but helps on the farm when he can. Apart from William's help Richard runs the farm on his own, although he employs contractors to come in with their machinery to bale the silage when it is ready.

Because of its climate, soils and isolation, Richard's farm, like most farms in this area, is extensive. Most of the wetland on the farm has not been drained. Grassland has not been ploughed and reseeded with high-yield grasses as has happened on much of the grassland in lower-lying areas. In fact a survey counted 112 different species of grass and wild flowers growing in just one of Richard's fields. This compares with only three or four species found in many much bigger, lowland fields that have been farmed intensively using weedkillers and fungicides.

The marshland along the rivers, the wild uplands and the rich flora of the pastureland attract an enormous number and variety of birds to the area. Many birds visit seasonally to feed or nest in the rich habitat. These birds in turn attract predators to the region, including buzzards and kestrels.

In 1985 this part of the Pennines was designated an Environmentally Sensitive Area (ESA), so Richard is eligible for grants to conserve the environment.

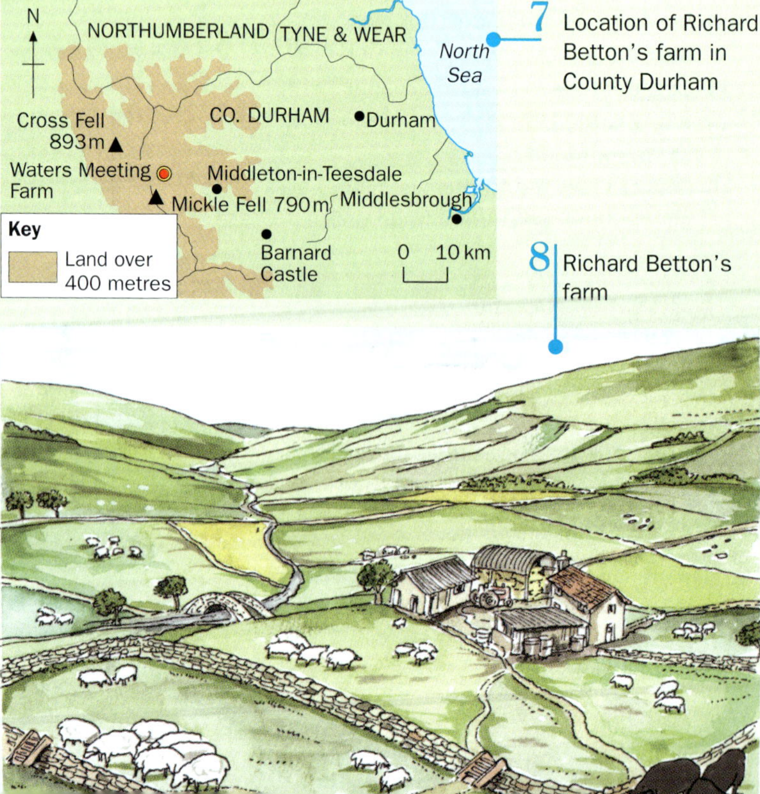

7 Location of Richard Betton's farm in County Durham

8 Richard Betton's farm

Most of the farmers around here love the natural environment. We're not environmental vandals but we do have to feed our families. The hay from the meadow here may contain 112 species but it has a low nutritional value. It will not support enough stock to make the farm profitable. So it's hard to make a living without going intensive. However, the ESA gives us a choice. Subsidies are paid to those who agree to farm an ESA in a less intensive way. The ESA does not force us to farm less intensively, it just makes it economically profitable. These subsidies are designed to make up the difference between the income that we actually earn and the income we could earn by farming the land intensively. But government policies change. Can we rely on ESA payments long term?

I love the life up here, but I'll never earn good money in farming. I do help my dad as much as possible when I'm not at school. My main jobs are taking feed to the stock in winter and helping with lambing and shepherding. I keep trying to persuade Dad to get a quad bike. It would be useful for carrying bales of hay and bringing sick sheep back to the farm, as well as being fun. The broad tyres are well suited to going across damp ground where tractors would cause ruts. Dad says it's not worth it. So many get stolen off farms – it's the new kind of rustling up here.
My ideal future would be to go to uni, get a well-paid job and then to live lower down the valley. It would be warmer, and more accessible there, but I'd like to have a small piece of land and to keep a few sheep so that I'm still involved in farming.

If William does not want the farm when I retire, it might be abandoned. If that happened, the 'traditional' landscape and the wildlife it supports would soon be lost. Grassland would be replaced by scrub vegetation; the drystone walls would collapse; footpaths would no longer be maintained; and the sheep and cattle that are part of the scenery would be lost. A lot of the attractive features of the countryside would be gone.

Tasks

1 Describe Richard Betton's farm, using the **NEWS** technique: describe the **N**atural environment, **E**conomic environment, **W**ho has power (the political environment) and the **S**ocial environment.

2 Copy this table.

Intensive farming option	Environmental impact	ESA alternative
Put drainage pipes in flat, wet fields.		
Plough and reseed grassland.		
Spread lime on fields to raise pH value of soil and improve its fertility.		
Graze more sheep on the open moorland.		
Replace drystone walls with cheaper fences.		

a) In column 2 match one of the following environmental impact statements to each option.

b) In column 3 match one of the ESA alternative statements to each option.

Environmental impact statements

- Overgrazing leading to bare soil patches, causing soil erosion and gully formation
- Loss of moorland birds
- Diversity of species lost
- Loss of flower meadow
- Soil erosion after ploughing
- Marsh habitat would be lost
- Acid-loving plants would be lost
- Loss of 'traditional' appearance of the farmed landscape

ESA alternative statements

- Accept lower stocking levels
- Accept poor yield of hay
- Keep fewer animals
- Buy in hay from lowland
- Stock levels must be kept low because of poor grazing
- Grants paid to farmers to maintain stone walls
- Marsh can be used for summer grazing but not for hay or winter grazing

3 What would you advise Richard Betton to do? Go the intensive route, go the ESA route, or give up farming altogether? Write a paragraph explaining your choice.

'Niche marketing' and 'direct marketing'

The ESA alternative is much more sustainable than intensive farming. Intensive farming on Richard Betton's land might bring a bigger short-term profit, but eventually it could destroy the soil and the environment. One way in which Richard can increase his income while still taking the ESA route is to specialise in quality food production and NICHE MARKETING. Niche marketing means producing something to sell to a fairly small, specialised market. Some farmers can make bigger profits by producing high-quality products to meet the special demands of certain types of consumer. They can charge a higher price for excellent products.

If they can also cut out the 'middleman' – such as the supermarket – by selling direct to the customers they may be able to make more profit, or keep prices down, and also make sure that their product gets to the customer as fresh as possible.

Richard sells most of his cattle to a supermarket chain. They pay a premium for Aberdeen Angus cattle that have spent all of their lives on the farm of their birth, because they know that they produce high-quality meat. Now, though, he has become a partner in a company that is starting to market high-quality meat over the internet. The company began on a farm near Penrith in Cumbria. They guarantee that their meat comes from Herdwick sheep and Aberdeen Angus cattle – breeds well known for producing high-quality meat. They also state that the animals are raised:

> *on farms which rear to the highest standards of welfare under totally natural conditions. Feeding and rearing conditions are as stress-free as possible, and transport to the abattoir and care prior to slaughter are carried out to standards in excess of EU recommendations.*

The company finds that people will pay more for their food if it:

- tastes excellent
- is very well presented
- has been produced by environmentally friendly and cruelty-free farming methods.

Some other farmers have set up 'farmers' markets' in their nearby towns or cities. Here they actually meet their customers and build up a close relationship with them. This direct contact means that they can find out exactly what their customers want, and they can explain any special features of their food – such as ESA farming or cruelty-free animal rearing. Other farmers have moved over to ORGANIC FARMING (see case study 2) and can charge a premium price for food that is guaranteed free from artificial chemicals.

Case study 2: Organic farming

Richard Betton (see case study 1) runs his farm with low inputs and tries to conserve the environment. However, he is *not* an organic farmer. He still uses some chemical inputs on his farm. To be recognised as organic, farmers must be able to prove that they do not use any inputs of artificial chemicals (except for a very few which are acceptable to the Soil Association). Once a farmer achieves organic status he can charge a premium for food guaranteed free from artificial chemicals

Oliver Dowding owns Aviaries Farm, near Wincanton in Somerset. In 1989 he started to convert the farm and it achieved organic status in 1993. You can find full details of this farm on the internet:

www.nfu.org.uk

or your teacher may provide you with a summary of the farm.

There were many reasons why Oliver went organic:

- **Environmental reasons:** By stopping the addition of chemical inputs into the farming system, Oliver realised that he would be: helping to conserve the environment; producing food that contained fewer impurities; providing better working conditions on the farm.
- **Economic reasons:** He would be able to sell his products for more money. He would also cut the costs of chemical inputs, although this would be offset by increased labour costs (for weeding, etc.) and reduced outputs.

Eleven years after taking the decision to go organic, Oliver describes the results (see opposite).

Tasks

1 Write two sentences to explain how:
  a) ESA subsidies
  b) niche marketing
  can increase a farmer's income.

2 *Either:* Write a leaflet for taxpayers explaining how their taxes have been well used to help farmers preserve the environment.

Or: Write an advertisement for meat from Richard Betton's farm. Convince people that there are several good reasons to pay extra for this meat.

3 Discuss: 'Are there other ways Richard could diversify his farm to bring in a better income without destroying the ESA?'

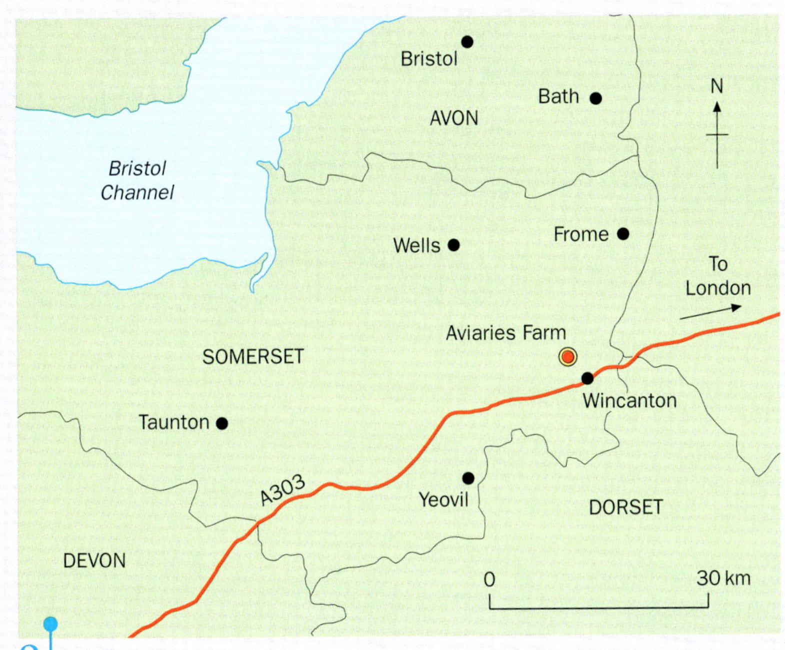

9 Oliver Dowding's farm: Aviaries Farm, Somerset

Tasks

4 Draw a farm system diagram for Aviaries Farm (see Source 3 on page 3 for a similar diagram). Mark on: Physical inputs, Labour inputs, Other inputs, Processes, Outputs.

5 What problems have been caused for Oliver Dowding by turning organic?

6 Write three paragraphs to answer the question 'Can organic farming offer a solution to the farming crisis in the UK?'

You should put forward some evidence that it can, and some that it can't; and then write a summary paragraph giving your balanced view.

The current farming crisis has not hit organic farms as badly as non-organic farms. Our crop and milk prices have remained steady.
Unfortunately the prices for our calves and old cows have fallen. Prices are so low that it is often not worth the transport costs of sending them to market. Sadly we sometimes just have to shoot them. But overall the change has been an economic success.

We've improved a lot of the old footpaths over the farmland. However, they're not all in sensible places. We'd like to re-route some of them, to make them more user-friendly both to walkers and to the farm.

Conversion has had a very good effect on the wildlife on the farm. Numbers of birds have grown and these are mostly very helpful. However:

- We now have about 1,000 rabbits on the farm. They eat grass. 320 rabbits require the equivalent of 1 hectare of grazing land.
- We have several hundred badgers on the farm. They eat slugs (good) and worms (bad). They also drop their dung in the crops, scrape the soil and make holes in dangerous places. Their numbers are now excessive.
- Birds of prey – buzzards, hawks, etc. – have increased and now kill many songbirds. It is difficult to get the balance right.

It's impossible to measure the benefits like improved job satisfaction and better health. However, I'm still worried about the future. If farm prices continue to fall it will mean redundancies and the collapse of businesses that depend on farmers, like vets and machinery dealers. It will also mean that farmers are forced to cut back on non-productive expenditure, like environmental work. If that happens, the nation will lose a significant part of its heritage.

Case study 3: Precision farming

ESA farming or organic farming (see Case studies 1 and 2) are important for conserving some of the most beautiful and wild areas in the country. However, it is unrealistic to think that this type of farming will ever become widespread in the arable farming areas of eastern and southern Britain. The food produced from these areas is so important that intensive farming will be essential there for the foreseeable future. Methods are being developed that could make intensive arable farming more environmentally friendly. These methods are known as PRECISION FARMING or 'COMPUTER-ASSISTED FARMING' (CAF). They use the most modern technology available to help farmers to work with the natural environment rather than against it. Precision farming depends on knowing precisely what is happening in each part of each field. Then each part has to be farmed to gain the biggest yield for the least amount of inputs (i.e. fertilisers and pesticides).

In this way precision farming both cuts costs and reduces damage to the environment: win–win! By ensuring that fertilisers and pesticides go in just the right places, farmers reduce their inputs without reducing their yields and indeed may increase their yields.

By knowing which parts of a field produce high or low yields they can:

- plan their inputs more carefully
- know which parts of the land should be 'set aside'
- improve profits
- reduce damage to the environment.

Table 10 compares precision farming with conventional high-input farming, and the technology of precision farming is described in detail on pages 264–65.

Let's first find out why and how precision farming helps farmers. This composite case study is based on two neighbouring farms in Oxfordshire. Lucy Greene is representative of a small but rapidly growing band of farmers in eastern and southern England and in eastern Scotland who are becoming deeply committed to using precision farming.

High-input farming	Precision farming
Farmer knows field's average yield	Farmer knows yields of each part of field
Adds fertiliser evenly over field to maximise average yield	Puts carefully measured amounts of fertiliser on parts of field where it is most needed
Some parts get too much fertiliser so it is wasted. Other parts get insufficient, so yields are lower than they could be	All parts of field get enough fertiliser to maximise yield without wasted inputs
Farmer fears pests in field and sprays it all with pesticide	Farmer knows where pests are so sprays areas where it is needed, reducing costs and causing less environmental damage

10 Comparing high-input farming and precision farming

Lucy Greene is 29 and runs a farm in Oxfordshire. Her farm covers 160 ha and mainly produces wheat, oilseed rape and barley. It was built up by her father who ran the farm for 30 years. Lucy went to agricultural college and always hoped to take over from her father, but she would have liked to have worked alongside him for longer. Unfortunately he died suddenly and she had to take control.

I realised that most of Dad's knowledge of the land died with him. I had to learn more about the potential of the soil if I was to farm it efficiently, so I started yield mapping straight away. I had studied precision farming at college and it seemed to suit my needs perfectly. The companies that make the systems were keen to help me to start. They hoped to use me as a demonstration farm for the systems.

As soon as I got the first yield maps I realised how complex things were. Here are six of the things that showed up:

1 Some of the land Dad had bought from neighbouring farms was very poor, even though the soil seemed exactly the same as ours. It had not been well cultivated, so it needed a lot of manure added to improve the soil structure. Once the soil was improved the crops could develop better root systems and the yields started to increase.

2 Land beside some of the hedges had much lower yields than land in the middle of the fields. This may have been due to shade, to birds from the hedges eating the crop, or to the roots of the hedges using up the soil minerals. The obvious answer was to leave wider strips down the sides of the hedges and claim set-aside payments for them. It stops wasting time and money on low-yielding land, and creates a really good wildlife haven.

3 Some parts of the fields were producing very high yields. High outputs can only be sustained with high inputs, so we concentrated fertiliser on those areas. I knew the soil was good, so the fertiliser would not be wasted and washed down to pollute the groundwater.

4 Some patches of land seemed to be producing low yields because there were lots of weeds. So last year we downloaded satellite images of the fields which had different shades of colour to show patches of different plant growth. The weeds were growing in clearly defined patches. Once again we could target our spraying precisely. We caught the weeds early and used the minimum of chemicals. We cut inputs and saved money. We used fewer chemicals so the risk of pollution was reduced.

5 Other areas had low yields, but we did not really know why – so we paid some students from the college to take soil samples. Some of these patches had a very low pH value. They needed lime to be added, because acidic soil cannot absorb fertiliser. If the fertiliser is not absorbed by the soil it is not available for the crops, and it can pollute groundwater and rivers. By learning exactly where to add the lime we saved money and helped the environment.

6 Some small low-yield patches turned out to be filled-in ponds. They had been filled to make it easier to use machinery – but the soil was poor, and the surface was marshy. We let a local conservation group reclaim two of the ponds and plant them up. That gave us two more small conservation areas at almost no cost. We hope to stock them with a few fish soon. They're not big enough for fishing ever to be a source of income, but my brother can enjoy fishing there.

So you can see how the system is working now . . . and in future it will become even more precise. The techniques are being developed to provide ever more accurate detail about the land and its crops. It may all sound too good to be true, but we can produce food more cheaply in money terms, and with less damage to the environment. It seems as though everyone wins – farmers, customers, machinery manufacturers, the water board and even the birds and small mammals. They say only 5–10 per cent of farmers are using precision farming at the moment. By 2010 I would guess that only 5–10 per cent will *not* be using it in some form!

Tasks

1 List all of the ways in which Lucy Greene's precision farming has helped:

- to improve the environment
- to cut her costs
- to increase her yields.

2 Design a poster to display to farmers who are *not* using precision farming to convince them that it is in their interests to try it.

3 Precision farming is still a fairly new idea. It is not specifically mentioned in any GCSE Geography specifications. Therefore you will not be asked a whole question about this topic. However, it illustrates very well some of the key ideas involved in studying farming. As a class, discuss the types of questions where you can use your knowledge and understanding of precision farming.

How does technology help the precision farmer?

Precision farming needs a lot of detailed information about each field. This is where modern technology comes in. Precision farming can use:

- satellite technology, including:
  - GLOBAL POSITIONING SYSTEMS (GPS)
  - INFRARED IMAGING of the field
- sensors in the combine harvester to measure yields
- computers in the cab of the tractor or combine harvester to collect data
- computers in the farm office to analyse this data
- precision controls on seed and fertiliser spreaders, linked to computers in the tractor cab.

The exact location of farm vehicles can be established by GPS. Satellite signals can be picked up by a receiver and relayed to the driver via a computer in the cab.

★ **FACTFILE**

Precision farming

- **Global Positioning Systems (GPS)** are small computers which can receive signals from satellites which show you *exactly* where you are on the Earth's surface. They can be used for finding your position to within a metre or two.
- **Infrared imaging** is a form of satellite data collection. The camera on the satellite picks up infrared rays, rather than light rays, and records these. It can 'see' the surface even when it is covered by cloud. Precision farmers can use infrared images to get data on types of crop growing in a field, growth rates, amount of weeds, insect damage, soil moisture levels, etc.
- **YIELD MAPPING** records the amount of grain picked up by a combine harvester as it harvests a field. Data are recorded every few metres to give a very precise map of the yields from different parts of the field.
- **Application maps.** The computer works out exactly how much seed, fertiliser etc. is needed on each part of the field.

Precision farming outcomes

Increased yield → increased food supply, reduced cost of food
Reduced inputs → less environmental damage
Better targeted inputs → less environmental damage
Higher profit margins → farm's survival.

Specimen yield map

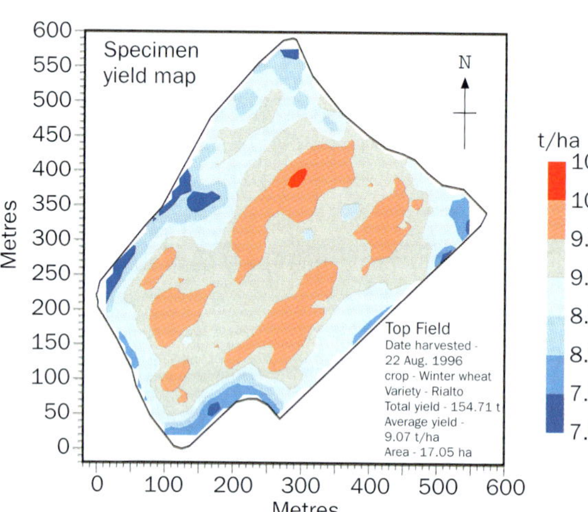

Yield maps produce information for further research (e.g. precise soil mapping)

Tasks

SAVE AS...

1 Draw a simple flow diagram to show how precision farming works. Include the following:

- satellite
- tractor with GPS and computer
- combine harvester with yield mapper
- farm computer
- fertiliser spreader with computer.

Example

Here GPS is being used to give precise soil mapping at a rhino sanctuary in Kenya.

GPS provides location data

Infrared imaging

Infrared imaging of fields to measure growth rates, amount of pests, moisture, etc.

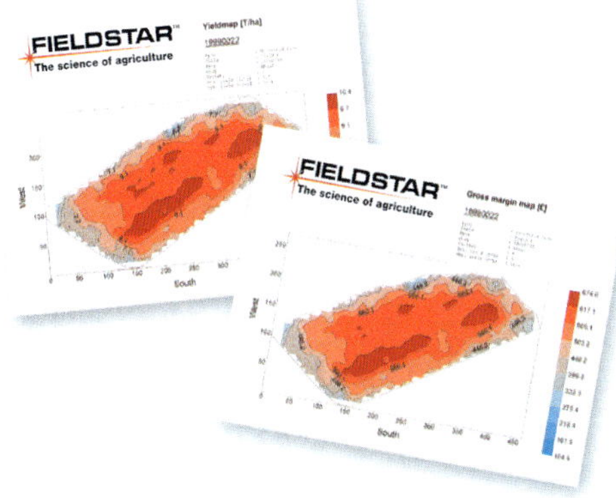

Satellite maps of field yields. Red shows high yields; blue shows low. Note the low yields along the edges of the field where hedges are home to wildlife.

Location data are transmitted simultaneously to the farmer's computer. The farmer can use this data and information provided by infrared imaging to create yield maps and application maps. The data collected can then be applied to farming methods.

Yield maps

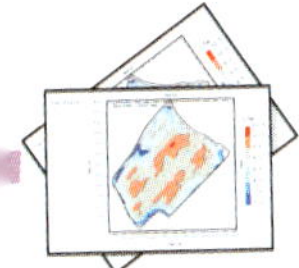

Application maps

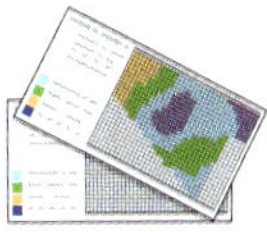

Cultivation

Seeding

Fertilisation

Plant protection

Example

Variable seed rate plan – Pond field (Autumn 1996)

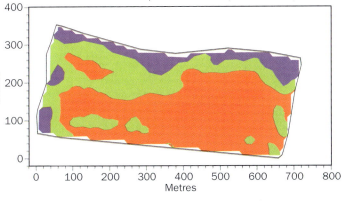

Seed rate plan is fed into communication unit which transmits instructions to the seed drill.

A brief history of plant breeding technology

1 Since prehistoric times, soon after people first started trying to grow crops, they have tried to breed better strains of plants.

Let's eat the wheat from the small plants but save the seeds from the good plants so we'll have more next year.

2 They realised they could improve their seeds if the best characteristics of different plants were combined. They must have relied on a lot of trial and error to make fairly small improvements.

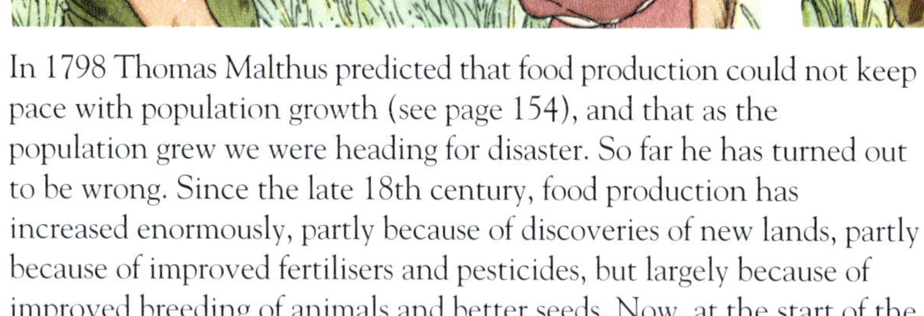

If we plant wheat with **a lot** of grains next to wheat with **big** grains, we might get new plants with **a lot of big grains!**

In 1798 Thomas Malthus predicted that food production could not keep pace with population growth (see page 154), and that as the population grew we were heading for disaster. So far he has turned out to be wrong. Since the late 18th century, food production has increased enormously, partly because of discoveries of new lands, partly because of improved fertilisers and pesticides, but largely because of improved breeding of animals and better seeds. Now, at the start of the 21st century, the technology is taking another leap forward.

The Green Revolution in India

The GREEN REVOLUTION is the name given to changes in farming in LEDCs during the second half of the 20th century. The key factor in this revolution was the introduction of new, high-yielding varieties of seed.

In the 1950s scientists in Mexico developed a new variety of wheat which doubled its yield compared to existing crops. From Mexico this Green Revolution spread to many other LEDCs – most famously to India.

In the 1950s India had suffered regular famines. In 1965 there was a major drought which led to another serious famine with many casualties. Then improved Mexican wheat was planted in North India and the results were amazing – yields improved by two to four times. This encouraged research into improved varieties of rice and other crops. Food production in India increased quickly. India has not suffered serious famine since. It was a real success story. However, as you will see from Tasks 1–4 opposite, there were some controversial issues as well.

The most obvious was that the revolution depended on a lot of expensive inputs which only rich farmers could afford. So the improvements were mainly taken up by the rich commercial farmers (who got still richer through it) rather than by the small subsistence farmers who make up the vast majority of India's farmers. Currently only about 10 per cent of India's rice fields are planted with improved varieties. The rest are still using traditional seeds and methods.

The Green Revolution also increased India's dependence on TRANSNATIONAL or foreign companies who supplied the seeds, fertilisers, pesticides or farm machinery.

People involved in the Green Revolution

12 Scientist checking rice plants at the International Rice Research Institute (IRRI) in Manila, in the Philippines

13 A technician takes a methane sample as part of a project to measure growth rates in rice crops. The project, based in Laguna Province in the Philippines, is organised by the IRRI.

3 Seeds were developed in many different parts of the world, including the Middle East (wheat), Asia (rice) and Central America (corn or maize). The improved seeds spread outwards from their areas of origin. In Europe, from the 18th century onwards, plant breeding became more scientific and developed rapidly.

If we transfer the pollen from these plants to those plants we should get even better plants next year.

4 In the second half of the 20th century, plant breeding became a big business. The development of new strains, through carefully planned cross-breeding programmes, led to the Green Revolution. This still involved cross-breeding between two strains of the same crop, e.g. rice.

We've got lots of different strains of rice. We'll carry out a controlled cross-breeding programme to ensure we get high-yielding varieties of rice.

5 In the late 20th century, and on into the 21st century, genetic engineering became possible. In this system scientists can add genetic material from one species to another completely different species. This is known as GENETIC MODIFICATION (or GM). (See page 268–69.)

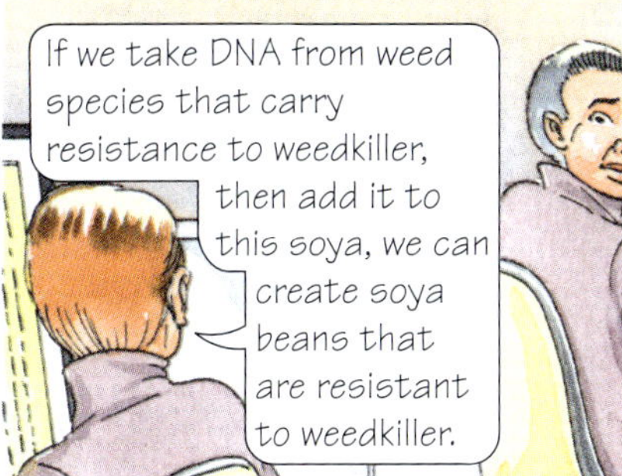

If we take DNA from weed species that carry resistance to weedkiller, then add it to this soya, we can create soya beans that are resistant to weedkiller.

Tasks

Was the Green Revolution a good thing for India?

1 Place each of statements A–L on the scale below.

+2	+1	0	−1	−2
A very good change	A quite good change	Neither good nor bad	Quite a bad change	A very bad change

Record your answers in a table.

2 a) Some changes are so important that they override other changes. Which ones?

b) Decide on a weighting. For example, you could double or treble the score of the really important factors, and/or halve the score of the less important ones.

3 Write a paragraph explaining your view of the Green Revolution. Support your answer with evidence from your table.

4 The Green Revolution affected different groups in India in different ways. Write a second paragraph explaining why and how the following groups might differ in their view of the Green Revolution:

- rich landowners
- poor, landless rural workers
- large transnational corporations with Indian factories
- the urban poor
- owners of small farms
- scientists working in Indian universities.

A Cross-bred strains of rice were sterile. Farmers could not save seeds from one year's crop to plant the next. They had to buy new seed from the seed company.

B Higher outputs took more minerals from the soil, so bigger inputs of fertilisers were needed.

C Imported seed varieties were higher-yielding than indigenous seeds but they were less resistant to local pests, so farmers increased the use of pesticides.

D Efficient farmers made big profits with which they could buy more land.

E Bigger farms encouraged mechanisation, which increased unemployment.

F Demand for fertiliser and farm machinery helped India to industrialise.

G The push of rural unemployment and the pull of industry increased urbanisation.

H Increased food production meant that India could feed its rapidly growing population.

I Some people felt that rice from the new higher-yielding varieties tasted bland and boring.

J In areas where farmers grew richer, more was invested in schools, clinics, small-scale industry, and so on.

K Most farmers cannot afford the new imported seeds. They still follow a mainly subsistence system.

L Farmers no longer save their seeds for planting next year, so some traditional strains are dying out.

Genetic modification of crops – a solution to the farming crisis, or a new crisis in the making?

The latest revolution in plant breeding is a result of genetic modification (GM) of seeds. All living things contain DNA. This is a complex combination of proteins which contains the 'genetic code' for that plant or animal. It contains the instructions, inherited from the previous generation, for building the new organism. Genetic modification involves taking some of the DNA from one species and adding it to another species. When a plant is genetically modified, one or more characteristics of the donor species is transferred to the new plant.

This can produce powerful new varieties of crop, for example:

- By adding the relevant genes of a weedkiller-resistant weed to a wheat seed, you can make a type of wheat that is not harmed by herbicides. So a field of wheat can be sprayed to kill all the grass-like weeds in the field, but the wheat will not be affected.
- By adding a gene from a pest-resistant species to soya bean seed, you can make soya beans which are resistant to pest species. So crops are not damaged by pests which used to destroy old crops.
- By adding a gene from a plant that grows well in arid conditions to the DNA of a rice plant, you can produce rice which needs less water than traditional rice crops, so it can be grown successfully in areas with lower rainfall.

14 Protest about a GM crop field trial

Some advocates of genetic modification claim that newly engineered crops could solve the problem of world hunger for many years to come, plus reduce the input of chemicals. They say that the advantages of genetic engineering will be as great as, or even greater than, the advantages of the Green Revolution in the 1960s.

In trials in the USA, GM soya beans and maize have done especially well. Already much of the soya imported into the UK – which is used in animal feeds and in many processed foods – is GM.

Apart from the USA, China has done most to develop GM crops. It has put large amounts of money into GM research into improved rice and cotton. Rice is the staple diet for most of China's 1 billion people and cotton is an essential raw material for its rapidly growing cloth-making industry. GM crops are therefore seen both as a help in feeding China's huge population and as a way out of poverty. In China there is no opposition to GM crops, and environmental groups opposing them are not allowed into the country. By 2010 much of China's food will be produced from GM crops.

In the UK GM field trials have been going on since 1999. They have been very controversial. Critics say that such trials are being rushed. They worry that:

- the pollen from GM plants *may* pollinate nearby plants and crops, spreading the modifications in an uncontrolled way
- crops on nearby organic farms *might* be contaminated by pollen from GM crops, then the farms would lose their 'organic' status
- GM crops *may* be attacked by new diseases and pests, which *may* increase the need for pesticides and not decrease it
- altering the genetic structure of plants *may* be a long-term threat to the health of people who eat the crops.

15 GM crop on 'field trial' being destroyed by one of a group of protesters

Such uncertainties have led some protesters to destroy GM crops undergoing field trials in the UK (see Photo 15).

There are two extreme views on the future of GM crops in the UK.

An opportunity

GM crops represent such a great opportunity for progress that we must get involved. Potential profits for UK firms and UK farmers are so great that we cannot afford to wait and let other countries rush ahead of us.

A threat

GM crops represent such an unknown threat to health and the environment that we cannot afford to have anything to do with trials of the seeds. They should be banned completely.

The government is currently taking a middle view.

A compromise

GM developments represent both a big opportunity and a big threat. We should be involved but we must be very cautious. More laboratory tests could be carried out. When scientists can prove that the crops are safe in the laboratory, some limited and very carefully controlled field trials could be carried out. In the meantime any GM crops in foods sold to the public should be very clearly labelled. Long-term impacts should be researched.

Whatever the UK decides, the GM revolution is happening anyway. Developments in China and the USA will have a global impact. The international seed companies and food manufacturers will not be influenced by protesters in any one country. Even governments of a single country will not be able to stand in their way. Governments may find it impossible to control the import of GM food or seed even if they wished to.

The problems caused for farmers in LEDCs may be even greater. GM seeds can only be obtained from the big seed companies. Small farmers who rely on saving their seeds from year to year will not be able to compete against rich farmers. This will be one more pressure helping to push traditional farmers off the land.

Tasks

1 Explain why China is one of the leading countries in the development of GM crops.

2 a) As a class, carry out a survey into people's knowledge of and views on genetic modification. Your teacher may provide you with a questionnaire that you could use, or you could draw up a questionnaire of your own.

b) Make sure that you question a large number of people. Among you, try to question at least 50 people. These should include different generations and, if possible, country dwellers and town dwellers. If you do not have access to both of these groups, you could contact a school in a different environment from yours by e-mail. You could each use the questionnaire and compare your results.

SAVE AS . . .

3 Write three paragraphs to record the main arguments for and the main arguments against GM crops and your opinion on whether the current British government policy is correct.

Focus tasks

What is the future for farming in the UK? Several possible futures have been suggested:

- Farming in the UK might become less intensive and concentrate more on conserving the environment rather than on producing large quantities of food.
- Farmers might go the scientific route – using GM seeds.
- They might go organic.
- They could combine precision farming techniques with any of the above scenarios.
- They might give up farming and abandon the land altogether.

4 Discuss each of these futures with a partner. What are the advantages and disadvantages of each for:

a) feeding the population

b) conserving the countryside?

5 Choose one of the farmers you have studied in this chapter.
Either: Write an outline for a short story in which you imagine the future for that person's farm.
Or: Write a letter to the farmer summarising the decisions to be faced and giving your opinion (backed by evidence) of how the farmer should develop the farm.

6.2 Is our use of energy resources sustainable?

A CHANGING ENERGY SUPPLIES IN THE UK

DO YOU KNOW?

- Modern economies depend on energy. Protesters caused havoc in the UK in September 2000 simply by stopping the supply of petrol.
- Each year, the world consumes an amount of fossil fuels that took 1 million years to produce.
- The amount of energy used in a country is a good indicator of the level of economic development. Each person in Tanzania uses the equivalent of 0.1 tonnes of coal per year. Each person in the USA uses the equivalent of 12 tonnes.
- Electricity generating is a major source of atmospheric pollution. In the 1990s over 300,000 tonnes of sulphur were deposited in Scandinavia, most coming from waste products from power stations in other parts of Europe and North America.

AIMS

- **To find out about the changing sources of energy in the UK during the last 50 years and the reasons for these changes.**
- **To consider some of the pollution problems caused by the use of energy sources and to look at the development of alternative energy.**
- **To assess the problems caused by relying on fuelwood as a source of energy in many regions in LEDCs.**
- **To consider how alternative energy sources that are being developed in MEDCs could be used in LEDCs.**

In the early 20th century a Geography textbook author wrote:

> *Britain is fortunate because it is set on an island of coal surrounded by seas full of fish.*

These were thought to be the two most important natural resources of the country. In particular the 'island of coal' provided the power to heat the UK's homes, run its factories, power its railways (still the main form of transport at the time), make its steel, pots, etc. Almost all of life depended on coal as its energy source and this was still true in 1950. Since then the UK's energy sources have shifted dramatically, as Table 1 shows.

Percentage energy from:	Coal	Oil	Natural gas	Nuclear power	HEP	Renewable energy
1950	91	7			2	
1960	78	20			2	
1970	50	42	3	3	2	
1980	37	36	21	4	2	
1990	33	34	23	6	2	2
2005 (est.)	31	34	25	6	2	2
2020 (est.)	31	31	27	3	2	5

1 Changing energy sources in the UK. Most of this energy is used to make electricity. Electricity drives most domestic and industrial machinery

The story of Paul Roberts and his family helps to explain how this change came about.

Case study: The story of Paul Roberts

2 Paul Roberts and his family

'Taffy' Roberts – Paul's grandad

I moved to Barnsley, in South Yorkshire, in the 1930s when my pit in Wales closed. I worked in the Barnsley area until the late 1950s, then they closed my pit again. It was almost worked out and the coal that was left was too expensive and dangerous to mine. In 1960 the National Coal Board moved me to a new 'superpit' over to the east, near Doncaster. We always knew that coalmines wouldn't last for ever, so miners were used to moving. I always wanted my son to work with me, although soon after he started mining in 1961 I was moved to a surface job because of my bad lungs.

Tasks

1 Look at the photos in Source 3, which show an old pit and a new pit. Compare them. Which do you think provides a more pleasant environment to work in?

2 Using the information in Figure 4, suggest why the older pits were sunk in the west of the Yorkshire coalfield and the newer ones in the east.

3 Why do you think Taffy suffered from lung disease as a result of mining?

SAVE AS . . .

4 a) Using an appropriate ICT program convert the data in Table 1 into a graph.

b) Take each of the energy sources in turn. Describe the main changes in the importance of each energy source between 1950 and the present.

3 Old and new coalpits in Yorkshire

Reasons for change
- Loss of markets
- Dirty, dangerous work, unattractive to labour force
- Smoke pollution and Clean Air Acts

19th-century collieries
- All pits now closed
- Most coal seams exhausted
- Those remaining tend to be thin, faulted and liable to flooding
- Extraction becomes very costly
- Equipment out of date
- Little money invested
- Mining towns decaying, e.g. bad housing, spoil heaps

Late 19th/20th-century mines
- Most pits now closed
- Mechanisation gives higher output per man in new pits
- Coal used locally in coal-fired power stations
- Less industrial decay

A few large, modern collieries
(few mining towns)
- Advanced mechanisation gives very high output per man
- High investment
- Conservation of the environment important

Fall in demand due to imports of cheaper fuels, e.g. oil from Middle East, coal from USA and Poland, gas from the North Sea

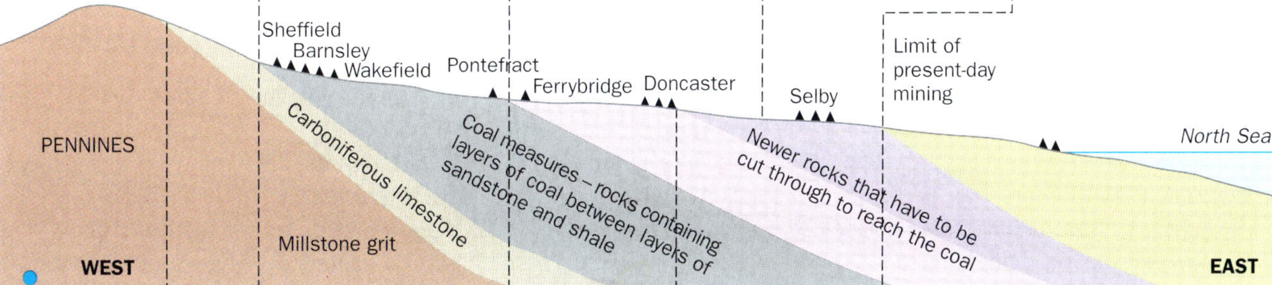

4 Cross-section of the Yorkshire coalfield

Dave Roberts – Paul's dad

When I left school I went to work down the pit. It was the reason this village existed at all. It was the only work here.

I joined at a bad time. In the '60s coal couldn't compete with cheap oil imports so we had to take low wages in order to keep the pits working. In the 1970s things got better. The price of oil went sky-high so coal was in big demand again. Our wages increased, and there was plenty of investment in the pits. Lots of miners bought their own homes at this time. I did too in 1978!

In the 1980s things became really bad again. There were rumours that the Coal Board was going to close a lot of pits. They said it was because coal was running out – but we knew that it was not. The real reason was that the government and the electricity generators wanted to import cheaper, foreign coal. If they did close the pits, it would be the death of communities like ours and we had to do something to save them. We also knew that once a mine was closed it could not be opened again. It seemed wrong to abandon perfectly good resources, when there was supposed to be a world energy crisis.

All the miners could do to protect their jobs was to strike. In 1983 we thought we were powerful. The government thought we were *too* powerful. They were determined not to give in whatever the cost. They won. We lost. The Union in our area stayed united, but men in other areas gradually trickled back to work. Our pit was closed two years later.

One reason why we lost was the decline in the country's heavy industry. Steelmaking and other heavy industries need coal. If the demand for steel falls, the demand for coal falls too. I lost my job in my 40s. I've never had a full-time job since. I would have had to leave the area, and I was not willing to do that with all my family here. My health was poor after over 20 years underground, so I swallowed my pride and became a house-husband. My wife found work. Her IT skills can earn far more than my old skills!

Miners know that coalmines don't last for ever. As you dig deeper it gets more expensive and dangerous to mine the coal. But the government didn't plan to run down the industry carefully. In Germany they give four years' notice of plans to close pits, and they subsidise the areas to bring new jobs and to clear up the environment. That never happened here, even though we were in the EU, like Germany. The pits closed suddenly. People had no time to adapt or plan. One thing I did make sure of was that our son Paul finished school and went to college. I'd never let him work in the mines and be treated like I was.

5 The miners' strike, 1983

★ FACTFILE
OPEC and oil

- OPEC stands for Organization of Petroleum Exporting Countries. It includes Saudi Arabia, Kuwait, Iran, Libya, Algeria, Venezuela, Nigeria and several other major oil exporters.
- OPEC does not include some major producers such as the USA, Russia and the UK.
- In the 1970s OPEC organised itself very well and all members raised the price of oil and cut production at the same time. Their income increased but they also conserved their reserves of oil.
- The industrial countries were so dependent on OPEC oil that they had to pay whatever price was asked. It wasn't just cars that needed oil – many power stations, factories and homes used it too.
- The price of crude oil quadrupled in 1973 and doubled again before the end of the decade.
- For a while in the 1970s and 1980s the UK became less dependent on OPEC oil because it discovered oil in the North Sea, and the high prices made it commercially viable to exploit it.
- Rising oil prices mean that petrol and diesel prices also rise (although the price of oil is only a tiny part of what we pay for petrol). In the UK more than 80 per cent of the petrol price is tax levied by the government. This tax is levied partly to raise money for the government, and partly to make petrol expensive so that people use less and so create less pollution.

Tasks

1. The closure of coalpits was controversial at the time. On page 274 you will consider whether it was the right decision. To start with you should gather data. Read Dave and Paul's accounts. Use them to list:
   a) reasons why the closures were necessary
   b) reasons the closures were a disaster.
   Keep your list.
2. Were the closures the right decision? Write a sentence giving your view. You will return to this later on page 274.

Paul Roberts

When I was little everyone I knew worked at the pit and I always expected that I would too. Then the pit closed and the community was almost completely destroyed. It had a bad effect on a lot of my mates. Some gave up hope, gave up school and ended up in dead-end jobs, or unemployed. Others went to college and left the area. I had always liked Geology at school and Sheffield Uni had a good Geology department, so I went there. I felt a bit like a traitor because I specialised in the geology of oil. In 1989 I got a job with an exploration company developing natural gasfields in the North Sea.

As the price of other fuels went up in the 1970s, natural gas became very important. By the end of the 1980s the UK was starting its 'dash for gas' – lots of medium-sized, gas-fired power stations were built. They produced electricity cheaper than coal-fired power stations and they also produced much less pollution. This 'dash for gas' was one reason why the coal industry never recovered after the strike. Still, my dad doesn't bear me a grudge. I had to find work somewhere.

The fields in the North Sea won't last forever, but when it runs out I hope the effect won't be like the death of a coal pit. Communities don't grow up around oil and gasfields. It's not labour-intensive like coalmining. We also try to make sure we don't leave scars on the environment like the coalmines did. There are no big waste heaps like there were around the mines, and the industry has invested in trying to stop spills from the drilling platforms. It's also usually careful about disposing of the old oil rigs and drilling platforms.

The industry is dangerous, though. Working in the North Sea with a force 9 gale blowing snow across the rig, with ice on every untreated surface, is tricky. And as the easy fields are worked out we're having to search for smaller and deeper fields. Still, the pay is better than it was in the mines. I've been able to help out my parents with the mortgage. They struggled to pay it after the pit closed.

6 North Sea gas platform

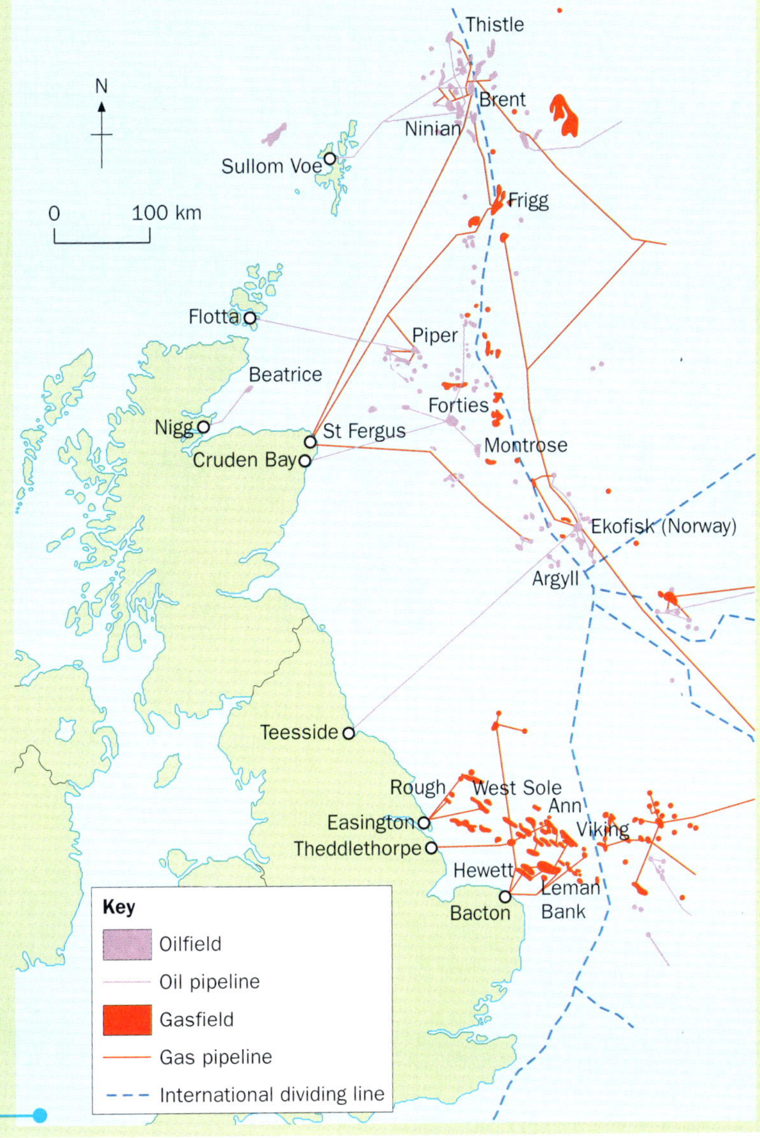

7 North Sea oil- and gasfields

Mary Roberts – Paul's mum

During the miners' strike we women set up support groups. At first we made tea and sandwiches for the picket lines, but gradually we became involved in other ways. I had learned to type at school, and so I started writing leaflets, which we gave out in nearby towns to try to get support. Word-processors were just coming in then, and someone gave us a simple one. I learned how to use it, and I've never looked back!

After the pit closed I got a job in the Benefits Office in 1986, as a clerk. They helped me to learn more IT skills and then I got a job at the Earth Centre near Doncaster soon after it opened. It's a sort of 'alternative theme park' which teaches people all about caring for the Earth and encouraging SUSTAINABLE DEVELOPMENT. It has sections on recycling – even making energy from recycled household waste.

Now I realise that burning coal was harming the environment. It increases the amount of carbon dioxide in the atmosphere, leading to global warming. It can also cause ACID RAIN and smog. But it is not just coal; natural gas also adds to the problem.

The Earth Centre has made me quite committed to clean energy. If we can develop new energy sources in this country, and then share them with people throughout the world, it will bring jobs and a better environment for all of our children and grandchildren.

The miners' strike changed the lives of the women around here for ever. The pit closures did enormous damage to our community but maybe, at last, some good is going to come out of it.

8 The Earth Centre in Doncaster opened in 2000

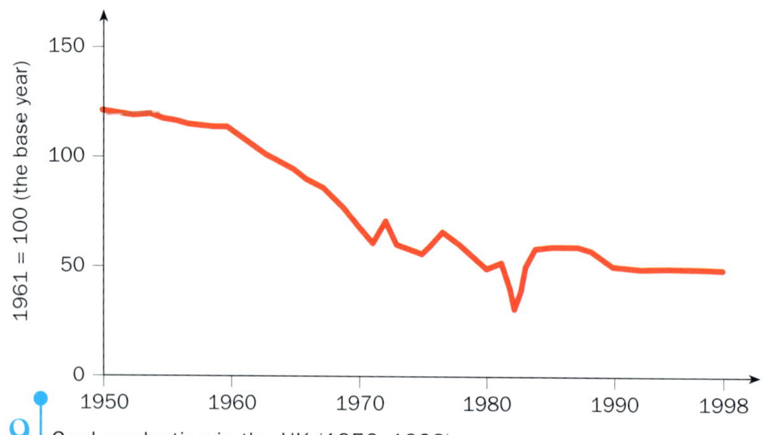

9 Coal production in the UK (1950–1998)

Tasks

SAVE AS . . .

1 Draw a living graph, based on Figure 9, to show the changes that have affected the Roberts family during the period since 1950. Your teacher may give you an outline graph to help you.

a) Add labels to your graph in the correct places which refer to:

- OPEC increases price of oil
- miners' strikes
- dash for gas
- pit closures.

b) Add other labels giving details of changes in the lives of the Roberts family. For example:

- Taffy moves to a surface job
- Dave takes out a mortgage
- Dave is made redundant
- Paul gets a job in North Sea gas
- Mary gets worried about global warming.

2 Look back at your work on page 272.

a) Add any more arguments to your lists from Task 1.

b) Classify your arguments into

- economic
- environmental
- social.

3 Now think about your view on pit closures. You may want to change your view. You may not. Whatever view you take, write up your viewpoint as a paragraph, giving evidence to back up your judgement.

When is a resource not a resource?

Which of the following are resources for me?

- the pile of rubble that the builders left in my garden
- the hollybush covered in red berries in mid-December
- the song thrush eating some of the holly berries
- my vegetable patch with Brussels sprouts growing in it.
- the layer of coal about 3 cm thick and 1 metre below the surface
- my son, Vik
- the tools that Vik is carrying.

A resource is something that is needed, and can be used, by people. So which of these are resources for *me*?
ECEMBER

The answer is that they all are, except for the coal.

- I'll use the rubble for the base of my new drive.
- I am going to cut and sell some of the holly. I do that every year. It is a renewable resource.
- I don't mind the thrush eating some of the berries because, in summer, he eats snails from my vegetable patch. He also sings beautifully. He is a natural resource.
- We will eat the sprouts for Christmas dinner. The soil is a carefully managed, natural resource.
- Vik is just about to break up the rubble to make it usable. He is a human resource.
- The tools are capital resources.
- But the coal is not a resource *for me*. Even if I dug it up, I could not burn it because I live in a smoke-free zone. During the Depression, in the 1930s, some of the people in the area did dig up their coal and burn it. Who knows what will happen in the future? One day it may become a useful resource again.

Why do we use less coal?

You saw from the story of the Roberts family (pages 270–74) how the importance of the UK's coal resources changed. In the early 1900s coal was *the* vital resource. By the 1990s it was just one of a number of possible resources. Many coalmines, with good reserves of coal, were abandoned. Coal itself has not changed but both attitudes to it and demand for it have. A resource is only a resource if there is demand for it. Pages 271–73 highlighted mainly economic factors that led to reduced demand for coal. You are now going to look at environmental factors.

Tasks

4 Make your own large copy of the classification chart below. Add your own examples that have not been mentioned in the text or diagram already.

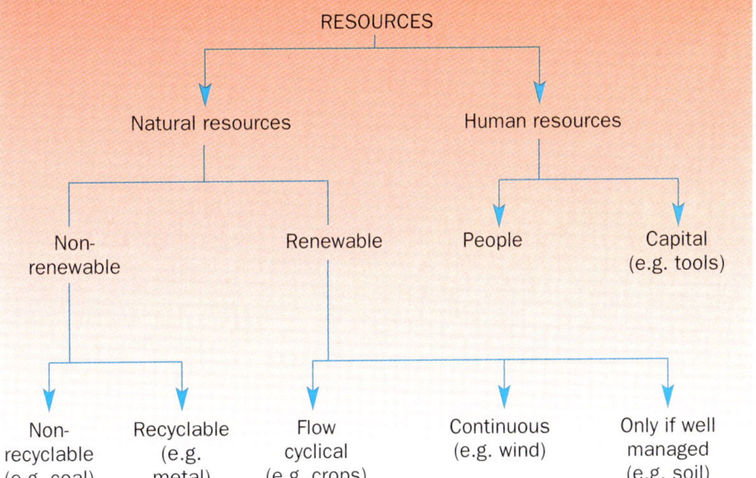

B WHAT ENVIRONMENTAL PROBLEMS ARE CAUSED BY BURNING FOSSIL FUELS?

Coal is a finite, or non-renewable, resource. Once it has been used, it is gone forever.

Coal is made up of important hydrocarbon elements which are used in the chemical industry. Some people think that coal reserves should be conserved for use in the chemical industry rather than burnt in power stations. Figure 10 shows the current world reserves of coal and other fossil fuels.

The current focus for concern is the issue of pollution. Burning coal produces two very damaging forms of pollution: carbon dioxide (CO_2), which contributes towards global warming, and sulphur and nitrogen, which are components in ACID RAIN.

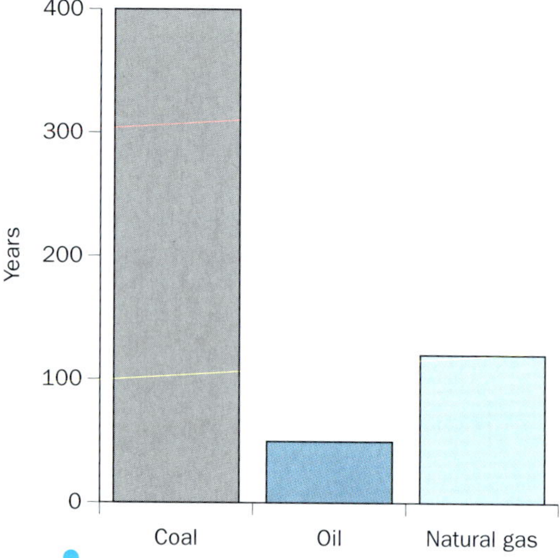

10 This graph shows how long the world's known supplies of fossil fuels will last if we continue using them at the current rate. It excludes supplies which would be so expensive to exploit using today's technology that they are not economically recoverable at today's prices. But remember, as the supply of a resource runs low its price goes up. As the price goes up the less-accessible supplies become more attractive to exploit.

Global warming

Global warming is probably caused by a build-up of greenhouse gases (such as CO_2) in the atmosphere. These trap heat and so the temperature on Earth rises. (See pages 91–92 for a more detailed analysis of this process.)

Concern about global warming has been one of the reasons for the reduction in the use of coal to generate electricity. Coal-fired power stations emit more CO_2 than any other source of energy. On a global scale this forms only a small part of CO_2 emissions, but on the principle that environmental action starts at home it is vital that the UK tries to reduce its emissions of carbon dioxide to the lowest levels possible. At the same time it is also vital that we plan for the consequences of global warming because, even if CO_2 emissions are reduced, warming is likely to continue. In the UK this could include, for example:

- building more sea defences, like the Thames Barrier, to protect vital areas from flooding
- researching into drought-resistant strains of crops to grow in areas of reduced rainfall (see page 268)
- planning cities and buildings so that they can cope with more extreme weather conditions.

The exact impact of global warming is unknown. No one can say for sure what will happen. Ironically it could make the UK colder or wetter. It will almost certainly mean more extreme weather. However, two common factors in most predictions are a rise in sea temperatures and a rise in sea levels.

Global warming seems a good idea to me: hotter summers... more holidays... no cold feet in winter...

Sadly, it's not as simple as that.

11 Global warming – is it all bad news?

Case study: How might global warming affect Bangladesh?

Bangladesh often suffers disastrous flooding. Several factors combine to make the country very vulnerable to floods (Map 12).

Global warming can only make this situation worse. A recent study estimated that global warming by 2° C would lead to a 30 cm rise in sea-level. Warming by 4° C would lead to a 100 cm rise. In Bangladesh higher sea-levels and higher sea temperatures would lead to:

- more and stronger cyclones
- accelerated coastal erosion
- reduced river flow because of inceased evapotranspiration
- saltwater pollution of freshwater supplies in wells and under ground
- destruction of mangrove forests and other habitats along the coast
- increased pressure on land and resources inland.

Cyclones sometimes coincide with river floods caused by monsoon.

Flat land and poor infrastructure make evacuation to high land difficult.

The land is very low-lying – 80% of the country lies on the river floodplain.

Poor farmers are unwilling to leave their land – it is all they have.

Country is poor, so cannot afford best flood defences.

Rivers now more likely to flood because of deforestation in Himalayan foothills (see pages 284–87).

N

BANGLADESH

INDIA

MYANMAR

Bay of Bengal

0 50 km

Shape of Bay of Bengal concentrates force of waves and surges on coast of Bangladesh.

Delta soils are fertile so these areas are densely populated.

Frequent cyclones bring high waves, storm surges and heavy rainfall.

Many distributaries, so many hundred km of flood barriers needed to protect the land.

12 Bangladesh at risk

Action Plan

The Bangladesh Flood Action Plan (FAP) has been set up by the government, the World Bank, and other foreign aid donors, to try to reduce the damage caused by floods both now and in the future. The main solution suggested by the FAP is to strengthen and raise the sea defences along the coast and river banks. It is hoped that this will protect the country from flooding, even if the sea-level rises by 100 cm. It will cost billions of dollars, which will have to come from the World Bank and other organisations based in MEDCs.

13 Major engineering work in progress on flood defences in Bangladesh

Tasks

SAVE AS...

1 a) Look at the labels on Map 12, which explain the causes of the floods which affect Bangladesh. Place them correctly in a diagram like the one opposite.

b) Which of these causes might be made worse by global warming? Underline or highlight them on your diagram.

2 a) Choose two additional problems that are likely to be caused in Bangladesh by global warming.

b) Explain how global warming could cause each of these two problems.

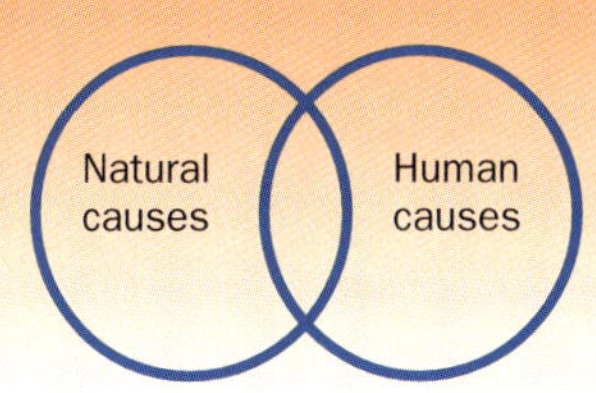

Acid rain

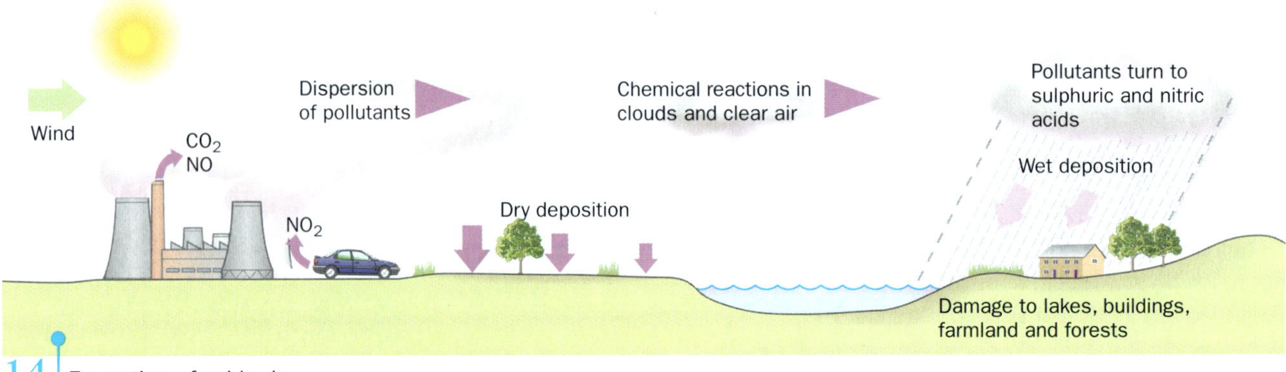

14 Formation of acid rain

The main causes of acid rain are explained in Figure 14.

All rain is slightly acidic because water in the atmosphere dissolves some carbon dioxide to form a weak solution of carbonic acid. This slight acidity is not always a bad thing. Small amounts of sulphur and nitrogen in rainfall can even help plant growth. However, acid rain can have a pH as low as 4.5, and this can have very damaging effects in areas where it falls. These are summarised in Table 15.

Human health can also be affected by acid rain:

- Inhaling acids causes lung and breathing problems.
- Acids help to transfer metals like lead and mercury from the soil into drinking water. These can have serious, long-term effects on health.

The effects of acid rain	Measures to reduce the effects
In lakes and rivers acidity can destroy fish stocks and amphibians like frogs and toads. This also leads to loss of birds that feed on these creatures.	The acidity can be reduced by adding lime (which helps to neutralise acid). In some remote lakes in Scandinavia lime is dropped by helicopter.
Increasing soil acidity can cause leaching of minerals and loss of valuable nutrients from farmland.	Increased addition of fertiliser may be needed, adding to farmers' costs.
Acidity can also damage plants. Forests in Germany and the UK have lost many plants. This is partly due to the effect of acid on the leaves. Acid may also stop trees taking up nutrients.	Commercial forests may have to be replanted with different species which are more resistant to acid rain.

15 The consequences of acid rain

The best way of reducing all these problems is to reduce emissions into the atmosphere. This can be done by:

- reducing the amount of coal burnt in power stations
- using alternative energy sources
- removing sulphur from coal – it must either be removed from the coal before burning (desulphurisation) or during burning (fluidised bed technology); or the SO_2 could also be removed from the waste gases before they are released (scrubber technology). All of these methods are expensive
- fitting catalytic converters to cars to reduce nitrogen dioxide emissions.

Tasks

SAVE AS . . .

1 Study Figure 14. Make a simple written summary of this process.

- Summarise each stage on no more than one line.
- Each line should contain just one key point.

★ THINK!

It is easier to revise if you summarise topics, as suggested in the 'Save as . . .' task. Underline one key word on each line of your summary. Learn the key words in their correct order. These words should act as triggers to remind you of the whole sequence of ideas.

Case study: Acid rain in Scandinavia

The Scandinavian countries realised that their forests and lakes were being severely damaged by acid rain. In 1979 a survey showed that one-third of all trees in Scandinavia had lost at least 25 per cent of their leaves because of this pollution. Salmon had almost ceased migrating up several rivers where they had been common for hundreds of years.

Scientists found that most of the pollution came from outside Scandinavia, as shown in Table 16.

This led to the signing of the Convention on Long-Range Transboundary Air Pollution (CLRTAP) by most major industrial countries. They agreed to reduce their sulphur emissions by 30 per cent between 1980 and 1993. At first, the UK refused to meet this target. Then, as more and more evidence of the damage from acid rain was produced, the UK did start to reduce its emissions. The UK now has one of the best records for reduction of sulphur emissions.

By 1999 total acid deposition on Scandinavia had fallen by over 25 per cent. And measures such as adding lime were introduced to help to repair damaged rivers (see Figure 17). However, acid rain leaves a 'residual effect', which means that pollution from previous years is still damaging Scandinavia's lakes and forests. Today, a high proportion of the pollution comes from the former communist countries of eastern Europe. Their economies are not prosperous, so they cannot afford anti-pollution measures.

Source	Annual deposition of sulphur (thousands of tonnes)
Germany	50
UK	48
Sweden and Norway	33
Poland	21
Russia	15
Denmark	14
Ocean trade	7
Natural emissions	6
20 other countries	40
Unknown	76
Total	**310**

16 Sources of sulphur falling on Scandinavia

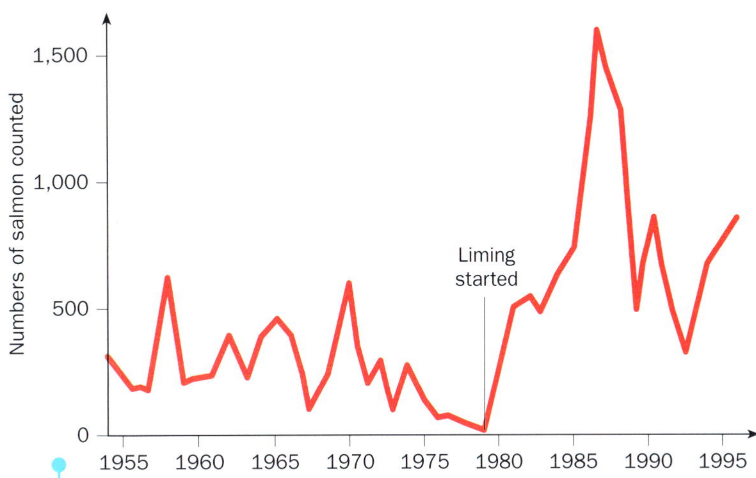

17 Counts of upstream migrating salmon in the River Högvadsån, south-west Sweden (1954–1996). Rivers in this area had lime added to reduce acidity

2 As a class, discuss these questions:

- Should the polluter pay to clear up the problems caused by acid rain?
- Should there be a tax on polluters? If so, how could this tax be made as fair as possible? Bear in mind that most research identifies only a country's pollution output, not an individual power station's output. Should this tax be waived if the extra cost might put a company out of business and cause job losses?

3 a) Describe what Figure 17 tells you about the number of salmon migrating upstream between 1954 and 1970.

b) Describe what happened to the number of salmon between 1974 and 1979.

c) The management of the river changed after 1979. What evidence is there that liming was successful?

d) What evidence is there that reduced sulphur emissions following CLRTAP affected salmon numbers in the river?

SAVE AS . . .

4 Write a summary of the history of acid rain pollution in Scandinavia.

- Give two pieces of evidence suggesting that Scandinavia was suffering pollution.
- Name three major polluting countries.
- Name, and give the initials of, the treaty which tried to reduce acid rain pollution.
- Describe the targets of this treaty.
- List two facts about how acid rain pollution had changed by 1999.
- Summarise the present situation regarding salmon migration in one Swedish river.

C HOW CAN OUR USE OF ENERGY BE MADE MORE SUSTAINABLE?

Electricity generation is the main cause of the pollution which leads to global warming and acid rain. Reducing the pollution will be expensive, and the companies will pass on the cost to the electricity consumer – your family and *you*.

Coal, oil and gas (the fossil fuels) cannot provide all our energy in future. They are non-renewable resources. Even if they do not run out they will become more and more expensive as all the easily accessible supplies are used up. These fuels also cause pollution, which cannot continue without causing problems. Continued use of these fuels at current levels is not sustainable. There are several possible solutions to the problem. Three of these are:

- Reduce energy demand
- Develop NUCLEAR POWER
- Develop renewable energy.

Tasks

1 As a class, brainstorm ways that energy demand can be reduced in your homes, schools or leisure facilities, without reducing your quality of life. Discuss:

a) how you would benefit from such a reduction.
b) how the environment would benefit.

SAVE AS . . .

2 Copy this table.

Solution	Advantages	Disadvantages
Reduce demand		
Develop nuclear power		
Develop renewable energy sources		

As you read pages 280–81 make notes to record the advantages and disadvantages of each solution.

Solution 1: Reduce energy demand

The MEDCs are able to reduce demand for energy. By saving energy in the home, in industry and in transport, the total demand has been cut in some countries, and can be cut further.

However, this is *not* a solution in LEDCs. Most people in LEDCs use only tiny amounts of energy anyway. The low use of energy is a sign of their low standard of living. In fact, if their living standards are to improve they need to use *more* energy – in industry, agriculture, transport and the home.

Solution 2: Develop nuclear power

In the 1950s one expert said 'nuclear power will produce electricity so cheaply that it will not be worth charging people for it!'

Nuclear power uses small amounts of radioactive material to produce heat in a nuclear reactor. This is then used to drive a turbine to make electricity. Some very successful nuclear reactors have been developed, but there are two major problems with nuclear energy that have still not been completely solved.

1 **Nuclear reactors are dangerous.** If they are not very carefully controlled explosions can result, like the Chernobyl disaster in the Ukraine. In 1986 a nuclear reactor at Chernobyl went out of control and caught fire. It caused very serious NUCLEAR FALLOUT in large parts of Europe, including the UK. Today, much of the countryside at Chernobyl is still badly contaminated with radioactive material that can cause cancers, or deformities in babies. Crops and animals there cannot be eaten. People in the nuclear industry claim that nuclear power stations in the UK are far safer than the Chernobyl plant – but they still face the second problem . . .
2 **Radioactive waste from the reactors has to be disposed of somehow.** Nuclear waste remains dangerous for hundreds or even thousands of years. Attempts to reprocess waste to make it completely safe are proving difficult and costly. There is still no agreement on the best way to dispose of it safely. Should it be buried deep underground? Or at sea? Or should it be stored where people can keep watch over it and constantly monitor its safety?

Until research can develop foolproof, long-term ways of dealing with the waste, development of nuclear power cannot really proceed. It is a source of great potential benefit, but also potential disaster. The claims of the 1950s for this form of energy have not yet come true.

Solution 3: Develop renewable energy resources

Renewable energy resources are not finite. Renewable energy comes from processes that go on all the time in the atmosphere or on Earth. Most renewable resources are powered by the Sun. Renewable energy sources include:

- solar power
- wind power
- hydro-electric power (HEP)
- wave power
- tidal power
- BIOMASS energy (the burning of agricultural or other wastes for energy)
- geothermal energy (comes from hot rocks beneath the surface of the Earth).

Most of these forms of energy:

- do not produce any waste material
- have low running costs once they are set up

but they:

- have not yet been developed sufficiently to be used on a large scale
- will need a lot of investment in research and/or start-up costs
- may produce visual or noise pollution
- are only suitable in certain, limited sites.

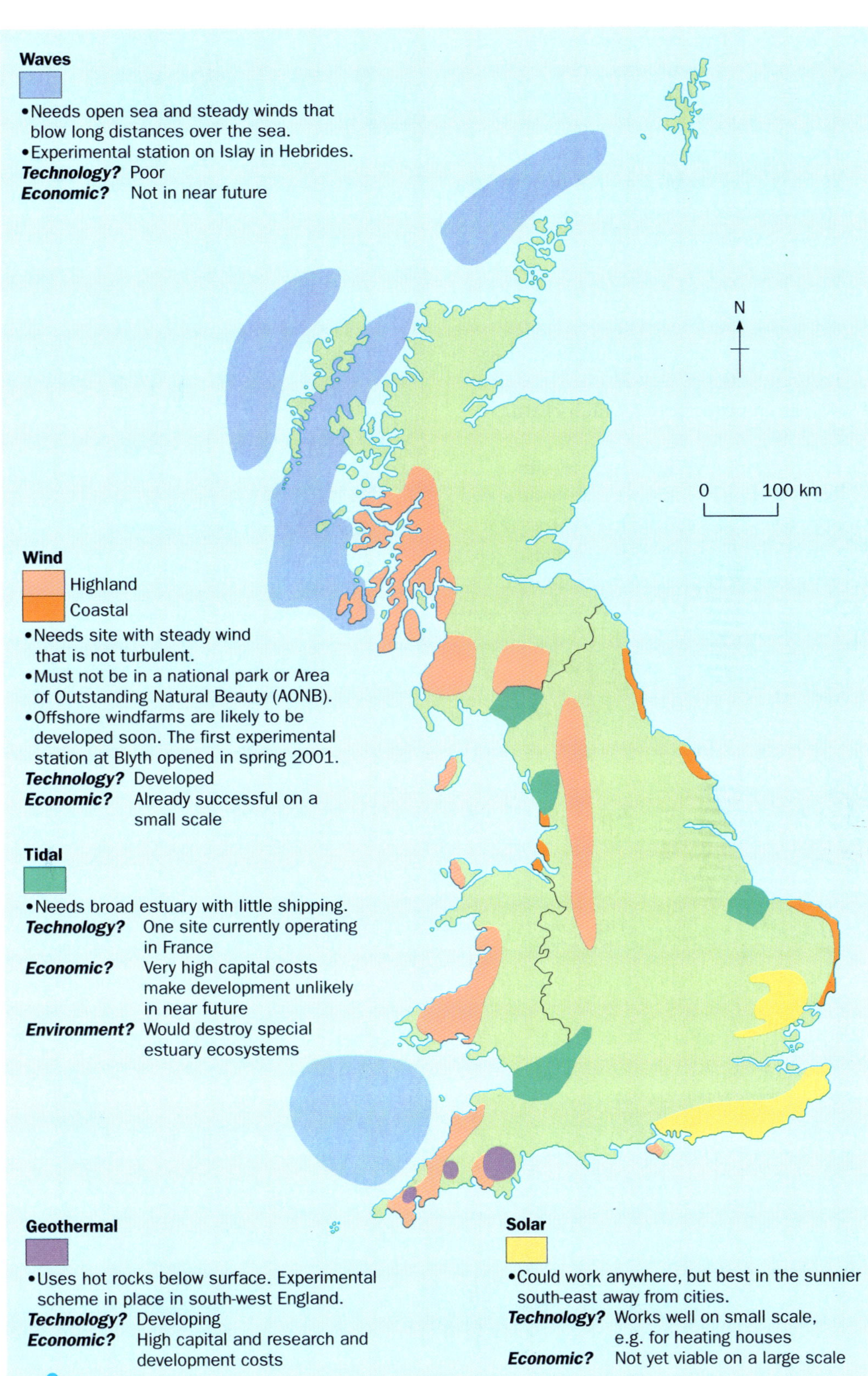
Waves
•Needs open sea and steady winds that blow long distances over the sea.
•Experimental station on Islay in Hebrides.
Technology? Poor
Economic? Not in near future
N
0
100 km
Wind
Highland
Coastal
•Needs site with steady wind that is not turbulent.
•Must not be in a national park or Area of Outstanding Natural Beauty (AONB).
•Offshore windfarms are likely to be developed soon. The first experimental station at Blyth opened in spring 2001.
Technology? Developed
Economic? Already successful on a small scale
Tidal
•Needs broad estuary with little shipping.
Technology? One site currently operating in France
Economic? Very high capital costs make development unlikely in near future
Environment? Would destroy special estuary ecosystems
Geothermal
•Uses hot rocks below surface. Experimental scheme in place in south-west England.
Technology? Developing
Economic? High capital and research and development costs
Solar
•Could work anywhere, but best in the sunnier south-east away from cities.
Technology? Works well on small scale, e.g. for heating houses
Economic? Not yet viable on a large scale

18 Potential sites for the production of renewable energy in Britain

Case study: Wind power in an MEDC

In the UK the most widely used form of renewable energy is wind power.

Wind energy is harnessed using a tall pylon fitted with blades to catch the wind. The energy of the moving blades is turned into electricity in a turbine housed in the 'nascelle'. The nascelle turns so the blades face into the wind to catch maximum energy. The electricity that is generated is carried to the electricity grid, or directly to the user. Turbines are usually built together in groups, called WIND FARMS.

Wind farms need to be sited:

- where the wind is strong and reliable – the average wind speed must be no less than 5 metres per second
- away from obstructions which obstruct the flow
- on flat ground, because hilly terrain causes turbulence and reduces wind speed
- away from National Parks or Areas of Outstanding Natural Beauty where planning permission might be refused.

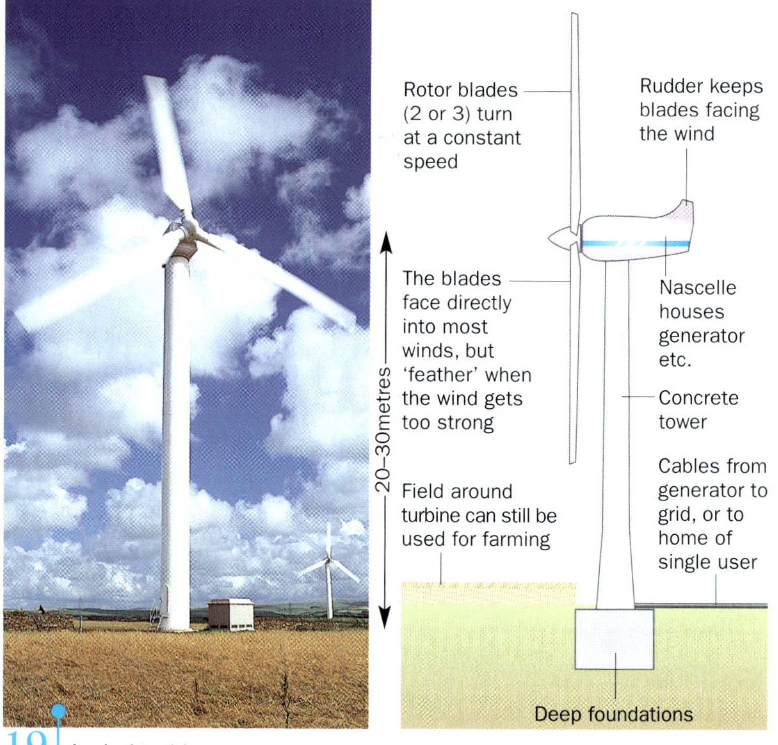
Rotor blades (2 or 3) turn at a constant speed
Rudder keeps blades facing the wind
The blades face directly into most winds, but 'feather' when the wind gets too strong
Nascelle houses generator etc.
Concrete tower
20–30metres
Field around turbine can still be used for farming
Cables from generator to grid, or to home of single user
Deep foundations

19 A wind turbine

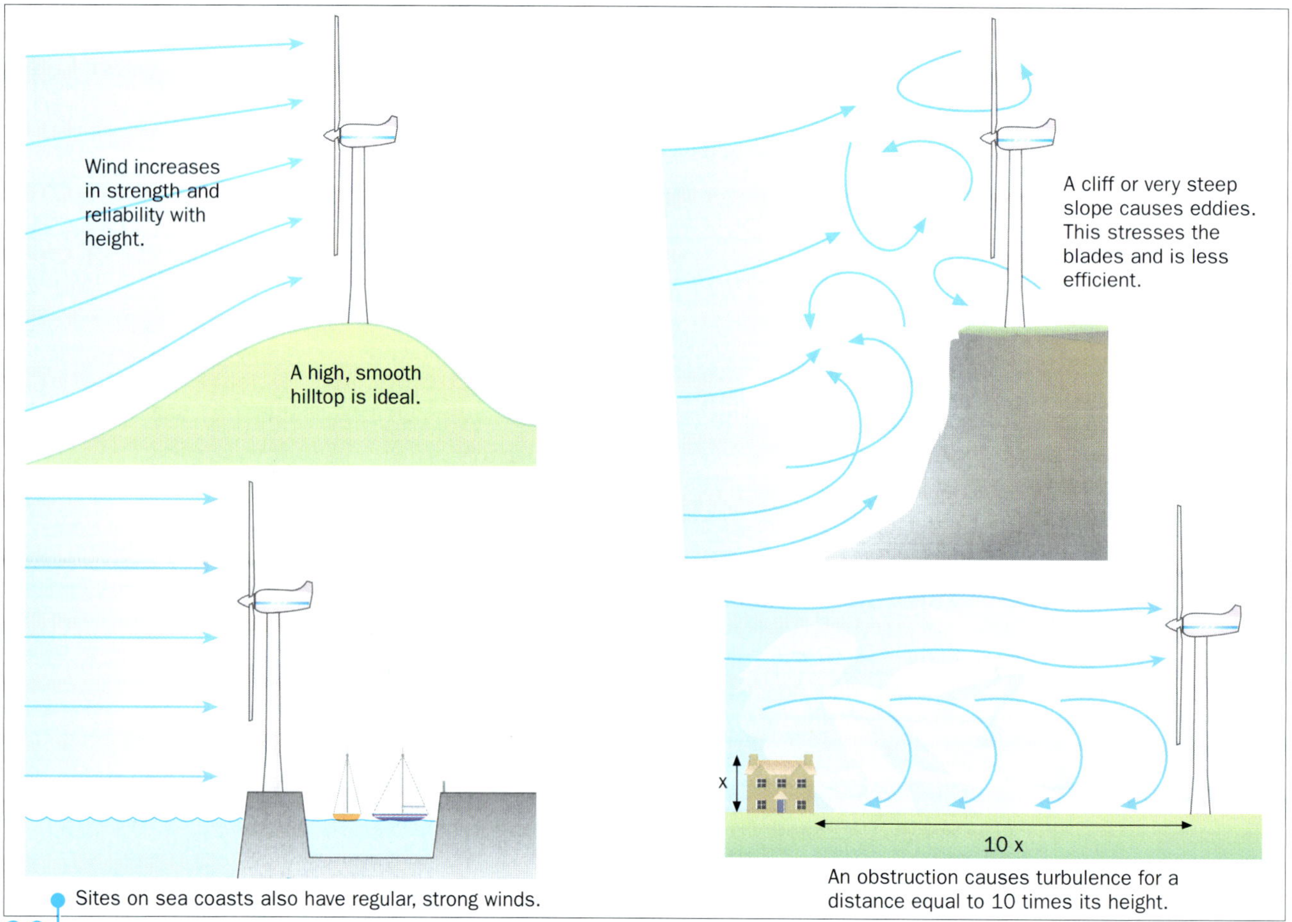

Wind increases in strength and reliability with height.

A high, smooth hilltop is ideal.

A cliff or very steep slope causes eddies. This stresses the blades and is less efficient.

Sites on sea coasts also have regular, strong winds.

x

10 x

An obstruction causes turbulence for a distance equal to 10 times its height.

20 Good and bad sites for wind turbines

Tasks

Decision-making exercise: wind energy

1 a) Imagine that you are a surveyor who has been asked to choose a site for a new wind farm by a company developing wind power. Eight possible sites are shown on Map 21. Study the map and make reference to some or all of these factors, for each site:

- reliability of wind
- accessibility
- distance from settlement
- value of land
- environmental value of site.

Fill in a table like the one below. (Your teacher may provide you with a copy of this.)

Site	Costs	Benefits
A		
B		

b) Choose the best site. Write a summary of your reasons for choosing this site, with brief comments on your reasons for rejecting the others.

ICT

2 Many wind farm developments have been strongly resisted by local residents who are worried about the possible noise and visual pollution caused by a wind farm.

a) Visit the British Wind Energy Association website at:

www.bwea.com

Study the information there about surveys that have been done in areas before and after wind farms have been built.

b) Use information from these surveys to write a leaflet for local residents aimed at setting their minds at rest about the impact of the development. Mention:

- the general reasons why the wind farm is better for the environment than fossil fuel electricity generation
- specific measures that may be taken locally to minimise visual or noise pollution
- any benefits that might come to the area from the building of the wind farm.

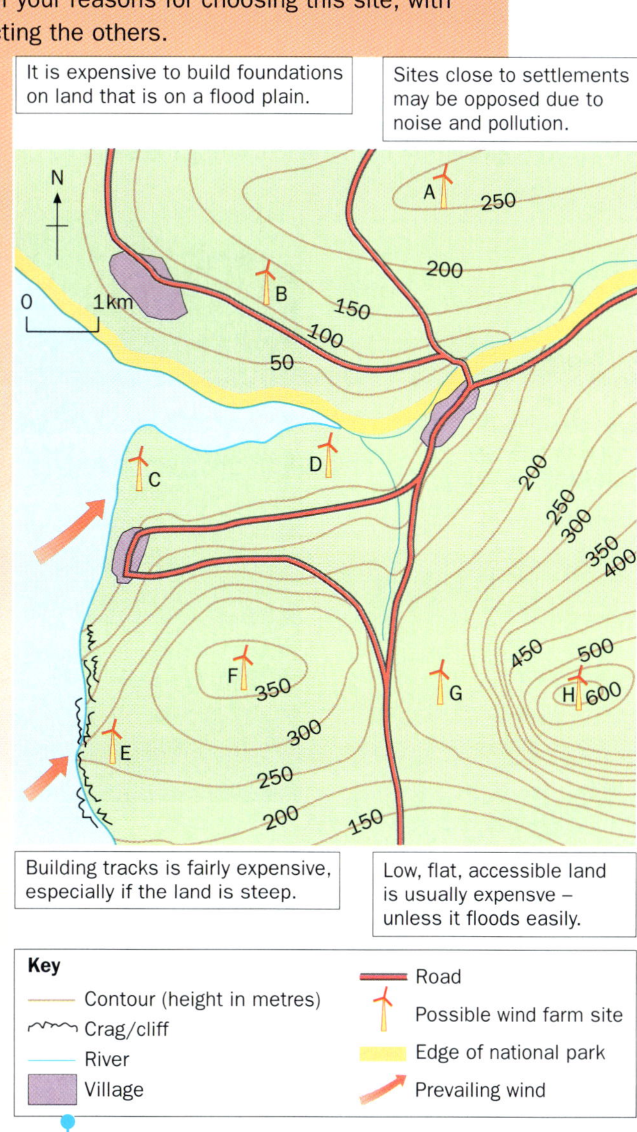

21 Possible sites for a wind farm in mid-Wales

Case study: Nepal – How can energy use in LEDCs be made sustainable?

If people in LEDCs are to improve their standard of living it seems inevitable that they will have to increase their use of energy. In many areas even the present low levels of energy use are damaging the environment. How can environmental damage be reduced at the same time as energy use is increasing?

Nepal is a small country situated high in the Himalayas (Map 22). It can be divided into three broad sections:

- **The mountain zone** is over 3,000 metres high. The land is extremely inhospitable. Population densities are low.
- **The hill zone** lies between 900 and 3,000 metres. Population densities here are much higher, averaging 120 per km^2. There is acute pressure on the flat, agricultural land on the valley floors. Farming is mainly subsistence rice-growing on flat valley bottoms and on terraced hill slopes. Wheat is grown in drier valleys and potatoes on higher, cooler slopes.
- **The Terai** is low, flat land on the northern part of the Ganges plain. It used to be sparsely populated, because it was marshy and malaria was widespread. In the last 50 years people have moved into this area to cultivate the land. Cash crops of sugar and oilseeds are grown, as well as subsistence crops. This area has an acute shortage of fuelwood. Much is brought in from the remaining forests in the hill zone.

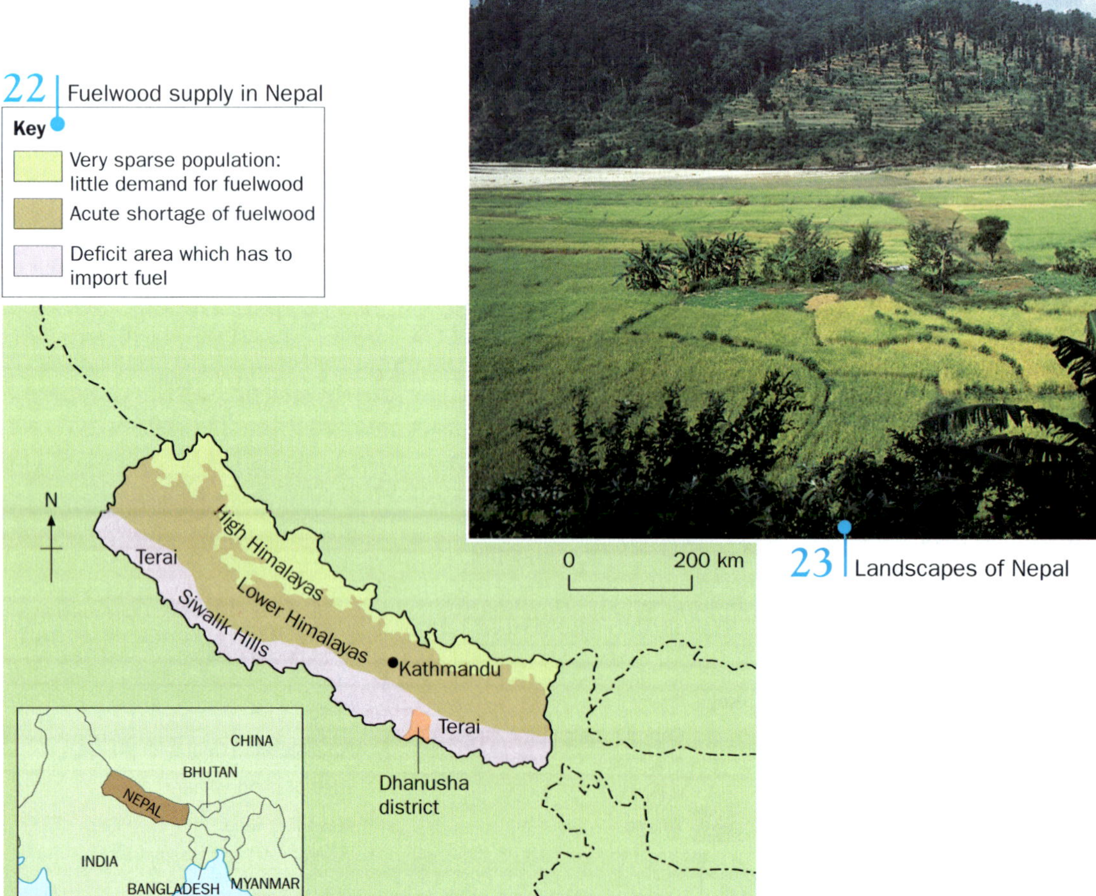

22 Fuelwood supply in Nepal

23 Landscapes of Nepal

★ **NEPAL FACTFILE**

- Population (2000) = 24.3 million
- Birth rate = 31‰, Death rate = 15‰, Infant mortality = 98‰
- Annual population growth rate = 1.6%
- People living in cities = 9% (but increasing rapidly)
- Life expectancy = 51 years
- GNP per capita = US $179
- Employment: primary = 58%, secondary = 10%, tertiary = 32%
- Population/doctor = 22,312
- Literacy rate = 26%
- Main city = Kathmandu (population 250,000 approx.); six other cities between 50,000 and 160,000
- Long history of supplying Gurkha soldiers to the British army
- Climate is dominated by the monsoon, but strongly influenced by varying altitude
- Most areas have adequate precipitation, most of which falls during the summer monsoon season – May to October
- Industry dominated by food processing and cement manufacture. However, tourism, particularly trekking in the Himalayan foothills, is growing in importance.
- Electricity supply only available in and around the major cities. Uses imported coal and some HEP. Increasing use of small-scale HEP stations in more remote regions.

Map 24 shows Dhanusha, a district which includes part of the Siwalik Hills and part of the Terai. The Siwalik Hills are foothills of the Himalayas. Until about 30 years ago, they used to be forested, and the people relied on the wood to provide them with all their fuel for cooking and heating.

Since then, an increasing population has increased demand for fuel and also for land on which to grow crops. Forest near the villages has been cleared rapidly and women (whose job it is to collect the wood) are now having to travel further and further to gather supplies.

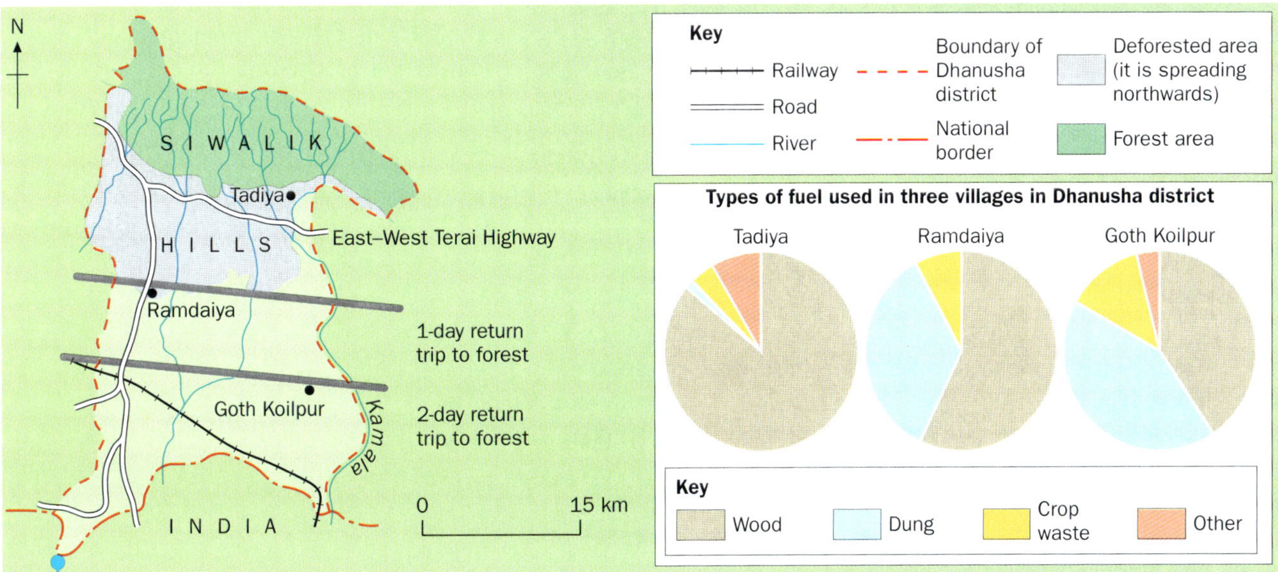

24 Fuel supplies in Dhanusha district of Nepal

Is the use of wood as fuel sustainable?

Nepal is typical of many LEDCs. Most people still rely on fuelwood as their main source of energy for cooking and heating. When population densities are low, this is a sustainable form of energy because the wood replaces itself naturally. However, if population density increases the amount of land available for gathering wood is reduced.

Then there is a problem. The use of wood as fuel is no longer sustainable. The wood is used more quickly than it can be replaced. The shortage of wood causes a range of human health and environmental problems:

- burning valuable dung instead of using it on the land (see Map 24)
- burning of crop waste that used to be used for grazing animals, instead of burning wood (see Map 24)
- reduction of protein intake because fewer animals can be grazed and then used for food
- reduced cooking time, leading to disease
- increased back problems among women who are forced to carry their wood supplies greater distances
- conflict with neighbouring peoples over wood supplies
- scarce money spent on old trucks to bring in wood
- malnutrition, poverty and urbanisation
- deforestation with all its associated problems (see Factfile).

★ FACTFILE

The problems caused by deforestation

- **Soil erosion** because:
  - there are no leaves to intercept rainfall and reduce its impact on the soil
  - there are no roots to bind the soil together
  - there are no roots to make passages through the soil to allow INFILTRATION, so RUN-OFF increases
  - there is nothing to slow down run-off.
- **Flooding** because:
  - run-off is faster
  - there is more run-off and less throughflow, so water arrives in the river more quickly
  - eroded soil can be deposited downstream, blocking the river.
- **Drought** because:
  - there is less evapotranspiration to return water vapour to the atmosphere
  - if run-off is faster there is less chance of evaporation from the soil
  - reduced evapotranspiration leads to reduced rainfall.
- **Loss of soil fertility** because:
  - topsoil is eroded (see above)
  - leaf fall no longer returns nutrients to the soil to be recycled.
- **Loss of habitats for wildlife**
- **Loss of attractive scenery**

Task

ICT

1 Study Map 24. Does distance from the forest have an effect on the type of fuel used in the villages of Dhanusha district in Nepal?

a) Devise a way to show how fuel use changes with distance from the forest. Here are the exact percentages of energy use for the three villages:

	Tadiya	Ramdaiya	Goth Koilpur
Wood	87	57	40
Dung	1	35	43
Crop waste	4	8	13
Other	8	0	4

Use the map to measure the distance from each village to the forest.

Now use your two sets of information to construct a graph. This could be a line graph or a stacked bar chart.

b) Use your graph to answer the question above.

c) Write a paragraph to explain the strengths and weaknesses of your method of displaying the data.

★ THINK!

Geography is about interconnections in the environment. In particular it concerns connections between the human and physical environments. You may think it odd that a Factfile on deforestation is here, in the chapter on resources and energy. However, it shows that all parts of the environment are interconnected.

How can Dhanusha's energy needs be met?

Various measures have been tried, including:

- planting new, quick-growing trees (pine and eucalyptus) on steep slopes
- giving plots of land to village communities, on condition that they grow wood for a sustainable harvest
- developing techniques for COPPICING to harvest rather than chop down trees – see Figure 26
- buying reconditioned trucks to bring in supplies of wood from the north of the country
- setting up 'mini-hydro schemes' to generate electricity supplies (see Figure 25) – Dhanusha has some sustainable sites
- setting up local craftspeople to make stoves, using scrap metal, which will cut demand for wood by about 50 per cent.

Wind power has not yet been tried, except on a small scale. Nepal's physical geography means that it does have some very windy places but they are mainly inaccessible and mountainous. However, some parts of Dhanusha are exposed and windy and might be suited to a wind turbine.

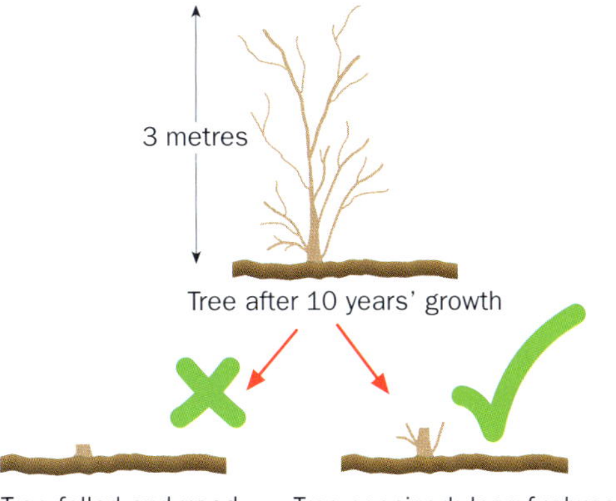

26 Coppicing – sustainable wood harvesting

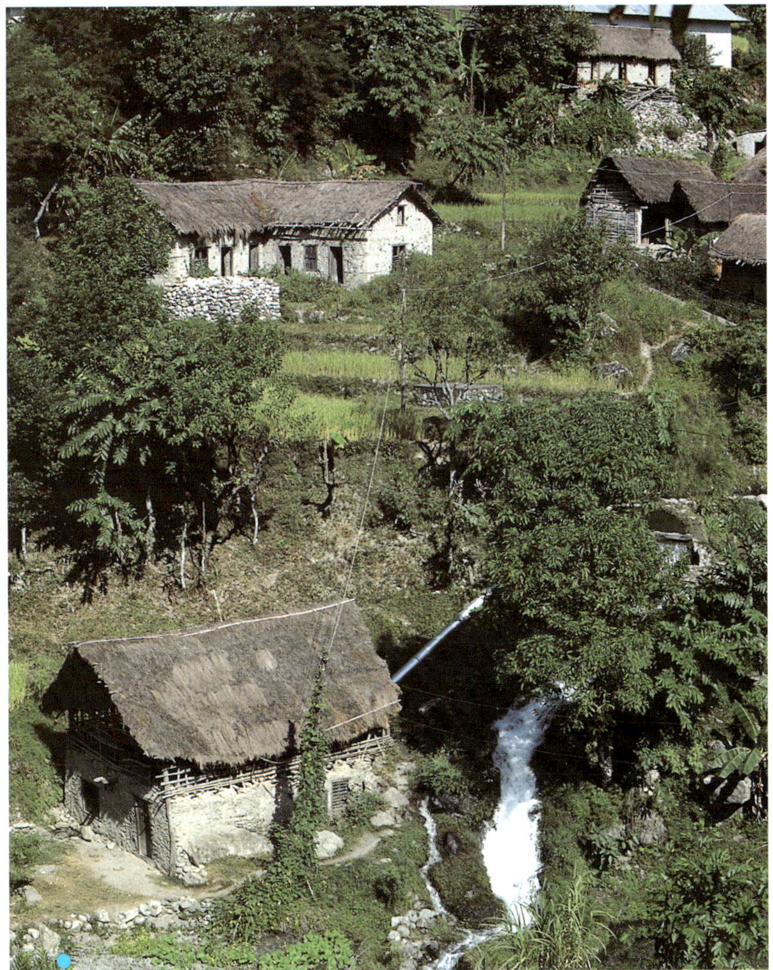

25 A small-scale HEP scheme in Nepal supplying sustainable electricity and so helping to reduce the need to use fuelwood

Tasks

2 Study the measures mentioned in the text to meet Dhanusha's energy needs.

a) One of these solutions is not sustainable. Which one?

b) Why might it still be essential to use this as a short-term solution to the fuelwood problem?

c) Consider whether wind power might be suitable for this area. Should MEDCs give the technology to LEDCs?

3 Bearing in mind all you now know about Dhanusha, work in groups to produce an action plan for the villages. It should:

a) summarise the differing needs of the three villages

b) suggest one measure each village can take *individually*

c) suggest two measures that the villages could take *together* to meet their long-term energy needs.

Your measures should be *affordable* (the villagers are mainly subsistence farmers although some have a little income from putting up and guiding trekkers); *environmentally sustainable*; *practical* rather than idealistic (so the villagers will actually be able to do what you suggest).

6.3 Can we meet the challenge of industrial change?

DO YOU KNOW?

- In 1967 there were 24 integrated iron and steel works in the UK. By 2000 there were only 4.
- Between the two World Wars the UK was the world's leading shipbuilding nation. Now the industry has almost completely disappeared from many areas where it used to be the main employer.
- In the 1960s many areas competed to attract car assembly plants – 'the new growth industry'. Now many of these plants have closed or are threatened with closure.
- The computer and micro-electronic industries hardly existed in the 1970s. Now they are the major industries in some parts of the country, and all parts of the country try to attract these industries.

AIMS

- **To understand how heavy industry developed in the UK in the 19th century, and to look at some of the reasons for its decline in the 20th century.**
- **To consider how Reading has developed new high-tech industries to replace its old declining industries.**
- **To look at the motor industry, and to compare its recent growth in north-east England with the motor industry in Malaysia, a NEWLY INDUSTRIALISING COUNTRY (NIC).**
- **To consider the effect of industry on the environment.**

1	**Central Scotland**	Iron, steel, ships, machinery, cotton
2	**North East**	Iron, steel, ships, machinery, chemicals
3	**Cumbria**	Iron, steel
4	**Lancashire, north-east Cheshire**	Cotton, machinery, chemicals
5	**West Yorkshire**	Woollen goods, machinery
6	**South Yorkshire**	Iron, steel, cutlery
7	**Staffordshire**	Pottery
8	**West Midlands**	Metal goods, machinery
9	**East Midlands**	Iron, steel, textiles, leather
10	**South Wales**	Iron, steel, metals
11	**London**	Various industries
12	**Belfast**	Ships, textiles

A THE GROWTH AND DECLINE OF HEAVY INDUSTRY

During the Industrial Revolution in the 19th century, heavy industry developed in many parts of the UK. Some of the main areas are shown in Map 1.

The most important factor which helped to decide where heavy industry was located was access to coal. This was important because coal:

- provided the energy to produce the steam that powered the industry
- was bulky and expensive to transport
- was transported by canal and by railway. The cost of coal rose rapidly with distance from the coalfield, so most heavy industry was built close to a coalfield, canal, or railway.

The main heavy industries that used coal are shown in Table 2.

The factory system

Every factory or industry, whether it was producing textiles in the 19th century or is producing computer components in the 21st century, can be seen as a system (Figure 3), with inputs, processes and outputs.

When people are deciding where to locate a factory, they try to choose a site where the costs will be low. They try to reduce collection and distribution costs.

- **Collection costs** – have to be paid to bring the main inputs into the factory: raw materials, power supply and labour.
- **Distribution costs** – have to be paid to transport the outputs to the market. The outputs are the product – either a finished product for sale, or a partly-finished product for use in another factory.

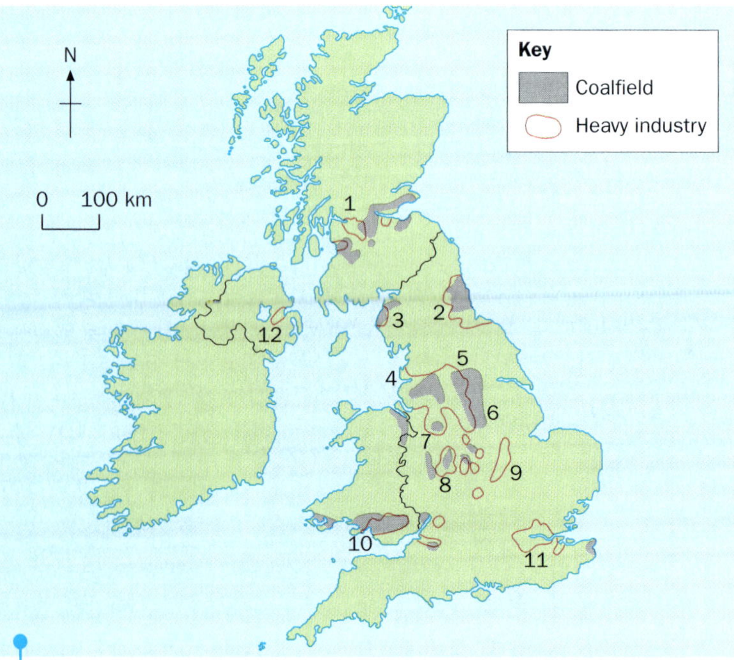

1 Coalfields and traditional heavy industrial areas in the UK

Industry	Location
Iron and steel	Mostly on coalfields which also had iron ore deposits. The finished product was bulky, so many users of iron and steel located nearby too.
Shipbuilding	On estuaries, to provide shelter close to deep water, and close to iron and steel supplies, to make the ships.
Textiles	Originally close to uplands, where sheep were reared for wool, and in areas with soft water (away from limestone areas) for washing the cloth. Later: • Cotton manufacturing developed close to west-coast ports where raw cotton was imported from the USA. • Wool industries remained further east in Yorkshire, where cotton was less accessible.
Heavy chemicals	Close to salt deposits and limestone (two raw materials) and near the textile industry which provided a market for its products (soap, dyes, etc.).
Heavy engineering	Produced machinery for the factories, railway engines, girders for bridges, etc., so were located close to iron and steel works.
Other metals	Near ports for imports of copper, lead and other metal ores.

2 Location of heavy industries using coal

Task

SAVE AS...

1 Look at Figure 4.

a) Complete two more diagrams like this to show the factors important in the location of other heavy industries described on these pages. Your labour box can be the same for each diagram. Use the information in Map 1 and Table 2 to help you.

b) Find out what has happened to each industry now, in the early 21st century, and write a paragraph to explain this.

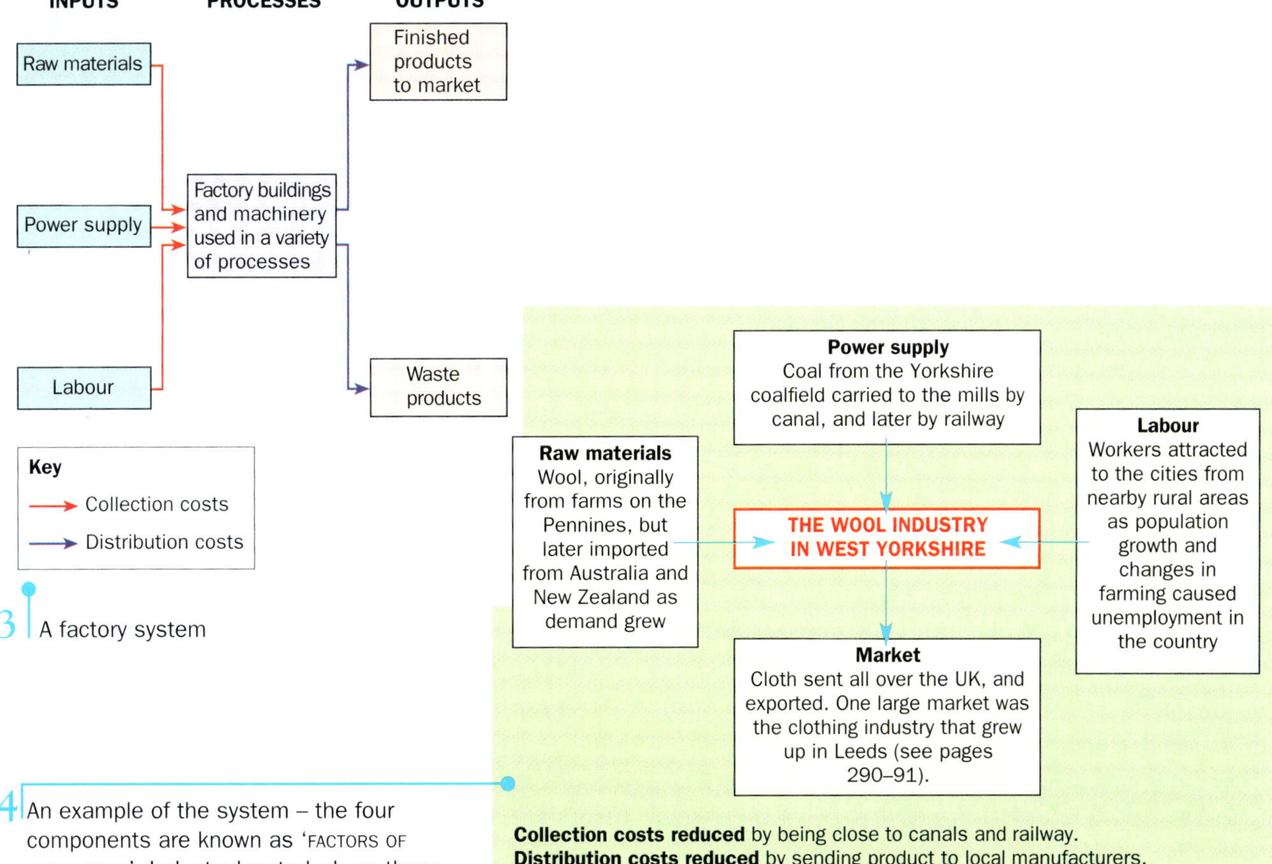

3 A factory system

4 An example of the system – the four components are known as 'FACTORS OF PRODUCTION'. Industry located where these four factors were available.

Aire Street workshops: then and now

One major market for the Yorkshire woollen industry was the clothing manufacturing industry in Leeds. Aire Street Workshops were built in the centre of Leeds, in about 1880, to make army uniforms, suits, trousers, jackets and other woollen clothing. The workshops were owned and run by Jed Bishop and his spirit has recently been disturbed . . .

In my day this location was just right. It had all we needed close by. Our raw material was wool from sheep grazed on the Pennine Moors, made into cloth in Bradford. The cloth came by canal. Power came from coal. A steam engine in the basement pumped all day. Great drive belts spread the power around the building to drive the cutters and presses. We employed boilermen to keep the machine working smoothly. All the coal came here by canal. Now the canal is used by tourist boats, and I hear that most of Yorkshire's pits have closed down. How can factories run their machines?

As for my workers, they were grateful for any job. Many of them moved here from country areas because of the depression in farming, so we provided them with housing as well as work. Hours were long in those days so folk had to live near the job. We had to crowd as many little cottages as possible close to the town centre. The folk had to walk to work too – there was none of those new omnibuses when my workshops were first built.

The smoke from the chimneys, the dust from the cloth and the damp houses meant that some of my workers had bad chests. Of course that worried me. You can't get a proper day's work from a sickly worker, but we had to put up with the problem. We couldn't replace workers with machines back then. We just had to get on with the job and meet demand. How about my market? Well, I sold to the British Army. They needed uniforms to wear in all parts of the Empire. We made tailored suits for gentlemen too. We could transport our goods to anywhere in the country by railway.

I made good profits and built myself a fine house on the edge of Leeds, away from the smoke and noise. I tried to put something back into the town too. I provided jobs for people but I also made generous contributions to charity and helped to build the local hospital and grammar school.

Those were the days! But they're gone forever now. No one uses coal, no one moves goods by canal: in fact manufacturing hardly needs workers now – just machines. So why are people poking about in Aire Street again? What's going on?

By the 1990s the Aire Street workshops were derelict. Then a property development company bought them to develop them for renting out to modern ICT-based companies.

2000

Many ICT companies favour inner city locations especially in university cities, because young workers like working in city centres, particularly exciting ones like Leeds today. ICT industries don't use many raw materials. For power, all they need are electric sockets. Public transport into the centre of Leeds is good so workers can travel easily.

All around the centre of Leeds there are small, but growing ICT-based companies as well as other businesses in the SERVICE SECTOR, such as banks, insurance firms and television companies.

5 The back of Aire Street Workshops in the 21st century

Tasks

1 Aire Street Workshops were built at GR 293333 on Map 15 on page 321. Why was this a good location for a factory in 1880? Refer to access to energy, raw materials, labour and markets.

2 Why do you think this location was no longer suitable for a factory using bulky raw materials in 1990?

3 Why do you think this might be a good site to develop for modern service or ICT industries?

SAVE AS . . .

4 a) When a site like this is redeveloped it is called **brownfield development**. It is often contrasted with **greenfield development** on land at the edges of towns and cities. Explain the meaning of these two phrases. You can also refer back to pages 215 and 238–39.

b) Explain why Aire Street Workshops is a good example of a brownfield development.

c) Explain why the government and conservation organisations are keen to encourage development on brownfield sites.

5 Many service companies benefit from locating close to each other and close to city centres. Why might the following all benefit from being close together:

- TV studio
- costume designer
- lighting hire
- fancy dress hire
- children's party planners.
- entertainment agency
- music shop
- outside catering agency
- keep-fit and dance studio

Draw a diagram to show how these services might depend on each other.

The decline of heavy industry in the late 20th century

The Aire Street story could be repeated thousands of times for locations all over the UK. During the 20th century many of the UK's old industries went into decline. The following were some of the causes and consequences of that decline.

Power
- Coal reserves began to run out in some areas.
- In other areas all the most accessible coal had been mined, and what was left was too expensive to mine.

Outdated products
- The demand for many of the products fell. For example, steel was replaced by plastic in many industries. Fewer railway engines were needed as cars and lorries were used more.

Poor sites
- Factories and warehouses were on congested inner-city sites, surrounded by housing that was often condemned as unfit for human habitation. The surrounding land was often seriously polluted by the waste from over a century of uncontrolled industrial development.

6 Derelict warehouse in the Sheffield Canal Basin

Lack of investment
- The owners of these factories were unwilling or unable to invest and modernise. They struggled on, gradually losing their markets in the face of growing competition from abroad.

Low wages
- Industry survived in old, dilapidated buildings only by keeping wages down.
- A lot of industries relied on immigrants who were prepared to take poorly-paid jobs in the inner city for as long as it took to establish themselves. They worked hard, saved money, made sure that their children received a good education, and then followed the earlier labour force (whom they had replaced) to better jobs with better pay.

Foreign competition
- Many other countries developed strong manufacturing industries which took over UK export markets and sold products to the UK itself.
- Some of these competitors had a much cheaper labour supply than in the UK, so made goods cheaper.
- Many of the competitors had newer, more efficient factories and machines than the traditional UK industries.
- In the 1950s the competition came from other European countries (such as Germany) that had invested in new plant and equipment after the war; then it came from the newly industrialising countries (NICs), such as Malaysia, which had a cheap labour force and used modern methods.

Renewal
- Many early factories had been located near canals and railways. As road transport took over in the second half of the 20th century, these sites were less attractive than accessible sites on the edges of cities.

Entrepreneurs
- The most eager businessmen and women in the UK and the best workers were attracted to new industries with more pleasant working conditions, better pay and a brighter future. Usually these businesses were located well away from traditional manufacturing areas. Often they were in the south east of England.

Strikes
- At times industrial relations in the UK were very poor, with lots of strikes, poor management of the workforce, and inflexible working practices (for example, one worker refusing to do another's work to help out in busy times). This made UK manufacturing inefficient and unreliable.

B READING: A CASE STUDY OF INDUSTRIAL CHANGE

The rapid decline in the UK's manufacturing industry has caused great economic, social and environmental problems. However, the UK is now going through a new industrial revolution. New industries are growing which are less harmful to the environment. Society has to adapt to the needs of this new revolution.

The town of Reading in Berkshire illustrates many points about the growth and decline of the old industries. It is also a very good example of a town which is enjoying a new industrial 'boom' in the 21st century. Both of Reading's periods of industrial growth have been largely due to its position and to excellent communications, although the means of communication were very different in the two periods.

Tasks

1 How did each of the industries in Table 8 reduce collection costs and distribution costs?

2 Annotate a copy of Map 7 to show the reasons for the growth of each of these industries.

3 Suggest why Reading's position between London and Bristol helped the growth of the town's industry in the 19th century.

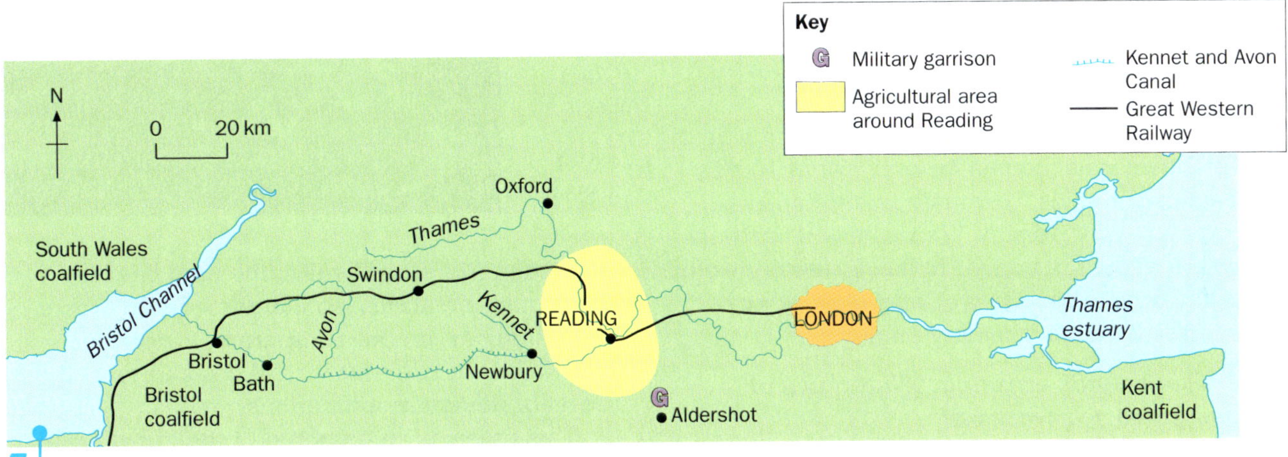

7 Reading in 1870

	Raw materials	Energy	Labour supply	Market
Beer The brewing industry developed from the malting industry. Originally, local barley was made into malt in Reading, then the malt was sent down the Thames to breweries in London. Later, the local industry was able to use the profits from malting to build its own breweries.	Local barley. Hops from Kent.	Coal to heat vats.	Moved to town from nearby countryside.	London. British Army.
Biscuits Huntley and Palmer's biscuits were made using local wheat. They were packed into tins to be sent to London. Then the firm got a contract to supply the army, and their tins of biscuits were sent to the forces in India and many other parts of the British Empire.	Local wheat and butter. Sugar from Bristol. Tins from South Wales.	Coal to power ovens and mixers.	Moved to town from nearby countryside.	London. British Army at home and abroad. National.
Bulbs and seeds Suttons mainly produced bulbs and seeds for market gardeners, and later for house gardens.	Fertile soil.	Coal to heat greenhouses.	Moved to town from nearby countryside.	London. National.

8 Reading's '3Bs': the first industrial boom

The new industrial boom in Reading

By the middle of the 20th century, Reading's industries were in decline, just like those in many other towns in the UK. However, the great advantages of Reading, its position and its accessibility, helped to attract modern, high-tech industry. The importance of these factors is emphasised in this interview with Tony Sanders, Marketing Manager for SGI, an American-based TRANSNATIONAL CORPORATION with a large operation in Reading.

SGI was originally called Silicon Graphics Inc. It was one of the new ICT companies that developed in Silicon Valley, California, in the late 20th century. Reading is the company HQ for Europe and Africa.

9 *Star Wars* special effects prepared with SGI equipment

Q ***I believe you make the equipment that was used to produce the special effects for Star Wars and many Stephen Spielberg movies?***

A That's true. Many film production companies use our equipment but that side of the business accounts for only about 10 per cent of our sales. Our main business is in virtual reality design. We make 360° screens which allow the viewer to feel in the centre of the product they are working on. They can see the product from any angle.

Q ***That sounds amazing. What types of design do you work on?***

A We do a lot of work on car design. We also do a lot of work for the oil industry. We can run a simulation of the conditions inside an oilwell, so that the engineers can work out the best method of extracting the oil. They trial pumping the oil under various pressures and other different conditions so that they can see how to get the maximum amount of oil from the well. It means that they can save money and also make the limited supplies of oil last longer. Our system cuts out a lot of the waste. It's an example of a very 'green' application.

Q ***Do you design the software in Reading?***

A No. Most is designed in California. They develop the hardware too. In Reading we develop applications. We work with our clients to develop new ways of using the software. For instance, designers and engineers from VW came here to develop the new Beetle design applications, so that they could create a 'virtual car' on the computer. Designers could 'get inside' this virtual car and test all sorts of design modifications.

This really is an example of the knowledge economy. We put together our know-how with that of our partners to produce applications that they can use for their very specific needs.

Q ***Where is the hardware manufactured?***

A Our hardware is relatively easy to manufacture – once a few very specialised components have been made. The chips have to be made in specialised factories, in sterile conditions. Most of our chips are made in

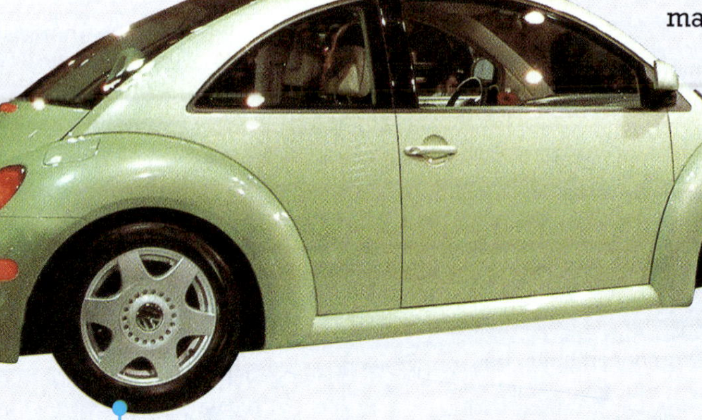

10 The new VW Beetle, designed with the help of SGI applications

California, although new factories were built in many countries in the late 1990s. The chips are usually sent to areas of low labour costs to assemble the finished product. Assembly takes place in two main types of location: areas of high unemployment in more developed countries – like South Wales, where there's a large pool of semi-skilled workers who are used to industrial work; and accessible places in stable, less developed countries where the labour is even cheaper and where they are developing skills very quickly. Malaysia is an excellent example of this. They are developing their skills and infrastructure so fast that many companies are opening research facilities there as well as manufacturing the hardware.

Q ***So why did SGI choose Reading?***

A Two main reasons. First, it's very close to Heathrow Airport which is vital for a firm like ours, doing business with all of Europe and Africa, and with a parent company in the USA. The M4 and the railway both provide excellent links to Heathrow for us. Then, second, Reading is not London! We have all the advantages of being close to the big London market, but prices are lower and the environment is better. Costs to rent premises are much less here, and housing is cheaper. What's more, we're trying to attract staff who want to live in a good environment. Here they are close to the Thames Valley and the Chilterns. London, Oxford and Bristol are all accessible cities with important universities which help with research and recruitment.

Our proximity to other ICT firms is important too. Here in Reading there are Microsoft, IBM, Compaq, Fujitsu and many smaller companies specialising in services like marketing and human resources. We're close to many research centres too. We've no direct links to places like the Meteorological Office in Bracknell and the Aldermaston nuclear research centre (Figure 13) but all these places combine to create an exciting mix of ideas and people in the area. This leads to what we call linkages. By providing services and support for each other, all in the same area, we help each other to grow. In fact it supplies a MULTIPLIER EFFECT (see the Factfile on page 299) for the whole area. Don't you geographers talk about the 'M4 corridor'? Well, it is a very important region and the fact that so many firms are there already makes the area more attractive to new firms – like a snowball growing as it rolls downhill.

11 The SGI site at Reading

Tasks

1 a) Read through the interview and find at least 5 reasons for locating SGI in Reading.

b) SGI is located at GR 650713 on Map 12 on page 296. Using information from Map 12, explain why this was a good site for this firm to choose.

2 a) Explain why manufacture of hardware for high-tech companies takes place in areas like South Wales rather than places like Reading.

b) Read about Shah Alam on pages 302–303. Does this support what the representative from SGI said about Malaysia?

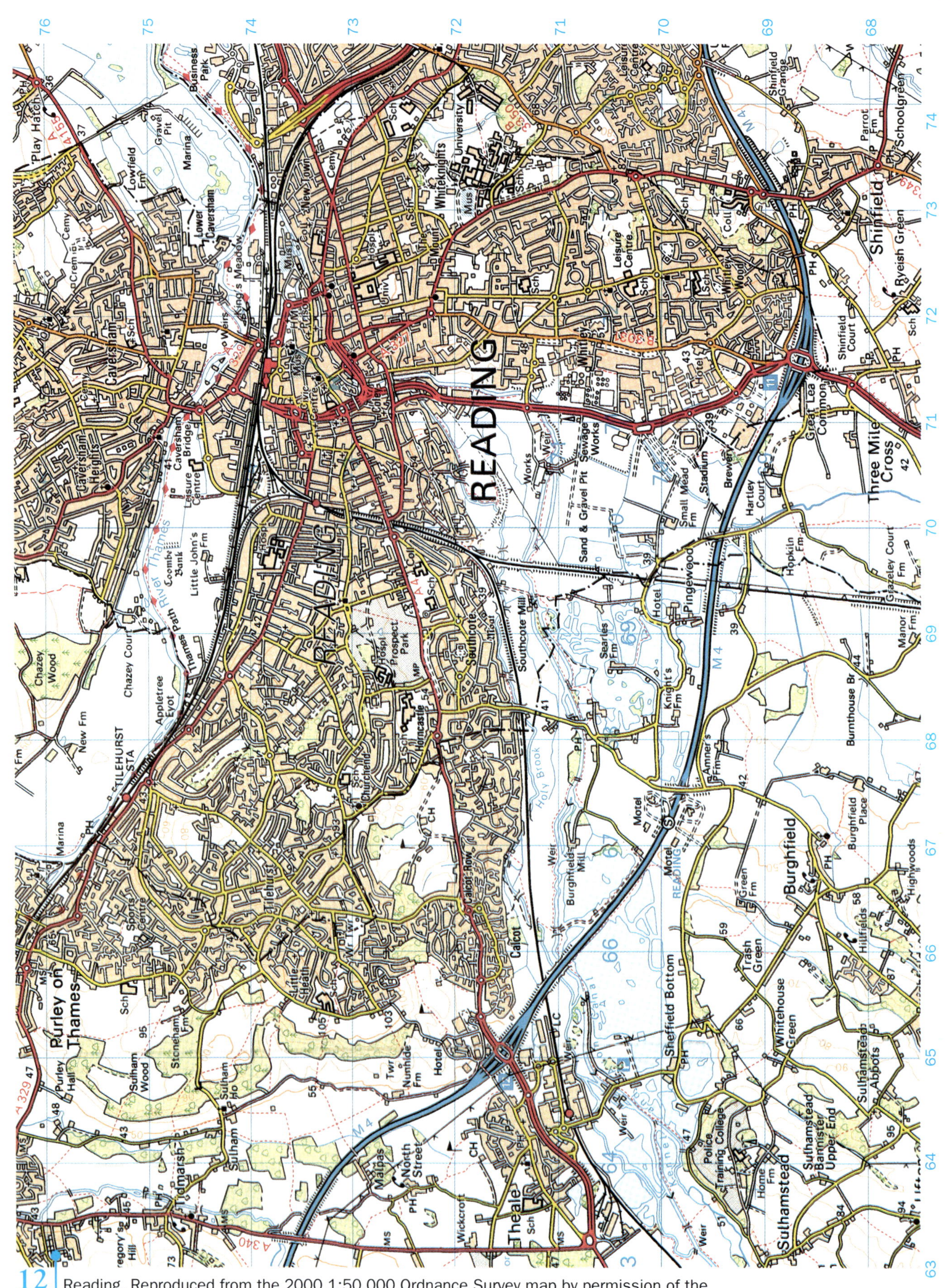

12 Reading. Reproduced from the 2000 1:50,000 Ordnance Survey map by permission of the Controller of HMSO © Crown Copyright.

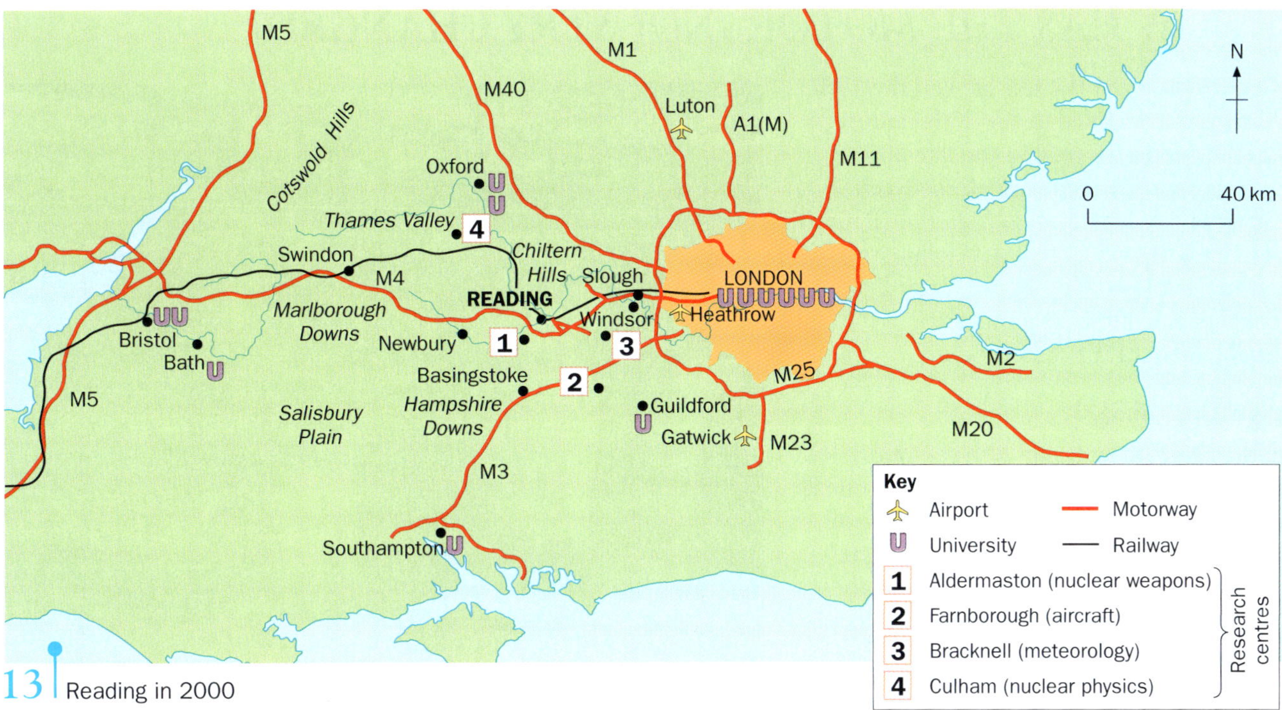

13 Reading in 2000

Tasks

1 'The M4 affects the geography of Reading today just as the railway affected it in the late 19th century and the canal affected it in the early 19th century.' Explain what this statement means.

2 The old Huntley and Palmer's biscuit factory used to be at GR 721734 on Map 12.

a) Why was this a good site for Huntley and Palmer's in the 19th century?

b) Why would it not be suitable for SGI now? Look at Photo 11 to help you.

3 Imagine that another big transnational ICT firm is thinking of building an office in Reading similar to SGI's. Their managing director has said that they need 'access and environment'! Give grid references for two places (other than the industrial estate at Theale) which might be suitable for them. Give reasons for your choices, explaining why each site is good for 'access' and 'environment'.

4 Look back at the case study of Cybercity in Kuala Lumpur (page 196). Which of the features which make Reading a good location could you also find in Cybercity?

5 Reading also provides a good example of how environmental damage done by earlier industrial development can be repaired. Reading was – and still is – a centre for gravel extraction. Gravel is used mainly in the construction industry.

Study Map 12 carefully and list all the evidence you can find of past or present gravel extraction. Your teacher may give you a copy of the map which you could use to record the evidence.

SAVE AS . . .

6 Write a paragraph to describe how former gravel pits are now being used. You could illustrate this on another copy of the outline map.

ICT

7 An annual survey of the working environment in offices in UK towns and cities claimed that Reading is the best place to work in England! Check this fact on the website:

www.reading.gov.uk

The website also shows what the planners think Reading will be like in 2020. They hope that Reading will grow as a 'sustainable city'.

Use the website to answer the following questions:

a) Why is Reading a good city to work in now?

b) How will Reading grow between 2000 and 2020?

c) How will the planners try to make this growth sustainable?

d) What will be the advantages and disadvantages of this growth for Reading's present inhabitants?

Prepare your answers to these questions as a word-processed report. Give your document the title: 'Reading's Future'. Cut and paste images from the website to illustrate your report.

C HOW HAS THE MOTOR INDUSTRY CHANGED?

Heavy industry in the UK grew in the 19th century and declined in the 20th century. The ICT industry grew in the late 20th century and is still growing. In between these two 'industrial ages', from the 1950s to the 1990s the motor industry dominated the country's industrial economy. Pages 298–302 look at some of the stages in the development of the motor industry.

14 A factory production line in the 1920s

Pioneer stage

In the late 19th and early 20th centuries a few motor enthusiasts started to make cars in small workshops. Each workshop produced a small number of vehicles built to its own designs. These workshops were scattered around the country – wherever the owner chose to locate and wherever skilled workers were available. Building a car required skilled craftsmen for almost every job – people who could work metal, wood, rubber, glass or fabric, as well as engineers to build the engines. One of the most successful of these workshops was owned by Mr Morris, in Oxford.

Expansion stage, 1910–1930

Some of these small workshops were turned into successful large-scale factories.

In the USA Henry Ford transformed car making with mechanisation and the introduction of the production line. Machines replaced many of the traditional skills. For example, a machine rather than a craftsman shaped the body panels, repeating the same process – perfectly – every time. Meanwhile, the production line technique meant that the worker needed to be able to perform only a single task, for example, to weld one panel to another as the car moved past. At this stage, the chief skill lay in making the machine to make the component. On the shop floor itself, cheap, semi-skilled labour became more important than highly skilled but very expensive labour. This allowed mass production of cheap cars and led to massive expansion of the industry.

The same thing happened on a smaller scale in the UK. Morris turned his small workshop into a Ford-style production line factory in Cowley near Oxford. Other manufacturers did the same, particularly in the Midlands where there was a long tradition of engineering and therefore a skilled workforce was available. The Midlands area was also accessible to the main markets in the country.

15 The motor industry in the early 1960s

Concentration stage, 1930–the 1950s

As the successful factories grew, they attracted other industries. Manufacturers no longer built cars from scratch but became assembly plants where they put together components made in other specialist factories. The suppliers could cut their transport costs by locating close to the assembly plants, so factories producing steel body-parts, tyres, instruments and dials, seats and upholstery and many other components were attracted to the Midlands. As a result, any manufacturer thinking about setting up or expanding was almost certain to choose the Midlands because the suppliers were close by. By the 1950s the Midland area had come to dominate the UK motor industry (see Map 15).

Dispersion stage, 1960s and 1970s

In the 1960s the government saw that the motor industry could create prosperity in areas where it developed – the so-called 'multiplier effect' (see Factfile). They decided to encourage firms to set up assembly plants in areas of high unemployment. They hoped that this would attract other businesses and stimulate the economy of those areas. Tax incentives, grants and subsidised energy were used to attract existing car manufacturers to set up plants in Merseyside and central Scotland, where there was plenty of cheap labour (see Map 16).

Some of these new factories were successful for a time, but many struggled. Their collection and distribution costs were higher than those of their competitors in the Midlands, which remained the heartland of the motor industry. In difficult periods when sales fell, the parent companies were more likely to close down these branch factories than to close their main factories.

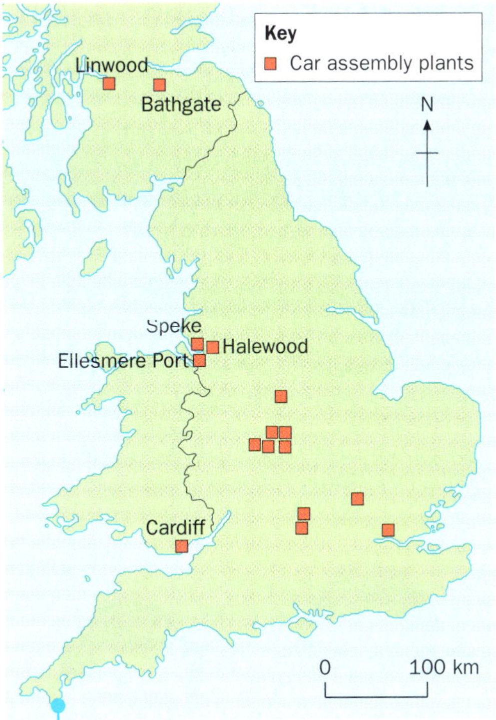

16 The motor industry in 1972

Inward investment stage, 1980s onwards

Inward investment is the term used when a transnational company invests money in a country other than the one where it is based. Inward investment from Japan, the USA and mainland Europe played an important part in the next stage of development of the UK motor industry, as you can see from the case study of Nissan on page 300.

★ **FACTFILE**

The multiplier effect

The motor industry in the West Midlands is a good example of the 'multiplier effect' in action. A strong industry attracts other satellite industries. The main industry and the satellite industries all need workers. The workers need housing and services (shops, clubs, banks, gas installers, electricians, local newspapers, street sweepers, and so on). So a whole new buzz of economic activity builds up around the original industry. The effect multiplies. For example, for every person employed in a West Midlands car plant another two jobs were created in the area in component manufacture or in services to the businesses or the workers.

Tasks

1 Name one location of motor manufacturing in each of the following stages:

- expansion stage
- concentration stage
- dispersion stage.

2 For each stage choose one or more factors from the following list which you think made this a desirable location for car making. Explain your choice.

- close to suppliers of components
- cheap energy supply
- labour supply
- closeness to market
- government policy
- an existing tradition of car making.

3 For each stage choose one factor which was *not* important. Explain your choice.

Remember:

- Car assembly mostly uses inputs called 'components'. These include gear boxes, body parts, seats, electronic equipment, brakes – and in recent years plastic and computer chips – which are made in other factories.
- Energy is provided in the form of electricity from the National Grid.
- The market exists wherever there are large populations – especially in the big conurbations, and especially where the population is more wealthy than average.

Case study: Nissan, north-east England

In the 1980s and 1990s a number of car plants were built in the UK by Japanese companies, to build Japanese cars for the European market. One of these was built in Washington in north-east England. This was a long way from the traditional heart of the motor industry. In some ways it seemed an odd location, but by the end of the 1990s it had become the most efficient car plant in Europe. Why was it built there?

1 We needed to build a factory in Europe because the EU was threatening to put up tariffs on imported Japanese cars. These tariffs were like an extra tax on cars made outside the EU and imported into Europe. We could avoid those tariffs by building in the UK.

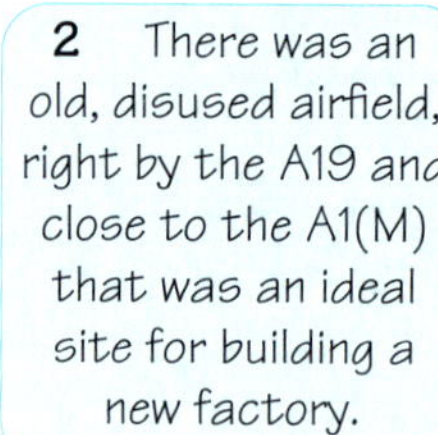

2 There was an old, disused airfield, right by the A19 and close to the A1(M) that was an ideal site for building a new factory.

Government official

3 We offered help with start-up costs, including grants towards the buildings and the training of workers, because this was an area with high unemployment due to the decline of local industries like coal, steel and shipbuilding.

Nissan European sales manager

Estate agent

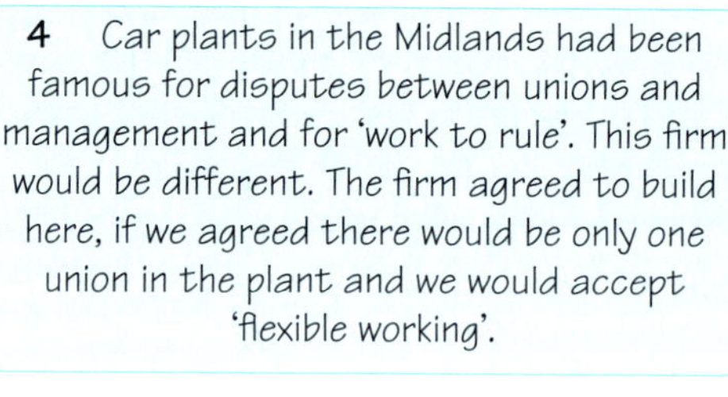

4 Car plants in the Midlands had been famous for disputes between unions and management and for 'work to rule'. This firm would be different. The firm agreed to build here, if we agreed there would be only one union in the plant and we would accept 'flexible working'.

Trade union official

Electricity company manager

5 When Nissan were thinking about building here they took nothing for granted. They spent a long time talking to us and to the gas suppliers, making sure that we could guarantee reliable energy supplies at competitive prices – we wanted their custom so we had to give those guarantees.

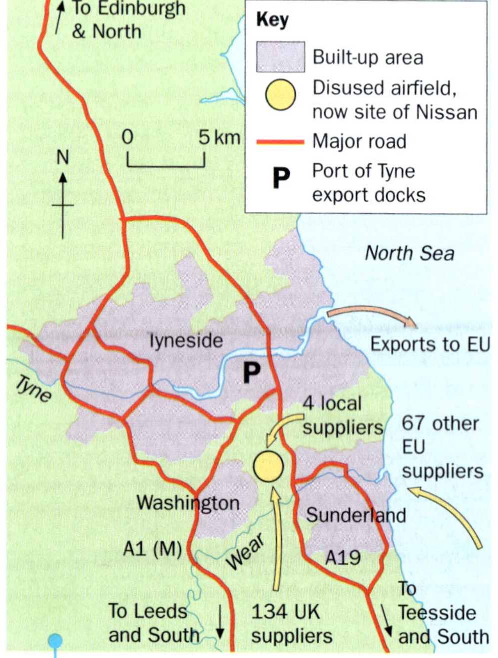

17 North-east England and the Nissan plant

Nissan production manager

6 This factory was one of the first in the UK to use 'just-in-time' deliveries. This is a computerised system for the delivery of components that means we do not have large, expensive reserves of supplies at the plant. We just order them when we need them, but then we expect immediate delivery. This saves money and storage costs. Because the plant is highly automated it is also easy to predict what we will need and when we will need it. Of course it also means that we need good road access all the time to ensure that things do arrive 'just in time'.

7 We had to agree that the roads around the site would be adequate for the plant. For instance, when Sunderland FC applied to build a new stadium nearby we had to veto it because it could have caused congestion that would have slowed down deliveries to the plant.

Local planner

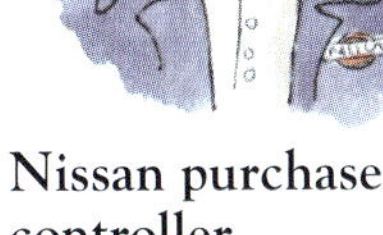

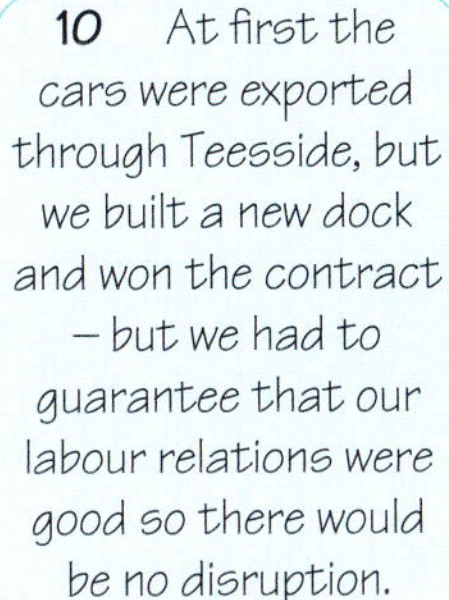

Nissan purchase controller

8 At first we imported most of the parts from Japan, but we have steadily built up a network of suppliers in Europe. These suppliers have been trained to Japanese standards of efficiency. We had to agree to produce parts in Europe, but it saves money because labour costs are lower here than in Japan.

9 The Nissan plant brought a great boost to the steel industry in this area. They provided a new market for a traditional industry, and their demands forced us to modernise our methods to become more reliable.

Steel supplier, Redcar

10 At first the cars were exported through Teesside, but we built a new dock and won the contract – but we had to guarantee that our labour relations were good so there would be no disruption.

Port of Tyne representative

18 Nissan workers on the production line in Washington

Tasks

SAVE AS . . .

1 a) From the comments on these two pages find evidence that Washington was a good location for a new car plant in terms of access to:

- materials
- energy
- workforce
- markets.

b) Find evidence which shows that Japanese working methods were different from traditional UK working methods. Why is this important?

c) What does 'inward investment' mean?

d) Why did Nissan need to invest in Europe?

e) Has the effect of this transnational investment been good or bad for the region? Explain your answer using evidence from these two pages.

11 I suppose we were a bit suspicious of the Japanese at first. They wanted everything done their way. Then they took us to Japan to train us and we saw how efficient their factory was and how dedicated the workforce was. One thing that struck me was the fact that the bosses wore the same uniforms as the workers and all ate in the same canteen. That felt good, and then I started to realise that they would listen to our suggestions and our complaints. It made everyone try to pull together. Now I have become a trainer for new workers and I have been given more and more responsibility. I feel as though I've got a career here, not just a job. Now this is the most efficient car plant in Europe, and I'm proud to be part of it.

Nissan worker

12 At first Nissan sent managers over from Japan, or took Europeans to Japan for training. Now we've established many links with the firm and we are helping to train their managers.

Lecturer at Sunderland University

Problems in the UK car industry

The story of the UK car industry is ongoing. The car industry is regularly in the news. For example, July 2000 saw major setbacks including the following:

- BMW, the German firm that owned Rover, threatened to close the factory at Longbridge altogether. They were persuaded to sell it instead. Eventually a group headed by a former manager at Longbridge bought the plant. They paid £1 for it! Both BMW and the UK government agreed to make payments to help to keep the Longbridge plant open. Even so, its prospects do not look good:
  - There had not been enough money invested in modernising the plant in the past. It is out of date.
  - The UK market does not buy enough Rovers to support Longbridge on its own, but the strength of the pound has made it expensive to export Rovers from the UK, so exports are falling not rising.
- The American-owned Ford announced that they were cutting production at Dagenham and were going to build their new model in their German factory instead. Like Longbridge, the Dagenham plant was very old-fashioned and had not had enough investment to compete with other European plants.
- Nissan announced that even though their plant in Washington was the most efficient and productive car plant in Europe, it was still not very profitable because the strong pound made exports expensive.

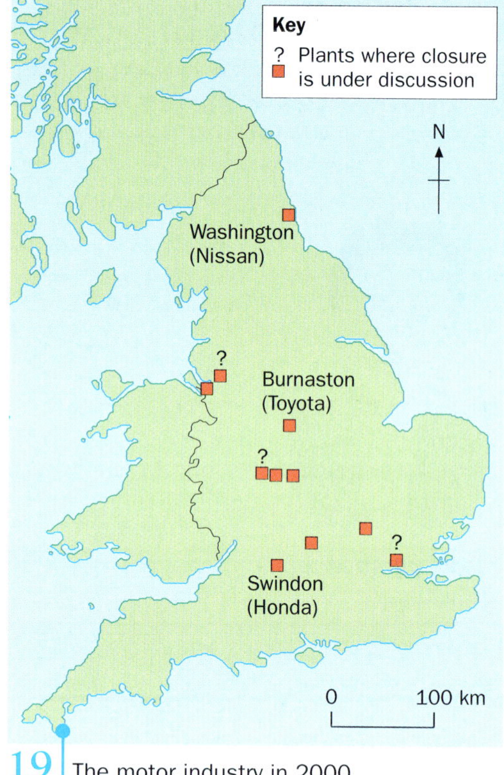

19 The motor industry in 2000

You can use ICT to keep yourself up to date with developments in the car industry, for example at www.ford.co.uk.

Case study: Proton, Shah Alam, Malaysia

Shah Alam is a new town in Malaysia. It lies just south-west of Kuala Lumpur, near the port of Kelang and the international airport at Subaya. It is being built on land where the rainforest had already been badly damaged by logging, so little further ecological damage was done as a result of the building work. Firms that locate there are given assistance by the government (for example, they do not have to pay any taxes for the first ten years). This is part of the government policy to modernise and enrich Malaysia by developing strong industries. At first firms sold goods to the home market, replacing expensive imports. Now they are encouraged to export to the world market.

The largest factory in the town is Proton, a Malaysian firm. It was set up with massive grants from the Malaysian government. They assemble cars in partnership with the Japanese firm, Mitsubishi. Their first models were produced in 1987 from parts that were all made in Japan. The workforce was trained in Japan too. By 1994 some parts were being made in factories nearby, and by 2000 all parts were made in Malaysia. In fact by 2000 Proton had started building a factory in China to assemble cars using some parts made in Malaysia.

Their tyres and some other components come from a local rubber manufacturing firm set up by Proton. Rubber was already one of Malaysia's most important raw materials but in the past raw rubber was exported for manufacture in MEDCs, especially the UK. Now that Malaysian companies control the Malaysian rubber industry, more of the profits stay in Malaysia itself. There is much more 'added value' in processing raw materials than in simply exporting them.

At first Proton cars were built for the South East Asian market, but they are now being exported to Europe and the USA. They compete well in the low-cost end of the market because they are cheap due to Malaysia's low labour costs. However, the management realise that they cannot rely on cheap Malaysian labour for long. Already the firm is bringing workers in from Bangladesh and Indonesia to do jobs that the Malaysians will not do because they are dirty or badly paid.

As Malaysians become more skilled and educated they look for higher wages. As Malaysia's industry grows there will be more competition for workers. The only way that Proton can compete in the world market is by being as efficient as any other motor manufacturer. The firm is investing in the most modern plant and machinery available, and the government is helping by investing in education, to ensure that the workers are skilled and efficient too. The company will also have to develop more stylish, up-market models so that they can compete in all parts of the world market.

The government's policy of encouraging industrialisation has been extremely successful so far. In the early 1990s the country's economy grew at about 8 per cent per year. (In most European countries the growth rate was around 1–2 per cent.) There was a slow-down in growth at the end of the 1990s because of recession in South East Asia, but by 2000 the rate was above 5 per cent again. Unemployment is below 4 per cent in the country. In fact the rapid growth has led to labour shortages in some industries.

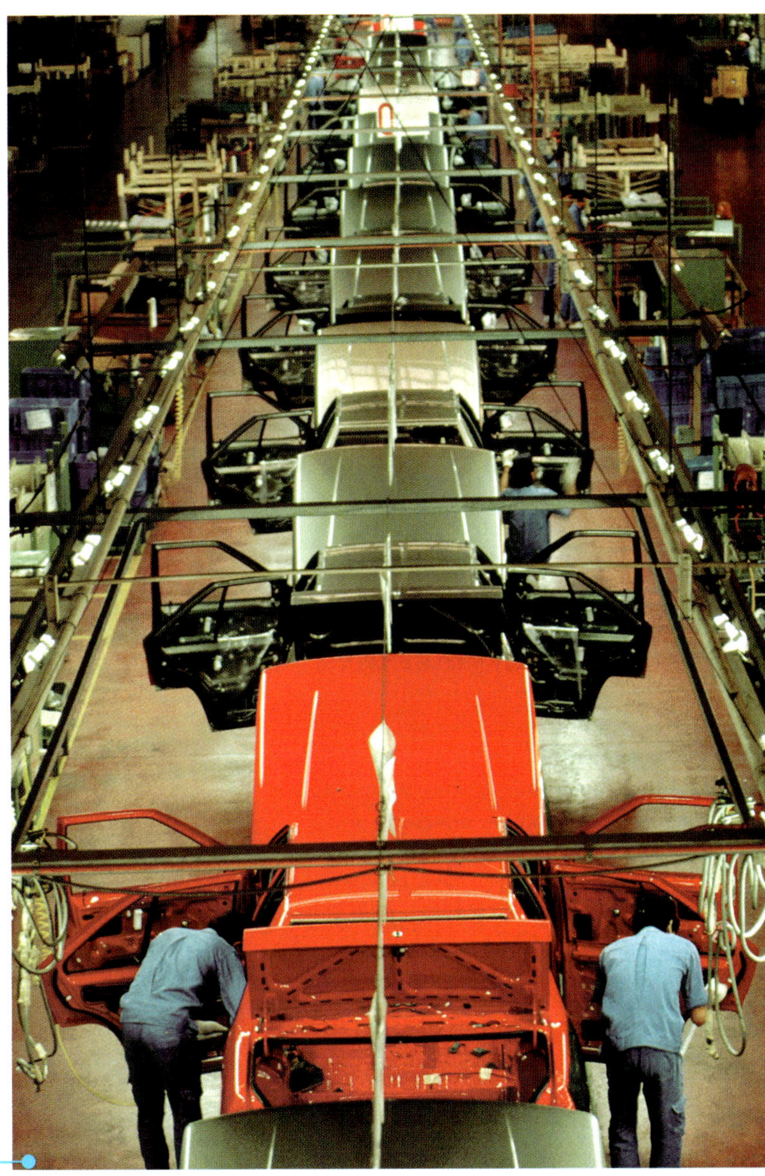

20 Proton factory at Shah Alam

Tasks

1 The Malaysian government chose to support the motor industry because they believed that it would encourage the development of other industries. Explain whether this has happened in Malaysia. Give examples.

2 A newly industrialising country (NIC) is one that has moved from depending mainly on the primary sector of the economy to one with an important secondary (manufacturing) sector. Malaysia is an NIC. Study the case studies of Proton at Shah Alam and of Cybercity on page 196. Summarise:

- the types of industry found there
- where the money comes from to develop them
- the markets where these industries are aiming to sell.

Focus task

3 Compare the development of the Proton factory in Shah Alam with the development of the Nissan factory in Washington. In what ways are they similar? In what ways are they different?

★ THINK!

Exam questions which ask about 'the advantages of locating factories in LEDCs or NICs' are quite common. Many candidates give answers about 'cheap labour'. Think how much better an answer would be if it referred to Malaysia and then said: 'Although Malaysians may originally have worked for low wages, this is no longer the case. Now the government is allowing cheap labour from Bangladesh into the country because . . .'

6.4 What is the future of work in the 21st century?

DO YOU KNOW?

- People who work in services (tertiary sector) do not make anything. Their work provides services that make life better for people.
- Employment in services increased particularly quickly in the last 20 years of the 20th century.
- In the UK more people now work in Indian restaurants than in shipbuilding, steel and coalmining combined.
- Some people who work in services (e.g. lawyers) can earn huge amounts of money while others (e.g. workers in McDonald's) earn very little.

AIMS

- **To find out how employment structure changes as economies develop.**
- **To consider why manufacturing industry has declined in the UK and been replaced by tertiary employment.**
- **To investigate tourism as an example of employment in the tertiary sector.**

Tasks

1. Read the advice. Who do you think is giving the best advice? Who do you think is giving the worst?
2. Have you got any other good advice for Phil?
3. Write three more speech bubbles that might go with the following advice:
   - **a)** Start at the bottom, Phil, then work your way up.
   - **b)** Whatever you do, Phil, do the best you can.
   - **c)** Get a job helping people, Phil.
4. Think Geography! How is work changing? List all the changes you can think of. This page will get you started but you will have other ideas.
5. How might these changes affect:
   - **a)** where people live?
   - **b)** where they work?
   - **c)** how they work?

A CHANGING PATTERNS OF EMPLOYMENT

Phil is doing A-levels and is thinking about the future. Everyone is giving different advice.

Get a steady job, Phil. Your grandfather worked in the same factory from leaving school until he retired in 1991. He had security, a steady income and a company pension scheme. Why don't you look for the same kind of career?

But Gran, I'm not sure it's like that any longer.

Phil's granny

Get into ICT, Phil. Flexible ICT skills are needed in many different businesses. That way you'll be able to move around easily. The workplace is changing so fast that no-one can be certain what they'll be doing ten years from now.

ICT! I'm good at that!

Careers Officer

Do whatever you most enjoy, Phil. I liked media studies so I'm working in media. Basically admin at the moment but it's a start and I'm learning on the job. We video the Everton matches for the Club. And a few mates and I have set up our own production company. We've made a few promotional videos to get experience and contacts.

I won't know what I enjoy until I've tried it. Then it might be too late!

Phil's sister, Ruth

Phil's cousin, Joe

Go where the money is, Phil. No doubt where that is: banking. I ran up £20,000 debt at uni but I'll pay it off with my first bonus. The top finance jobs are here in London. That's why I've moved here. But remember, big rewards come with big risks. If performance slips here they call you up to the fifth floor to be sacked. They escort you from the building. They don't even let you return to your desk or else you might steal or sabotage computer files.

All right if you're one of the successes. But what if you're one of the failures?

Get a trade, Phil. Be an electrician or something like that. Everyone will always need electricians. You can be your own boss; work when you want; go on holiday when you want; play golf when you want.

Everyone tells me I should do something more than that with my education.

Phil's friend, Andrew

Phil

Travel the world, Phil. Don't join the rat race. Stay free. Just earn what you need. I work two jobs in bars and fast food through the summer. That earns me enough to go travelling using cheap tickets from the internet. I did the Far East straight after college, and I'm off to South America in three weeks.

For a couple of years maybe, but what then?

Daisy, the barmaid at the 'Dog'

Get qualifications, Phil. I messed up my exams by not working and I found that you're worth nothing in the job market without training and qualifications. I tried shelf-stacking and fast food then got a job where they promised me training. I got NVQs the hard way – studying while I was working – but now I've got qualifications and five years' work experience.

Better get on with my coursework then!

Phil's friend, Richard

Auntie Cath

Work from home, Phil. When I had young children I left my job in publishing. Now my youngest has started school I've taken up freelance work. I can work almost entirely from home, using the internet, e-mail, faxes and so on. My partner is a graphic artist and he's thinking of becoming a home worker too. That would mean we could move away from the London area. We could probably save about £50,000 on the house, cut down on travelling costs, and improve our quality of life.

That would be boring, only myself and a computer for company!

Think global, Phil. After I did my engineering degree, I did a stint in the Middle East, then off to an American firm of management consultants. Six years' brilliant training and great contacts – mainly Japanese companies working throughout Europe. I could work anywhere in the world with my experience.

You make everything sound so easy!

Phil's cousin, Sarah

Classification of employment

Employment and industry can be classified according to what people (or firms) do.

- In the *primary* sector people produce raw materials. It includes farming, mining, fishing and forestry.
- In the *secondary* sector people take raw materials and partly finished products and turn them into finished goods. For example, steelmaking and cheese manufacture are both in the secondary sector. This sector is sometimes called the manufacturing sector.
- In the *tertiary* sector workers do not produce things, but provide a service for people, or for industry. For example, hairdressers, shop keepers and lorry drivers are all in the tertiary sector.

In recent years the tertiary sector has become so complex and wide-ranging that it has been subdivided. A fourth sector has been added to the classification:

- In the *quaternary* sector people provide information and expertise. It includes the information and communication technology (ICT) industry. This sector often provides information and research services to all the other sectors.

It is hard to obtain reliable statistics at present for the quaternary sector. Government statistics do not separate it out as a different category. In the rest of this chapter the figures for the tertiary sector or service sector will include workers from both tertiary and quaternary industries.

Tasks

1. Look at each of the pictures in Figure 1. Into which sector should each of these jobs be classified?
2. Think of a product that you know something about. Draw a flow diagram to show its life story from start to finish. Show all the jobs involved in getting it from the raw material stage to the consumer. Highlight primary, secondary, tertiary and quaternary jobs in four different colours. Add a key.

1 Workers in the different sectors of employment: primary, secondary, tertiary and quaternary

Changing employment structure

As countries develop economically their employment structure changes.

Stage 1: Primitive stage

In primitive economies almost all of the people work in the primary sector, mainly in subsistence farming. The UK was in this phase in the Middle Ages. There are very few countries, if any, where this is still the case, although there are regions of some LEDCs where economies are still like this.

Stage 2: Early developing stage

In the early developing stage agriculture becomes more efficient and produces a surplus. This can support some people in service employment. When the first cities were set up in the Middle East and Mediterranean Europe, people in the cities became administrators, priests, rulers, soldiers, scientists, teachers, domestic servants, and so on. All of these people were in services. Some people also started to work in small-scale industries. The UK was in this phase in the Tudor period. Many LEDCs still have most of the population working in the primary sector but also have a reasonably large tertiary sector.

Stage 3: Industrial stage

During the Industrial Revolution in the UK and Europe, the secondary (manufacturing) sector increased rapidly. In the UK this happened between 1750 and 1900. The same thing can be seen happening today in newly industrialising countries (NICs) like Malaysia. These countries are in the industrial stage.

Stage 4: Post-industrial stage

The UK and many other countries are in a post-industrial stage. Employment in both the primary and secondary sectors is falling, or has fallen, and employment in the tertiary (service) sector is increasing still further. Employment in the quaternary sector is increasing and these countries may be entering a new ICT-driven industrial stage.

Task

SAVE AS . . .

3 The table below shows the percentage of the working population in different sectors of employment for six countries.

	Primary sector	Secondary sector	Tertiary sector
China	42	27	31
Germany	6	28	66
India	70	11	19
Jamaica	36	17	47
Malaysia	29	19	52
UK	2	24	74

a) Draw horizontal divided bars for each country, to show the percentage employment in each sector of the economy. Draw your bars 10 cm long and 1 cm deep, so that 1 mm of the bar will then represent 1 per cent of the workforce in that country.

Make sure that your bars lie directly underneath each other. Use the same three colours to shade each bar. Add a key to show what each colour represents.

b) Next to each bar note which stage the country is in: primitive, early developing, industrial, or post-industrial.

★ THINK!

NICs today are going through changes in employment structure that MEDCs went through some time ago. The process is similar but the NICs are experiencing it later and usually much, much faster than MEDCs did in the 19th century.

This is a pattern that you may have seen elsewhere in the book. For example, LEDCs are experiencing rapid urbanisation, rapid population growth and worrying environmental pollution just as MEDCs did in the last century.

The decline of manufacturing in the UK

Figure 2 shows how employment structure in the UK has changed since 1841.

In the late 19th and early 20th centuries the UK was one of the world's leading manufacturing countries. But, as you learned in Chapter 6.3, many of the country's leading industries have declined rapidly. Others have grown, but almost all manufacturing employs fewer people than the old industries did. Mechanisation means that manufacturing, even where it has survived, has moved from being labour-intensive industry (i.e. needing a lot of workers) to being capital-intensive (i.e. needing a lot of machinery but not many workers). Figure 3 shows the loss of manufacturing jobs in Britain between 1981 and 1995. Since then employment in manufacturing has continued to fall, but less rapidly.

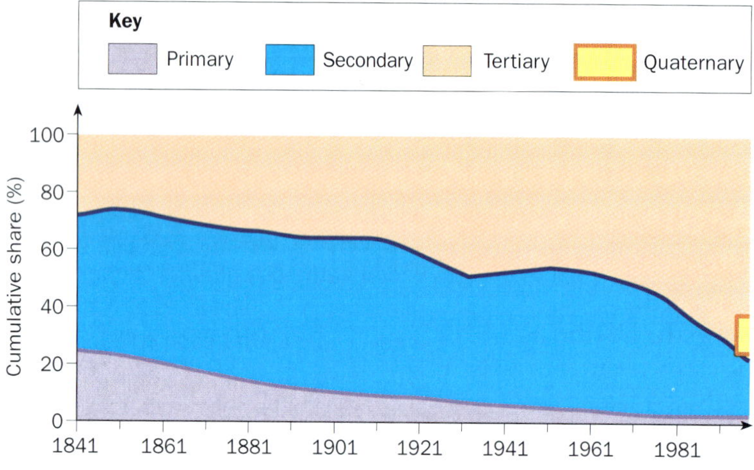

2 Changing structure of employment in the UK, 1841–1999

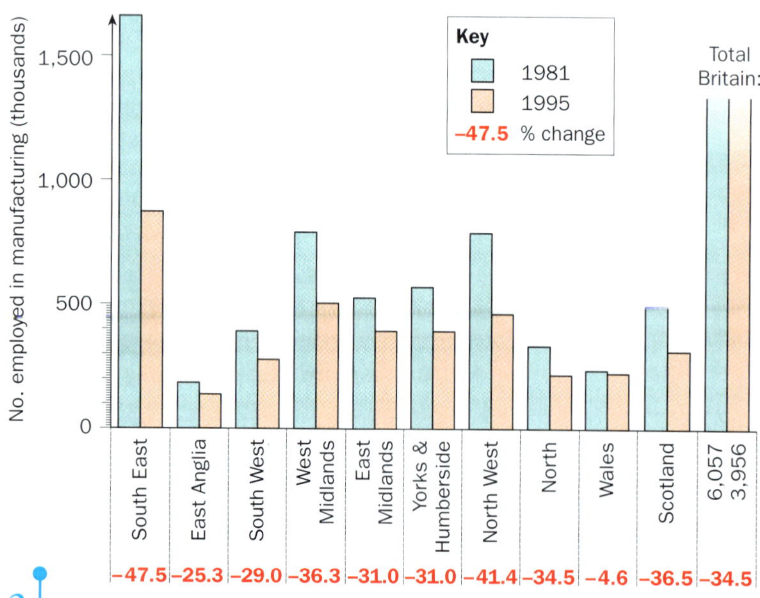

3 Decline of manufacturing employment in the regions of Britain, 1981–95

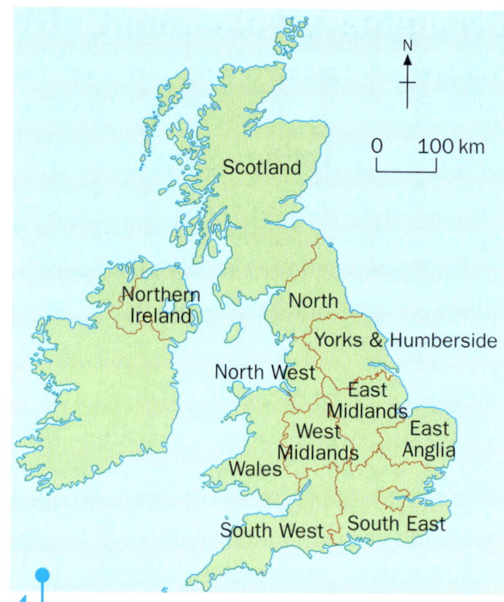

4 The 'Standard Regions' of the UK

Tasks

1 Study Figure 2.

a) Describe what happened to secondary employment:

- between 1841 and 1911
- between 1911 and 1931
- between 1931 and 1951
- between 1951 and 1999.

b) Suggest how each of the following factors helped to cause these changes: mechanisation, foreign competition and changing demand.

c) What other reasons can you think of?

2 On a copy of Map 4:

a) Shade each region to show the percentage loss of manufacturing jobs. You will need to work out a key first.

b) Draw a bar in each region – the length of the bar should represent the total number of jobs lost. You will need to work out a scale for this.

c) Add a scale and key to your map.

d) Describe the patterns shown by your map.

During this period unemployment also fell from 3 million to 1 million. Suggest why the loss of manufacturing jobs did not lead to increased unemployment.

Who makes your clothes?

The manufacture of cloth and clothing used to be one of the UK's biggest employers. In the Leeds area alone the industry employed 50,000 people in the heyday of Jed Bishop (see page 290). It now employs one-tenth of that number, and these jobs are almost exclusively in small workshops producing small-run, high-quality items. Mass production has largely gone from the area. It's a similar story elsewhere in the UK – for example in the Lancashire cotton towns.

But we still wear clothes! In fact most people in the UK now own more clothes than their grandparents would ever have dreamed of 50 years ago. So who makes all these clothes? And where have all the clothes-making jobs gone?

SAVE AS . . .

3 a) State the total loss of manufacturing jobs in the UK between 1981 and 1995.

b) Name the three regions that lost the most manufacturing jobs.

c) Give three reasons for these job losses.

d) Give the most important reason for job losses in each of the three regions.

You are going to carry out some research on your own and then pool your results with others to make a class database which you can analyse.

Stage 1 – on your own

4 Working on your own, copy and complete this table to analyse the clothes you own.

Item	Cost when new	Country of origin	Category
Jeans	£40	Vietnam	Leisure-wear

- Try to include at least five items. The more you do, the more useful your results will be.
- If you don't know what the item cost, make a sensible guess.
- So that your results can be combined with others and analysed more easily, in column 4 use one of the following categories: underwear, footwear, sportswear, school uniform, leisurewear.

Stage 2 – with others

5 Pool your findings with others in your class and enter your results into a spreadsheet or a Word table.

6 Using the sort and formula functions you can now analyse your results to find out which country produced:

a) the most items in total

b) the most items in each category

c) the highest total value

d) the highest value in each category.

SAVE AS . . .

7 On your own, write a paragraph under the title 'The fashion industry is a global industry'. Use your graphs from Tasks 5 and 6 to illustrate your paragraph.

Stage 3 – discussion

Just because clothes are made in another country does not mean that all of the jobs involved in selling them move overseas. Nor does it mean that all of the best jobs have gone overseas.

8 a) Draw up a list of all of the jobs you think would be involved in getting clothing from raw cloth to finished garment. You can get a sheet from your teacher to help you.

b) Highlight those that still take place in the UK even if the garment is imported.

c) Discuss those highlighted jobs. Do you think they are skilled or unskilled? Highly paid or low paid? Desirable jobs or undesirable jobs?

SAVE AS . . .

9 Write a second paragraph, under the title 'The clothing industry as a tertiary employer in the UK', summarising the conclusions of your discussion in Task 8.

The regional pattern of changing employment in the UK

	% of the national workforce in:		
	primary employment	secondary employment	tertiary employment
1971	3	40	57
1999	2	24	74

5 Changing employment structure in the UK

	% of the regional workforce in:			Total regional workforce (millions)
	primary employment	secondary employment	tertiary employment	
1976				
South East	2	36	62	8.5
East Anglia	7	30	63	1.0
West Midlands	5	51	44	3.5
1999				
South East	1	23	76	8.0
East Anglia	5	27	67	1.0
West Midlands	2	36	62	2.9
Scotland	2	25	73	2.3
North	4	27	69	2.1
Yorks & Humb	3	30	67	2.9
North West	1	26	73	3.3
East Midlands	3	32	65	2.1
Wales	4	21	75	1.2
South West	3	22	75	2.7

6 Changing employment structure in regions of Britain

Tasks

1 You are going to decide on the best sort of graph to use to present the data in Table 6. Different members of the class should try to use different techniques, for example:

- divided bars
- pie charts
- bar graphs.

You will need to draw pairs of graphs for each region, to represent the two different years. You will find this easiest if you use an ICT graphics program, but you can use hand drawn graphs. Whichever method you choose you should:

a) make the total size of your bar(s) or circle represent the total employment in that region

b) make the different sections of each bar or circle represent each sector of employment

c) shade the sections so that each sector of employment is represented by the same colour on each graph.

You could fix your completed graphs on to a base map showing the regions.

2 Once you have drawn your graphs, compare them with other people's.

- Which was easiest to do?
- Which shows the information most clearly?
- Which looks the most striking and eye-catching?

SAVE AS . . .

3 Write a summary of what the graphs show.

- Have all regions seen a decline in primary employment?
- Where has the decline been fastest?
- Have all regions seen a decline in secondary employment?
- Where has the decline been fastest?
- Have all regions seen an increase in tertiary employment?
- Where has the increase been fastest?
- Using the information on pages 308–10, give reasons for the differences between the regions.

★ THINK!

In your coursework you will have to use a variety of techniques to show the data that you collect. Think very carefully about which ones to use. You must show the information clearly and in an interesting way. You will not gain extra marks just for using difficult techniques. The most marks are always awarded when you choose the best technique for the purpose, and can explain clearly why you chose that technique.

Case study: The growth in service industries in Edinburgh

The following conversation was overheard in a sauna in a health club in Newcastle. A tall man (TM) – a TV producer – and a short man (SM) – a lawyer – were talking.

TM: *I haven't seen you here lately. Where have you been?*

SM: I'm working in our Edinburgh office a lot these days. The company has rented a flat for me up there, so I stay three nights a week.

TM: *Nice flat?*

SM: Great! You should see what the firm is having to pay for it. There's so much competition with all the business that's being attracted there.

TM: *So, why Edinburgh?*

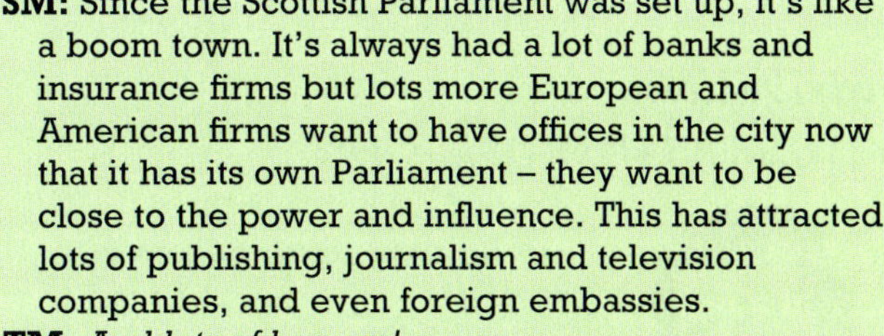

SM: Since the Scottish Parliament was set up, it's like a boom town. It's always had a lot of banks and insurance firms but lots more European and American firms want to have offices in the city now that it has its own Parliament – they want to be close to the power and influence. This has attracted lots of publishing, journalism and television companies, and even foreign embassies.

TM: *And lots of lawyers!*

SM: Exactly! They all need lawyers to make sure things are done properly. The combination of Scots law, English law and European law makes lots of work for firms like ours.

TM: *So, is it nicer than Newcastle?*

SM: It's magnificent. Good night life. All the money has attracted lots more good restaurants, clubs and bars. It's too crowded in the summer when the Edinburgh Festival is on – but all the tourists make it an exciting place to be. The tourist season goes on all year round. They've attracted lots of good shops – which all bring more work. There's rugby at Murrayfield, and plenty of other sports, and it's just an hour or so's drive to the hills for climbing and walking. Almost everything that I need.

TM: *So you're moving the family up there I presume?*

SM: Maybe! The schools up there have a good reputation. They help attract businesses. Trouble is all the children are well settled here. But Dominic is applying to Edinburgh University. It has a good reputation for medicine and it has a lot of links with high-tech firms in the area. His long-term ambition is to work in medical research, and Edinburgh is as good as anywhere for that. So he might join me.

7 Edinburgh

Tasks

SAVE AS . . .

4 Read the 'conversation in the sauna'. List all of the different kinds of service employment that are mentioned.

5 Draw a concept map (see page 32) to show how all of these are linked in Edinburgh.

6 Give your diagram the title: 'The multiplier effect in service employment'. Then below your diagram write a clear definition of the term 'multiplier effect'. (see page 299.)

B CASE STUDY: EMPLOYMENT AND ECONOMIC ACTIVITY IN LEEDS AND WEST YORKSHIRE

In the 1990s Leeds was the most successful UK city outside London in creating new jobs. As a result, it was the fastest growing city in the UK. You are going to investigate the kind of jobs that people do in Leeds today. Figure 8 gives you the main statistics. It features four representatives from Leeds' main employment sectors.

Distribution

Employment: 14% up (1991–1998)

I drive a truck distributing books and stationery all over the north of England. Leeds is good for distribution because of the motorways. Leeds has lots of purpose-built warehousing estates – most of them close to the motorway junctions. Royal Mail have a big warehouse for their Automated Processing Centre. I say 'automated' but it still needs a surprising amount of workers. More than 3,000 people in Leeds work for Royal Mail. Warehousing can be automated but transport can't be – yet! The M62 is one of the busiest motorways in the country – half a million tonnes of freight every day: that's nine juggernauts a minute all day every day, every week, every year! No wonder they need a lot of lorry drivers.

Retailing

Employment: 3% up (1991–1998)

There is always work in shops. Lots are part-time and flexible. People from far around come to shop in Leeds. There are 189 different big-name companies trying to rent shop space in Leeds at the moment – that's the highest figure in the country. The Council have helped – they have improved the city centre and the latest development is Trinity Square. It's a big covered shopping centre right in the middle of Leeds. You can shop out of town as well – the White Rose Centre had 16 million shoppers in its first year.

Engineering

Employment: 18% down (1991–1998)

Everyone thinks of industrial Leeds as a wool town, but the wool industry needed machinery and so engineering grew alongside wool. That industry developed so much that by 1900 Leeds' biggest industry was engineering – everything from copperworks, forges and engine-makers to fine-engineering. Everyone will tell that you that manufacturing has declined in the UK but it's still important in Leeds – we supply items like oil-rig equipment and engines all around the world. In fact, engineering still contributes more to Leeds' economy than any other sector except finance. It's just that with mechanisation we don't need so many workers to do it. We have a highly skilled workforce – most of the unskilled work is done by machine.

Business and financial services

Employment: 40% up (1991–1998)

Businesses like ours are driving Leeds' growth. There are 32 national or international banks here including two HQs – Yorkshire Bank and First Direct. First Direct was the first 24-hour telephone only bank – it opened in 1989. Since then 20 call centres have located here. Figure 10 shows you some of the things that you can phone Leeds for! The experts say that Leeds has reached 'critical mass' – so many financial services have moved here that now all the others that aren't here want to come as well! It's England's third most important financial centre after London and Birmingham. Financial services still need a lot of workers. First Direct alone employs 3,500 people. Our business is good for the area in other ways. Our workers develop very transferable skills – such as confidence, presentation, ICT – and these spread through the whole region helping other industries to develop. For example, the internet service provider Freeserve is based in Leeds. Over 30 per cent of all of the e-mail and internet messages in the UK go through Leeds. Many people who worked for Freeserve have left to set up their own ICT companies in the area.

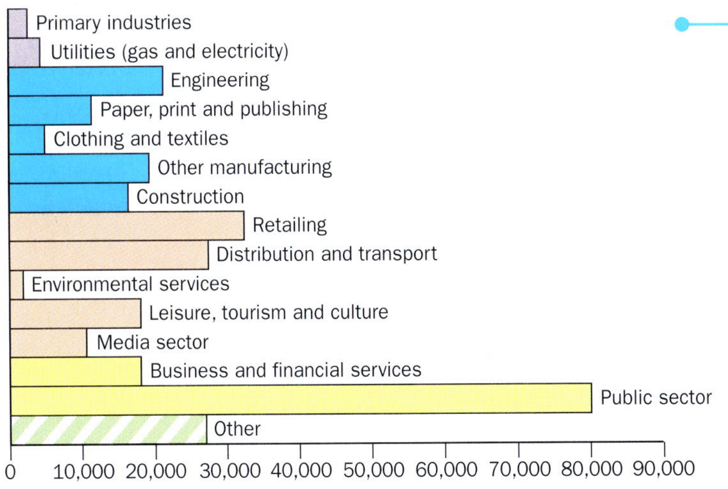

8 Employment structure in the Leeds area, 1998. (Primary industries include farming, forestry, mining, quarrying and fishing)

Engineering £2 bn
Other manufacturing £2.4 bn
Business and financial services £2.6 bn
Tourism £0.6bn

9 The Leeds economy: revenue earned in 1998 by four sectors

You might phone Leeds when you:	Company	Call centre employees
open a bank account	First Direct	3,500
order clothes	Ventura (for Next)	2,000
buy insurance	Direct Line Insurance	1,200
sort out your store card	GE Capital Bank	900
break down on the road	Green Flag	700
trade shares	TD Waterhouse	620
top up your mobile phone	BT Mobile	500
buy a mortgage	Barclays Mortgages	400
place a bet	William Hill	220

10 Some of the major call centres in Leeds

The call centres are attracted to Leeds by:

- a cheap but well-educated workforce
- the Yorkshire accent – which callers see as trustworthy. Yorkshire people are famous for being honest and speaking their mind. For the same reason, people with Yorkshire accents are often used in TV adverts
- the 'synergy' that comes from similar industries being located together.

Tasks

1 Read what the workers say about Leeds. Make a list of all of the different reasons that employers choose Leeds.

2 Compare Figures 8 and 9. The sectors of the economy that make the most money don't always employ the most people. Why do you think this is?

3 a) Which sector in Figure 8 would you most like to work in? Why?

b) Which employment sector do you think has the best prospects? Why?

c) Which sector do you think requires the most skills? Why?

4 Here are some local newspaper headlines: how might these affect employment in Leeds?

- ICT teaching in Leeds schools is criticised by inspectors.
- Pland Stainless move sink production to Malta.
- Visiting celebrity opens new city centre theatre.

5 Education is a big employer in Leeds – there are 40,000 higher education students and well over 100,000 in schools.

a) List all of the different types of job generated by education. (Use what you know about your own school as a starting point.)

b) What factors might affect the number of people employed in education in the future? Give reasons.

SAVE AS . . .

6 Imagine that you work for Leeds City Council and have to prepare a homepage for the employment area of their website. Write a snappy paragraph for the website describing employment opportunities in Leeds. Make sure that it is accurate, i.e. use the facts given on these two pages, but also make sure that it 'sells' Leeds.

Leisure and tourism as a major employer

More people in the world work in leisure and tourism than in any industry apart from farming. It employs 7 per cent of all the world's workers. In the UK tourism and leisure employs 2 million people and the figure is rising. In the 21st century tourism is therefore a vital part of any economy.

Tourism also cuts across the neat thematic divisions used in Geography. Through this book have already studied many different aspects of tourism. For example, as part of:

- agriculture – how the EU is encouraging farmers such as Richard Betton to diversify into tourism as one way of cutting down on overproduction (see page 258)
- urban geography – how redevelopment of urban areas has made inner city sites such as Brindleyplace, Birmingham attractive to tourists (see page 246)
- settlement – how tourism has transformed towns such as Ambleside or created new towns from scratch such as Matalascañas in Southern Spain (see page 198)
- coastal processes – how the value of tourism has influenced decisions about coastal defences in towns such as Lyme Regis (see page 142)
- development – in Chapter 7.3 you will investigate the impact of tourism in an LEDC, Jamaica, for which tourism offers a means of economic development.

This reflects the growing importance of tourism in Geography. Tourism is a labour-intensive industry. It creates lots of jobs. Why? What kind of jobs are they? What skills are needed? What is the future for tourism?

Tourism, leisure and employment in West Yorkshire

Leeds is not traditionally thought of as a tourist destination. Yet tourism and leisure industries employ around 17,000 people in Leeds alone and many more if you include the nearby countryside. Tourism is growing at 4 per cent per year. Yorkshire is the third most popular tourist destination in the country. Leeds itself has 18 million visitors a year. Many more use Leeds as a base for touring the surrounding area.

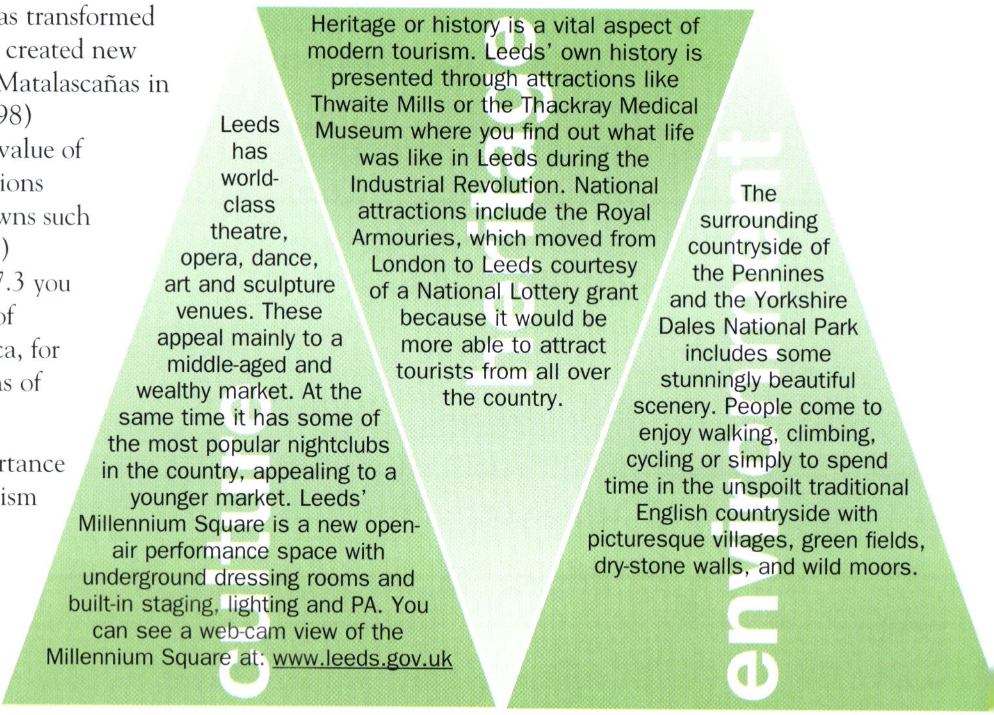

Leeds has world-class theatre, opera, dance, art and sculpture venues. These appeal mainly to a middle-aged and wealthy market. At the same time it has some of the most popular nightclubs in the country, appealing to a younger market. Leeds' Millennium Square is a new open-air performance space with underground dressing rooms and built-in staging, lighting and PA. You can see a web-cam view of the Millennium Square at: www.leeds.gov.uk

Heritage or history is a vital aspect of modern tourism. Leeds' own history is presented through attractions like Thwaite Mills or the Thackray Medical Museum where you find out what life was like in Leeds during the Industrial Revolution. National attractions include the Royal Armouries, which moved from London to Leeds courtesy of a National Lottery grant because it would be more able to attract tourists from all over the country.

The surrounding countryside of the Pennines and the Yorkshire Dales National Park includes some stunningly beautiful scenery. People come to enjoy walking, climbing, cycling or simply to spend time in the unspoilt traditional English countryside with picturesque villages, green fields, dry-stone walls, and wild moors.

★ **FACTFILE**

UK tourism

- 25.7 million overseas tourists visited the UK in 1999.
- The average UK family spends 15.6 per cent of its income on holidays.
- 2 million people in the UK are employed directly in the tourist industry.
- Of all the UK people who stayed in the UK for their holidays:
  31 per cent visited the countryside
  64 per cent visited towns
  55 per cent visited the coast.
- The West Country, East Anglia, and Yorkshire are amonst the most popular regions in England for holidaymakers.

Tasks

1 a) Match up each of the photos on page 315 with one of the categories of tourism in the diagram above. Explain your choice.

b) Who might visit that attraction and why?

2 On a copy of a map of Yorkshire:

a) Mark all the attractions visited by Connie and Ray.

b) Add symbols to your map to show attractions which are linked to:

- farming
- traditional manfacturing industry.

Connie and Ray visit Leeds

Connie and Ray come from Winston-Salem in North Carolina in the USA. Connie's mother grew up in Yorkshire. She married a GI (an American soldier) in 1945 and moved to the USA. Connie was keen to visit the places that her Mum talked about and to rediscover her heritage. So when Ray had to attend a conference in London, they took the chance to visit the area. This is Connie's diary:

Yorkshire Sculpture Park

Friday

Arrived in Leeds by train. It only took two and a half hours. Coffee was served at the table by a steward.

Visited the Yorkshire Sculpture Park (see photo). Plenty of security guards in evidence – although you'd need a crane to pick up one of these giant stone sculptures!

Went to see 'Macbeth' at the West Yorkshire Playhouse. After 400 years Shakespeare still keeps dozens of actors and theatres in work!

Dinner at the Sheesh Mahal, a top Indian restaurant with Indian dance performance.

Stayed the night in a hotel overlooking the river Aire, right in the heart of Leeds.

Pateley Bridge

Saturday

Hire-car brought round to the hotel by a pleasant young woman who seemed to know all the attractions of Yorkshire inside-out. Made lots of suggestions for places we could visit.

Went to Harewood House. One of the four car-park attendants kept beckoning me without looking so I nearly ran into him – that would have served him right. Had guided tour. Bought souvenirs, including a skirt for sister Amy – hand-made from traditional Yorkshire wool. Lunch in café at Harewood. Avoided the bird gardens since there seemed to be a thousand children and masses of red-shirted workers on one of these specialist children's holiday camps.

Drove to Pateley Bridge. Visited the old wool mill at Glasshouses where my grandfather used to work. It is now a design studio on one side, on the other a riverside restaurant – very chic.

Evening meal at farmhouse near Pateley Bridge featured 'Dales Spring Lamb' cooked by Betty, the farmer's wife. She told us all about the changing times in farming. It's only her B&B that keeps them afloat – without it the farm is losing money.

Brewery Wharf

Sunday

Visited Stump Cross Caverns. They were in the middle of making a promotional video – there were film crews and microphones. One of them interviewed me.

Lunch in nearby pub. Landlord used to be an engineering worker. Said he got a big redundancy payment when he was laid off in 1984 – bought the pub just at the right time. Trade is really good. Said his pub now employs more than work in farming in the whole dale!

Drove to Wensleydale to see traditional cheese-making in action. They supply all the nearby restaurants as well as exporting the cheese to places like North Carolina!

Stayed in B&B in Hawes. Evening meal was roast beef and Yorkshire pudding.

Monday

Drove back to Leeds. Did some shopping.

Ray went off to the Thackray Medical Museum – he is a doctor so was very interested in medical history. I downloaded one of the recommended walks from the Leeds website. The walk started at Brewery Wharf. They've rebuilt the 19th century canal-side around the brewery with shops and pubs. Bought souvenirs: hand-made cloth; local beer and traditional tankards. There was still lots of building going on. There were dozens of workmen eating lunch by the canal. I had a steak and kidney pie in a canal-side pub.

Finally we caught a train back to London.

What kind of jobs?

One of the attractions visited by Connie and Ray was Harewood House. Each year 300,000 people visit Harewood and it employs 130 people in the main tourist season. You can find out more about it at:

www.harewood.org

Figure 12 shows you the range of jobs that people do at Harewood.

Tasks

1 You are now going to think about jobs created by tourism.

a) List as many jobs as you can think of which are needed within leisure and tourism. Connie's diary (page 315) will give you some ideas. You will think of more. Try to go beyond the obvious.

b) Pool your ideas as a class. Now classify these jobs into different categories:

- part-time and full-time
- seasonal or all-year-round
- managerial or menial
- mainly for males or mainly for females
- ones you would like to do and ones you would not like to do.

SAVE AS . . .

2 Write an article (about a page long) which could be posted on the Leeds website explaining how tourism is helping people in the traditional industrial and farming area of West Yorkshire to make a living.

Break it down like this:

- Explain why traditional industrial and farming areas are experiencing problems today.
- Explain, with examples, how tourism can run alongside agriculture.
- Explain, with examples, how tourism is linked to the area's industrial past.
- In conclusion, explain why jobs in tourism are vital in areas that have lost more traditional sources of employment.

Jobs at Harewood
Curator
The house is full of beautiful things. We show them off but we have to look after them too.
Education Officer
Here is your worksheet. Now start researching . . .
Chief Executive
This is our five-year development plan for Harewood. In Year 1 we will . . .
Tour guide
Let me show you round.
Accountant
We still have not received that money.
Property maintenance
The stables need a new roof.
Catering
We do excellent wedding receptions here.
Gardener
It's a full-time job just mowing all the lawns.
Receptionist
Welcome to Harewood.
Bird Garden Keeper
Rare birds take a lot of looking after.
Events
I can arrange everything except the weather.
Marketing and promotions
Must keep the website up to date.
Ticket office
That's one ticket for the whole family.
Market research
What did you most like about your visit?
Press office
Yes, we are open this week despite the weather.
Cleaner
300,000 visitors a year leave a lot of mess.
Shop Manager
Everyone likes to buy something before they go.
Head of Security
We watch this place 24 hours a day, 365 days a year.

12 Some of the team at Harewood House

What kind of future for tourism jobs?

The optimists will say that, as disposable income increases, so does people's use of leisure and tourism and so employment in the tourist industry can continue to increase. It will always be a labour-intensive industry – you can't get machines to do many of the things that people do in leisure and tourism – so, as long as wealth is increasing, people can be sure of a future in tourism.

The pessimists will say that tourism is a very fragile sector – subject to fashion and variables such as the weather which are well beyond human control. Many jobs in the tourist industry are unskilled and therefore insecure. Many are seasonal – summer only. There are also longer-term cycles: tourism can decline as well as grow!

There is sometimes a lot of hype surrounding tourism, which leads to over-ambitious planning. When the visitors don't materialise then the jobs disappear very quickly. Only one year after opening, with lots of publicity, three of Yorkshire's 'national' tourist attractions – the Earth Centre in Doncaster (see page 274), the Royal Armouries in Leeds and the National Centre for Popular Music in Sheffield – were all laying off workers after disappointing visitor numbers. The attractions had drawn in local people but had all failed to attract enough people from around the country despite masses of promotion.

In 2001 rural tourism took an enormous knock from the foot and mouth epidemic (see page 253). Whole areas of countryside were closed to walkers. Rural hotels and restaurants reported bookings down by half. However, one person's pain can be another's gain – seaside resorts around the UK reported increased bookings as people opted for areas unaffected by foot and mouth disease.

Tasks

3 How might each of the four attractions be affected by each of the following?

- unusually bad weather
- a national rail problem that makes the trains unreliable
- a sudden rise in the cost of petrol
- a proposal to charge an entry fee to cycle or walk in a national park
- the screening of a popular television drama set in the Yorkshire Dales
- a visit from a member of the royal family which is featured in the local and national newspapers
- a scare about health in cows
- being featured on a nationwide travel programme
- an enthusiastic pupil telling his class what a fantastic time he had last weekend when he visited . . .

SAVE AS …

4 Are you an optimist or a pessimist about tourist sector jobs? Write two paragraphs – one optimistic and one pessimistic – predicting the future for employment in the tourist industry. Whatever your predictions, base your ideas around evidence on pages 314–17.

DALES CYCLING AND WALKING TOURS
Cycling and walking tours made simple. Accommodation, bikes, routes, luggage transfer all arranged. Extended and mini-breaks from April till end of October.

WENSLEYDALE CHEESE VISITOR CENTRE, HAWES, NORTH YORKSHIRE
www.wensleydale.co.uk
See Wensleydale cheese being handmade. Cheese-tasting, viewing gallery, exhibition, video. Restaurant, shops. Free parking.

THWAITE MILLS WATERMILL, LEEDS
www.leeds.gov.uk
Restored water mill on the River Aire. Workers in 19th-century costume. Restored Victorian mill-manager's house. Interactive exhibits.

THE NATIONAL CENTRE FOR POPULAR MUSIC, SHEFFIELD
www.ncpm.co.uk
Interactive visitor attraction with 3D arena telling the story of popular music. Hosts a programme of special events and exhibitions including live music and workshops.

13 Tourist attractions in Yorkshire

Review case study: location of a high-tech industry in Leeds

Where should Tara Multimedia locate?

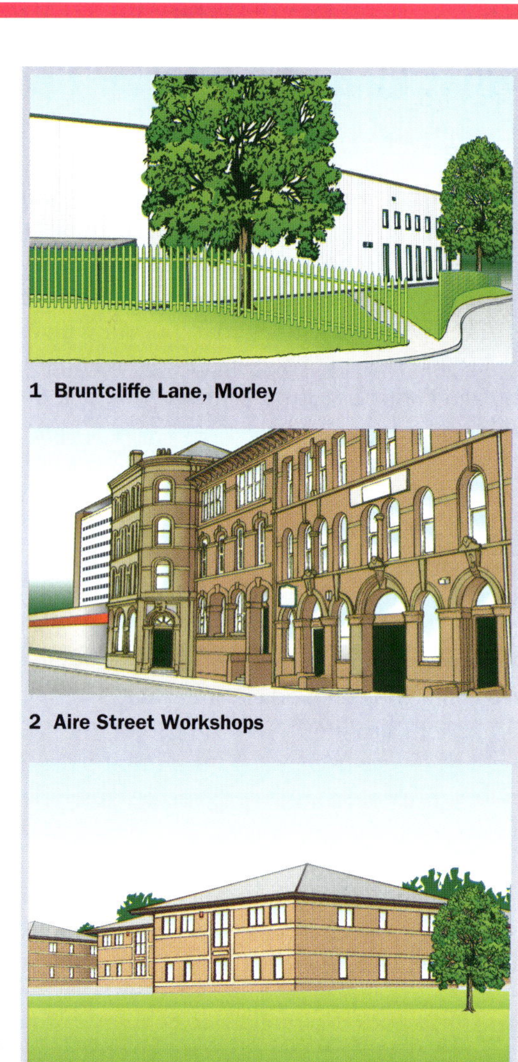

1 Bruntcliffe Lane, Morley

2 Aire Street Workshops

3 York Park

4 Lawnswood Park

The company

Tara Multimedia (TM) writes, designs and makes CD-ROMs. TM was set up in 1989 by a group of people who worked in computing and education at Newcastle University. It soon developed an international reputation for its CD-ROMs for education and industry. TM has won many awards for its products.

TM developed close links with a television production company based in Leeds. The television company invests money in helping TM to develop new products. TM can therefore be more ambitious. It is much better to get investment from the TV company than from a bank which would charge interest.

As the links became closer it was decided that the whole firm would move to Leeds. TM chose Leeds because the city already has a dynamic high-tech and media sector. The directors feel that TM will be able to expand there, because of the region's dynamism and its growing pool of skilled and trained labour.

The workforce

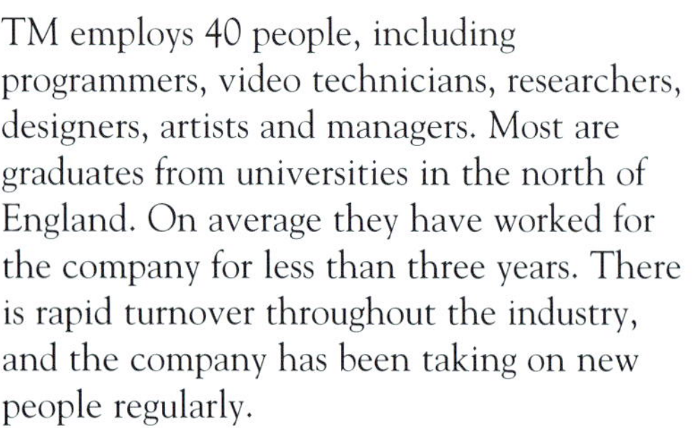

TM employs 40 people, including programmers, video technicians, researchers, designers, artists and managers. Most are graduates from universities in the north of England. On average they have worked for the company for less than three years. There is rapid turnover throughout the industry, and the company has been taking on new people regularly.

The ideal premises

At present TM has 700 m^2 of office space. It wants to double this, at least, because the present office space is cramped already, and the firm hopes to expand even further. It wants a site that:

- is *accessible* to the TV company, just north of the A65, close to Leeds centre
- is accessible to Leeds University and Leeds Metropolitan University, so that they can have close contact with university staff and students for future recruitment
- is *well designed* and up-to-date, so that it will be convenient to work in
- presents a *good 'image'* of the company to customers and other contacts
- is surrounded by a *pleasant environment*
- is reasonably close to *pleasant housing* areas, which will attract the present workforce to move down from Newcastle.
- has *good links* to the rest of the UK by motorway, rail and air
- is in a good location for making *contacts* with other high-tech and media companies.

Interviews

This firm is its people. I built it up by employing the best creative talent that I could find, and letting them develop their ideas. We're going to Leeds because there'll be lots of creative people there. The firm can only keep on growing and developing if we're located in an environment that allows people to think and create.
We must not be too close to the company that's financing us. That would mean that the 'men in suits' could start to dominate the company. We must be in a nice area, which will encourage our present staff to move, where we can easily recruit new people, and where we can make contacts with other firms in the high-tech and media sectors of the economy.

Tara Taylor, Managing Director

Jamie Weslowski, Sales Manager

Don't be fooled by talk of easy access for lorries which carry our goods. We produce <u>ideas</u>. CDs are so light and easily transportable that our actual transport costs are minute. What we need is access for people – particularly visitors from other firms which we work for, or which retail our products. Road and rail access are both important, and airport access will be increasingly important. Then, when people arrive, the office must impress them and allow them to relax.

In the Focus task on page 322, you must decide where is the best place for TM to locate its new office, bearing in mind what you know about Leeds and TM's needs.

All four possible sites (see page 320) have different strengths and weaknesses. There is not a single 'correct' answer. Even the best location is not perfect.

The shortlisted premises

TM is considering four sites. All fall within TM's price range, have enough space, have full services laid on, and should be ready for the firm in six months' time when they plan to move. All are accessible by car to pleasant housing.

Site name	Description	Rent/m^2/year	OS Grid Ref.
1 Bruntcliffe Lane, Morley	Industrial estate	£48.45	252280
2 Aire Street Workshops	Converted former factory	£52.20	293333
3 York Park	Business park	£48.45	323337
4 Lawnswood Park	Science park	£54.90	256382

1 Bruntcliffe Lane, Morley This prime area of warehousing and distribution space lies in the Leeds 'Golden Triangle' between the M1, M62 and M621. Built in 1982, it has a proven record of serving the needs of local industry. Premises can be re-designed to meet the needs of clients. Any business needing quick, easy access to the motorway network could not be better placed. Its out-of-town position means that rush-hour congestion can be almost completely avoided here.

2 Aire Street Workshops are superbly situated with excellent access to motorway and rail transport. These converted Victorian warehouses (which used to be training workshops for ex-prisoners) lie close to Leeds CBD, on the edge of the rapidly developing 'Armouries' tourist area. Units of around 500 to 1,000 m^2 are available, but they could be combined to give more space. The rents are comparatively low for property so close to the city centre. The chance to rent at these prices is unlikely to last long as the redevelopment of the area gathers pace.

3 York Park Built in the early 1990s, this development was carefully landscaped to provide an attractive environment for science-based companies. As the area has matured it has changed its nature and now mainly provides premises for general office and business development. Self-contained two-storey units range in size from 1,000 to 2,500 m^2. Large spaces are available in the three-storey blocks on the site. Ample landscaped car parking is available with all units.

4 Lawnswood Park There are very few office or industrial developments in the north-west sector of Leeds, and yet this is an ideal location for modern science-based developments. It has easy access to the city's two universities and to the attractive housing areas in the suburbs of Headingley, Horsforth, Meanwood and Moor Allerton. The development is exceptionally attractive and prestigious, with unit sizes ranging from 1,000 to 2,500 m^2. Strict controls have been placed on the type of client in order to ensure that the high status and the scientific emphasis of the development are maintained.

14 The four possible sites in Leeds

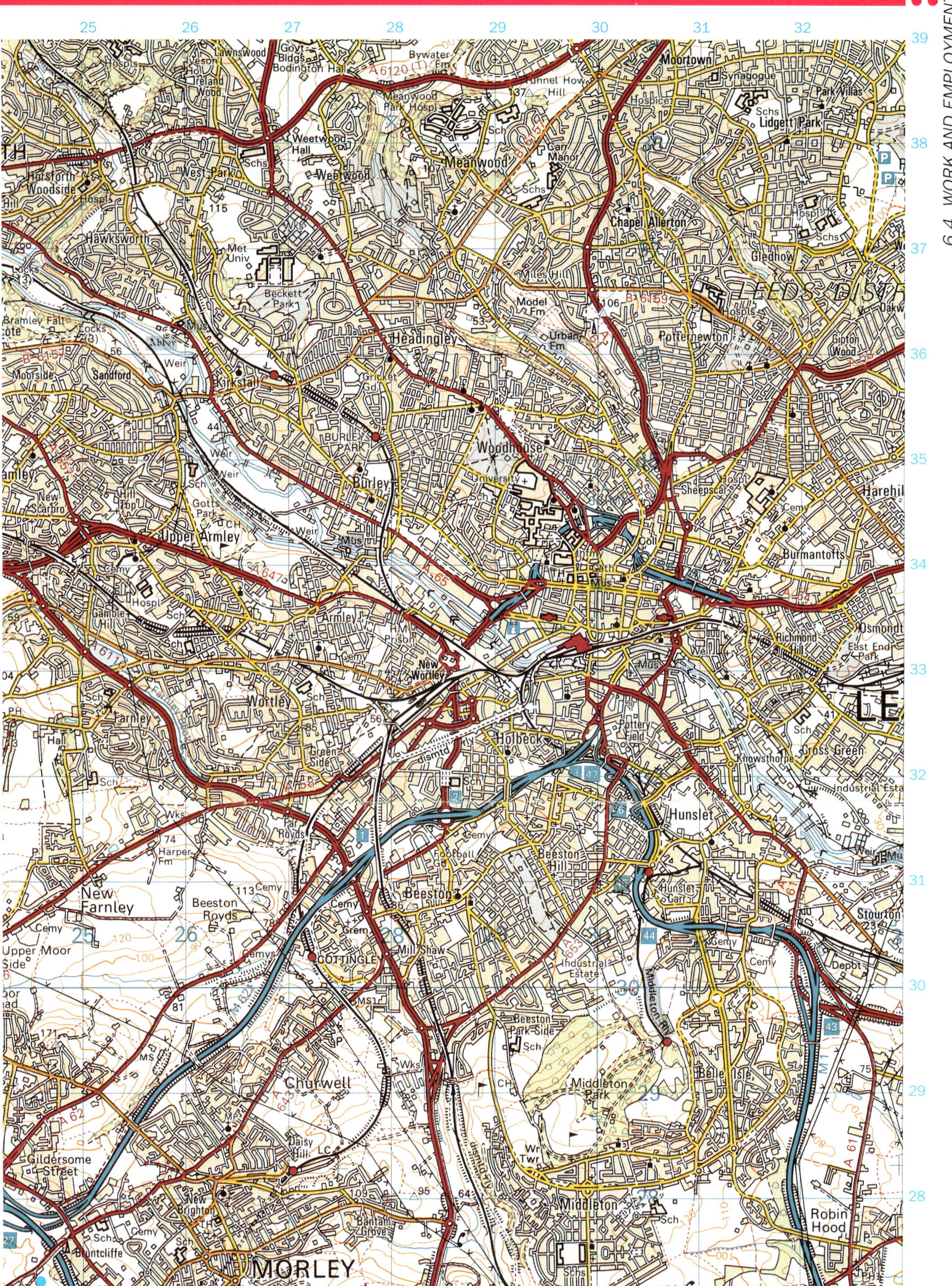

15 Leeds. Reproduced from the 1998 1:50,000 Ordnance Survey map by permission of the Controller of HMSO © Crown Copyright

Focus task

1 Imagine that you are advising Tara Multimedia. Four sites have been shortlisted. You have to decide which is the best location for the firm.

Stage 1: Consider TM's needs

a) Make a list of six important criteria that must be considered when choosing a site for TM in Leeds.

b) Arrange these criteria in order of their importance, with the most important at the top.

c) Explain how you decided on your ranking.

Stage 2: Assess the possible sites

d) Study each of the four suggested sites (Figure 16). Award them marks out of five for each criterion. Record these marks in the column marked R on a chart like this:

Stage 3: Make your decision and write your report

e) 'Raw' marks are not good enough. You need to 'weight' some of the marks to show importance. Multiply the most important factor's score by four, working down so that the score of the least important factor is multiplied by one. Record the scores in column W.

f) Decide on the best site and write a report for Tara Multimedia recommending a site and explaining why you chose this site. Attach your matrix to your report, so that you do not need to repeat figures from the matrix. Instead, your report should focus on explaining the main strengths of your chosen site. You may want to add brief comments on a 'reserve' site, and even briefer comments on your reasons for rejecting the other two sites.

Criteria	Bruntcliffe Lane		Aire Street		York Road		Lawnswood	
	R	W	R	W	R	W	R	W
1	3		3		1		4	
2								
3								
4								
5								
6								
Total score								

R = raw score W = weighted score

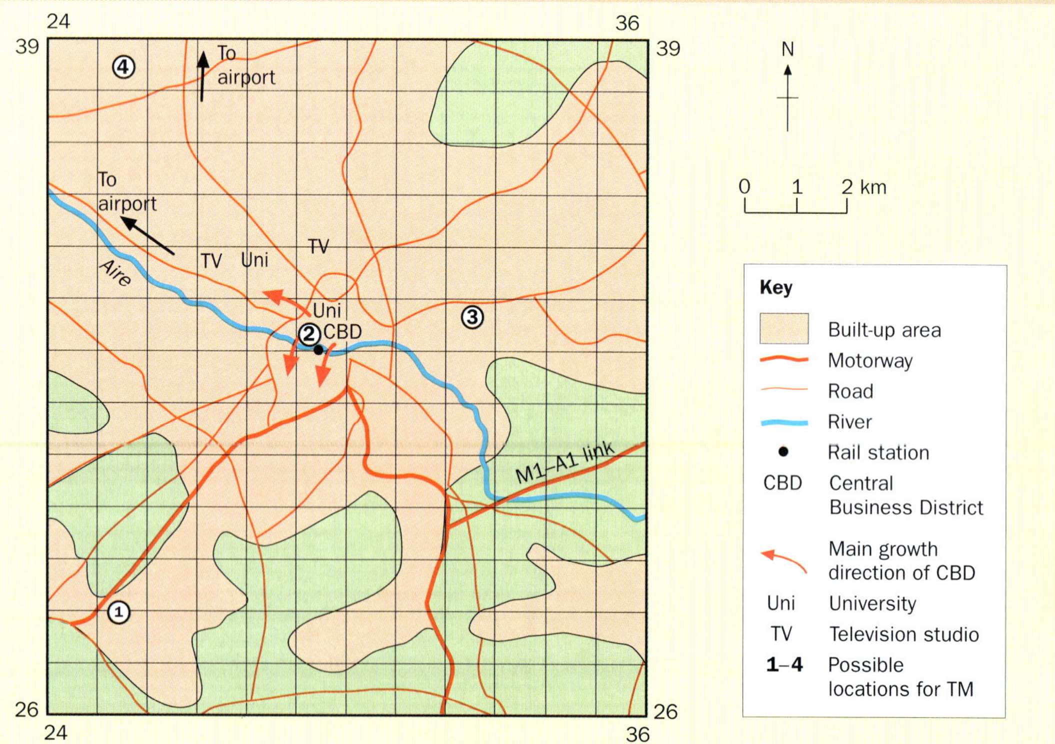

16 The four possible sites for TM's new premises

Quality of life and development

How can we get rid of poverty?

Wealth in an LEDC

All men are created equal.

Thomas Jefferson, former US President (and slave owner!)

The life of a single human being is worth a million times more than all the property of the richest man on Earth.

Che Guevara, a revolutionary leader from South America

Where do you think the photos were taken? What do they tell you about the quality of life in each place? What questions would you like to ask people in the photos?

What do you think each quotation means? Which ones do you agree with and which do you disagree with? Why?

A fur hat is no good on an empty stomach!

Vladimir Ivanov, owner of a Moscow hat stall

It is easier for a camel to go through the eye of a needle than for a rich man to enter the kingdom of heaven.

Jesus, founder of the Christian faith

Poverty in an MEDC

7.1 What is quality of life?

DO YOU KNOW?

- People are considered poor if they get less than half the average income of people in the country where they live. In the UK, 3 million children live below the POVERTY LINE.
- The number of Americans who describe themselves as 'happy' peaked in 1957, even though living standards in the USA are now twice as high as they were then.

AIMS

- **To consider what things are important for a good quality of life.**
- **To decide whether there is a North/South divide in quality of life in the UK.**
- **To identify patterns of inequality in Europe.**

What affects people's quality of life?

Quality of life is a measure of human well-being and of what people's life is like. It tells us how happy, or content, people are with their lifestyle and environment. This sounds quite straightforward, until you start to ask people what makes them happy and content! Everyone is different. There are probably as many definitions of quality of life as there are people. It is important to remember this when we try to compare quality of life in different places. It is even more important when we consider DEVELOPMENT – which is the way that people's quality of life improves.

Here are two contrasting examples. What affects Erica and Gregory's quality of life?

1 Life on the Easterhouse estate in Glasgow

Erica lives on the Easterhouse estate in Glasgow with her husband, Ivor, and children, Charlotte, Deirdre, Gilbert and Mary. This is an extract from her diary (taken from *Faith in the Poor* by Bob Holman).

MONDAY

I get up and get the kids dressed for school and give them some tea and toast. I give Deirdre £2.80 for two days' bus fares to school and back out of the child benefit. Kids go to school, I go and buy:

potatoes £0.69, loaf of bread £0.69, gas token £5.00, cigarettes £5.03, two tins beans £0.60, soap £0.22, washing-up liquid £0.49.

When they come home from school, Gilbert and Mary need money for school Christmas trips, but I don't have it to give to them. I can't even start thinking about Christmas presents for the kids. I am still in debt paying for last Christmas. I am still paying for the kids' clothes. Sometimes you have to borrow to eat. We watch some TV and then we all go to bed.

WEDNESDAY

Get up with the kids. Make them tea and toast and get them off to school. Peace at last. Charlotte goes to the Post Office to get our social money, £74.49. Spend money on:

gas token £10.00, electricity token £10.00, water token £5.00, milk £5.18, bread £2.76, potatoes £3.57, margarine £0.49, two meat pies £1.96, two tins beans £0.60, Deirdre's bus fare £1.40.

I have spent £40.96 and have £33.53 left. Gilbert is off school at the moment because he needs new shoes and that is £10.00. I can't afford them just now so I will have to wait and see what I can do for him. I'll go to the second-hand shop as soon as I can.

The kids come home from school. Thank God they get a meal at school. Ivor is making tea while I get the kids ready for bed. Deirdre and Charlotte are going out with some friends. So me and Ivor are on our own at last for the first time in ages. The kids come back and we all go to bed.

SATURDAY

Get up as usual but my head hurts bad. The kids are running about so I have to shout at them so they are not pleased with me at all. Make them some toast. I have to send for some bread plus some cigs. Must have some cigs or I can't get through the day. I hate it when people say that because you are on social security you shouldn't smoke. I wish I could give it up. I've tried and failed. I spend:

bread £0.69, cigs £2.93, electric token £5.00.

I have £5.75 left. I have asked the kids to help clean up for me. They moaned but they did it. We all sit and watch TV. The TV set is on its last legs and I don't know what I'd do without it. The kids go to bed. I want to go for a bath but I have to keep the hot water for Sunday for the kids.

Gregory, a boy from the UK, visits his grandparents in Tobago for the first time. Source 2 is from *Gregory Cool*, a story by Caroline Binch.

2 Life on the island of Tobago in the Caribbean

Sitting in the taxi from the airport, squashed tightly between his grandparents, Gregory wished he was back home with his mum and dad. Why did he have to come to Tobago?

The air was stifling and strange smells disturbed him. Gregory shut his eyes. All of a sudden he felt very tired. The taxi stopped outside a very small house. 'Do you really live here?' asked Gregory. His grandparents just laughed as they took him inside and showed him his room. The last thing he saw before he fell asleep was a lizard looking down at him from the ceiling.

Gregory woke up next morning with just a sheet over him. It was hot! Sun poured in through the open window. There were no toys, no books, no carpet – not even a proper door. Gregory scratched at his arm. Something had bitten him during the night. Was he really expected to stay here for four weeks?

In the kitchen, Grandma was cooking breakfast and Grandpa sat at a small table with a boy Gregory hadn't seen before. This must be his cousin. His mum had told him about Lennox, and how he lived with his grandparents. Gregory sat down and looked at his breakfast plate. Scrambled eggs – he could deal with that. But it wasn't eggs... Gregory spluttered and spat out the salty stuff as politely as he could.

'Hey, you don't like your bake and buljol?' said Grandpa. 'It's just bread and saltfish.' 'It's cool,' said Gregory. 'I'm just not hungry.' He drank a glass of fruit juice and followed Lennox outside. Lennox was a year older than Gregory, but much smaller. 'What do you do around here?' asked Gregory. 'Got a bike?'

Lennox grinned shyly at him. He had bare feet – Gregory looked at them, then looked away quickly. 'Come, I'll show you the river,' said Lennox. The air was shimmering hot. Gregory sat down in the shade. 'I'd rather stay here,' he said. 'It's cool.' 'Well, I'll go feed the goats, then dip in the river,' said Lennox, and off he ran.

Granny appeared carrying a big basket. 'Right children,' she said. 'Grandpa an' me is taking you for a sea-bath.' 'Wicked!' shouted Lennox, leaping across the yard.

'Cool,' said Gregory politely. Cool was the last thing he felt, but he wasn't going to say so. At least he might get a fizzy drink and an ice-lolly at the beach.

The bus they caught was like an oven, crammed with people. When they finally got there the beach had palm trees and sand, just like a travel poster. But there wasn't anywhere to get ice cream or chips – and Gregory had missed out on breakfast, so he was feeling very hungry.

Tasks

SAVE AS...

1 Study Figures 1 and 2. Select *all* the information that gives evidence about quality of life, both positive and negative. List this in two columns under the headings Positive and Negative.

Don't miss things that seem obvious! For example, *friends* help to improve our quality of life. Some things could be listed in both columns. For example, *child benefit in the UK* is positive because it shows that the government helps people with children – but the fact that some people are so poor that they depend on it, is negative.

Add your own ideas to each list.

2 a) Choose five things from each list in Task 1 that you think are most important in affecting quality of life.

b) Rank them in order of importance.

c) Compare your priorities for quality of life with those of other people in your class. Are any two people's ideas the same?

d) Talk about the differences. Why are some things more important to some people than others?

e) How does wealth affect quality of life? Which important things in your lists are affected by wealth, and which are not?

3 Work in a group. Think about your local area. What is the quality of life like there? Where are the best parts and where are the worst?

a) *Either:* Use a digital camera to take five photos that show examples of good quality of life in your area, and five more that show poor quality of life in the area.

Or: Find five photos and/or articles in your local newspaper that show good quality of life in your area and five more that show poor quality of life.

b) Use your photos and articles to make a display about quality of life in your local area. Annotate your photos or news cuttings to explain why you chose them.

Does the North or the South of the UK have a better quality of life?

Many people would say that there is a North/South divide in the UK, with people in the South enjoying a better quality of life than people in the North. There is some evidence to support this argument. The economy in the South East of England is booming, while much of the North has suffered industrial decline. People from the North migrate to the South to find better-paid jobs. Does this mean that everyone in the South has a better quality of life than those in the North?

Focus task

1 Work with a small group. Your task is to devise a hypothesis about quality of life in the North and the South of the UK. You should then use the information on pages 326–29 to test your hypothesis.

Stage 1: Think about what North and South mean. Where would you draw a line on a map of the UK to divide the country into North and South?

Stage 2: Think about the images that you have of the two parts of the country. Do you think there is a difference in the quality of life between North and South? If so, which is better?

Stage 3: Now write a hypothesis that you can test. For example: *Quality of life is better in the North than it is in the South.*

Stage 4: Study Source 3. List all of the evidence it gives for or against your hypothesis. As you work through pages 326–29 do the same with all the other Figures. At the end you will weigh up this evidence and decide whether your hypothesis is true or not.

Seeking Southern Comfort

Northerners are migrating down the M1 in their thousands, reports Richard Reeves

Midnight in a Toxteth pub. A dispute has broken out. For once it's not the usual Liverpool *v.* Everton rivalry. The issue behind the locked door and closed curtains is this: Who wants to live in Liverpool?

'There are no jobs, and the weather's terrible,' says Nick Allen. 'And the drugs and crime, it's like a cancer spreading across Liverpool.' His wife, Thelma, has seen all her children leave the city. 'Good for them,' she says. 'They're bettering themselves.' 'Nonsense,' retorts Billy Hill, a lifelong resident of the area made infamous by its 1980s riots. 'This is God's Little Acre, Liverpool 8. Around here you're never alone. People look out for each other. It's like a family.'

But the Allen family's downbeat view is backed by the facts. Recent population figures show our northern cities are being hollowed out by an astonishing exodus to the South. The North loses 23,500 people every year to the South; Liverpool alone has lost 19,000 since 1990. Our friends in the North are becoming our neighbours in the South.

The trend is set to accelerate in the new millennium, with a government projection for Merseyside showing the loss of 8.5 per cent of its people by 2021. 'There has been a significant shift in the levels of migration from North to South,' says Alan Holmans, an expert in population and housing at Cambridge University. 'Before about 1993 there was a drift southwards, but only at a rate of around 9,000 a year. We are now running at well over double that rate.'

Since the beginning of the 1980s, a third of a million people have been added to the population of the South, almost entirely at the expense of the North. The result is a southern housing market so inflated that nurses cannot afford to live in London, while estate agents in Liverpool are offering 'buy one, get one free' deals to shift terraced houses in deserted areas.

Holmans blames the job market. 'The South was harder hit by the recession of the early 1990s, but now the effects of that downturn have worn off, the attraction of the South has emerged stronger than ever.' There are, of course islands of prosperity north of the Wash, such as Leeds and York, but these are exceptions.

Back in the Toxteth pub, Allen agrees that the migrants are leaving for economic reasons. 'There are no long-term jobs round here, so it's difficult to consider raising a family. There's no stability.' He also suggests there has been a shift in attitudes towards future prospects. 'When all the docking and factory jobs went in the 1980s, people sort of thought they might come back. Now they are realising those days are gone for good, so they're packing up and getting out.'

There are other explanations on offer. 'It's the bloody weather, isn't it?' the barmaid offers, serving another round of 'bitter and browns'. 'I lived in Malta for four years and it was great. I'm back here because my gran died but there's no work around here. My husband is a baker and has no chance of getting a job. At least out there it's warm.'

One refugee from Merseyside, dancer Kelly Matthews – here on a weekend visit to her old home – says lack of creative arts has driven her out. She says, 'The best teachers, directors and choreographers are in London and the South. It's a shame because there's at least as much talent in the North, but lots of it goes to waste.'

Nine out of ten of the migrants are in their twenties or thirties. Matthews is 22 and believes that, for women, there is an additional incentive to join the march south: New Men are scarce in the North, so women are expected to adopt more traditional roles. 'The further north you go, the greater the pressure to cook, clean, iron and generally look after your man. It makes me shudder. I have no plans to cook a Sunday roast in my life, ever.'

3 Extract from the *Observer*, 7 July 1999

★ **FACTFILE**

North East of England

- Average house prices in the North East are lower than in any other region of the UK.
- The unemployment rate in the North East is higher than in any other region.
- The North East has the poorest GCSE results among 16-year-olds of all the regions.
- Almost three-quarters of households in the North East participate in the National Lottery – a higher proportion than any other region.
- The number of violent crimes in the North East is lower than in any other region.

4 Home worth £100,000 (at 2000 prices) in Jesmond, Newcastle

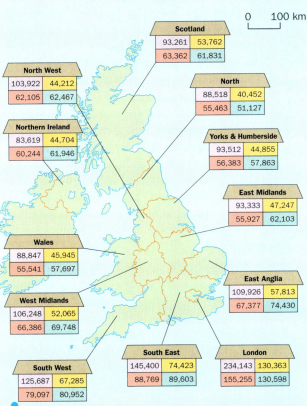

Key

	UK average (£)		
Detached house	114,364	59,230	Terraced house
Semi-detached house	71,670	74,638	All homes

N

0 100 km

Region	Detached house	Terraced house	Semi-detached house	All homes
Scotland	93,261	53,762	63,362	61,831
North West	103,922	44,212	62,105	62,467
North	88,518	40,452	55,463	51,127
Northern Ireland	83,619	44,704	60,244	61,946
Yorks & Humberside	93,512	44,855	56,383	57,863
East Midlands	93,333	47,247	55,927	62,103
Wales	88,847	45,945	55,541	57,697
East Anglia	109,926	57,813	67,377	74,430
West Midlands	106,248	52,065	66,386	69,748
South East	145,400	74,423	88,769	89,603
London	234,143	130,363	155,255	130,598
South West	125,687	67,285	79,097	80,952

6 Average house prices in the UK regions, January 2000. House prices are a good guide to both the wealth of a region, and whether people want to live there

★ **FACTFILE**

London

- Housing costs in London are higher than in any other region, but people are less likely to be satisfied with their homes.
- More than 4 per cent of people in London have an income of £50,000 or more – the highest proportion in the UK.
- The crime rate in London is higher than in any other region.
- 29 per cent of young people in London admit to illegal drug use – more than in any other region.
- The road accident rate in London is nearly three times higher than in any other region.

5 Homes worth £100,000 (at 2000 prices) in Forest Gate, London

Tasks

2 Look at Figures 4 and 5. Think about the links between house prices and the quality of life. How do you think that house prices:

a) affect the quality of life in a region?

b) reflect the quality of life in the region?

3 Two of the authors of this book live in Jesmond, Newcastle and Forest Gate, London. As teachers their incomes are roughly the same (teachers in London receive an extra allowance worth about 10 per cent more). In London the cost of living is much higher. Not only house prices but also things like council tax, car insurance, transport and food cost more.

a) Which author do you think would enjoy the best quality of life? Give reasons for your answer.

b) Does this help to prove, or disprove, your hypothesis?

Region	Unemployment rate (%)	Average weekly earnings (£)	Proportion of people who own their homes (%)	Proportion of pupils achieving grades A–C at GCSE (%)	Homes with central heating (%)	Number of crimes per 100,000 people	Proportion of households with one car or more (%)	Proportion of people satisfied with the area they live in (%)
North East	8.2	339.2	63	33.9	94.1	9,713	58	49
North West	6.6	361.6	68	39.2	82.9	9,422	69	47
Yorkshire & Humberside	7.0	344.9	64	35.4	82.5	10,857	66	53
East Midlands	4.9	350.4	71	39.0	92.3	9,187	73	54
West Midlands	6.3	358.8	68	37.3	82.2	9,055	68	52
East	5.0	378.6	73	43.6	92.3	6,643	77	57
London	8.1	500.9	56	39.3	87.5	10,527	61	40
South East	4.3	405.5	75	46.4	91.6	7,276	81	56
South West	4.5	354.0	73	44.6	88.4	7,490	76	61
Wales	6.7	343.9	72	38.6	–	8,111	69	–
Scotland	7.4	350.3	–	48.9	87.2	–	65	59
Northern Ireland	7.3	332.6	70	46.8	88.7	–	70	–

7 Regional variations in the quality of life in the UK, 1999. Up-to-date information for each region can be found at www.statistics.gov.uk

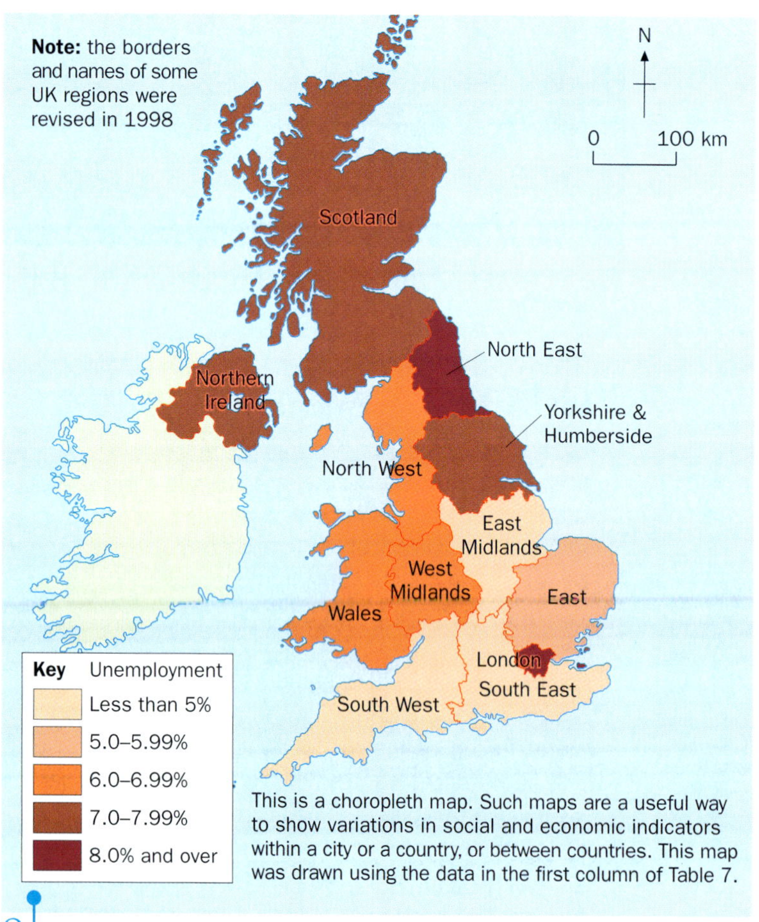

This is a choropleth map. Such maps are a useful way to show variations in social and economic indicators within a city or a country, or between countries. This map was drawn using the data in the first column of Table 7.

8 Unemployment in the UK regions, 1999

Task

1 Work in a small group.

a) Choose at least four things from Table 7 that you think are important indicators of quality of life. Each person should use a map of the UK to draw a choropleth map showing the regional variation for at least one indicator.

b) Compare the maps that you have drawn. Are there any similarities or differences? Do the maps help to prove, or disprove, your hypothesis?

★ THINK!

Tables are useful sources of data. You can get a lot more information into a table than into a paragraph of ordinary writing. But tables don't show any obvious geographical patterns and they are hard to remember for exams. However, if you can turn the data in a table into a map, it will show the patterns more clearly and make the information easier to remember. Compare the data in Table 7 with Map 8 and the maps you have drawn for Task 1. Which is easiest to remember?

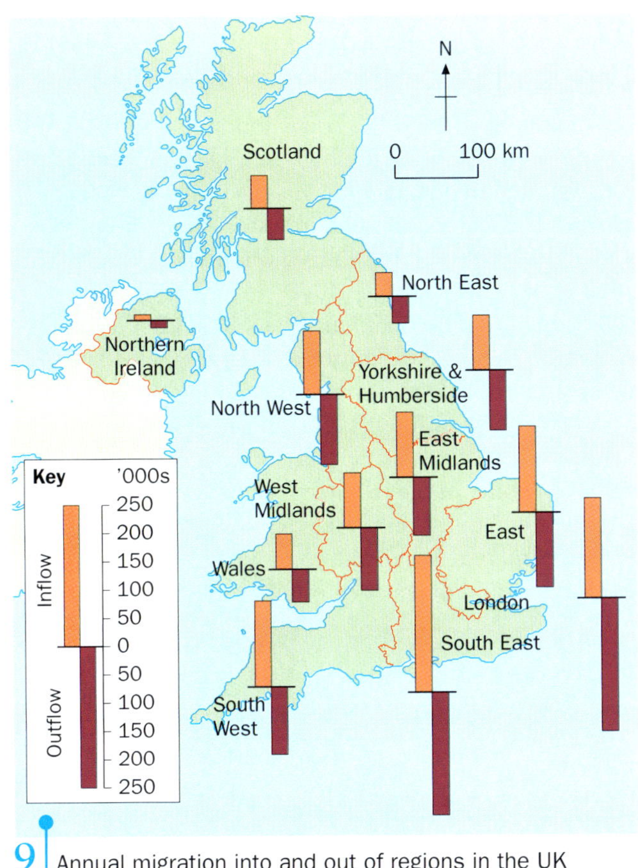

9 Annual migration into and out of regions in the UK

Tasks

2 Look at Map 9.
  a) Describe the pattern of migration shown on the map.
  b) What evidence does this give of a North/South divide?
  c) Compare Map 9 with the maps you drew in Task 1. Explain any links that you can find between the maps.

3 Look at Figure 10. How would this advertisement help to attract people to the South West region?

4 a) Find out more about quality of life in your own region. You could look at the website of the Office for National Statistics at: www.statistics.gov.uk
  b) Use the information to produce an advert for your own region, attracting people to live there. Emphasise all the positive aspects of quality of life in your region.

5 Complete the list of evidence for and against your hypothesis from the Focus task on page 326. Check that you have used all the evidence on pages 326–29. You can also use evidence from the maps you drew in Task 1.

6 Look again at your hypothesis. On balance, does the evidence support it or not? Explain your answer. If the evidence does not support it, how would you rewrite your hypothesis now?

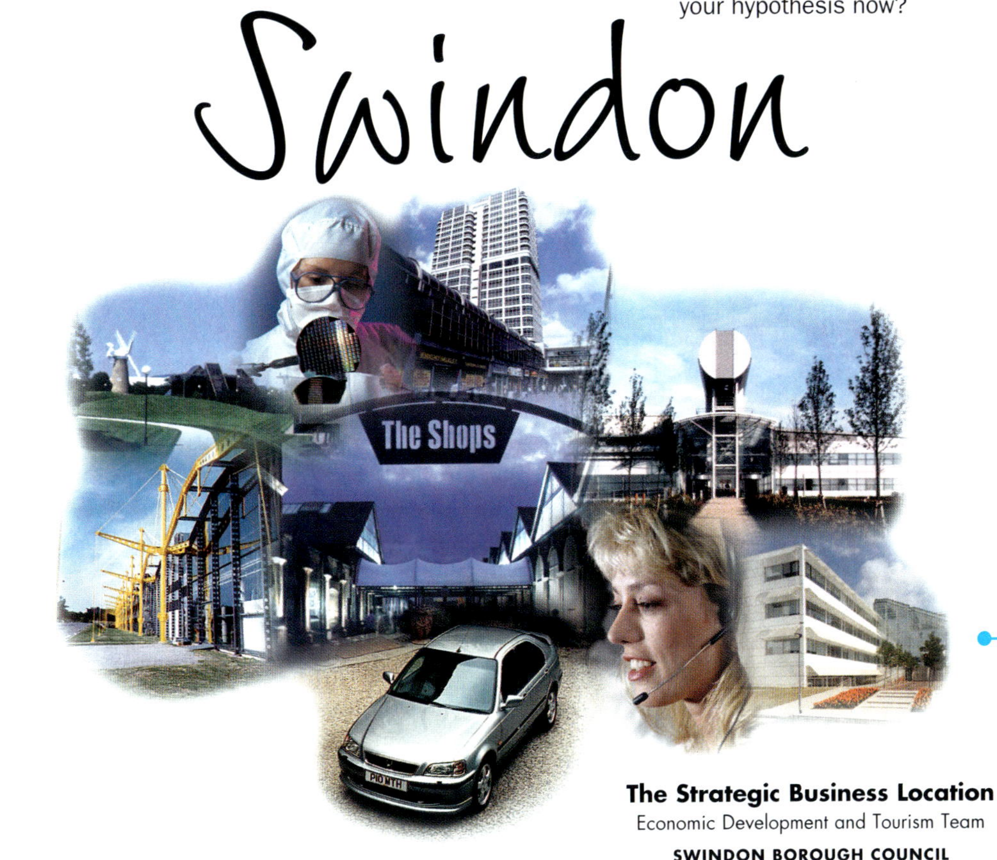

10 An advertisement for Swindon, a growing town in the South West. What does it tell you about the quality of life there?

Are there divisions within Europe too?

You have found that there are differences in the quality of life between regions in the UK. But how do the UK regions compare with other parts of Europe? Can you find any patterns in Europe equivalent to the North/South divide in the UK?

When you looked at quality of life in the UK you compared several indicators, including unemployment, weekly earnings, crime, and people's satisfaction with where they live. Geographers try to simplify this sort of comparison by using just one indicator for quality of life. One they often use is wealth, measured by GROSS DOMESTIC PRODUCT PER PERSON, or GDP/CAPITA. This is the value of all the goods and services produced in an area, divided by its population. Map 11 shows how GDP/capita varies in the regions of the European Union (EU), including regions within the UK.

Regions with a high GDP/capita are those that produce more goods and services and where people earn more money. This means that they are likely to enjoy a higher LIVING STANDARD and, in theory at least, this will give them a better quality of life. But is GDP/capita always a good indicator of quality of life? It has three weaknesses.

- It is an average measurement that tells us nothing about how the wealth is actually shared.
- It tells us nothing about how wealth is used. For example, GDP/capita can rise because the government spends a lot of money fighting a war against another country! It is possible that as GDP grows there will be an increase in environmental problems or crime within a region or country.
- It doesn't allow for non-economic factors.

Tasks

1 Look at Map 11. You should be able to name all the countries. If not, use an atlas. Your teacher may give you a map that names the regions too.

a) Describe the pattern on the map showing variations in GDP/capita in the EU.

b) Compare Maps 11 and 15. Name the countries and/or regions that are part of the core and periphery.

2 a) Look at Photos 12–14. What do they tell you about quality of life in each of the three regions?

b) Look at Map 11 to find out the GDP/capita in these regions. What is the link between GDP/capita and quality of life?

SAVE AS . . .

3 Write two paragraphs explaining:

a) Why GDP/capita is an indicator of quality of life.

b) What problems GDP/capita might have as an indicator of quality of life.

Core and periphery

Look carefully at Map 11. You will notice that the regions of the EU with the highest GDP/capita are concentrated in an area known as the core. It includes many of the major cities and industrial areas of Europe. You may also notice that the regions around some of the other major cities, outside this core, have a higher GDP/capita too. Regions with the lowest GDP/capita away from the core are known as the periphery. These are areas with little industry and declining agriculture. Often their population is falling as people migrate to the cities. The EU, as part of its economic policy, tries to improve quality of life in these regions by bringing new jobs so that people don't have to move. It has given the periphery more regional aid, to provide better transport and services and incentives for new industry.

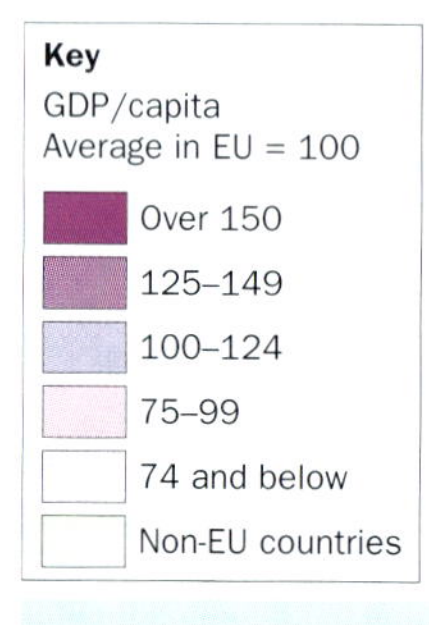

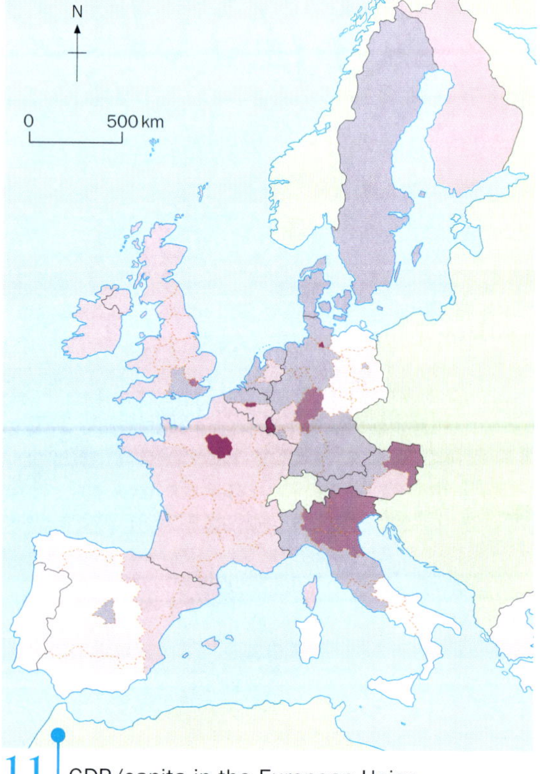

11 GDP/capita in the European Union

12 The CBD in London, in the core of Europe

13 Farming in rural Spain, part of the European periphery

14 This steelworks in northern France has closed. Areas of industrial decline are often found in the intermediate area between the core and periphery

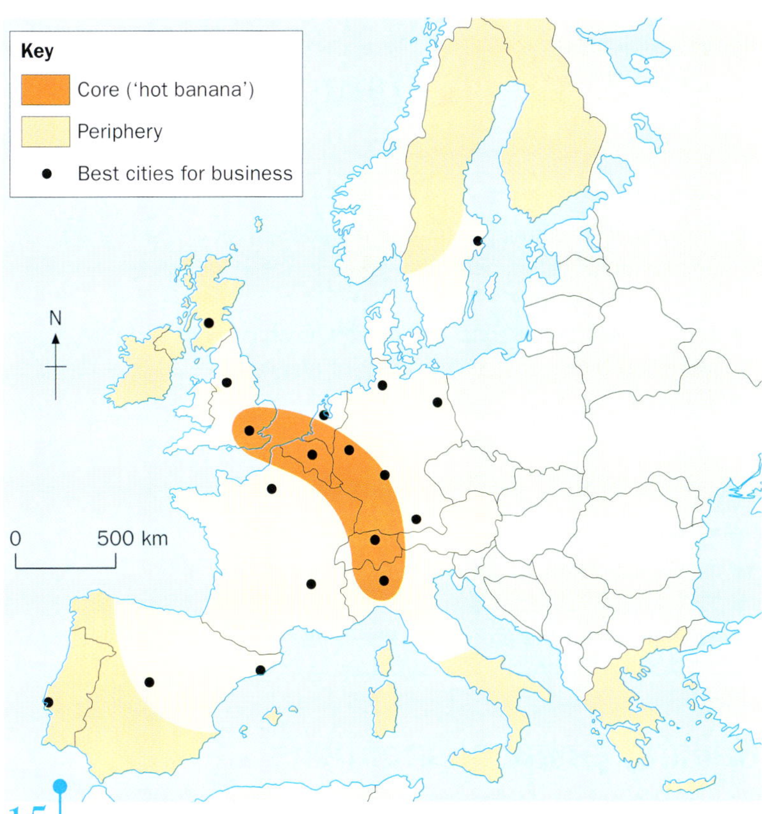

15 The core and periphery in Europe – the core is sometimes called Europe's 'hot banana'

Focus task

4 You have been asked to prepare a case arguing for regional aid from the EU for one of the regions outside the core of Europe. It could be one of the regions in the UK – possibly the region that you live in yourself. You have to persuade the EU that your chosen region needs financial support to improve the quality of life for people living there. Write a report to include:

a) Problems in the region that lead to a poorer quality of life. Give as many examples as you can. Try to explain how these problems have arisen. Be as specific as possible. Obviously, this will be easier if you live in the region as you will have local knowledge. If you choose another EU region you could get more information on the following website:

www.europa.eu.int/

b) Suggest how EU money could be used to improve the quality of life in your chosen region, for example to build new roads, create new jobs, reduce environmental pollution. Remember, there will be many other regions asking for money, so you must make your case as convincing as possible.

7.2 How are global patterns of development changing?

DO YOU KNOW?

- The combined annual income of the poorer half of the world's population (3 billion people) is equal to the wealth of the world's richest 225 people.
- The largest transnational companies, such as Ford and IBM, have an annual income larger than that of an average-sized country like Malaysia.

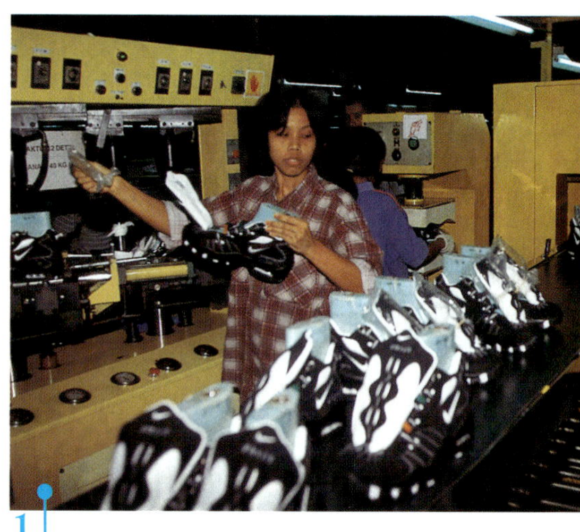

1

AIMS

- **To identify global inequalities based on wealth.**
- **To consider how to measure development.**
- **To compare levels of development between countries.**
- **To understand the processes that make global inequalities grow.**

Making global connections

Have you heard the expression, 'It's a small world'? People have been saying it for years, but now it's true! These days you need look no further than your classroom. Find the labels on your own clothes – almost certainly they have been made in another part of the world. Turn on your computer and the internet will give you access to websites almost anywhere. And when the school term ends, will any of you be jetting off to distant corners of the globe? If so, you won't be alone, because more people than ever go abroad for their holiday.

Throughout history there have been links between people around the world, but in recent years these have increased at an astonishing rate. It has even added a new word to our vocabulary – globalisation. Driving the process of globalisation is the growth in the GLOBAL ECONOMY – the goods and services produced and traded around the world. Not only are people in the more wealthy countries such as the UK consuming more, but more of what we consume is produced outside our own country, and very often on a different continent. Globalisation brings us the benefit of a wider variety of goods and lower prices. And it's not just goods that move around. Money and information flash around the world today in ways that would have been impossible just 10 years ago. But are the benefits of globalisation equally shared around the world?

2

3

4

5

6

Tasks

1 Work with a partner. Look at Photos 1–6 and analyse them. For each photo ask these questions:

- Where do you think the photo was taken?
- If you can see any people, who are they?
- What are they doing?
- Why are they doing it?

Think of other questions that you would like to ask about each of the photos. (Later, your teacher may give you an information sheet about each photo that may help you to answer some of these questions.)

2 There is a connection between all of the photos on these two pages. With your partner, can you work out what it is?

Create a story that links all the photos. Tell your story to the rest of the class. After you have heard all the stories, decide which story you think is most convincing.

3 Read each of the following statements to do with globalisation. Discuss the statements with your partner.

A Globalisation benefits everybody around the world.

B Consumers in the UK pay less in the shops for goods that are produced abroad.

C Producers of goods and services in other countries are paid too little.

D Everybody would be better off if they produced all the things they need in their own country.

E People around the world now depend more on each other, so wars are less likely.

F Both producers and consumers are the victims of large companies that control the global economy.

a) Decide which statements you agree with and which you disagree with. You may be able to use the evidence in the photos to help you.

b) Place the statements on a line, like the one shown below, with the statement you most agree with furthest to the left.

Statement I most agree with — *Statement I least agree with*

Keep your diagram until you have finished Unit 7. Then look again at your opinions to see if they have changed.

How is the world divided?

The simplest way to examine global inequality is to compare wealth country by country. National wealth is usually measured by GROSS NATIONAL PRODUCT PER PERSON or GNP/CAPITA. This is the total value of the goods and services produced by the people of one country, divided by its population. It is similar to GDP/capita (which we came across in Chapter 7.1), but while GDP is based on goods and services produced *in the country*, GNP is based on goods and services produced by its population *in the country and abroad*. Both GNP and GDP are usually given in US$ to make comparisons between countries easier.

Geographers use GNP/capita to divide the world into two areas: MEDCs (More Economically Developed Countries) with high GNP/capita, in the North, and LEDCs (Less Economically Developed Countries) with a lower GNP/capita, in the South. Map 7 shows the line that can be drawn to divide the world into 'North' and 'South'. The map also shows four types of countries grouped by income levels. If you look carefully you will see that not all the countries in the North are rich and not all the countries in the South are poor.

So is there a better way to divide the world? As the global economy changes, some countries grow richer at a faster rate than others. A few countries have even become poorer. If you compare Maps 7 and 8 you will notice some of the changes that have happened since 1960 and the ways that patterns of global inequality are changing.

Tasks

1 Look at Map 7.

a) Name five countries in each of the four income groups. Alternatively, label them on a copy of the map. Use an atlas to help you if you need to.

b) Write a paragraph explaining to what extent the North/South divide is an accurate way to describe global inequality. Name some countries that don't fit this pattern.

c) Describe any other patterns that you can see on the map.

2 Study the information in the Factfile. On Maps 7 and 8 locate the five countries given as examples

a) For each country, state:

i) which income group it is in (Map 7)

ii) how its relative wealth has changed since 1960 (Map 8).

b) Find at least one more example of each type of economy as defined in the Factfile.

c) Divide an outline map of the world to show the areas where you would find each type of economy. For example, north-west Europe is a rich industrial area. Draw lines on your map like the North/South divide on Map 7. Shade or label each area on your map. What pattern emerges?

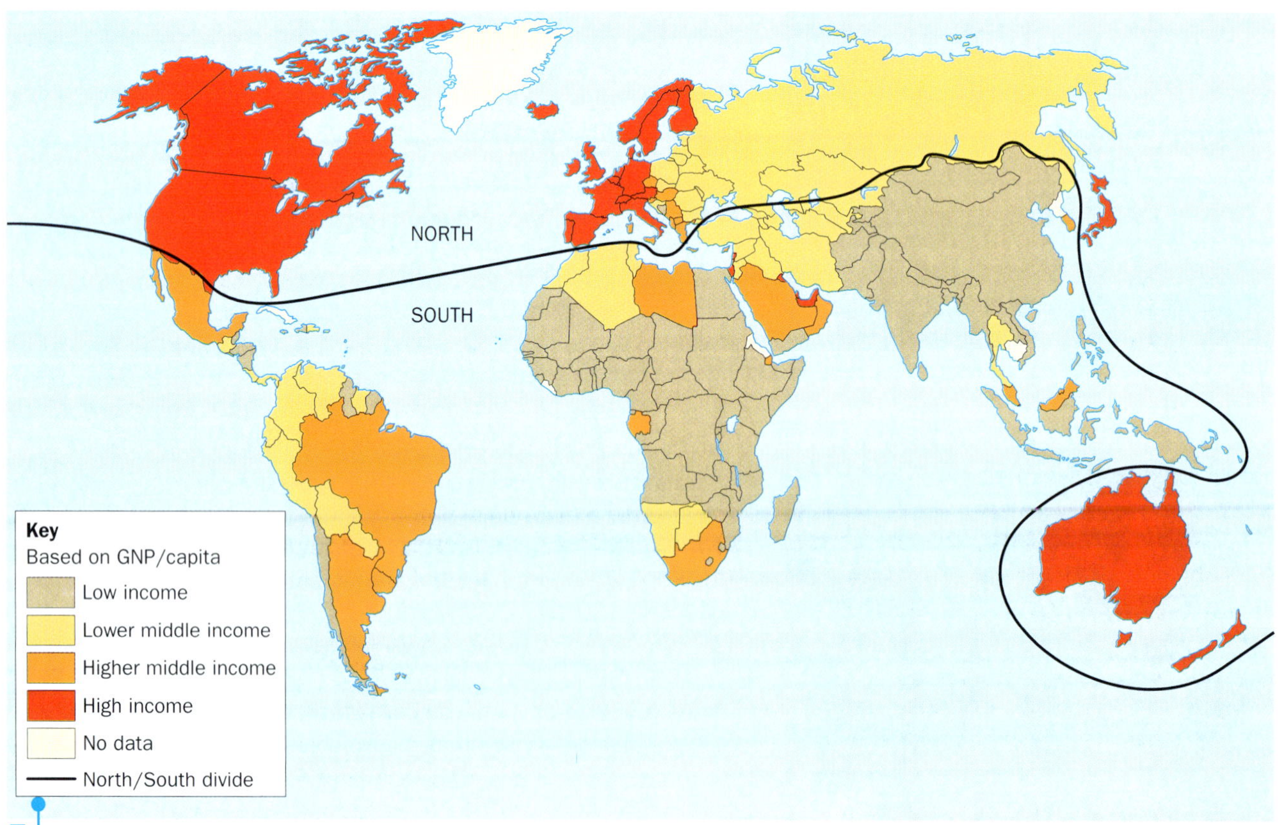

7 Global inequality in 1999

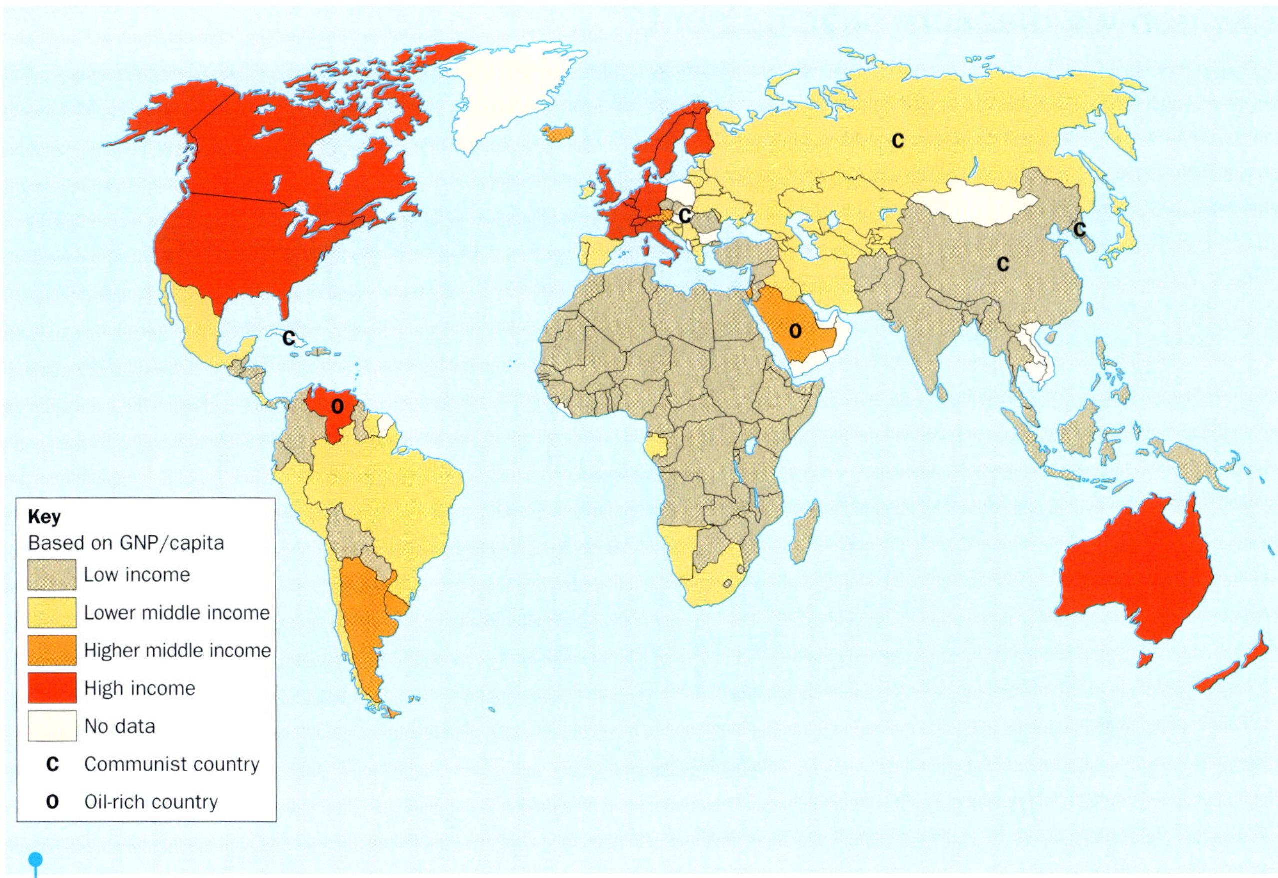

8 Global inequality in 1960

★ **FACTFILE**

Five types of economy

- **Rich industrialised economies** *e.g. Germany*
  These were the first countries to industrialise in the 19th and early 20th centuries. This gave them an advantage over the rest of the world since they were able to make more money from selling manufactured goods than by selling primary goods. Many of them have now shifted from manufacturing (secondary activity) to new service industries (tertiary activity), while manufacturing has relocated in newly industrialising countries. GNP/capita in these countries remains high.
- **Oil exporting economies** *e.g. Saudi Arabia*
  During the latter part of the 20th century the world economy became increasingly dependent on oil. While the price of most primary products was falling, the price of oil increased. The few countries that are rich in oil were in a strong position to negotiate with the rest of the world. GNP/capita in these countries is also high.
- **Newly industrialising economies** *e.g. Malaysia*
  Some poorer countries have been able to break out of dependence on primary products and low incomes through rapid industrialisation and economic growth. Since the 1960s their governments have encouraged local industry and foreign investment. Many companies from the rich industrialised countries were attracted by cheaper labour and so began to manufacture goods in these countries. GNP/capita has risen quickly.
- **Former centrally-planned economies** *e.g. Russia*
  For much of the 20th century some countries were ruled by communist governments. Industry was under government control rather than in the hands of private companies. Since 1989, and the decline of communism, most of these countries have re-entered the global economy. For some, this has been followed by a difficult period of readjustment. GNP/capita has remained average or even fallen as their industry struggles to modernise and compete in the global market.
- **Heavily indebted poorer economies** *e.g. Jamaica*
  These countries have been unable to industrialise and still depend mainly on the export of primary products for their income. Most primary products (apart from oil) have fallen in value on the world market, so the countries have become poorer. Many borrowed money from international banks or richer governments to help them to develop, but now have to repay large debts. They are trapped in a cycle of poverty and debt. GNP/capita in these countries was low and has fallen.

How can we measure development?

In general, people in the richer MEDCs enjoy a better quality of life than those in poorer LEDCs. However, a country's wealth does not paint the whole picture. To judge how developed a country really is you need to look at other factors, too. Geographers use a variety of DEVELOPMENT INDICATORS, of which wealth is just one, to compare the level of development in different countries.

9 New York in the USA, an MEDC

10 Farming in rural India, an LEDC

Task

1 Work with a partner.
Compare Photos 9 and 10. Imagine that these photos are the *only* information that you have about the USA and India. What do the photos tell you about the level of development in each country?

In which of the two places shown in the photos do you think that people are more likely to:

- Have more *wealth*?
- Enjoy better *health*?
- Have better *jobs*?
- Take more care of the *environment*?
- Have greater *equality*?
- Receive a better *education*?
- Consume more *energy*?
- Enjoy greater *freedom*?
- Live in *peace*?

In each case, try to give evidence from the photos to support your opinion.

How can countries develop?

People disagree as to which development indicators they think are the most important. This has led to different attitudes to development. For example, in countries that think wealth is the most important indicator of development, leaders have followed different policies from countries where leaders think that equality is most important. So how a country develops is related to the political IDEOLOGY of its government. An ideology is a set of beliefs and values that determines the political and economic system in the country.

The most widely-held ideology is the capitalist free market view (see Figure 11): for a country to develop, it must have ECONOMIC GROWTH. Each year it produces more goods and services and is able to trade these around the world. Most MEDCs have developed in this way, but it is not the only way that a country can develop. Here are the three main views.

Capitalist free market view
A country can only develop through economic growth. It creates wealth by producing goods and services through its industry and exporting these to other countries. This can happen most easily within a FREE MARKET where countries can trade easily with each other without government interference. Most free market economies have elected democratic governments that cannot dictate to people, or to companies, how they should behave.

Socialist planned economy view
A country can only develop by curbing the market and planning or controlling the economy for the good of all. Development is not just about economic growth. It should also be about how wealth is distributed, or shared, among the population and how it is used. When the free market is left to work uncontrolled, the rich get richer and the poor get poorer. Governments need to intervene to ensure that wealth is used to benefit everybody by providing decent homes, health and education. They should plan their economy rather than leave it to the free market.

Sustainable development view
A country can only develop as far as natural resources will allow it, or else the global environment will be harmed. Development that is based simply on economic growth – whether it is planned or in the free market – is not sustainable. Sustainable development does not destroy the natural environment or deplete the world's resources. Both capitalist and socialist economies have been guilty of damaging the environment. The worst damage is caused by the richest countries with the highest levels of consumption.

11 Three ideological views on development

Tasks

SAVE AS . . .

2 Read the list of development indicators in the box on the right.

a) Write a sentence to explain the meaning of each indicator.

b) Match each indicator with one of the *italic* words in Task 1. For example, 'Adult literacy' and 'Children enrolled in primary school' both indicate *education*. Which words have no indicators to match them? How could these be measured?

- Access to drinking water
- Adult literacy
- Car ownership
- Carbon emissions
- Child malnutrition
- Children enrolled in primary school
- Employment in agriculture
- Forest loss
- Fuel consumption
- GNP/capita
- Infant mortality
- International debt
- Life expectancy
- People per doctor
- Population growth
- Telephone ownership
- Unemployment
- Urban population
- Women in the workforce

3 Work in a group of three. Each person should represent one of the three ideologies in Figure 11. How would you measure development according to your ideology? Choose the six most important development indicators from the box. Explain why you chose each one.

Development indicators	Brazil	China	Colombia	Germany	India	Jamaica	Kenya	Malaysia	Russian Federation	South Africa	UK	USA
Access to drinking water (% of population)	69	83	75	100	85	93	45	89	n.a.	59	100	90
Adult literacy (% of people 15 and above)	83	82	91	99	51	85	78	84	99	82	99	99
Car ownership (number of cars per 1,000 people)	79	8	38	528	7	50	13	152	158	134	399	767
Carbon emissions (tonnes per capita)	1.7	2.8	1.7	10.5	1.1	4.0	0.2	5.6	10.7	7.3	9.5	20.0
Child malnutrition (% of children under 5)	7	16	8	0	66	10	23	23	3	9	0	0
Children enrolled in primary school (% of 5–11-year-olds in school)	90	99	85	100	n.a.	100	n.a.	91	100	96	100	96
Employment in agriculture (% of total workforce)	23	72	27	4	64	25	80	27	14	14	2	3
Forest loss (% annual deforestation)	0.5	0.1	0.5	0	0	7.2	0.3	2.4	0	0.2	−0.5	−0.3
Fuel consumption (kg oil equivalent per person)	1,012	902	799	4,267	672	1,465	476	1,950	4,169	2,482	3,992	8,051
GNP/capita (US$ per person)	4,790	860	2,180	28,280	370	1,550	340	4,530	2,680	3,210	20,870	29,080
Infant mortality (per 1,000 live births)	34	32	24	5	71	12	74	11	17	48	6	7
International debt (US$ per person)	1,180	119	794	n.a.	98	1,565	223	2,146	854	615	n.a.	n.a.
Life expectancy (years)	67	70	70	77	63	75	52	72	67	65	77	76
People per doctor	729	642	1,124	365	2,494	2,157	7,358	2,847	215	1,597	623	408
Population growth (% growth per year)	1.4	1.1	1.9	0.5	1.8	0.9	2.8	2.5	–0.1	2.0	0.4	1.0
Telephone ownership (per 1,000 people)	107	56	148	550	19	140	8	195	183	107	540	644
Urban population (% of total population)	79	31	73	87	27	54	30	54	76	50	89	76
Women in the workforce (% of total workforce)	35	45	38	42	32	46	46	37	49	38	44	46

12 Development indicators for 12 countries

Task

ICT

1 Your task is to use the data in Table 12 to compare levels of development in different countries. You could continue to work in the same group of three students as you did for Task 3 on page 337. Each person could use the six development indicators that they chose for that task.

You can change the table below to a spreadsheet on a computer to do this task. This will also help when you do the investigation on page 340.

a) Decide which countries you want to compare. You don't need to compare all 12. For example, you could choose the four countries that are featured in case studies in this book – Germany, Jamaica, Malaysia and the UK. Write their names on a large copy of the table below, in the left column. List the six development indicators you chose, in order of importance, across the top of the table.

You can update the data in Table 12 by visiting the World Bank website:

www.worldbank.org

b) Complete your table. Give the countries scores for each of the indicators you have chosen. Give maximum points to the country that does best on that indicator (4 points if you compare four countries) down to one point for the country that does worst. For example, on GNP/capita, Germany scores 4 points, the UK 3, Malaysia 2, Jamaica 1. Write the scores in the 'R' columns.

Country	**Development indicators**												**Total development score**
	1		2		3		4		5		6		
	R	W (x6)	R	W (x5)	R	W (x4)	R	W (x3)	R	W (x2)	R	W (x1)	
Germany													
Jamaica													
Malaysia													
UK													

R = raw score, W = weighted score

Now give a weighted score for each indicator in the 'W' columns to show importance. For example, in the first column the most important indicator should be multiplied by six. Add up the weighted scores for each country to find the total development score.

c) Compare development scores. How can you explain the similarities or the differences? Are they what you expected? Write a paragraph to explain your results.

Comparing levels of development

Until 1990, GNP/capita was used as the standard way of comparing levels of development between countries. But clearly, on its own, GNP/capita is an inadequate measure of development. It does not take into account the way that wealth is shared, or the other factors that influence development. In many LEDCs, GNP does not even provide an accurate measure of the wealth in a country. The subsistence activities that support many of the people in these countries do not contribute to the country's GNP.

In 1990 the UN came up with the HUMAN DEVELOPMENT INDEX (HDI) to measure development. The HDI combines three important development indicators:

- Income, measured by GDP/capita, adjusted to take into account what the money can actually buy in each country – this is known as purchasing power parity
- Knowledge, measured by adult literacy and the proportion of children attending school
- Health, measured by life expectancy.

HDI is calculated from a country's average score for each of these three indicators. For each country, the HDI is given a value between 0 and 1 – the closer to 1 the score is, the more developed the country is. The UN calculates the HDI for every country each year to compare rates of progress.

★ THINK!

Development is one of those issues in Geography about which there are many opinions. Each person in your group probably worked out a different development score for each country. None of them is wrong! They are each based on different development indicators which, in turn, were chosen from different ideological viewpoints.

It is important to recognise different viewpoints when you answer issue-based questions in Geography. You can use phrases like 'some people think . . .', 'while others believe . . .', 'alternatively . . .'. This helps to show that you realise that there is more than one viewpoint. Don't be afraid to express your own view, but remember that it is not the only one. In Geography there is frequently no one right answer. In exams most marks will be given for presenting a range of opinions, not for choosing the right one.

What is the link between wealth and development?

Tasks

What is the correlation – or link – between wealth and the other indicators of development?

1 Look back at Table 12 on page 338. Working on your own or with a partner, choose one of the indicators in the table to compare with GNP/capita (wealth). As a class, make sure that all of the indicators are allocated.

Rank all the countries in the table according to their GNP/capita: the country with the highest GNP/capita should be ranked 1 and the lowest, 12. Now do the same for your chosen indicator, ranking the country that performs best as 1 and the country that performs worst, 12. Plot the results on a scatter graph, like the one below. This one shows what happens if you plot life expectancy against GNP/capita. Can you see any pattern?

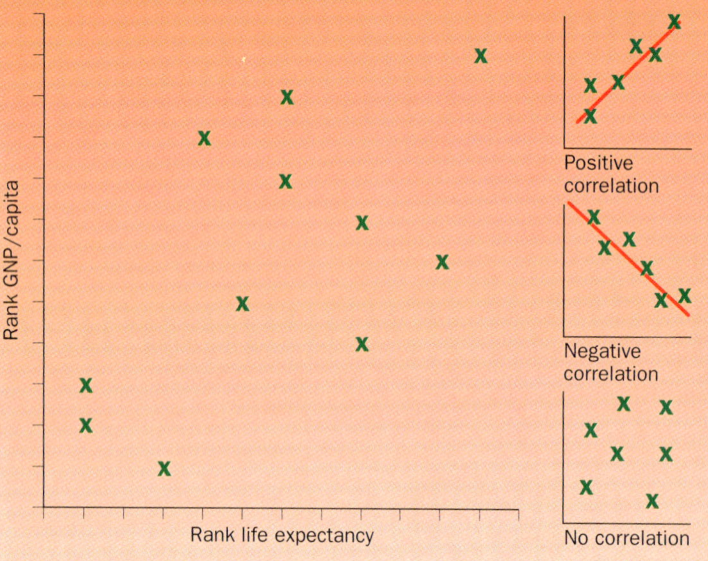

2 Interpret the graph that you have drawn:

- If the points that you plot on your scatter graph form a linear pattern then there is a correlation between wealth and the indicator you chose. The closer that the points come to an imaginary straight line, the better the correlation – or the closer the link.
- If the line slopes *up* there is *positive correlation*. This means that as one indicator increases, so does the other.
- If the line slopes *down* there is *negative correlation*. This means that as one indicator increases the other one decreases.
- If there is no obvious pattern on the graph and the points are spread randomly, there is *no correlation*, or link.

3 Share your results with other members of the class. Your teacher may need to photocopy your graphs to help you to do this.

Which indicators show a correlation with wealth? In each case, is this a positive correlation or a negative correlation? How can you explain the links?

Using the results from the whole class, write a conclusion to explain what the investigation showed.

Your investigation should show that the correlation between GNP/capita and other development indicators can vary. There is strong correlation between wealth and indicators like energy consumption, whereas the correlation between wealth and life expectancy is weaker. This suggests that for some countries, improving health has a low priority as their wealth increases. On the other hand, some poorer countries have managed to increase life expectancy faster than many richer countries.

Compare Map 14, showing HDI for countries around the world, with Map 7 on page 334. You will notice that the pattern is similar. Countries that are wealthy often have a higher HDI than poorer countries. However, if you look closely you will notice some differences. South Africa, a country that falls in the higher middle income group for GNP/capita, is in the medium low category for HDI. Its wealth has not been translated into development for all its people. Countries like this can be called 'underachievers'. In contrast, Russia, which is in the lower middle income group for GNP/capita, has a medium high HDI score. Even though its wealth is more limited, it still has a higher level of development. Countries like this can be called 'overachievers'.

13 South Africa has a fairly high GNP/capita but its wealth is not shared equally. Most of the black population, which make up 76 per cent of the total population in South Africa, remain poor, and have lower life expectancy and literacy rates than the rest of the population

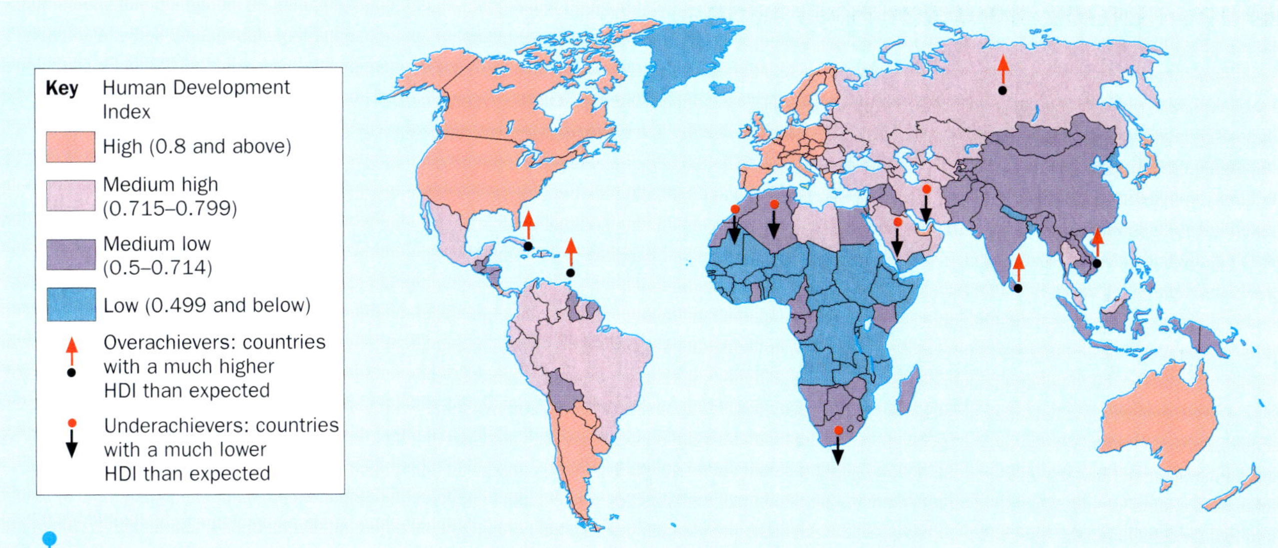

14 Human development around the world

	GNP/capita (US$)	World rank	Human Development Index	World rank
USA	28,740	8	0.943	4
Germany	28,260	9	0.925	19
UK	20,710	18	0.932	14
Brazil	4,720	37	0.809	62
Malaysia	4,680	38	0.834	60
South Africa	3,400	49	0.717	89
Russian Federation	2,740	58	0.769	72
Colombia	2,280	65	0.850	53
Jamaica	1,560	79	0.753	84
China	860	104	0.650	106
India	390	128	0.451	139
Kenya	330	135	0.463	137

15 GNP/capita and HDI compared

SAVE AS . . .

4 Compare Map 14 with Map 7 on page 334.

a) Explain why the patterns on the two maps are similar.

b) Name five countries that are underachieving (where the rank for HDI is lower than for GNP/capita) and five countries that are overachieving (where the rank for HDI is higher than for GNP/capita).

Focus task

5 You are going to write a development report about an LEDC.

Look at Table 15. Choose one country where the rank position for both GNP/capita and HDI is low (over 70). Do some background research on this country. You could begin by looking at Table 12 on page 338.

For which development indicators does the country do well, and for which does it do badly? Try to explain why. You could use these websites to find out more about how the country is developing:

World Bank: www.worldbank.org
United Nations Development Programme: www.undp.org
US Central Intelligence Agency: www.odci.gov

Write your own development report for the country. Recommend what priorities it should have in order to improve its HDI.

Why is global inequality growing?

Global inequalities are no accident of history. They have developed over hundreds of years as a result of unfair patterns of trade. The countries of Europe, North America and east Asia, with just 20 per cent of the world's population, have 85 per cent of international trade. The vast majority of the world, living in the rest of Asia, Africa and South America, have just 15 per cent. You can see this pattern on Map 16.

This imbalance goes further than simply the way that trade is shared. Most MEDCs export manufactured goods and services with a high value, while LEDCs export mainly primary products that have a low value. What's more, the value of primary products has been falling in relation to other exports since the 1960s. LEDCs have to produce more and more of the same products just to keep up with the rest of the world. They lack both the technology and the investment that would be needed to switch to the production of manufactured goods.

A growing proportion of international trade is now controlled by large transnational corporations (TNCs), all of them based in MEDCs. Examples include Ford, Nike, Sony and McDonald's. The ten largest TNCs now have a combined income greater than that of the world's 100 poorest countries!

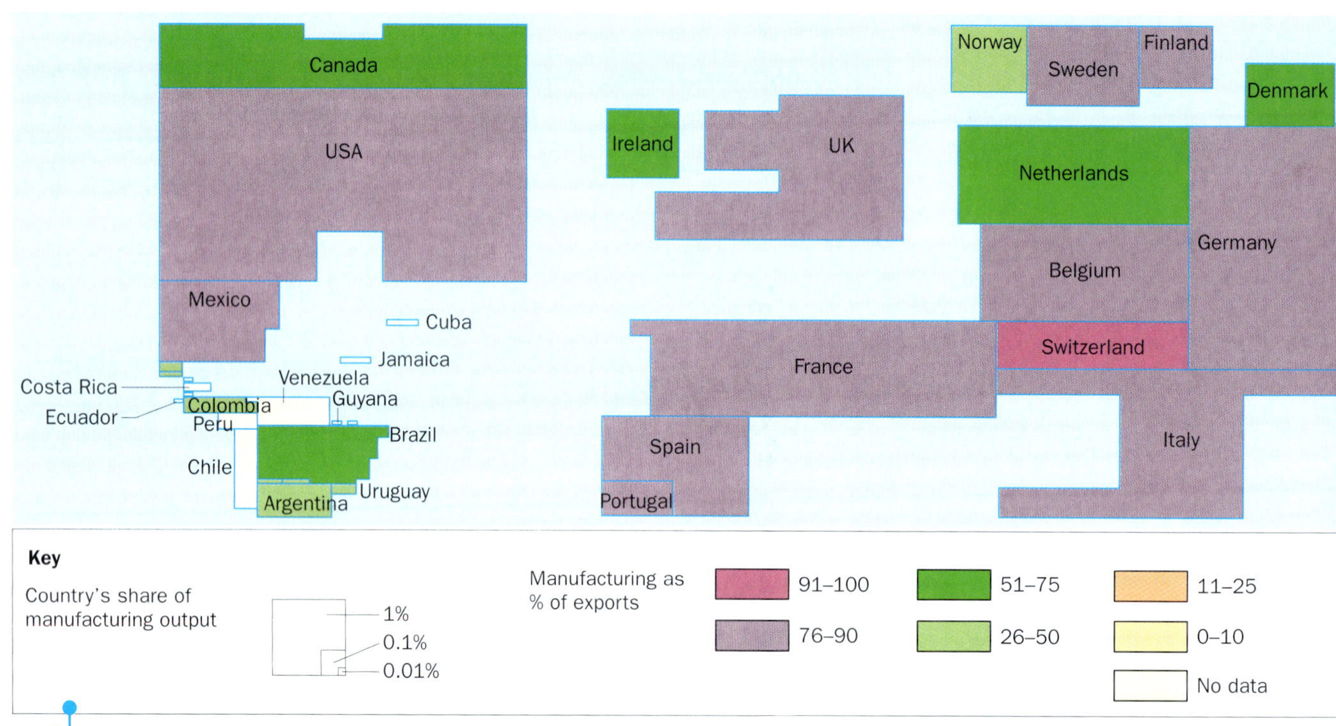

16 A world trade map adapted from *The State of the World Atlas*, 6th edition

★ FACTFILE

Benefits and problems that transnational corporations (TNCs) can bring to an LEDC

- TNCs bring new investment into a country's economy.
- TNCs provide jobs, often at higher wage levels than the local average.
- TNCs frequently import their own inputs rather than use local products.
- TNCs can damage the local environment and deplete resources.
- TNCs provide management skills that may be lacking.
- TNCs export profits, draining wealth from the country's economy.
- TNCs usually provide low-skilled jobs in LEDCs and retain skilled jobs in MEDCs.
- TNCs provide research and development that can assist local development.
- TNCs are mobile and can leave a country as quickly as they arrived.
- TNCs are powerful and can interfere in a country's political processes.
- TNCs' international links can gain a country access to world markets.
- TNCs can weaken workers' rights by looking for countries with cheaper labour.

Nike is a well-known US-based transnational corporation that specialises in manufacturing sports shoes and other sportswear. The company operates in many countries around the world and there are few countries where its products cannot be bought. The company was set up in Beaverton, Oregon USA in the 1970s by a couple of college graduates who had an idea for a new sports shoe. The company headquarters is still in Beaverton where most of its US employees work.

By 1998 the company had a market value of US$8 billion. Its annual sales were $6,135 million and its profit was $491 million.

Today, Nike directly employs around 20,000 people worldwide. Most of these people are employed in product design, marketing and administration, working in the USA. However, there are probably another 500,000 people around the world who are employed in factories making Nike products. Most of them work in Asia for companies to which Nike subcontracts most of its manufacturing. There are factories making Nike products in Taiwan, South Korea, Hong Kong, China, Indonesia, Vietnam, the Philippines and Bangladesh. Most employees are semi-skilled and unskilled assembly-line workers. They earn a small fraction of the wages of a factory worker in an MEDC. Some people argue that TNCs like Nike exploit the cheap labour in these countries. However, Nike argues that it is helping to provide jobs that are often well above the local wage rate.

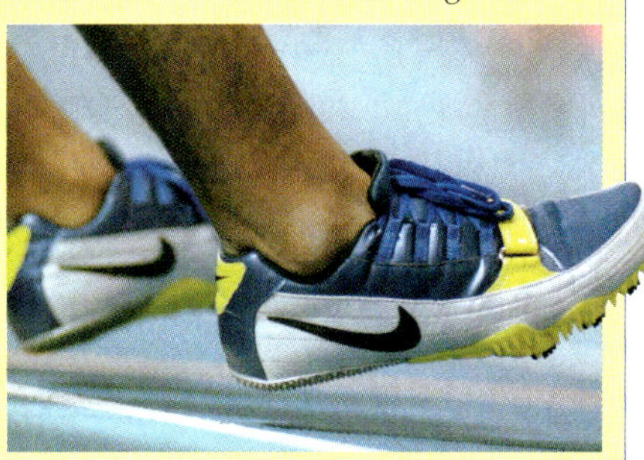

17 Nike – a major TNC

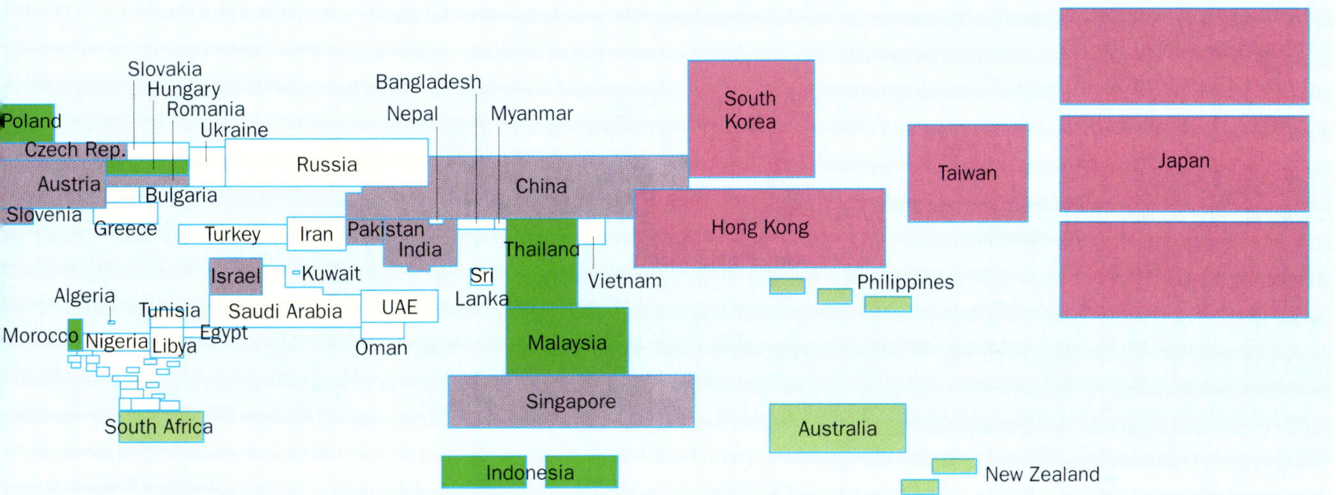

Tasks

1 Look at Map 16.

a) List the ten countries with the largest share of world trade. In each case, estimate the percentage of world trade, with the help of the key.

b) Compare the map with a similar map of population, which your teacher will give you. Name ten countries that have a smaller share of world trade than you would expect from their population.

2 Read the information in the Factfile.

Sort the statements into *benefits* and *problems* that TNCs bring to LEDCs. Write two lists. On balance, would you say that TNCs have a positive or negative impact on the economies of LEDCs?

3 Study the information in Figure 17.

a) Look back at Photos 1–6 on pages 332–33. Now that you know more about TNCs, describe what each photo shows you about the way a company like Nike operates.

b) What benefits and/or problems do you think TNCs like Nike bring to LEDCs and MEDCs? You could present your ideas in a table.

c) Have your opinions for Task 3 on page 333 changed? If so, how?

How can inequalities be reduced?

One way in which MEDCs can help LEDCs to break out of poverty and to develop is by providing AID. We often hear about aid going to LEDCs when disasters, such as floods or famine, strike. But we are not as generous as you might think. Only a tiny fraction of the UK's annual budget is spent on aid, and we are not alone. The United Nations has laid down a target of 0.7 per cent of the GNP of MEDCs to be spent on aid. Most countries come nowhere near this modest target (see Table 18).

But even where aid has been given, it does not always have the desired effect. Most of the poorest countries in the world are poorer now than they were 40 years ago, despite having received millions of dollars of aid. So why has all this aid been so ineffective?

Over the years there has been a change in the way that aid money is spent, away from health and education – two of the most important aspects of development – towards emergency relief and debt repayment. Much of the aid that is given is in the form of tied aid – that is, aid given on condition that it is spent on goods and services from the donor country. This reduces the control that the recipient country has over the way that aid is spent. Even worse, aid may be in the form of loans from MEDC governments or international banks, which have to be repaid over a number of years, often with interest. This has been one of the reasons for many LEDCs falling increasingly into debt.

	Aid as a percentage of GNP (%)
Denmark	1.04
Norway	0.85
Sweden	0.84
Netherlands	0.81
France	0.48
Luxembourg	0.44
Belgium	0.34
Finland	0.34
Switzerland	0.34
Germany	0.33
Canada	0.32
Ireland	0.31
Australia	0.30
UK	0.27
Austria	0.24
Spain	0.22
New Zealand	0.21
Portugal	0.21
Italy	0.20
Japan	0.20
USA	0.12

Source: UNDP Human Development Report 1998

18 Aid from the richest MEDCs

19 Some aid mistakes

The main obstacle to development in most LEDCs is huge INTERNATIONAL DEBT. Ironically, this has arisen as a result of loans provided by MEDC governments and banks during the 1970s and 1980s to bring about economic development. The total debt of all of the HEAVILY-INDEBTED POORER COUNTRIES (HIPCs) combined now runs into US$ trillions (million millions) and is still growing. It works out at a debt of $400 for every person in countries where average income is less than one dollar a day.

Of course, this amount of debt will be impossible to repay, but that has not stopped organisations like the World Bank and the International Monetary Fund (IMF), who lent part of the money, from trying to recover it. During the 1990s they insisted that poor countries adopt very strict economic policies to enable them to repay the debts. This usually meant that government spending had to be cut by closing schools and hospitals, or by charging people to use them. It also meant selling more exports so that the country could earn more money, often by growing cash crops (see page 184) at the expense of growing crops for the local population. These policies have led to great hardship for the poorest people in LEDCs but no reduction in the debt mountain, which continues to grow. Many people believe that the only solution to the problems of HIPCs is for richer countries, along with the World Bank and the IMF, to cancel the debt.

20 Food aid being distributed in Nepal. MEDCs can provide aid to overcome food shortage in a crisis, but often such crises occur because the country is in debt

Task

1 Study all of the information on these two pages. Which strategy – giving further aid, or cancelling debt – do you think is the best way to reduce global inequalities?

Write a letter to the UK government to argue *either* for more aid *or* for debt cancellation.

Use as much relevant information as you can to support your argument.

Focus task

2 Look back at the Focus task that you did on page 341. You wrote a development report for an LEDC to recommend priorities for future development to improve its Human Development Index.

Write a second part for this report, again about 500 words, to recommend how the country might develop.

Use the work you have done on pages 342–35 to help you. You should consider how **a)** trade **b)** aid **c)** debt cancellation would help the country to achieve its priorities for development. Suggest what sort of trade and aid should be encouraged, and whether TNCs should be involved.

	GNP/capita (US$)	Debt/person (US$)	Debt as % of GNP
Mozambique	89	336	376
Nicaragua	372	1,318	354
Somalia	92	290	315
Sudan	221	636	288
Congo	817	2,278	279
Guyana	829	2,039	246
Zambia	350	757	216
Congo, Dem. Rep.	138	292	212
Côte d'Ivoire	662	1,332	201
Ethiopia	105	178	168
Tanzania	189	245	130
Sierra Leone	205	259	126

21 Debt in the poorest HIPCs (data from Jubilee 2000)

7.3 *Jamaica – a case study of development*

Wish you were here?

The Southgate family went to Jamaica for a holiday. They stayed at an ALL-INCLUSIVE RESORT near Ocho Rios on the north coast of the island. It was the best family holiday they'd ever had. They're hoping to go again next year. But what did they find out about Jamaica while they were there? When they returned, Rebecca and Tom drew a mental map to show what they had learned about Jamaica (Map 1).

The people are so friendly. They must like tourists!
Why would anyone want to leave Jamaica to live in the UK?
This has got to beat camping in Cornwall. It never rains here!

DO YOU KNOW?

- Jamaica and the other Caribbean islands were inhabited long before they were 'discovered' by Christopher Columbus.
- Jamaica was ruled by the UK as a COLONY until it became independent in 1962.
- The population of Jamaica is 2.5 million. A similar number of Jamaicans live outside the country – mainly in the UK, the USA and Canada.
- These days, Jamaica has more links with the USA than with the UK. Americans sometimes refer to the Caribbean as their 'backyard'.

★ THINK!

Mental maps are maps that you draw using the information you have in your head. They are a useful way to show what you know about a place. They are also a good way to revise. Draw maps of some of the countries and case studies that you have done in your GCSE course to see what you remember. When you have completed this chapter on Jamaica, draw your own mental map of the country. With luck, you will do a bit better than Rebecca and Tom!

Montego Bay - we landed at the airport here. It's a popular resort.

Ocho Rios - the resort where we stayed. It's nearly all hotels.

The rest of Jamaica - they grow lots of sugar and bananas and the people are poor.

The Blue Mountains - where they produce Blue Mountain coffee

Kingston - Jamaica's capital. We were told not to go there because they say there's a lot of crime.

1 Mental map of Jamaica

AIMS

- **To consider whether or not Jamaica should be described as an LEDC.**
- **To understand how Jamaica's links with other countries influence its development.**
- **To assess the impact of transnational corporations and international debt on Jamaica.**
- **To suggest future development strategies for Jamaica.**

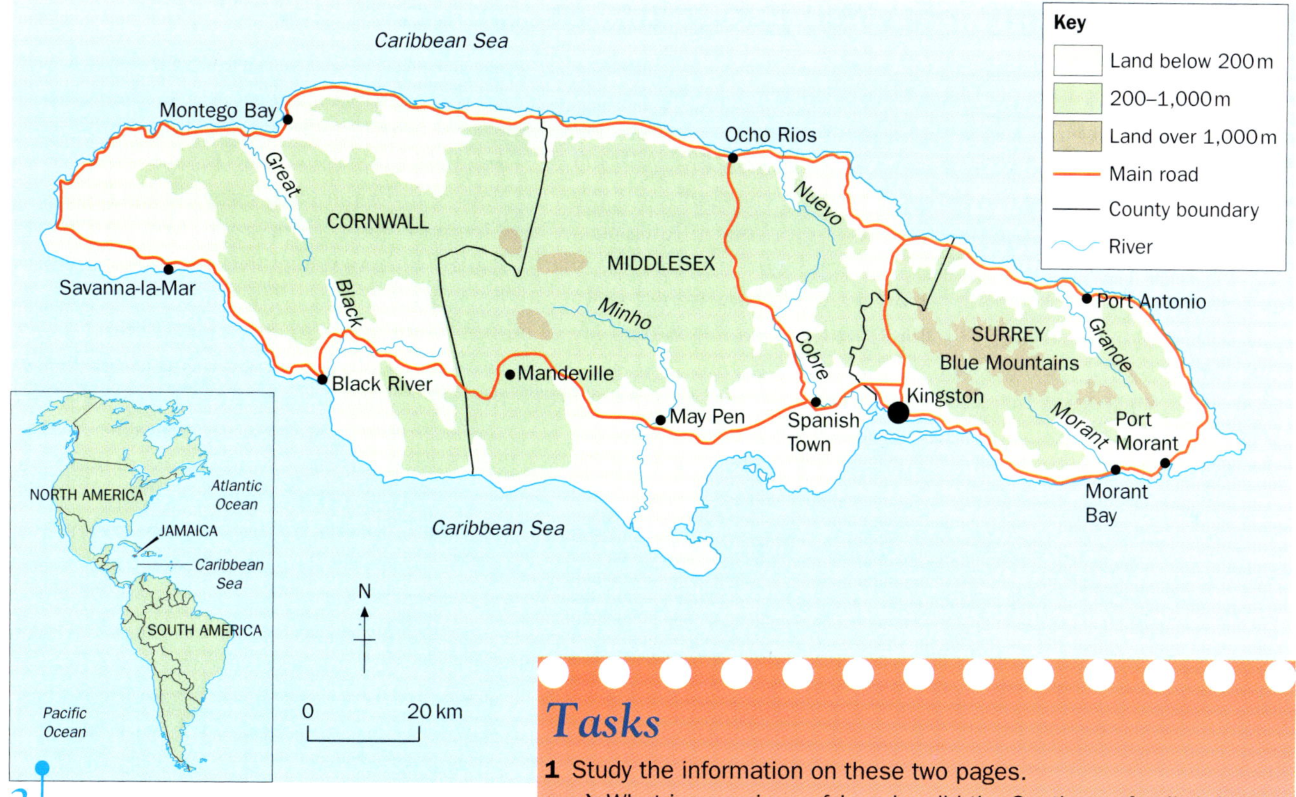

2 Jamaica

	UK	Jamaica
Area	241,600 km^2	10,800 km^2
Population	59,009,000	2,554,000
GNP/capita	US$ 20,870	US$ 1,550
Life expectancy	77	75
Adult literacy	99%	84%
Urban population	89%	54%

Tasks

1 Study the information on these two pages.

a) What impressions of Jamaica did the Southgate family bring back from their holiday?

b) How accurate do you think these impressions are? What evidence can you give?

c) What do *you* know about Jamaica? Write a paragraph to describe your own ideas. Keep your paragraph and come back to it when you have finished this chapter. Note how your ideas change.

2 Think of a place (at home or abroad) that you have been to for a holiday. Draw a mental map of the country that it is in, without the help of an atlas. Annotate your map with information. How accurate do you think your ideas are?

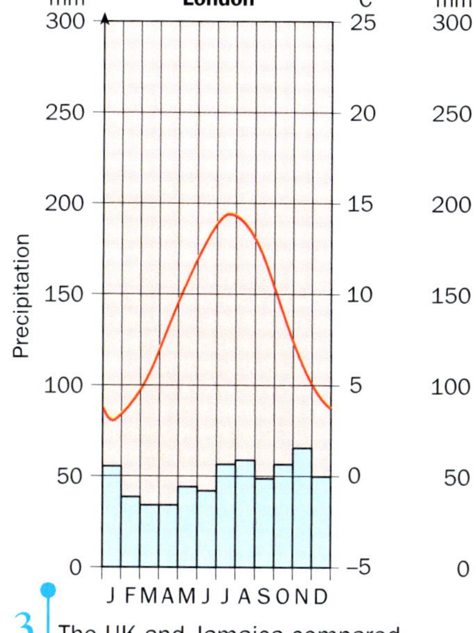

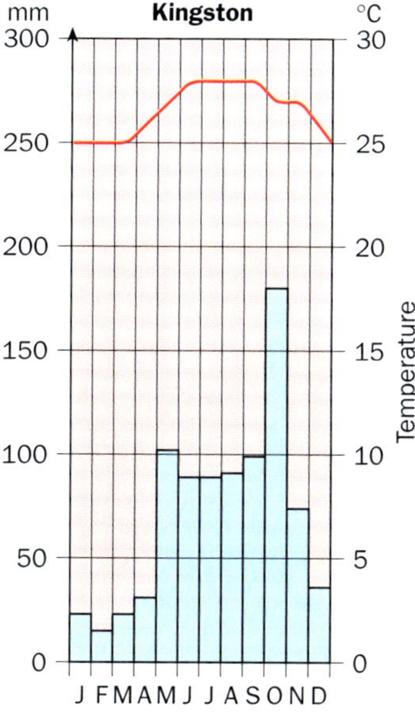

3 The UK and Jamaica compared

Focus task

3 Jamaica is sometimes described as an LEDC. How true is this? As you go through this chapter, collect evidence to support, or refute, this idea. You can start with this page.

Towards the end of the chapter (page 364), having considered all the evidence, you will decide whether you think Jamaica is an LEDC or not. If you don't think it is an LEDC, you will think of a more accurate way to describe it, using the evidence that you have collected to support your opinion.

How does history influence geography in Jamaica?

When Christopher Columbus landed on the north coast of Jamaica in May 1492 it was not so much a discovery as an accident! He had sailed west in search of India but, instead, he stumbled upon America. This is why the Caribbean is also known as the 'West Indies'. The islands were already inhabited by two groups of people – the Caribs and the Arawaks. The Arawaks gave the island its original name: Xaymaca.

In 1492 the period of European expansion began. Explorers from the countries of Europe set sail around the world in search of new wealth. The Spanish colonised Jamaica in 1510, and within a few years the native people had been wiped out by guns and disease. The colonists began to grow new crops, like sugar cane and bananas, on plantations. But Spanish domination was challenged by other European powers and in 1655 Britain took control of Jamaica.

4 The triangular trade between Europe, West Africa and the Caribbean

European countries needed labour to work on their Caribbean plantations. Their own people were unwilling to work in such harsh conditions, so slaves were brought from Africa. In Britain, ships from Bristol, Liverpool and London sailed to West Africa loaded with manufactured goods that could be traded for slaves. The slaves were piled on to the same ships to cross the Atlantic, chained together in terrible conditions below deck (see Source 5). Many did not survive the long voyage and died from disease or else committed suicide. Over a period of 200 years, until slavery was abolished, an estimated 5 million Africans were taken as slaves. It was the largest enforced migration of people in history.

In the Caribbean the slaves were sold to plantation owners and the ships returned to Europe laden with sugar, cotton and tobacco. This completed their round trip, which became known as the 'trade triangle' (see Map 4). It brought great wealth to plantation owners and slave traders, and laid the foundation for the Industrial Revolution in Europe.

★ THINK!

You may wonder what history is doing in a geography book. Pages 348–49 provide an important background to the rest of the chapter. The more links that you are able to make between the subjects that you study, the better you will be able to understand each one. There are many other examples of the way in which history helps us to understand geography, and vice versa. How many can you think of?

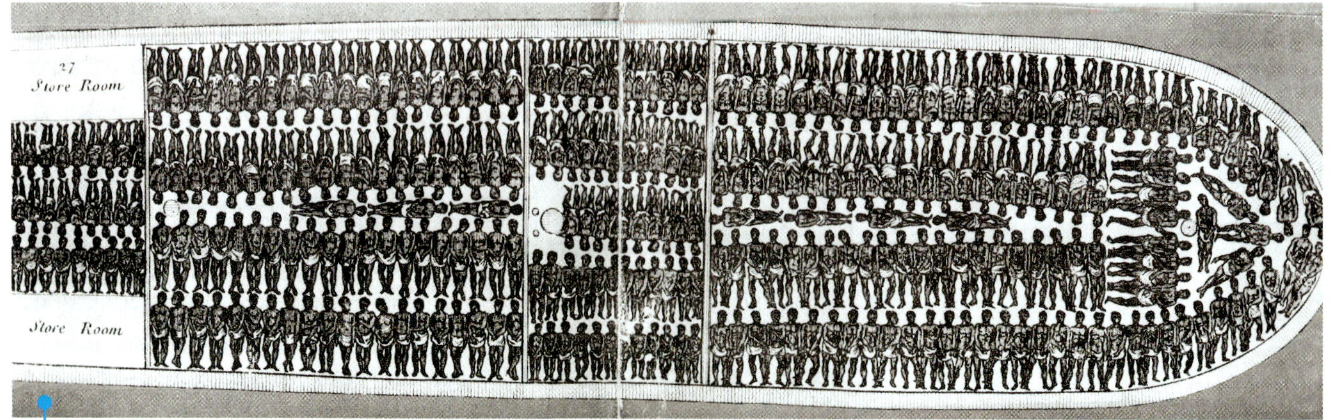

5 Plan view of a slave ship, 1790

6 People of Jamaica – the island's population today is made up of several different ethnic groups

For 200 years the Caribbean economy was based on slavery. Most slaves worked on plantations while others worked as domestic servants. Men, women and children were all expected to work. Not surprisingly, slaves resisted the conditions that were imposed on them and many tried to escape. In Jamaica a community of runaway slaves, known as the Maroons, lived in the mountains and from there they attacked their former owners. During the 18th century they periodically declared war on the British.

Plantation owners lived in fear of slave revolts and in Britain there was pressure to end slavery from those who thought it was immoral and unchristian. Slavery in Jamaica was eventually abolished in 1838. The plantation system began to decline and slaves had to be replaced by immigrant workers from Europe and, later, from Asia. Evidence of the people who came to the island can be found in Jamaican place names and the different ethnic groups who live there today (Photo 6). The national motto of Jamaica is 'Out of many, one people'.

Throughout the 20th century the importance of sugar and other primary products in the global economy declined, and the Jamaican economy declined with them. Colonialism over 400 years had left the countries of the Caribbean totally dependent on income from these products. The flow of immigrants to the Caribbean stopped and emigration began. Since 1945 there has been a huge exodus of Jamaicans to the UK, the USA and Canada to find work.

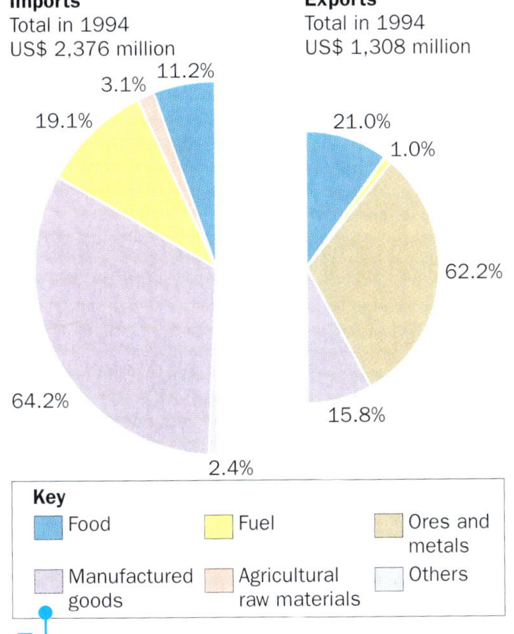

7 Jamaican imports and exports, 1994

Tasks

1 Study Map 4.

a) Using the map, describe the triangular trade.

b) Suggest what impact this trade would have had on:

- Africa
- Europe
- the Caribbean.

2 Look back at Map 2 on page 347, and at Photo 6. What do the place names on the map and the people in the photos tell you about the history of settlement in Jamaica? Explain your findings from the information on pages 348–49.

SAVE AS . . .

3 Look at Figure 7.

a) Describe what the pie charts show about Jamaica's imports and exports.

b) How does Jamaica's history explain this pattern of trade?

What is the future for farming in Jamaica?

Jamaica, like most other Caribbean countries, still relies on the export of primary products for a large part of its income. The pattern of trade established during the years of colonialism has been slow to change. Agriculture plays a vital role in the Jamaican economy, both as a way for the country to earn money from exports and as the main source of employment for almost half of the population.

Large plantations still contribute most to the country's agricultural exports, but it is small farms that provide food and work for the island. Most Jamaican farmers own small plots where they grow food for their family and produce cash crops which they sell to earn income. The pattern of farming is well illustrated by bananas: 75 per cent of the country's bananas are grown on large plantations on the fertile coastal plain, and the other 25 per cent by small-scale growers, often farming in the mountains.

8 Owen Smith, a Jamaican banana grower

I grow bananas on about 5 ha of land (the size of about ten football pitches) in the hills about 10 km from the coastal town of Port Antonio.

Bananas are ideally suited to conditions in the Caribbean – they like heat and plenty of rain. They're an all-year crop which provides farmers like me with a regular income. Most other crops are seasonal, so the income comes and goes. There is always a demand for bananas. Most of mine go for export to the UK, but I can also sell them locally.

Bananas are a hardy crop. They recover quickly from damage caused by the weather – which is just as well, because in recent years Jamaica has suffered more than its share of bad weather. In 1988 Hurricane Gilbert flattened most of the island's banana trees. Drought in the 1990s reduced our yields so that there were fewer bananas to sell. Then a storm in January 1998 produced unusually heavy rain and flooding. The bananas survived but much of the road to Port Antonio was washed away. For weeks we couldn't get our bananas to the boat. Even now, the road is still badly potholed so it takes an hour to drive there.

But it's not just the weather that's out of my control. Banana growers are affected by the world market, too. We used to be able to sell all our bananas to the UK without a problem, but now we face growing competition from other countries. Consumers always choose the bananas that look best. Jamaican bananas are good to eat but don't always look so good. When our bananas grow too short, or have any sort of blemish, consumers are less likely to buy them and we get a lower price.

When prices go down it's harder for me to survive. Right now I'm just breaking even. I employ eight people and I have to pay their wages (about J$300 or £5 a day). Then I have to buy pesticides, herbicides and fertiliser. If things get any worse I'll have to think of other ways to survive in future.

9 Bananas growing on Owen Smith's farm near Port Antonio

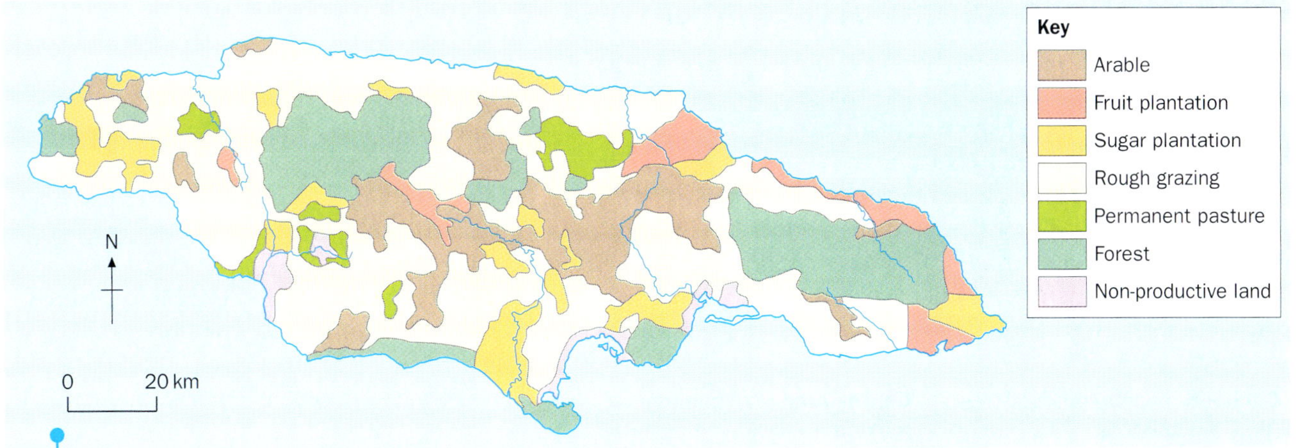

10 Agricultural land use in Jamaica

Intensify production
Increase the yield of bananas by using more chemical fertiliser, herbicide and pesticide. This will help to increase yields and to improve the appearance of the bananas. Large plantations use up to 30 kg of chemical/hectare to produce the best-quality bananas.

Diversify
Grow other crops in addition to bananas. This will protect farm income when the price of bananas falls. There is a developing export market for crops like mangoes, papayas and cut flowers. There is also an increasing market in MEDCs for crops that have been grown organically (without the use of chemicals).

Co-operate
Join together with other small-scale growers to form a co-operative. This benefits each grower by reducing costs through shared equipment and transport. Co-operatives can also help to sell the bananas by finding new markets and negotiating a better price.

Migrate
Move out of farming and find alternative work in the town or city. There are a growing number of jobs in the service sector in Jamaica, particularly in tourism.

11 Alternative strategies for banana growers in Jamaica

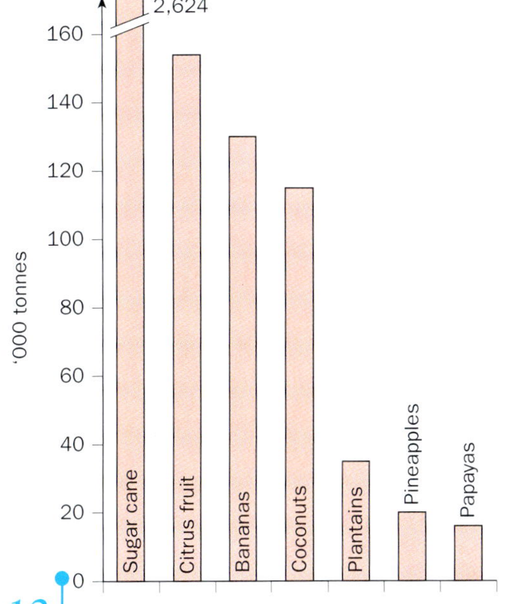

12 Production of export crops in Jamaica, 1996

Tasks

SAVE AS . . .

1 a) List the problems faced by small-scale banana growers in Jamaica.

b) Write notes on the problems that dependence on export crops could cause for the Jamaican economy.

2 Look at Figure 11. Consider the advantages and disadvantages of each strategy for small-scale banana growers like Owen Smith. How would each one help to solve his problems? Decide which would be his best option. Give reasons for your decision.

3 Compare the experience of Owen Smith with that of the British farmer Richard Betton on pages 258–60.

a) Which problems faced by the two farmers are similar? Which are different?

b) How might this affect the options that they would choose in order to survive as farmers?

13 Port Antonio harbour. Each week a boat sails to the UK with Jamaican bananas. Usually it leaves only half-full, so it calls in to ports in other countries for more bananas. Jamaican producers could earn more money if the boat was full

Does Fairtrade offer a better future?

From 1975, Caribbean bananas had a guaranteed market in the EU under the Lomé Convention. This was an agreement between European countries and their former colonies in Africa, the Caribbean and Pacific (ACP countries) to continue to import products from those countries even though their production costs might be higher than elsewhere and their products more expensive. The agreement came to an end in 2000 and is to be gradually phased out. It means that ACP countries will have to compete with other countries, many of which can produce the goods more cheaply.

Caribbean banana production (Euro-bananas) €	Central American banana production (Dollar-bananas) $
Land is hilly and mountainous – there is limited space	Large flat plains – there is plenty of space
Poor soil produces low yields, less than 20 tonnes/ha	Rich soil produces high yields, over 40 tonnes/ha
Hurricanes common	Hurricanes rare
Mainly small-scale independent producers	Mainly large-scale plantations owned by TNCs
Higher wages and social conditions (though not as good as Europe)	Lower wages and poor social conditions
High unit cost of inputs due to small scale of production	Low unit cost of inputs due to large scale of production
High shipping costs due to small volumes and more port calls	Low shipping costs due to large volumes and one-stop port calls

14 A comparison of banana production in the Caribbean and Central America

The World Trade Organization (WTO) sets the rules for international trade. In 1999 it ruled that it was unfair for the EU to continue to give preferential treatment to bananas from the Caribbean. It ordered EU countries to open their market to free competition. This will allow large American TNCs such as Dole and Del Monte, which produce bananas in neighbouring Central America, to sell more bananas in Europe. It could mean the end of banana production in countries like Jamaica where production costs are higher and growers are already struggling to survive.

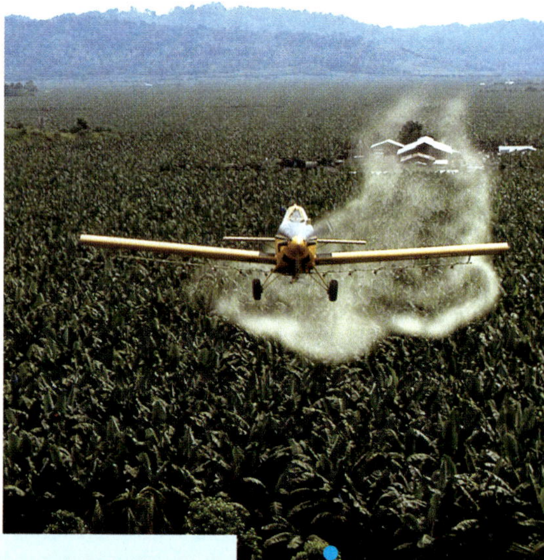

15 Spraying crops on a banana plantation in Central America

Key
Euro-bananas
Dollar-bananas
Atlantic Ocean
Gulf of Mexico
Caribbean Sea
Pacific Ocean
N
Trinidad & Tobago
Martinique
Guadeloupe
Dominican Rep.
Puerto Rico
Cuba
Haiti
Dominica
St Lucia
Jamaica
St Vincent
Grenada
Mexico
Belize
Honduras
Guatemala
El Salvador
Nicaragua
Costa Rica
Panama
Venezuela
Colombia
0 500 km

16 The Caribbean and Central America, showing countries producing Euro-bananas and Dollar-bananas

The Jamaican Banana Export Company (BECO) helps small-scale growers on the island to produce and sell their bananas. They must find new ways to market Jamaican bananas if they are to compete successfully with other countries. One way would be to obtain FAIRTRADE certification to show European consumers that more of the money they pay for their bananas goes to the grower. Most bananas now sold in the UK are produced by TNCs in Central America and do not have the Fairtrade certification. Normally, 10 per cent or less of the price that consumers pay goes to the grower (see Figure 19), but for fairly traded bananas it is double that. Some consumers in MEDCs are willing to pay more for their food if they know that their money goes to the people who produce it. In the UK, products that are fairly traded receive the Fairtrade mark from the Fairtrade organisation. This includes products such as tea, coffee, chocolate and bananas.

Fairtrade . . .

* Pays a fair price to producers, with a guaranteed minimum price.
* Works with co-operatives under the control of the producers themselves. They distribute income fairly and provide services for farmers.
* Brings the product directly to consumers and cuts out some of the intermediate costs. The benefits of trade are more likely to reach the producers.
* Encourages sustainable farming practices, particularly through the demand for organic products that do not use chemicals.
* Develops long-term trade based on trust and respect.

17 Benefits of Fairtrade for producers

18 Fairtrade bananas from a UK supermarket

19 Where the money for your bananas goes. Fairtrade bananas are more expensive but 20% of the money goes to growers

Tasks

1 Study the information in Table 14.

a) List the advantages that companies producing bananas in Central America have over banana growers in the Caribbean.

b) Would you rather be a banana grower in the Caribbean, or a worker on a Central American plantation? Explain why.

SAVE AS . . .

2 Compare Photo 15 with Photo 9 on page 350.

Draw sketches of the two photos. Annotate your sketches to show the differences between the two methods of banana production. Use all the information in Table 14.

3 Do some market research on Fairtrade products in UK supermarkets.

a) Produce a questionnaire about people's shopping habits and their views on fair trading.

b) Carry out interviews, using your questionnaire, with people you know.

c) Combine your results with those from the rest of your class. Do people in the UK support fair trade? Are they likely to change their shopping habits?

ICT

4 Imagine you work for BECO in Jamaica. You have been asked to devise a marketing strategy to sell Jamaican Fairtrade bananas in the UK. You need to consider:

- The benefits of fair trade for producers
- The reasons that UK consumers might prefer Caribbean bananas.
- The attitudes of UK consumers to fair trade (from Task 3).
- How you could persuade consumers to pay more for fairly traded bananas.

Prepare an information leaflet about Jamaican Fairtrade bananas for UK consumers. You could use a DTP program to help you.

Can TNCs help Jamaica?

When Jamaica gained independence from the UK in 1962, the government had to decide how the country was going to develop. Remember that after 450 years of colonial rule the country still had a mainly agricultural economy that depended on export crops for its income. Jamaica lacked the money that was needed to set up its own industries. However, it did have its own natural resources – in particular it had rich deposits of the mineral bauxite, used to produce aluminium, and it had a year-round warm climate and beautiful beaches that would attract tourists.

The government decided to invite foreign TNCs to invest in Jamaica to exploit these resources. For the TNCs, based mainly in the USA, it meant access to new resources and a supply of cheap labour. For Jamaica, it meant a share of the profits that the TNCs made and more jobs for its workforce. It was expected that this would make the economy grow and create wealth to invest in things like health and education. But the policy has not really worked as the government expected. Over the next four pages you will look at why this is so.

20 Part of an opencast bauxite mine near Mandeville. The area is mined by Alpart, a US-based TNC. Notice how close the mining is to people's homes

21 Bauxite mining and processing

Forest is cleared and the overlying soil is removed to expose the bauxite-rich rock beneath. The soil has to be replaced years later when mining is finished.

Bauxite is blasted loose and is excavated by huge power shovels. This creates clouds of red dust.

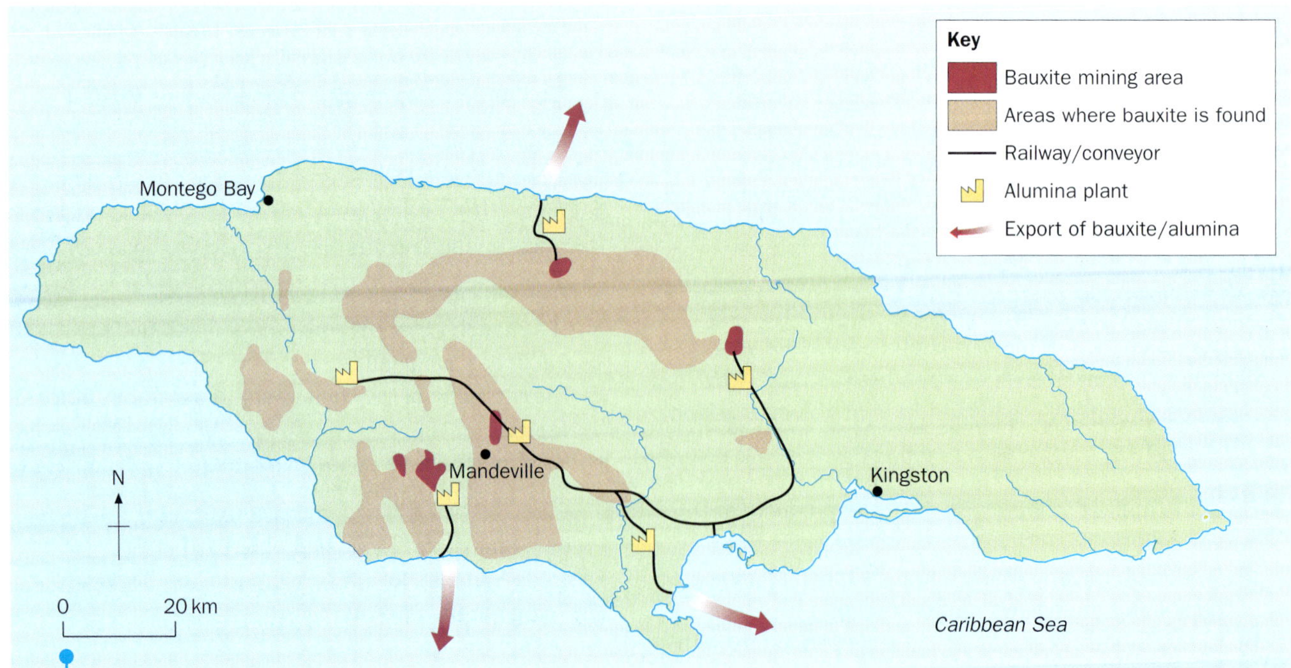

22 Bauxite mining in Jamaica

23 Norma Williams and her mother live near Mandeville in an area surrounded by bauxite mining

Alpart began mining bauxite in this area three years ago. There is hardly any part of our life that is not affected. The environment has changed beyond recognition. Trees have disappeared and hills have been pushed down. Farmers have had to move to new areas and it will be many years before the land can be used again. Huge trucks thunder past our house from 6 in the morning till 6 at night. They drop the red dirt as they pass and the dust covers everything. It gets in the house and ruins the washing on the line.

But worst of all the mining has brought on my asthma. It started two and a half years ago. I'd never had a problem until then. I think it must be all the dust. The company say there's no proof that they have caused it. They won't give me any compensation and I can't afford a lawyer to fight my case. There are many other people who are affected but the government won't help. They don't want to upset the bauxite companies because they earn money for the country.

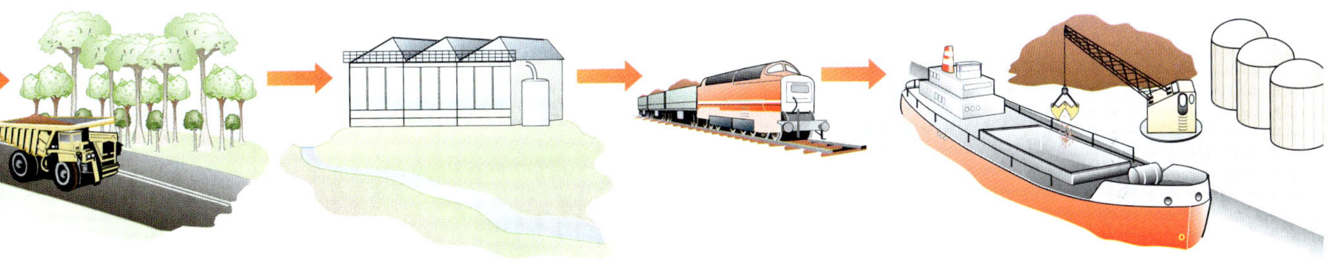

Lorries carry the rock either to a bauxite drying plant or to an alumina plant to be part-processed into aluminium.

Bauxite is crushed and washed in caustic soda to produce alumina. Effluent from the alumina plant may find its way into local rivers and lakes.

Trains or conveyors carry the alumina to the nearest port for storage and shipping.

Alumina is exported to the USA for further processing into aluminium. This requires huge amounts of electricity but is the most profitable part of the process.

Bauxite mining in Jamaica began in 1952. The country is now the world's third most important producer of bauxite, after Australia and Guinea. The mining is done by a few large TNCs based in the USA, Canada and Europe. The Jamaican government also has a share in some of them, which gives the government more control over their activities and helps to keep some of the profits in the country.

The bauxite is mined and part-processed into alumina in Jamaica (Photo 20 and Figure 21) but is then shipped abroad to produce the aluminium. Companies pay the government for the alumina they export but it is worth only 2 per cent of the value of the final product. Despite this, bauxite and alumina account for 42 per cent of Jamaica's exports and are the country's second most important source of income after tourism.

Tasks

SAVE AS . . .

1 Draw a large table, like the one below, to show the benefits and problems that TNCs bring to Jamaica. Divide them into two categories – social and environmental.

List all the benefits and problems that you find on these two pages. Leave space in your table for the benefits and problems of the tourist industry (pages 356–57).

	Benefits	**Problems**
Social		
Environmental		

2 Read the text in Figure 23.

a) Norma Williams believes that bauxite mining has caused her asthma. Do you think this is likely? Give reasons.

b) How could she try to prove the link between her asthma and the bauxite mining? What people could she call to give evidence? How could they help?

Paradise PLC?

Tourism is Jamaica's fastest-growing industry. It now earns almost half of the country's income and employs a quarter of the population. But, like the bauxite industry, tourism relies heavily on the involvement of foreign TNCs. Hotel chains like Trust House Forte, Sheraton and Holiday Inns, well known in the UK, are found in Jamaica too. TNC involvement in tourism does not end with hotels. Airlines carry tourists to Jamaica, construction companies build the hotels, and travel companies organise the holidays. Many are based outside Jamaica, draining profits from the country. This is sometimes called LEAKAGE.

However, not all TNCs operating in Jamaica are foreign-owned. 'Sandals' is a Jamaican company that owns all-inclusive resorts in all of the main tourist centres on the island. Most of them are concentrated on the north and west coasts (see Map 24). All-inclusives offer tourists the opportunity to experience a luxury holiday in the Caribbean without having to set foot outside the resort. Started in 1981, Sandals now has 12 resorts throughout the Caribbean with a total of over 3,000 beds.

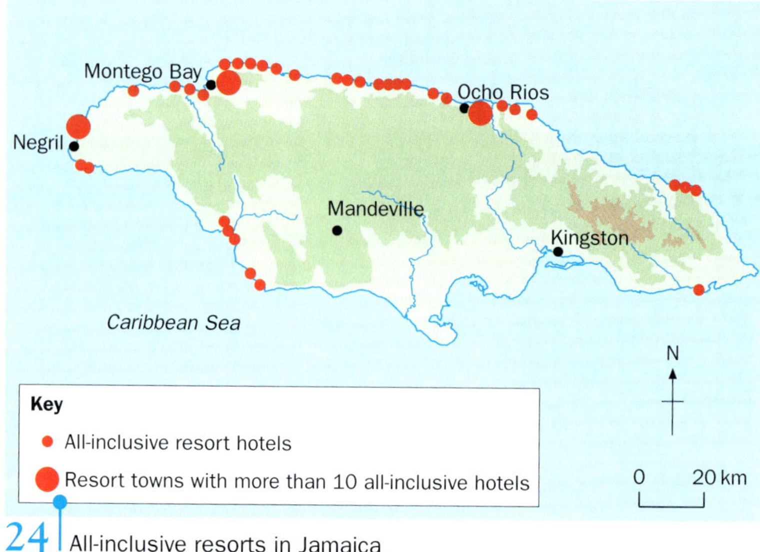

24 All-inclusive resorts in Jamaica

26 An all-inclusive resort at Ocho Rios

25 Only tourists are welcome at all-inclusive resorts. The local community is often excluded from the economic benefits that tourism can bring

Mister, don' feel up de fish
If you not buying it, leave it!
No Sir, sea egg price gone up.
No Sir, I ain't put it up
Is de government
What you say sir?
If you could take my picture?
How much you paying?
We natives doesn't pose
For free again!
Alright, But lemme
Fix up me face
All you move,
Move darlin', move little bit darlin'

Poem by Tim Tim

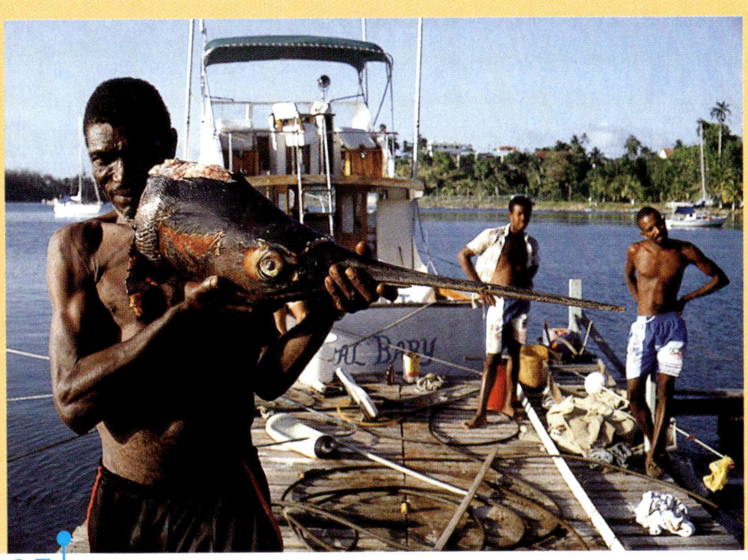

27 A Jamaican fisherman poses for a tourist camera

The resort has bedrooms for 350 guests, most of whom are foreign tourists from Europe or North America. Guests pay up to £2,000 per person per fortnight all-inclusive.

90 per cent of the food and drink at the resort is produced in Jamaica. This is not true of all resorts, which often import food and drink to meet tourists' taste.

The resort provides a high level of service – two members of staff for every room. This is only possible because of the low wage levels in Jamaica. Sandals employs 6,500 people in Jamaica.

Guests arrive by plane at Kingston or Montego Bay. The resort provides transport to and from the airport. Return flights from the UK cost around £600 per person.

The resort has a private beach which is fenced off to prevent local vendors from harassing tourists. As a result, tourists may never come into contact with the 'real' Jamaica.

Tourists coming from countries with higher living standards than Jamaica consume more resources than the local people. The average guest uses ten times as much water as a local person.

Inadequate sewage treatment and disposal by resorts pollutes the sea and causes damage to coral reefs. Larger resorts like Sandals can afford to install their own sewage treatment facilities.

If tourist numbers continue to grow there will soon be as many tourists visiting Jamaica each year as the number of people living there. This puts huge pressure on all the island's resources.

Tasks

1 Look at Photo 26.

a) Draw a spider diagram, with an all-inclusive resort at the centre. Using arrows and labels, show the social and environmental impact that this type of tourism has on Jamaica.

b) Use the information to complete the table that you started in Task 1 on page 355. List the benefits and problems that TNCs bring to tourism in Jamaica.

2 Read the poem in Figure 27.

a) What does the poem suggest the fish-seller feels about tourists?

b) Give reasons why she might feel this way. (You can use evidence from these two pages, but also think back to what you found out on pages 348–49.)

What role should foreign-based TNCs have in Jamaica?

3 Work in a small group to play the role of members of the Jamaican government. You have to make a decision about the role of TNCs. Use all of the information on pages 354–57.

a) Decide whether to encourage foreign-based TNCs to invest in bauxite mining and/or tourism. You could decide to encourage investment in neither or both. What conditions would you put on their activities?

b) If you decide not to encourage foreign investment, suggest alternative strategies for development. Write a one-page report to justify the decisions that you took. You can find out more about tourism in Jamaica in Chapter 7.4.

Is debt the new slavery?

In April 1999 there were riots on the streets of Kingston, the capital of Jamaica. People were protesting against government tax increases that put up the price of petrol. Many similar protests have happened in Jamaica in recent years. The petrol price rise was just one of many unpopular decisions that the Jamaican government has taken to deal with the problem of debt.

Task

1 What did the riots in Kingston (Figure 28) have to do with Jamaica's debt problems? With a partner, think about this question. Can you suggest an explanation?

RIOTS IN KINGSTON KILL EIGHT

20th April 1999

Thousands protest at petrol price rise

Violent scenes erupted on the streets of Kingston yesterday as thousands of angry people protested at government plans to put up the tax on petrol. Mobs overturned vehicles before setting light to them, and shops were looted as the protest threatened to get out of control. Eight protestors were shot and killed as police tried to restore order.

First to be affected by the tax increase will be taxi drivers and bus companies. They are expected to pass on the increases in the fares that they charge passengers. The government is trying to play down the riots, fearing they could damage tourism and make the difficult economic situation that the country faces even worse.

28 Adapted from a Jamaican newspaper

1996

Total debt:	US$ 4,000 million
Debt/person:	US$ 1,616
Total debt service:	US$ 682 million
Debt service/person:	US$ 273
GNP/capita:	US$ 1,714

Key: Private; Bilateral; Multilateral

1980 US$ 2,000 million

1996 US$ 4,000 million

29 Data on debt in Jamaica, 1996

★ FACTFILE

Glossary of debt

- **Private debt** Money owed by government as a result of borrowing from private banks.
- **Bilateral debt** Money owed by government to the government of another country.
- **Multilateral debt** Money owed by government to international organisations like the World Bank and the International Monetary Fund (IMF).
- **Debt service** Part of a larger total debt that is repaid during one year.
- **World Bank** A UN organisation that provides long-term loans to governments for development.
- **International Monetary Fund** Another UN organisation set up to keep international trade going. One way it does this is to provide loans for countries in debt.
- **Debt rescheduling** Spreading out debt repayments over a longer time period to help governments to repay the debt.
- **Structural Adjustment Programme** Strict economic conditions imposed by the IMF on governments that are allowed to reschedule debts.

CAUSES OF DEBT

1500–1838
Slave economy in Jamaica was geared to the production and export of raw materials to Europe. Products for local consumption often had to be imported.

1838–1938
After the abolition of slavery, people either continued to work on plantations or farmed their own small plots of land. Jamaica continued to export crops.

1938–62
Decline in the importance of primary products in the global economy led to unemployment in Jamaica. People emigrated to Europe and America for work. Many of the most able and productive people left.

1962–70
Jamaica gained political independence but had little money for development. It attracted foreign investment but had little control over decisions. Profits left the country and it remained economically dependent.

1970–80
World oil prices rose in 1973. The prices of other primary products were falling, so the gap between the value of Jamaica's imports and exports grew. The country had to borrow money to pay for development.

1980–90
Banks increased their interest rates, so Jamaica had to pay back more on its loans. The IMF loaned more money and the country's debts were rescheduled, but Jamaica had to agree to a structural adjustment programme.

Since 1990
Despite structural adjustment, Jamaica remains highly indebted. The gap between the value of imports and exports continues to grow, making it harder to repay the debt.

CONSEQUENCES OF DEBT

Employment
Unemployment rises and wage levels fall as structural adjustment forces the government to reduce spending. There are fewer jobs in public services such as health and education. As people have less money to spend, private companies cut back production and reduce their jobs too – 16 per cent of the workforce in Jamaica is unemployed.

Prices
As part of structural adjustment, Jamaica had to devalue its currency. This makes imports more expensive, and prices in the shops rise. In 1979 the minimum weekly wage of J$26.00 was enough to feed a family of five for a week. By 1999 the minimum wage of J$500.00 was barely enough to buy half of the food that a family of five needs.

Housing
House prices have also increased and so have the interest rates for mortgages. The price of a typical three-bedroom house in 2000 was J$2 million and a one-bedroom apartment was J$1 million. Few young people can afford to buy at this price, so they share homes with their family or build their own.

Public services
The biggest cuts in government spending have come in health and education. In 1975, 9 per cent of government spending went on health and 18 per cent on education. By 1996 these had fallen to 5 per cent and 8 per cent. The quality of these services has deteriorated and people now have to pay for healthcare and education.

Transport
Imported vehicles, spare parts and fuel are all more expensive, so not everyone can afford to buy or run their own car. Cuts in government spending mean less public transport, and the government also raises extra money through taxes on fuel, making all transport more expensive.

30 Causes and consequences of debt in Jamaica

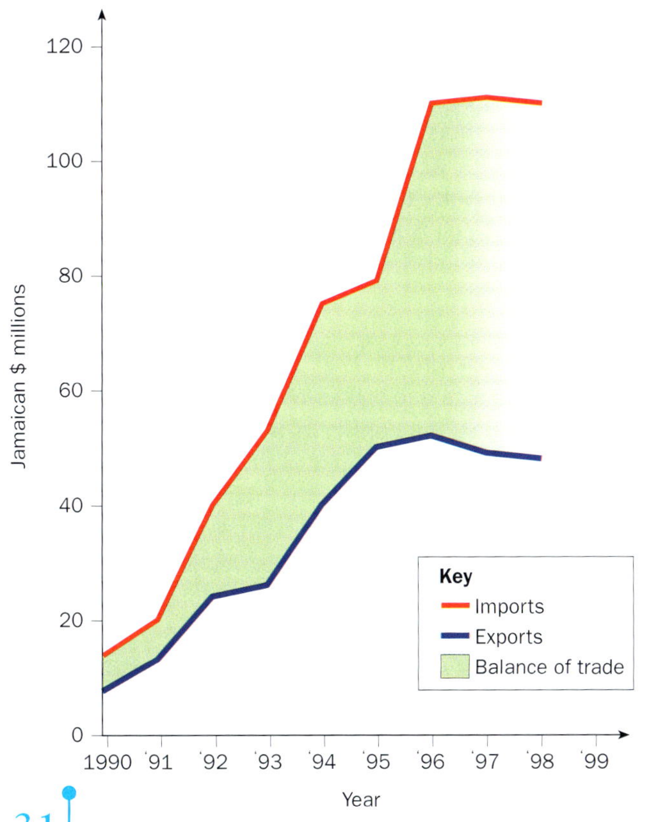

31 Jamaican imports and exports, 1990–99

Tasks

SAVE AS...

2 Study the *causes of debt* section of Figure 30.

a) Identify at least six causes of Jamaica's debt. Make a list of the causes.

b) Write your six causes around the edge of a sheet of paper. Draw lines to link the causes and label the links to explain them. For example, dependence on the export of raw materials meant that Jamaica suffered when the price of primary products fell. Identify as many links as you can.

3 Study the *consequences of debt* section of Figure 30.

'The impact of debt is greatest on the poorest people in a country.'

Do you think this statement is true of Jamaica? Discuss the statement with a partner. Use evidence to support your opinions.

4 Now look again at the question in Task 1. Would you answer this question differently now? Explain your answer.

What damage can debt do?

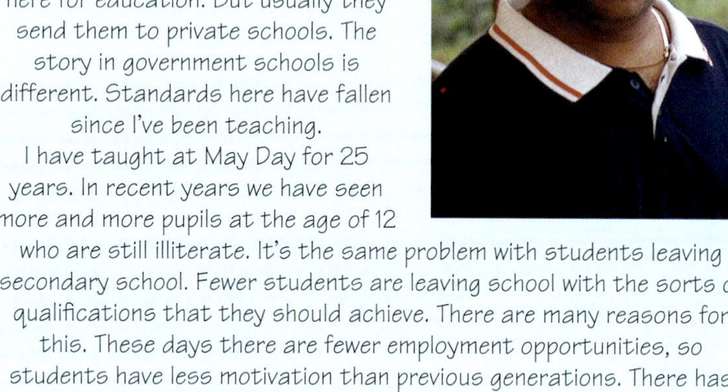

Jamaica rightly has a good reputation for its education. Parents of Jamaican background in the UK sometimes send their children back here for education. But usually they send them to private schools. The story in government schools is different. Standards here have fallen since I've been teaching.

I have taught at May Day for 25 years. In recent years we have seen more and more pupils at the age of 12 who are still illiterate. It's the same problem with students leaving secondary school. Fewer students are leaving school with the sorts of qualifications that they should achieve. There are many reasons for this. These days there are fewer employment opportunities, so students have less motivation than previous generations. There has been an increase in the informal economy and drug culture, which offer alternative ways to make money without finding a proper job.

School education in Jamaica is no longer free as it used to be. Parents at this school have to pay a fee of J$4,000 (about £60) a year to send their children to school. For poorer parents this can be a large part of their income, especially if they have more than one child at school. As many as 75 per cent of our parents have to apply for government assistance to pay for their children to come to school.

From the money that parents contribute, the school has to pay all of its bills, pay for the building and buy all its resources. We never have enough money to spend. Last year we fell J$242,000 short of our budget, so next year we shall need to be even more careful how we spend the money. The main problem we have is lack of space. Currently we have 23 classrooms, but really we need 26. This means that there are more students in each class than we want. The teacher : pupil ratio should be 1:25, but it's closer to 1:35. We're still waiting for the science laboratory to be completed for next term. If it's not, the school won't be able to open on time.

The government is responsible for paying teachers' salaries. The average salary is about J$27,000 (about £450) per month. This is low, even by Jamaican standards. Teachers are paid less than nurses, police and most white-collar workers in private industry. They used to be able to escape from teaching into other jobs but, with the economy struggling these days, there aren't any other jobs to go to. The only alternative is to go to the USA where they can earn more.

If education in Jamaica is to improve and get back to high standards, I believe we should have three priorities. First, make more human and physical resources available for schools. Second, pay teachers properly. Third, ensure that jobs are available for students when they leave school to provide them with the motivation to succeed. Unfortunately, none of these things is likely to happen while the economy is in crisis.

32 Stanley Skeene is the Principal of May Day High School near Mandeville, Jamaica. It is a government school for students aged 12 to 19. There are about 850 students in the school. Many of them come from poor homes in rural areas

Tasks

1 Ask for a copy of the annual budget for your school.

- How much money does the school spend each year?
- What is the money spent on?

Imagine that your school had to cut its budget by 50 per cent (like schools in Jamaica).

a) How would you reduce the spending? You should reduce spending on the things you consider least important. How would you spend the money that is left? Justify your decisions.

b) What effect would this have on your school? What difference would it make to the quality of education that the school provides?

2 Debate the motion 'Debt is the new slavery' with the rest of your class.

One half of the class should support the motion while the other half should be against it (for the purpose of the debate you should ignore your own personal view). Use all the information provided so far in this chapter and in Chapter 7.2 to support your argument. Each person should prepare a two-minute speech for the debate. One person should act as chairperson. Listen to all the arguments that are made.

At the end of the debate take a class vote. Having heard the arguments, you can vote according to the way that you really feel. Is debt the new slavery?

★ THINK!

Facts and figures about issues such as debt are hard to take in, let alone remember. It is through people's personal experiences that they begin to make sense. You could use the example of May Day School if you are asked a question about the impact of debt on a country. It also helps to relate issues to your own experience. That is why you are asked to think about your own school in Task 1.

What has been the impact of migration on Jamaica?

Thousands of Jamaicans living in the UK are going home. Having spent most of their working lives in the UK they have decided to enjoy their retirement in Jamaica. But the island they return to is very different from the one they left over 40 years ago.

Between 1955 and 1962 over 150,000 Jamaicans emigrated to the UK. That was about 10 per cent of the country's population at the time. They were invited to come and work in the UK at a time when industry and public services here were short of labour. Major employers, such as the National Health Service and London Transport, recruited workers in Jamaica and the other islands. Employment opportunities in the Caribbean then were limited. Migration was an alternative to being unemployed or, at best, earning a low income from farming.

33 Jamaican immigrants arrive at Southampton in 1962

The life that the Jamaican immigrants found in the UK was not always easy, but most stayed to raise their families here. They were also able to send money home and, in many cases, make long-term plans for their retirement. Some had modern new homes built on family land back in Jamaica. During the 1990s, as they reached retirement age, an estimated 1,000 people returned to Jamaica each year. Back home they are known as 'returnees'. They are marked out by their large, affluent homes which are often in contrast to the more modest homes of their neighbours. As a result some have become the victims of crime, and a few returnees have even been murdered for their wealth.

The lifestyle of returnees is just one aspect of the inequality that is dividing Jamaican society. It highlights the gulf between those who are part of the global economy and those who are left behind.

Task

3 Put yourself in the position of an older Jamaican person who has spent their working life in the UK. As you look back over your life, would you still have taken the same decisions:

a) To emigrate to the UK?

b) To return to Jamaica?

In each case, try to explain your feelings.

34 A returnee's home near Mandeville in the centre of Jamaica

Inequality in Kingston

Jamaica has become one of the most unequal societies in the world. The richest tenth of the population earn 32 per cent of the country's income, while the poorest tenth earn just 2.4 per cent. Nowhere are the contrasts more visible than in Kingston, the capital. The city has a population of 700,000 people – over 25 per cent of Jamaica's total population – and is still growing. Much of the country's business is located there, including the headquarters of several large foreign companies. These provide highly-paid jobs for the wealthy middle-class people who live in suburbs like Norbrook. Within a few minutes' drive is the inner city – tightly-packed areas, like Grant's Pen, where people build their own shacks on family plots that grow more crowded with each generation. Most people living there are likely to be unemployed or depend on work in the informal economy – anything from street-trading to drug dealing.

I was born in Kingston and have lived here all my life. I wouldn't really want to live anywhere else. In Norbrook we have fantastic views of the city, a cool sea breeze blows in from the Caribbean and there is a golf course at the end of my garden. What more could a man ask for? Of course, it's expensive to live here – house prices can be anything up to J$40 million (£750,000). Most of my neighbours are doctors, attorneys and foreign ambassadors. Until 1980 I used to run my own business. I owned a factory that produced underwear for the USA. But the economy got into trouble in the 1970s – taxes went through the roof and it was getting harder to find reliable workers. Many of my business friends left the country and took their money with them. But I stayed and began a new career in photography. People hear about the crime in Kingston and they are surprised that I still live here. But we see few problems in Norbrook. It's mainly in the inner city where crime is rife. The difference between rich and poor bothers me big time – and things are getting worse rather than better.

35 Brian Rosen, a well-known Jamaican photographer, lives in Norbrook, a suburb of Kingston

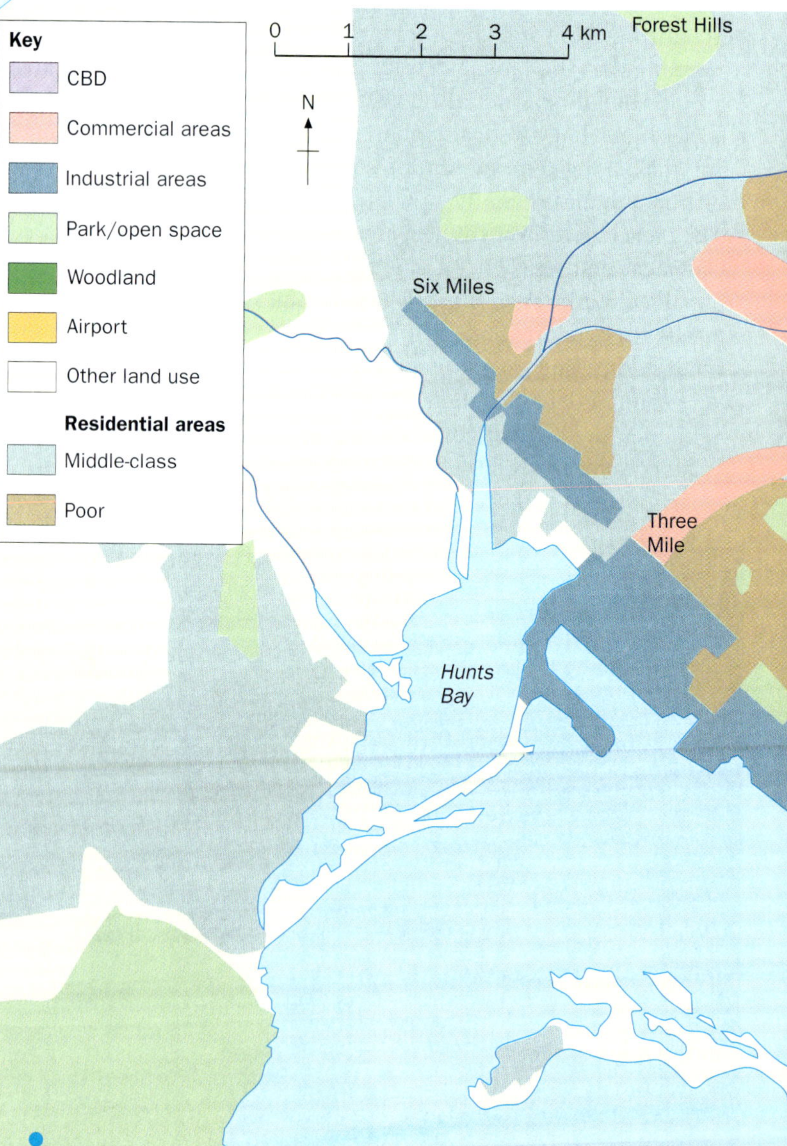

36 Luxury housing in Norbrook

37 Land use map of Kingston

I've lived in Grant's Pen all my life. Like most people here I'm struggling to survive. I have four children aged between 5 and 14 to look after. Their father doesn't live with us. It's too expensive to buy a house in Kingston so I've built my own. It has two small rooms and a kitchen. It's in the same yard as my mother's house so there's not much space. I left school at 15 and I've done all sorts of work since – in bars, factories and offices. I started doing a secretarial course a few years ago but I couldn't afford to finish. I've been lucky to find a job as a receptionist. I'm paid J$2,800 (£50) a week. Most of the money goes on food and school fees. It costs J$9,000 a year in Kingston to send each child to school. Then there are books, uniform, school lunches and bus fares to pay for.

But at least when they're in school they should be safe. Grant's Pen is controlled by gangs. They roam the streets with guns, shooting each other. It's easy for innocent people, including children, to be shot. The gangs threaten children into joining them and once they join they're trapped. I don't let my children out on their own, so they spend their spare time indoors. I really worry about them. I wish we could get out of here, but there's nowhere to go.

38 Sharon Palmer lives in Grant's Pen, part of Kingston's inner city (this is not her real name since she wants to protect her identity)

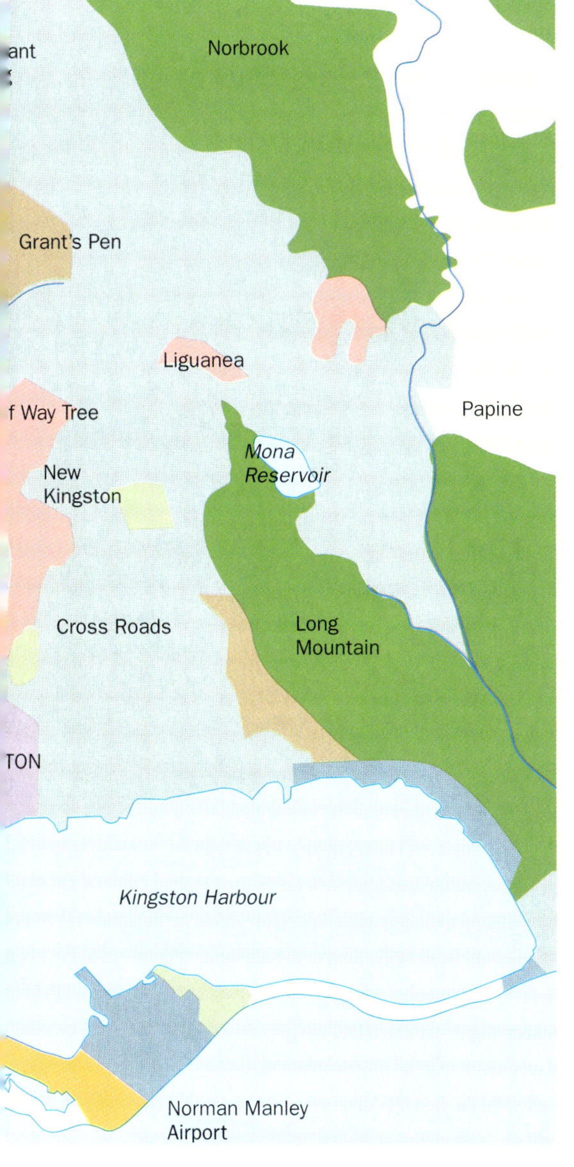

39 Housing in Grant's Pen – Sharon lives in a house similar to this

Tasks

1 Work with a partner. You are going to role play a conversation between Brian and Sharon. Choose who you will be.

a) Study all the information on these two pages. In particular, find out more about your character and where he or she lives.

b) Think about the views that your character might have on these topics: *inequality, crime, education, work*. Write these views down in the form of notes.

c) Together with your partner, create a situation in which these two people could meet. Role play the conversation that they would have. Try to cover the four topics in your conversation. You may be able to perform the conversation to the class.

SAVE AS . . .

2 Work in a small group. Research the reasons for inequality in Jamaica, using the information provided in this chapter. Each person could specialise in one area of research, for example:

- Jamaica's colonial history (pages 348–49)
- transnational corporations in Jamaica (pages 354–57)
- Jamaica's debt (pages 358–60)
- 20th-century migration (page 361).

Share the ideas that you have researched with the rest of your group. Discuss how important each idea is.

Write a joint report entitled *'Why is Jamaica so unequal?'* Each person should write one section of the report, based on their own research.

How should Jamaica develop now?

As you have read in this unit, there are different views about quality of life and development. Depending on which view you take, Jamaica's development since it gained independence might be seen as a success or a failure. If you think that development is about new industries, bigger cities and individual wealth, then Jamaica has been quite successful. However, if you think that development is about better health and education, shared wealth and less poverty, then it has been less successful.

Focus tasks

1 At the beginning of this chapter you were asked to consider whether Jamaica is an LEDC. You have collected evidence through the chapter, so what do you think now? Should Jamaica be described as an LEDC? If so, what evidence would you give to support this? If not, how would you describe Jamaica and how would you support this with evidence? Write a paragraph to evaluate development in Jamaica.

2 Look back at your ideas about quality of life (page 325) and development (pages 336–37). Have any of your views changed? What would be your priorities for development now?

3 Read the three models of development, Figures 40–42.

a) Which of the three models comes closest to the Jamaican experience?

b) Which of the three countries has developed most successfully, do you think?

c) Which would be an appropriate model for Jamaica to follow?

In each case give reasons for your opinion.

... LIKE PUERTO RICO

Puerto Rico is a Caribbean island similar in size to Jamaica. Like Jamaica it was a European colony that specialised in the production of sugar for export. By 1947 the sugar industry was stagnant and the island sought foreign investment to revive its economy. The country became a '*freely associated state*' of the USA – in effect a US colony. Many US-based companies, attracted by its low taxes and cheap labour, relocated their factories in Puerto Rico. The island benefited, not just from increased employment, but also from US subsidies to improve its infrastructure, including roads, power and ports.

The result was rapid growth of the island's economy. It soon had one of the highest per capita incomes in the region, with literacy and life expectancy approaching that of the USA itself. But the initial increase in employment was followed, in the 1970s, by a rise in unemployment. Companies began to reduce their dependence on labour as they mechanised, or else moved to countries where labour was even cheaper. Puerto Rico has become more and more dependent on subsidies from the USA to survive, but unemployment, at over 20 per cent, remains high.

Size:	9,100 km^2
Population:	3,522,000
Average income/person:	US$9,988
Life expectancy:	75
Adult literacy rate:	98%

40 The Puerto Rican model

... LIKE CUBA

Cuba, just to the north of Jamaica, is the largest and most populated Caribbean island. It has a similar colonial history and ethnic diversity to those of Jamaica. In 1959, to escape from the grip of its own political dictators and the influence of the USA, Cuba had a revolution. The country was taken over by a new socialist government, led by Fidel Castro, which changed the way the country was organised.

Before the revolution, 80 per cent of the island's income came from sugar, most of it exported to the USA. After the revolution the sugar industry was taken out of the hands of large companies and put under government control. The USA disapproved of socialism and began a trade BOYCOTT of Cuba (which still continues). Cuba had to look elsewhere for trade. It turned to the USSR (another socialist country) as an alternative market for its sugar and also as a source of manufactured imports and oil.

Meanwhile, the government made education and health a national priority. It organised a huge literacy campaign to teach everyone to read and write. Today the country still has levels of literacy and healthcare that match many MEDCs. But, with the collapse of the USSR in 1989, Cuba has been forced to open its door to trade with other countries. Without Russian aid it has also reformed its economy to allow foreign companies to invest there again. The country is now less dependent on sugar, and exports other primary products such as nickel and tobacco. Tourism is now the country's main source of income.

Size:	109,800 km^2
Population:	11,059,000
Average income/person:	US$1,170 (World Bank estimate)
Life expectancy:	76
Adult literacy rate:	95%

Socialist planned control

41 The Cuban model

... LIKE SINGAPORE

On the other side of the world from Jamaica, Singapore was another UK colony until it became self-governing in 1959. It was a small city state that had high population growth, few natural resources, little industry and high unemployment. In 1965 its average income per person was just US$700 – about the same as Jamaica at that time.

During the 1960s the Singapore government set up an Economic Development Board (EDB) which encouraged foreign investment. The country had the advantage that it was located on the Malacca Strait, an important trade route in the Far East. At first there was a growth in manufacturing industry and exports, but in the 1970s the EDB switched the emphasis to more skills-based industry and services. To make this happen, the government had to provide good-quality education and encourage companies to pay higher wages. Although the country relied on private investment, the government played a key role in the development of industry.

In many ways Singapore is a cross between the Puerto Rican model, dependent on foreign investment and the free market, and the Cuban model, in which the government played a central role. The result has been continual economic growth since 1961 at an average annual rate of about 10 per cent. Singapore is now one of the richest countries in the world – even richer than the USA, and far wealthier than its former coloniser, the UK.

Size:	648km^2
Population:	3,100,000
Average income/person:	US$32,810
Life expectancy:	76
Adult literacy rate:	90%

Carefully managed foreign investment

42 The Singapore model

What will Jamaica's place be in the global economy?

The global economy is changing fast. The rules that governed international trade in the past are breaking down, and countries that are slow to adjust may be left behind as the world moves on. The changes could have a huge impact on Jamaica in the 21st century.

The problem for countries like Jamaica is that they have little control over these changes. It is the MEDCs (in the form of organisations such as the IMF and the WTO) that make the rules, and it is TNCs based in these same wealthy countries that dominate trade in the global economy. Of course, this is not really new. For the past 500 years it has been rich countries that have dictated the way that countries like Jamaica have developed. It was the needs of the rich countries that decided what Jamaica should produce while, at the same time, the rich countries took the profits for themselves. They established the pattern of trade, which meant that Jamaica had to import goods that its people needed while it saw the value of its exports fall.

So can Jamaica do any better in the new global economy than it has in the past? Some politicians and economists argue that co-operation with other Caribbean countries is the way forward. Co-operation in the past has proved difficult, because each island wants to pursue its own economic interests and compete for trade with the others. For example, most of the Caribbean countries produce sugar, but overproduction of sugar around the world means that each country earns less money even though it is producing more. If they could agree to reduce sugar production and diversify into other exports, then they might overcome such economic weakness.

Advocates of economic co-operation point to the success of other groups of countries, such as the EU. Not only has the EU increased the amount of trade between member countries, it also provides individual countries with greater influence within the global economy. The framework for such co-operation in the Caribbean already exists in the shape of CARICOM (Caribbean Common Market and Community), but its impact has been limited due to disagreement between the individual governments.

Task

1 Look at Figures 43 and 44.

a) What do they tell you about the role of Jamaica and other Caribbean countries in the global economy?

b) What problems does this lead to in those countries?

Why do you encourage us developing countries to export food while we, the poor, starve?
IMF
Efficiency! Just think of the world as a giant global supermarket! The foreign …
IMF
… transnationals are producers of the goods, the rich developed countries their consumers!
IMF
But what about we the poor here, what are we in the global supermarket?
IMF
You know the people who carry groceries from the store to the people's cars …?
IMF

43 Poor countries' place in the global economy

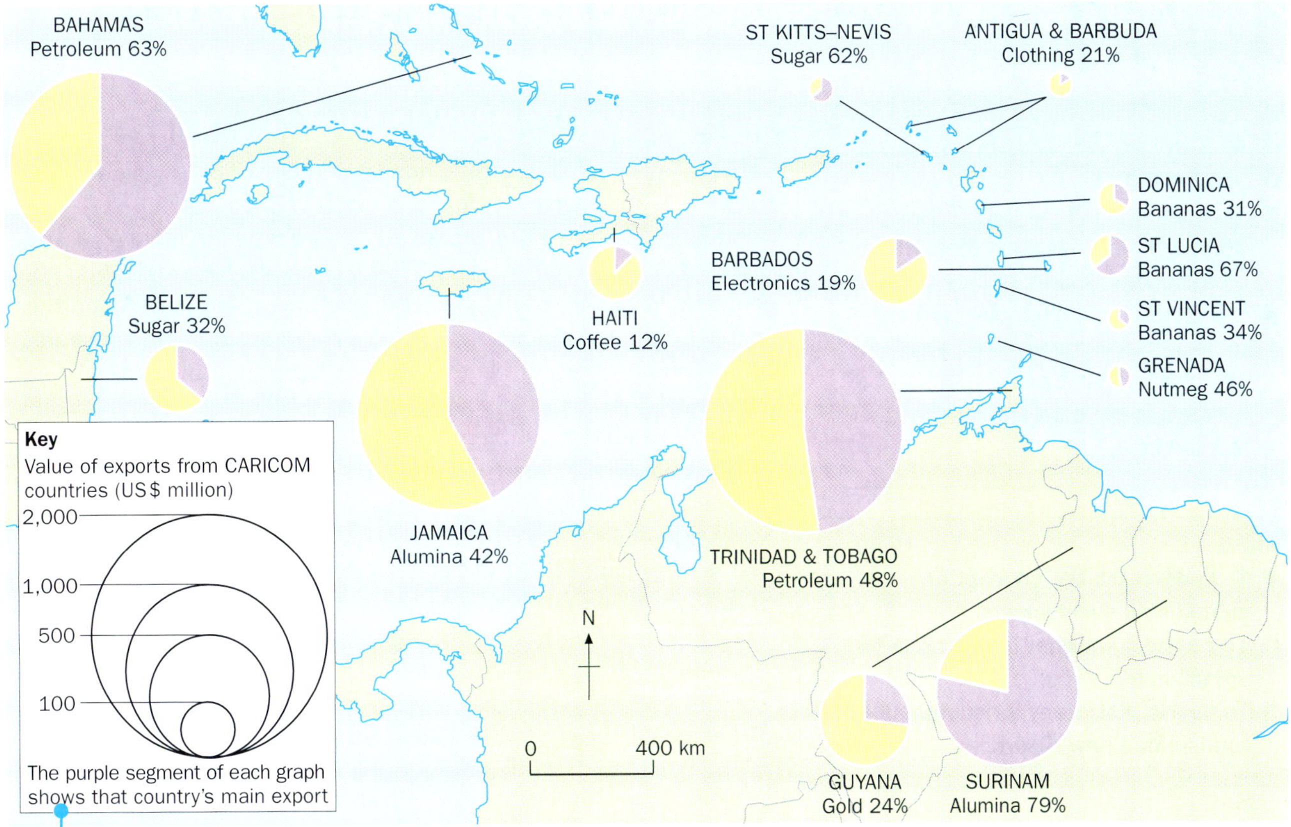

44 Caribbean exports from CARICOM countries

Task

SAVE AS...

2 Look at the box below. It suggests four economic strategies that Caribbean governments could follow. For each strategy, write down one advantage and one problem (think back to what you have learned about Jamaica in this chapter).

- Produce more primary products (like bananas or bauxite) to export.
- Encourage more foreign investment from TNCs to increase manufactured exports.
- Expand tourism to bring in more foreign visitors and money.
- Increase trade between the islands and reduce foreign imports.

Focus task

3 Work in groups of about ten. You are going to plan an economic strategy for the Caribbean. Each person should choose one CARICOM country to represent. (All the large countries, including Jamaica, should be represented.)

a) Together decide what the main economic strategies for CARICOM should be. Choose one or more of the four ideas in the box in Task 2, or suggest other strategies of your own.

b) Individually, consider how your strategies will affect the country you represent. Think about each of the following questions:

- How should production of your main export change?
- How should production of your other exports change?
- What other things could your country produce?
- Where will the investment come from?
- What will you be able to import/export within CARICOM?
- What will your country need to import from other countries?
- How will these decisions help your country to develop?

Report back to the whole group on what you decide. The group should take a vote on whether to accept or reject each country's report. If they reject it then you will have to decide whether to accept changes or leave the CARICOM group.

c) Write a report about the impact that your CARICOM economic strategy could have on *either* Jamaica *or* the country that you represented. Divide the report into seven sections dealing with each of the questions in **b)**.

7.4 Tourism – towards a sustainable future?

DO YOU KNOW?

- Altogether, people in the UK take over 60 million holidays each year. That is more holidays than there are people!
- ECOTOURISM is an alternative form of tourism that is meant to be more sustainable. But beware – all sorts of holidays these days are described as 'ecotourism'.

AIMS

- **To understand the concept of sustainable development.**
- **To consider whether tourism can be sustainable.**
- **To devise a strategy for sustainable tourism in a new resort.**

These days people have more holidays than ever before. Tourism is now the world's largest industry. It employs 200 million people worldwide. Through tourism we are able to travel to more places and find out more about the world. Great, especially for geographers! But holidays have a downside too. They can have a huge impact on the natural environment and on the local communities that it supports. Many people question whether the present growth in tourism is sustainable. And, if they're honest, how much do some people really learn about all of those wonderful places that they have visited?

Treasure Beach is a little-known resort on the south-west coast of Jamaica (Photo 1), far away from the tourist hotspots on the north coast that you saw on pages 356–57. It is a sleepy fishing village that has hardly changed within living memory and it is also the sort of place that would-be tourists dream of. Holiday brochures are full of 'Treasure Beaches' all over the world. But if tourism comes to Treasure Beach, and all the other unspoilt places in the world, will they ever be the same again? How will the environment change? What would happen to this sleepy fishing village and the people living there? Would it have a sustainable future?

★ SUSTAINABLE DEVELOPMENT IS . . .
Development that meets the needs of the present without compromising the ability of future generations to meet their own needs.

Come to where the only footprints you are likely to see on the beach are your own, and where the residents treat you as guests, not tourists. Nicknamed 'desert coast' because of the ever-present sunshine, Treasure Beach is a scenic 100-km drive from Montego Bay. This area is still unspoilt and largely unaffected by tourism, yet it offers the best of Jamaica's natural beauty and charm: deserted sandy beaches, colourful coral reefs, beautiful sunsets, starlit nights and mountains that reach the sky.

The natural beauty of Jamaica is only half of its magic. You will have the opportunity to learn about the rich culture and easy-paced lifestyle in which fishing and farming are done much as they were half a century ago. Relax and forget your worries – this is your holiday in paradise!

1 Treasure Beach and internet advertising

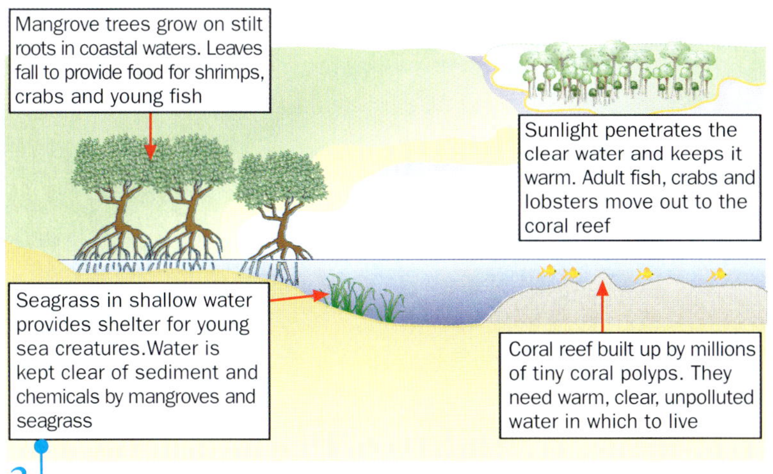

Mangrove trees grow on stilt roots in coastal waters. Leaves fall to provide food for shrimps, crabs and young fish
Sunlight penetrates the clear water and keeps it warm. Adult fish, crabs and lobsters move out to the coral reef
Seagrass in shallow water provides shelter for young sea creatures.Water is kept clear of sediment and chemicals by mangroves and seagrass
Coral reef built up by millions of tiny coral polyps. They need warm, clear, unpolluted water in which to live

2 The fragile environment on the Caribbean coast

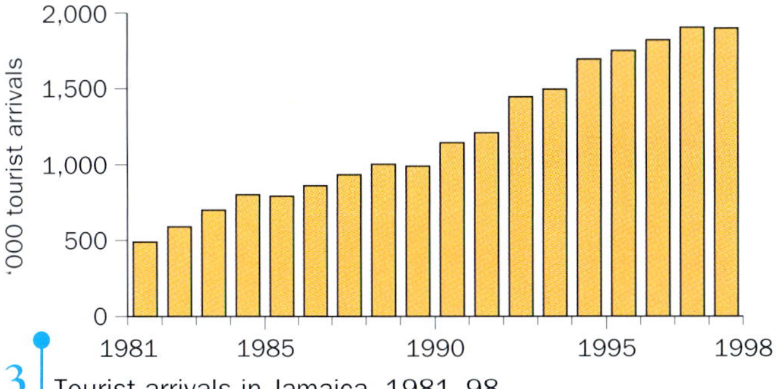

3 Tourist arrivals in Jamaica, 1981–98

Jamaica has a terrible reputation for violent crime – foreign documentaries flash images of poverty and gangsterism around the world. However, while the island's murder rate is undeniably high, its nightmare image is vastly exaggerated. The tourist board stresses that you are more likely to be mugged in New York than Montego Bay. At the same time robberies, assaults and other crimes against tourists do occur and you must take precautions that you'd take in any foreign city. Don't flaunt your wealth with fat rolls of banknotes and avoid walking alone late at night.

The chief irritation in Jamaica is hustling. The hard-sell pitches are wearisome, to be sure, but most of what is perceived as harassment is really nothing more than an attempt to make a living in an economically deprived country.

4 Extract from the *Rough Guide to Jamaica*, about crime and harassment

Tasks

1 Read the definition of sustainable development opposite. Write your own definition in language that anyone could understand. Share your definition with the rest of the class. Can you agree on one definition?

2 Look at Photo 1. Work with a partner to discuss the following questions.

- Who took the photo, and what might they have been thinking when they took it?
- What will people who see the photo think?
- What does the photo tell you about the place?
- What might the photo *not* tell you about the place?
- What might the place look like in the future? Why?
- What other questions would you like to ask about this photo?

Keep these questions. You may be able to answer them fully at the end of the chapter.

3 Look at Sources 2, 3 and 4.

a) For each figure, explain how it may mean that tourism in Jamaica is not sustainable.

b) For Figures 3 and 4, suggest what impact it could have on the scene in Photo 1.

Focus task

4 At the end of this chapter you will devise a strategy for the sustainable development of Treasure Beach. As you go through the chapter, consider the needs of different groups of people, the natural environment, and Jamaica as a whole.

Think about what to include in a brochure to advertise holidays in Treasure Beach.

★ THINK!

As you come towards the end of your GCSE Geography course your thoughts may already be turning to what you will do in the long summer holiday (don't forget your exams come before then . . .). But prepare for a shock! You can do Geography on holiday, too!

The purpose of this final chapter is not just to get you thinking about your holidays, exciting as that may be. It is also to bring together many of the themes and ideas in your GCSE Geography course, including coastal management, fragile ecosystems, employment, economic development, the global economy and, of course, sustainable development – all part of the Geography you have done. It may help to look back over your work, or the units in this book, to remind yourself about what you have learned. Then bring it together one more time.

Could tourism in Jamaica be more sustainable?

Situated on the north coast of Jamaica, Ocho Rios was originally a small fishing village – perhaps not unlike Treasure Beach today. Its development as a tourist resort began in 1948 when the first hotel was built there. Tourists were attracted by the white sandy beach and the nearby attraction of Dunn's River Falls.

Until the 1960s, Ocho Rios was an exclusive resort that catered for an elite group of tourists. Since then, with the development of MASS TOURISM and the PACKAGE HOLIDAY, it has grown dramatically. The original fishing village has been replaced by an artificial beach overlooked by hotel tower blocks. Ocho Rios is now Jamaica's second largest tourist destination with almost 30 per cent of the island's visitors. It is also the island's busiest port, where cruise liners bring over 300,000 passengers a year. Its residential population has grown to 110,000 with the number of Jamaicans who have come to work in the tourist industry.

Ocho Rios typifies the north coast of Jamaica where most of the island's beaches and resorts are located. The government is concerned about the social and environmental impact that mass tourism has along the coast. It has produced a report which states: 'Tourism often has severe environmental impacts which would reduce Jamaica's ability to sustain its reputation as a tourist destination . . . Long-term environmental protection must become a priority.' Although the number of visitors continues to grow, the government wants to ensure that tourism on the island is sustainable. It hopes to develop new resorts in other parts of the country which currently have no tourists, particularly on the south coast. This would be an opportunity to develop tourism that would bring greater benefits to the local community and also protect the environment. This type of tourism is sometimes called 'ecotourism'.

5 Ocho Rios, Jamaica

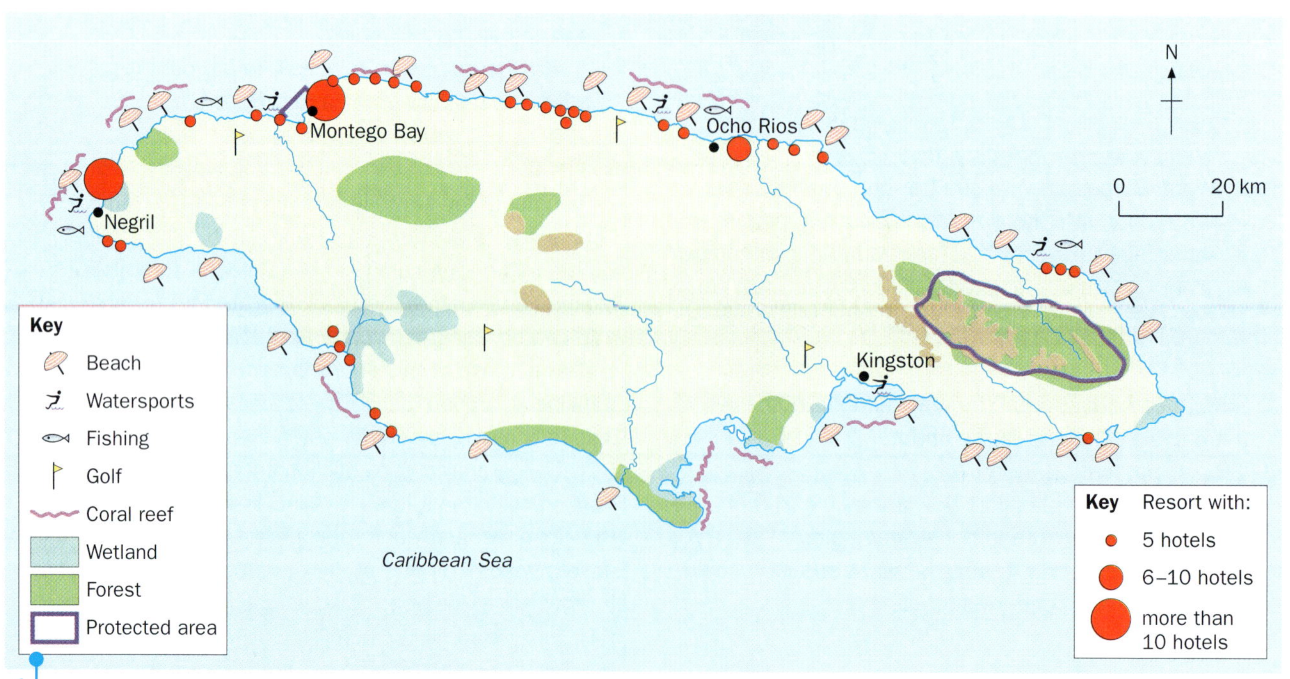

6 Tourism and the environment in Jamaica

In some years hurricanes hit Jamaica. These may become more frequent in future with global warming.

Storms wash away sand from the beach and carry it out to sea.

Tourists are attracted to Jamaica by its sunny climate and white sandy beaches.

Fishing boats and watersports can cause physical damage to coral reefs.

A growing number of tourists enjoy activities such as fishing, watersports and scuba-diving.

Hotel construction uses large amounts of sand taken from local beaches and sand dunes.

A coral reef is a fragile ecosystem. Once damaged it takes a very long time to recover.

Tourists expect high-quality hotels and a good natural environment when they go to Jamaica for a holiday.

Tourists consume more resources than local people. Hotels use more water and produce more waste per person than the average for the country.

The number of tourists visiting Jamaica has risen continuously for the past 20 years.

Fishing provides some local employment and is a source of food for local people and tourists.

Coral requires clear, unpolluted water in which to grow.

Hotels discharge untreated sewage into the sea, which increases the growth of algae and makes the water cloudy.

A coral reef forms a protective barrier against storms along the coast. It is also the source of white sand on the beaches.

Boats and divers can damage the reef, and some tourists remove coral to take home as souvenirs.

The coral reef provides a natural habitat for a huge variety of fish species.

7 The needs of tourism and the environment in Jamaica

8 Scuba-diving among coral reefs off the coast of Jamaica

Tasks

1 Look at Photo 5. Compare it with Photo 1 on page 368.

a) Where would you rather spend a holiday? Explain your answer.

b) Find out which place other people in your class would prefer to go to. What importance would this have for the Jamaican government if it wants to make tourism more sustainable?

SAVE AS . . .

2 Look at Map 6.

a) On a copy of the map, highlight areas that could be important for ecotourism. Give reasons for your choices.

b) Use the map to write a paragraph suggesting how tourism in Jamaica could be managed to make it sustainable.

3 Look at Figure 7. You will use these statements to create a model of *unsustainable tourism.*

a) Cut out copies of the statements and place them on your desk. Read each one. What links can you find between them?

b) Arrange the statements on a piece of paper so that those that are linked are close to each other. Some may be linked to more than one other statement.

c) Stick down the statements. Draw arrows to show the links between them. Explain how the model that you have created shows that tourism in Jamaica is unsustainable.

d) Create a similar model of *sustainable tourism.* Change the statements that make tourism unsustainable. For example, change 'Hotels discharge untreated sewage into the sea . . .' to 'Hotels treat their sewage before they discharge it . . .' How will this affect other parts of the model?

Sustainable tourism – is there such a thing?

Tourism is an industry, and the pattern of global tourism is little different from other parts of the global economy. Despite a growing number of visitors from MEDCs who go to LEDCs, most tourism is still concentrated in the richest countries. Eight of the top ten countries for tourist arrivals are in Europe and North America. Even where the pattern is changing, the role of TNCs ensures that the wealth from tourism finds its way back to the MEDCs.

Tourism is often seen by LEDCs as an alternative path to development – a way of breaking away from dependence on primary products and joining the global economy. Usually this involves foreign investment in mass tourism, based on the experience of areas like the Mediterranean. However, this type of development has been criticised as a new form of colonialism that is unlikely to bring any benefits to the majority of poor people in LEDCs. Rather, it may lead to so-called UNDERDEVELOPMENT, leaving poor countries vulnerable to changes in the world tourist industry. Places that, today, are hyped by travel companies as desirable holiday destinations may, tomorrow, lose their popularity and fall into decline (see Figure 10). This has happened to a number of Mediterranean resorts, not to mention seaside towns in the UK.

While some see tourism as a path to development and others see it as the cause of underdevelopment, there is a third view that tourism can be part of a strategy of sustainable development. Such tourism, under the control of the local community, would have a less damaging impact on both the natural environment and the people. Indeed it would bring benefits which would ensure that the community had an active interest in promoting tourism in their own area (see Figure 9).

Tourism as a path to development	Tourism as a cause of underdevelopment	Sustainable tourism
• Tourism enables LEDCs to *develop* in the same way that some MEDCs have done already.	• Tourism actually helps MEDCs to develop more than the LEDC because profits go back to the MEDCs. The LEDC becomes more dependent.	• Tourism can benefit both MEDC tourists and the LEDC local community. It should be part of a balanced economy where profits are retained locally.
• *Foreign investment*, expertise and technology are brought to the LEDC by TNCs.	• TNCs control tourism in their own interests, not those of the local community.	• Tourism is controlled by the local community and meets their needs as well as those of tourists.
• *Mass tourism* brings holidaymakers from MEDCs. They are attracted by advertisements showing idyllic resorts.	• Mass tourism may spoil the character of the resort and it will become a victim of its own success.	• The number of tourists should not destroy the character of the destination, so that visitors will continue to come.
• Traditional *local culture* is an obstacle to tourism. It is replaced by modern Western culture, symbolised by McDonald's, Coca-Cola, etc.	• The local community may be resentful if their culture is devalued. This can lead to crime and even terrorism.	• There must be respect for local traditions and culture. These should be part of the character of the tourist destination.
• Tourism requires development of the *local infrastructure*, like roads and airports. These may also benefit local people.	• Tourism can put too much pressure on local resources and reduce quality of life for people through pollution and congestion.	• The scale of tourism depends on the size of the resort and its resources, so that it does not damage the community or the environment.
• Tourism generates local *employment* and helps to increase wealth.	• Tourism exploits cheap local labour. The best-paid jobs often go to outsiders.	• Tourism brings a variety of well-paid jobs that encourage the development of local skills.

9 Three views of tourism in LEDCs

This model shows the stages through which a typical tourist destination may pass, from initial exploration through to eventual stagnation.

Exploration
A few tourists visit a newly discovered destination, seeking adventurous travel. This makes minimal impact.

Development
Improvements are made to local infrastructure, and tourism becomes a source of local employment. Foreign companies begin to get involved.

Consolidation
Control of the industry is taken over by foreign companies. Improved facilities and advertising lead to mass tourism.

Stagnation
The growth rate slows down as available resources such as beaches are fully exploited.

Decline or rejuvenation?
The area may lose popularity if the natural environment that first attracted tourists is spoilt. Alternatively, a policy of sustainable tourism could lead to rejuvenation.

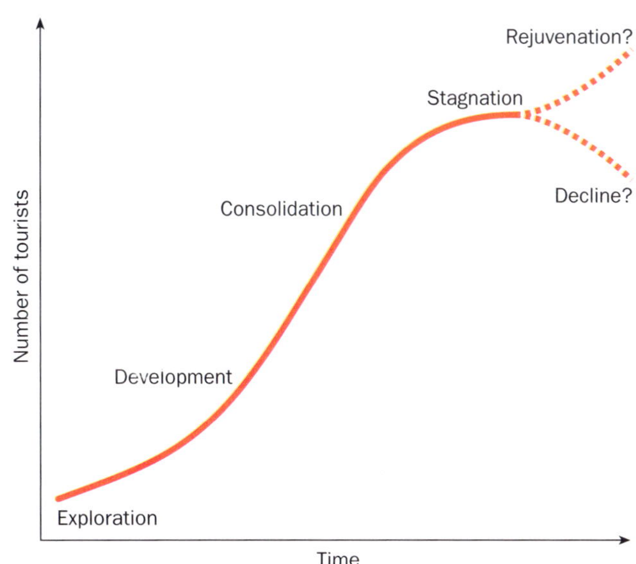

10 Model of development for a tourist destination

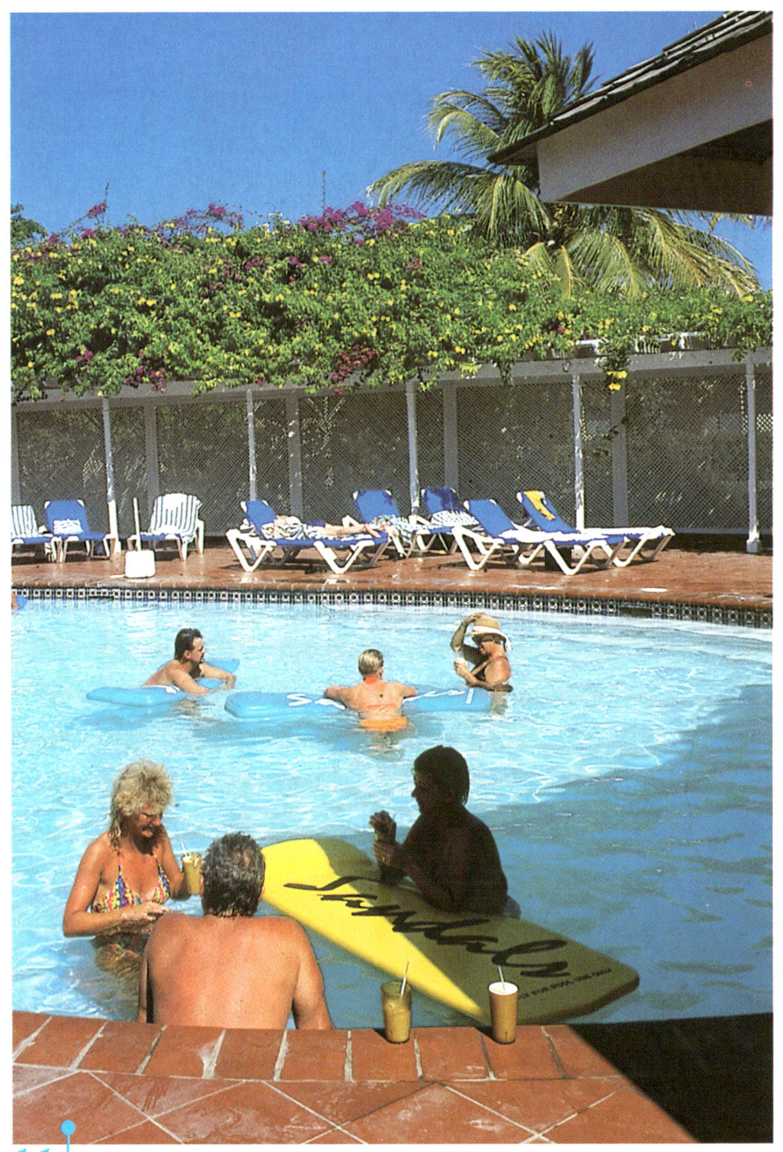

11 An all-inclusive resort at Ocho Rios, Jamaica

Tasks

1 Read the viewpoints in Figure 9. Use these views to evaluate the tourist resort in Photo 11. This is the same resort that you studied on pages 356–57. You will need to read those pages again to help you here.

- **a)** In what ways does the resort support each of the three views of tourism? List the ways under the three headings used in Figure 9. For example, the resort employs hundreds of local people, so this could go under 'Tourism as a path to development'.
- **b)** Overall, which of the three views do you think this resort supports? Why?

2 Think of a tourist destination that you have been to, at home or abroad.

- **a)** Would you describe that destination as an example of tourism as development, underdevelopment or sustainable development? Explain why.
- **b)** Which stage of the model in Figure 10 do you think that destination has reached? Again explain why. What do you think is likely to happen to it in the future?

Paradise for ever?

12 Joslin and Enid Barnett are farmers who are building a villa. They hope that tourists will come to rent it

We like the idea of tourists coming to our part of Jamaica, and they're always welcome. But we don't want to be taken over by tourism so that we lose our identity. About 20,000 people live in the Treasure Beach area. Most of us are farmers and fishermen. It's the way of life we know and we want to preserve it.

We could benefit from the trade that tourists might bring. They would buy our fruit, vegetables and fish, and tourism would create other jobs too. But this won't happen if they build all-inclusive resorts like they have elsewhere. They would come in and change the character of the area but provide no income for local people. They could also push up property prices so that ordinary families couldn't afford to live here.

We want to offer tourists a different sort of holiday – not the sort that you get in an all-inclusive resort where they never see the real Jamaica nor meet a real Jamaican. We hope to provide a range of accommodation, from traditional-style cottages to private beach-front villas. This will appeal to a more select group of tourists who are able to pay that bit more for their holiday. We hope they'll bring more income into the area without the large numbers of tourists that you get elsewhere.

13 The view from Jake's

14 Jason Henzell is a white Jamaican. He owns Jake's, a restaurant and holiday cottage complex on Calabash Bay.

We need to attract enough people to provide jobs for local people in tourism, but not so many that we attract hustlers and street-traders from other parts of Jamaica. I think we should limit tourist development to 25 rooms per hectare – not 75 rooms per hectare that you get in the busy resorts. That way, we can preserve the natural environment and character of Treasure Beach.

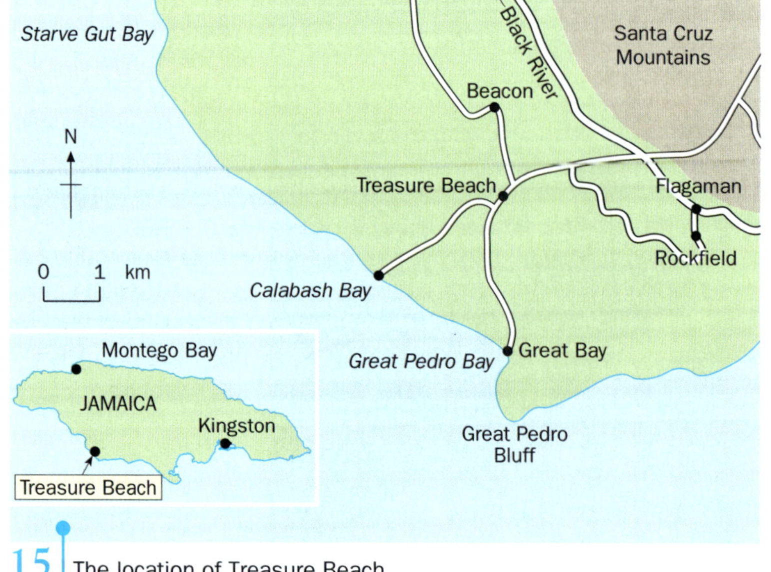

15 The location of Treasure Beach

16 Sign for Treasure Beach

I've been fishing here all my life. I keep my boat on the beach and bring my catch in here. The beach is as important to me as the farmer's land is to him. We don't want hotels and all-inclusives to take over Treasure Beach – that would put an end to fishing. One hotel tried to build a wall on the beach and the fishermen were angry. We wouldn't let them. Now my sons have taken over and will carry on the business. What would we do if we couldn't fish? I suppose we could use our boats to import drugs, but that's illegal!

18 Harold Stephenson, a fisherman from Great Bay

17 The fishing beach at Calabash Bay

This is the sort of place that you normally just dream about – deserted beaches, beautiful scenery, no stress. We found out about it on the internet. There are only a few tourists at the moment and we'd like it to stay that way, but people have begun to hear about it. The pop star Robbie Williams shot a video here, and the boxer Lennox Lewis comes here a lot. The danger is that soon everyone will want to come. We've already stayed in Montego Bay and that's just like America – KFC, McDonald's, that sort of thing. They shouldn't allow that to happen here. If you want fast food, music, nightlife, there are other places in Jamaica where you can go.

19 Jill and Debbie are from the UK, on holiday in Jamaica

Focus tasks

Having studied the ideas in this chapter, you now have to devise a strategy for sustainable development in Treasure Beach. But before you do, carefully read the opinions of the people on these pages. Remember, you have to consider the needs of different groups of people, the natural environment, and Jamaica as a whole.

1 Decide what the main objectives of development should be.

2 Think about how these objectives could be achieved. Here are some of the issues that you will need to consider:

- How many tourists do you want?
- Where will they stay?
- What accommodation needs to be built?
- How will they get there?
- What activities would they do?
- Where will the investment come from?
- Where will the money go to that tourists bring?
- How many jobs will be created?
- Who will do the work?
- What will be the impact on the community?
- What will be the impact on the environment?
- What will be the impact on the Jamaican economy?

ICT

3 Design a brochure to advertise holidays in Treasure Beach. Include any of the photos in this chapter to illustrate your brochure. You could also use a DTP package to make your brochure more effective and scan the photos you choose into the brochure. Alternatively, cut and paste photos from the internet. There are many Jamaican tourism websites, for example:

www.virtualjamaica.com
www.discoverjamaica.com

Your brochure should explain how these holidays will make tourism more sustainable.

Glossary

ABLATION The melting of snow and ice in a **glacier** or ice sheet.

ABRASION The action of river **sediment** rubbing along river beds and banks, causing the surfaces to be worn away.

ABSTRACTION The process of taking resources from the Earth.

ACCUMULATION The addition of snow or ice to a **glacier** or ice sheet.

ACID RAIN Rain polluted with chemicals, mainly sulphur dioxide and nitrogen oxides from coal-fired power stations and car exhausts, which damages buildings and forests.

AFTERSHOCKS Movements that occur after an **earthquake** as the Earth settles back again.

AGEING POPULATION An increase in the proportion of elderly people in a country or area.

AGROFORESTRY The combined use of land for agriculture and forestry.

AID Help in the form of money, food, medical supplies or technical expertise sent by international organisations or **MEDCs** to **LEDCs**. This may be disaster relief in response to an emergency such as a flood or **earthquake**, or may be in the form of a **loan**.

ALL-INCLUSIVE RESORT A holiday complex where guests are provided with hotels, restaurants, leisure activities and entertainment within one self-contained area.

ALLUVIAL SEDIMENT Deposits of **silt** left across a **floodplain** when a river has flooded.

ANDESITIC VOLCANOES **Volcanoes** in areas of andesite rock, which are particularly prone to violent eruptions.

ANTICYCLONE An area of high pressure, formed where cold air is moving from the **poles** towards the warmer tropical regions and sinks.

APARTHEID The policy of racial segregation in South Africa, by which non-whites were forced to live in certain areas and were denied many basic rights.

AQUIFER Underground rocks such as sandstones and limestones that store water.

ARÊTE A sharp mountain ridge formed by the erosion of two adjoining **corries**.

ARID ENVIRONMENT An environment where the rainfall is too low to support the growth of vegetation.

ASH Small particles of rock thrown out in a volcanic eruption.

ASPECT The direction in which a slope or valley side faces.

ATMOSPHERE The layer of air surrounding the Earth.

ATTRITION The wearing down of large rock particles as they collide during transportation in the river.

BACKWASH The return flow of water down a beach after a wave has broken.

BARRIER BUILDINGS Buildings specially constructed to withstand the fires that often follow **earthquakes**.

BARRIO A **shanty town**

BERGSCHRUND A crack in **glacier** ice caused by the separation of the ice inside a **corrie** from the mountain wall.

BIODIVERSITY The variety of plants and animals living in a particular **ecosystem**.

BIOMASS The amount of organic matter (plants and animals) contained within a habitat.

BIOME A natural **ecosystem** that covers a large part of the Earth.

BIRTH RATE The number of people born each year per 1,000 population.

BOULDER CLAY Sediments and rock fragments left on the ground after a **glacier** or ice sheet has retreated.

BOYCOTT To refuse to have relations with a person, group or country.

BROWNFIELD SITE Derelict urban land that is used for building.

CASH CROPS Crops grown to be sold.

CATCHMENT AREA The area drained by a river and its **tributaries**. Also known as a drainage basin.

CENSUS A population survey, carried out every ten years in most countries.

CENTRAL BUSINESS DISTRICT (CBD) The central area of a town or city containing mostly shops and offices.

CHALK DOWNLAND The **ecosystem** of a wide variety of grasses and flowering plants that grow in areas where the underlying rock is chalk. These plants support a large number of animal, bird and insect species.

CLIMATE The average pattern of **weather** over a long period of time.

CLOSED SYSTEM A **system** such as the **hydrological cycle** in which there are no net gains or losses.

COLD FRONT A line where cold air pushes up under warm air causing heavy rain.

COLONY An area of land both occupied and ruled by another country.

COMMON AGRICULTURAL POLICY (CAP) A European Economic Community policy which aimed to make agricultural production more efficient in order to end food rationing.

COMMUNIST GOVERNMENT A government that believes in a classless society in which the methods of production are owned and controlled by all its members and everyone works as much as they can and receives what they need.

CONDENSATION The process by which **water vapour** changes into liquid when it is cooled.

CONFLUENCE The point where a **tributary** meets the main river.

CONIFEROUS FOREST A **biome** found in northern **latitudes** consisting of evergreen trees with needle leaves that can withstand low temperatures.

CONSERVATION The preservation of landscapes, habitats or buildings so as to protect them for the future.

CONSTRUCTIVE WAVES Waves with strong **swash** and weak **backwash** causing material to be deposited on the shore and creating additional beach ridges.

CONTINENTAL CRUST The outer layer of the Earth, which forms major landmasses.

CONTINENTALITY The idea that inland areas of large continents suffer from more extreme temperatures and severe **drought** because of the increased distance from the sea.

CONTOURS Lines drawn on a map to join all places at the same height above sea level.

CONURBATION A large urban area that is formed when several settlements merge.

CONVECTION CURRENT A current of air that rises as it is warmed by contact with the Earth's surface.

COPPICING The process of regularly harvesting trees to provide posts or fuel etc.

CORRIE A rounded hollow eroded into a mountainside by the movement of ice at the source of a **glacier**. Also known as cwm or cirque.

CORROSION The wearing down of rock caused by a chemical reaction.

COUNTERURBANISATION The movement of people and industry away from large towns and cities.

CRAG AND TAIL A rock formation that is steep on one side (the crag) and gently sloping on the other (the tail), caused by the hard rock of the crag protecting soft rock on the other side as a **glacier** moved over it.

CREVASSE A deep crack found in **glaciers**.

CROP ROTATION A method of farming in which different crops are grown in succession so as to maintain the natural balance of **nutrients** without the need for artificial fertilisers.

CRUST The outer surface of the Earth. It is divided into **continental crust** and **oceanic crust**.

CYCLONE A **tropical storm** that occurs in the Indian Ocean.

DEPRESSION A **low pressure** system that develops when warm moist air from the south meets cold air from the north, causing winds to spiral inwards.

DESERT Area where the **climate** is too dry for vegetation to grow.

DESERTIFICATION The loss of vegetation in **arid** and **semi-arid** areas, often caused by human actions such as **overgrazing**, or by climate change.

DESTRUCTIVE WAVES Waves with a weak **swash** and a strong **backwash** that erode the beach and carry away material.

DEVELOPMENT The process of advancing and improving people's quality of life.

DEVELOPMENT INDICATORS A set of measurable characteristics which can be compared from country to country to assess their relative levels of development.

DEW POINT The point at which warm, rising air changes from water vapour to water droplets due to a fall in temperature.

DISCHARGE The volume of water flowing over the width of a river during a set amount of time.

DISPERSED SETTLEMENT Houses scattered over a wide area rather than being grouped together.

DISTRIBUTARY A branch of the main channel of a river that splits into several separate channels in a **delta**.

DIVERSIFICATION The process of increasing variety, e.g. using land for other activities in addition to farming.

DRAINAGE BASIN The area drained by a river and its **tributaries**. Also known as a river's catchment area.

DROUGHT A serious shortage of water.

DRUMLIN A smooth, oval low hill or mound formed by glacial **deposition**.

DEATH RATE The number of people dying each year per 1,000 population.

DECENTRALISED CITY A city where shops and industry have moved out from the **CBD** and spread over a wide area rather than being concentrated around one central point.

DECOMPOSITION The breaking down of plant or animal material so as to release the **nutrients** and return them to the **soil** as part of the **nutrient cycle**.

DEFORESTATION The clearance of forested areas by humans, causing danger to plant and animal species and damaging the natural **ecosystem**.

DELTA A low-lying area of land, formed when a river that is carrying a lot of material flows into a body of standing water such as a lake, sea or ocean. The **sediments** are deposited and gradually build up and extend from the **mouth** of the river.

DEMOCRATIC GOVERNMENT A form of government where political power lies in the hands of the people through elected representatives (politicians).

DEMOGRAPHIC TRANSITION MODEL A model of population growth, usually presented in graph form, to show changes in birth and death rates over a period of time.

DEPENDENCY RATIO A ratio representing how many economically active people support each non-economically active person.

DEPOSITION The process of **sediment** being laid down by water, wind or ice to create new land.

EARTHQUAKE A shaking or trembling of the ground caused by sudden movements within the Earth's **crust**.

ECOLOGICAL FOOTPRINT The area needed by a city to produce its resources and reabsorb its waste.

ECONOMIC GROWTH An increase in the amount of goods produced, leading to an increase in wealth.

ECOSYSTEM A community of plants and animals together with the physical environment in which they live.

ECOTOURISM **Sustainable** tourism which encourages the preservation of areas of natural beauty and local culture.

EMIGRATION Movement of people from their home country to live in another country.

ENVIRONMENTAL HEALTH A measure of the overall wellbeing of the environment.

ENVIRONMENTAL IMPACT ASSESSMENT (EIA) An assessment of the possible impacts on the environment of proposed developments such as **reservoirs** for **hydro-electric power** plants.

ENVIRONMENTAL QUALITY An evaluation of how pleasant the environment is.

ENVIRONMENTALLY SENSITIVE AREA (ESA) An area that is farmed using traditional methods in order to conserve a fragile **ecosystem**.

EPICENTRE The point on the Earth's surface that is directly above the centre (focus) of an earthquake.

EQUATOR An imaginary line which runs east–west around the middle of the Earth.

EROSION The process by which the land surface is worn away.

ERRATIC A large rock or boulder that has been carried by a **glacier** and dropped in a place far from its origin.

ESKER A long and steep-sided ridge that was formed by **sediment** deposited by streams flowing under a **glacier**. It is left exposed after the glacier has retreated.

ESTUARY The widest part of a river **mouth**, which is affected by tides.

EUROPEAN UNION (EU) A group of countries in Europe with a single market for goods.

EUTROPHICATION A process by which excess fertiliser is washed off the land and into streams where it causes plant life to grow rapidly, so that all the oxygen is used up and fish can no longer survive.

EVAPORATION The process by which water is turned into **water vapour** in the air when the temperature rises.

EVAPOTRANSPIRATION The release of water from **soil** and vegetation into the **atmosphere** by **evaporation** and **transpiration**.

EXTENSIVE FARMING Farming over a wide area and using few **inputs**.

EYE OF THE STORM The centre of a **tropical storm** where the air is calm as the winds spiral around it.

FACTORS OF PRODUCTION Four key issues which have to be considered when managing any industry: raw materials, power supply, labour and market.

FAIR TRADE Trade that offers a fair price and good working conditions to the people producing the goods.

FISSURE ERUPTION A volcanic eruption that occurs along a fissure (crack) where two **plates** are being pulled apart, rather than through a single vent.

FJORD A deep coastal inlet created by the sea filling a glaciated valley.

FLASH FLOOD A sudden and violent flood caused by heavy rain falling in a normally dry or arid place.

FLOODPLAIN Flat land on either side of a river where the river floods regularly and deposits fertile **alluvial sediments**.

FOOD CHAIN A simple system in which plants are eaten by herbivores, which are then eaten by carnivores.

FOOD WEB A more complex system of interacting **food chains** that together make up the **biodiversity** of an **ecosystem**.

FORCED MIGRATION An unavoidable move away from home, normally caused by factors beyond the control of the people involved.

FORMAL SECTOR Official jobs with specific hours, wages and regulated working conditions.

FOSSIL FUEL A fuel such as coal or oil that is formed in the ground by buried plants or animals.

FRAGILE ENVIRONMENT An **ecosystem** in which a small change can have a devastating effect on its **biodiversity**.

FREEZE–THAW A **weathering** process where frost causes rock to crack, and thawing water carries away the fragmented pieces.

FRONT The border between two contrasting air masses.

GENETIC ENGINEERING A scientific method of manipulating the genes and DNA that make every living thing different.

GENETIC MODIFICATION The process of changing or combining the DNA of different species to permanently alter individual characteristics.

GENTRIFICATION The action of wealthy people buying properties in deprived areas to renovate them and settle there.

GEOTHERMAL ENERGY Electricity generated using underground hot water and steam in volcanic areas.

GLACIER A slow-moving 'river' of ice.

GLOBAL ECONOMY All goods and services made and sold around the world.

GLOBAL POSITIONING SYSTEM (GPS) A system where a computer can receive satellite signals and pinpoint any position on the Earth.

GLOBAL WARMING The warming of the Earth's atmosphere, due to the increased production of **greenhouse gases**.

GLOBALISATION The process by which companies move around the world to find the most favourable locations.

GREEN BELT An area of land surrounding a city where development is restricted by law.

GREEN REVOLUTION The use of science and technology to increase crop productivity in **LEDCs** by using higher-yielding varieties (HYVs) of crops.

GREENFIELD SITE Rural land that is used for building.

GREENHOUSE EFFECT The natural effect of gases that trap heat in the lower layers of the **atmosphere** by allowing short-wave radiation from the sun to come through and warm the Earth's surface, but prevent long-wave radiation from escaping back out again.

GREENHOUSE GASES Gases such as carbon dioxide, methane, nitrous oxide and chlorofluorocarbons (CFCs) that contribute to the **greenhouse effect**.

GROSS DOMESTIC PRODUCT PER PERSON (GDP/CAPITA) The total value of all of the products and services produced in a country or area divided by the total population.

GROSS NATIONAL PRODUCT PER PERSON (GNP/CAPITA) The total value of all the products and services produced by a country divided by the total population.

GROUNDFLOW Water that collects in rock underground.

GROUNDWATER Water stored underground in porous rocks.

GULLY A narrow channel running into the ground, created by the action of running water.

HADLEY CELL A cycle of heat-driven air flows that starts at the **equator** where warm air is heated and forced to rise up towards the **poles**, where it cools and falls back down to the equator where it is heated and forced to rise again.

HANGING VALLEY A **tributary** to a valley that has been glaciated and eroded downwards, leaving the tributary suspended above the original valley and often marked by a waterfall.

HARD ENGINEERING The use of physical barriers such as concrete banks to provide protection against hazards such as flooding and coastal **erosion**.

HEAVILY INDEBTED POORER COUNTRIES (HIPCs) An **LEDC** where a large proportion of **GNP** is spent repaying debt.

HEAVY INDUSTRY An industry using heavy-weight materials and bulky machinery but producing a final product that is of low value compared to the weight, for example ship building.

HIGH PRESSURE An area of high atmospheric pressure characterised by fine and dry weather.

HUMAN DEVELOPMENT INDEX (HDI) A measure of **development** used by the United Nations to compare countries in terms of income, levels of education and health.

HUMUS Decaying organic matter in the **soil**, formed by the **decomposition** of plants and animals.

HURRICANE A **tropical storm** with wind speeds of 120 km per hour or more and very heavy rainfall.

HYDRAULIC PRESSURE The energy and power created by flowing water.

HYDRO-ELECTRIC POWER (HEP) Electricity generated by harnessing the energy of fast-flowing water, usually stored in a **reservoir** behind a dam.

HYDROGRAPH A graph showing the changes in the rates of **discharge** of a river.

HYDROLOGICAL CYCLE Also known as the water cycle, this is a **closed system** by which water circulates between the oceans, atmosphere and the land.

ICE AGE A period in time during which large parts of the Earth's surface were covered by vast **glaciers**.

IDEOLOGY A set of ideas which governs the social, economic and political organisation of a country.

IMMIGRATION Movement of people into an area or country.

IMPERMEABLE Material that does not allow water to flow through.

INDIGENOUS PEOPLE The original inhabitants of an area.

INDUSTRIAL REVOLUTION The process of industrialisation that occurred in Britain from about 1750 to 1900. New machines and techniques greatly increased the production of manufactured goods.

INFANT MORTALITY RATE The number of children who die in their first year of life for every 1,000 born.

INFILTRATION The movement of water from the surface of the ground into the **soil** or rock below.

INFORMAL SECTOR Unofficial jobs that don't have fixed wages, hours or working conditions.

INFRARED IMAGING A way of obtaining an image from satellites using heat rather than light waves.

INFRASTRUCTURE The facilities that provide the framework of an economy, such as transport, power supplies, and communications.

INNER-CITY AREAS The section of a city surrounding the city centre.

INPUTS Anything that goes into a **system**.

INTENSIVE FARMING The farming of a small area using many **inputs**.

INTERCEPTION The holding of rainfall by vegetation before it reaches the ground.

INTERCROPPING Growing a variety of crops in alternate rows, so as to protect the **soil**.

INTERLOCKING SPURS A series of projections from either side of a valley, that appear to cause the river to wind between them.

INTERNAL MIGRATION People moving from one place to another within the same country.

INTERNATIONAL DEBT Money owed by countries to international banks and to governments of other countries.

INTERNATIONAL MIGRATION People moving from one country to another.

INWARD INVESTMENT Money entering a business from an outside source.

IRRIGATION The supply of water to the land, usually for farming.

ISOBARS Lines on a weather map joining places with equal values of atmospheric pressure.

ISOTHERMS Lines on a map joining places of equal temperature.

KAMES A ridge or mound of gravel or sand deposited by a stream as it ran from beneath a **glacier**.

LABOUR-INTENSIVE INDUSTRY An industry which employs a large number of people.

LAG TIME The delay shown on a **hydrograph** between the time of peak rainfall and peak **discharge** of flow in a river.

LAGOON A shallow lake of salt water, usually in the area between a river and the sea.

LAHAR A mudflow of volcanic **ash** and water.

LATITUDE The position on the globe north or south of the **equator**.

LAVA Molten rock flowing from a **volcano**.

LEACHING The process by which rainwater travels down into the **soil** taking excess materials, e.g. fertiliser, with it.

LEAKAGE When profits are drained from a country, e.g. when the companies making money from tourism are based outside the country.

LESS ECONOMICALLY DEVELOPED COUNTRIES (LEDCs) Poorer countries with low GNP per capita.

LEVÉE A natural or artificial raised bank alongside a river.

LIFE EXPECTANCY The average age that people can expect to live to.

LINEAR SETTLEMENT Buildings in a line, usually along a valley or a road.

LIQUEFACTION The process in an **earthquake** when the movement of the waves through the rocks causes the ground to act like a liquid.

LIVING STANDARD A measure of the quality of life.

LOAN Money given by international governments or banking organisations to **LEDCs**, which has to be re-paid, often with interest.

LOMÉ CONVENTION An agreement made between European countries and their ex-colonies in Africa, the Caribbean and the Pacific to keep importing the latters' products although production costs may be higher and the products more expensive.

LONGSHORE DRIFT The movement of beach **sediments** along the shore in the direction of the **prevailing wind**.

LOW PRESSURE An area of low atmospheric pressure characterised by unsettled and changeable weather.

MAGMA Molten rock beneath the surface of the Earth. It flows from **volcanoes** in the form of **lava**.

MANTLE The layer in the Earth between the core at the centre and the **crust** on the surface.
MARKET TOWN A town that originally grew around a market, where people bought and sold goods.
MASS TOURISM Organised tourism for large numbers of people.
MEANDER A bend in the course of a river.
MELTWATER Water melting from snow and ice at the snout (lower end) of a **glacier**.
MICROCLIMATE The **climate** in a small local area such as around a building or within a wooded area.
MIGRATION The movement of people from one place or country to another.
MODEL A representation created by taking data from several different sources.
MONSOON A seasonal wind in India and South East Asia that brings heavy rain in summer.
MORAINE **Sediments** that have been transported and deposited by a **glacier**. These may be in the form of lateral (side) moraine, medial (middle) moraine, terminal moraine (at the end of a glacier) or recessional moraine (where a glacier's retreat has halted temporarily).
MORE ECONOMICALLY DEVELOPED COUNTRIES (MEDCs) Richer countries with a high GNP per capita.
MOUTH Where a river meets a sea, ocean or lake.
MULCHING Covering the soil between crops with a layer of organic matter in order to retain moisture and improve soil structure.
MULTIPLIER EFFECT The way in which an increase or decrease in an activity has direct consequences on other activities in the same location.

NATIONAL PARK An area of natural beauty and wildlife that is protected from development.
NATURAL POPULATION CHANGE The difference between the birth rate and the death rate.
NATURAL SUCCESSION The process by which agricultural land reverts back to woodland if grazing and other agricultural practices are stopped.
NÉVÉ Snow crystals that are compacted into ice by the weight of fresh snow above. Also called firn ice.
NEWLY INDUSTRIALISING COUNTRY (NIC) A country that has recently developed large-scale, machine-operated production.
NICHE MARKETING Creating and selling a product to a consumer group that is fairly small and specialised.
NORTH ATLANTIC DRIFT A warm air flow travelling across the North Atlantic Ocean and influencing Britain's weather.
NUCLEAR FALLOUT RADIOACTIVE NUCLEAR DEBRIS, LEFT AFTER THE DROPPING OF A NUCLEAR BOMB.
NUCLEAR POWER Energy produced by carrying out a nuclear reaction in a power station.
NUCLEAR SETTLEMENT Buildings grouped together in a cluster.
NUTRIENT CYCLE The movement of **nutrients** through an **ecosystem** by plants taking up the nutrients from the **soil**. When the plants die the nutrients are returned to the soil by **decomposition**.
NUTRIENTS Food required by plants or animals.

OCEANIC CRUST The thinner part of the Earth's **crust** which lies beneath the oceans.
OPEN SYSTEM A **system** such as a **river system** in which there are variations in **inputs** and **outputs** at different times.
ORGANIC FARMING Farming without using chemicals.
OUTPUTS Any products of a **system**.
OUTWASH PLAIN An area covered by material such as sand, gravel and clay washed out by the meltwater from glaciers.
OVERCULTIVATION Using the land too intensively for growing crops, so that the **soil's** natural fertility is reduced.
OVERGRAZING Grazing too many animals on the land and not allowing time for the vegetation to recover.
OXBOW LAKE A lake formed after a **meandering** river has straightened its course by cutting through the neck of the meander.
OZONE A gas that occurs in the **atmosphere**, produced from the breakdown of oxygen. It helps to protect the Earth's surface from harmful ultraviolet radiation.

PACKAGE HOLIDAY A holiday where travel and accommodation are organised by a holiday company.
PEAT A layer of black/brown part-decomposed and water-logged vegetable matter.
PERMANENT MIGRATION A movement of people who intend to stay in their new location for ever.
PERMEABLE ROCKS Rocks such as sandstones and chalk which allow water to pass through them easily.
PLANTATIONS Large farms planted with a single **cash crop** such as rubber, oil palm, sugar-cane or coffee.
PLATES The segments of the Earth's crust that are slowly moving due to **convection currents** in the **mantle**. At the margins, where the plates meet, **earthquakes** and **volcanoes** are likely to occur.
PLUCKING A form of **erosion** where the bottom of a **glacier** freezes onto rocks and pulls them away as it moves down the valley.
PLUNGE POOL A deep pool at the bottom of a waterfall cut out by rocks whirling in the water.
POLAR CLIMATE Very cold climate, such as that found at the **poles**.
POLES Imaginary points indicating the most northern and southern part of the Earth's axis.
POPULATION CHANGE The increase and decrease in the number of people in an area or country.
POPULATION DENSITY The number of people in a given area.
POPULATION DISTRIBUTION The way that people are spread within an area.
POPULATION PYRAMID A divided bar graph that shows the age and sex structure of a population.
POPULATION STRUCTURE What the population is made up of in terms of the age and sex of people.
POTHOLE A hole caused by erosion of bedrock by rocks swirling in water.
POVERTY LINE A certain level of income, below which people are considered to be poor.
PRECIPITATION Water that falls on the Earth's surface in the form of rain, snow, hail, sleet or dew.
PRECISION FARMING Farming using computers to work with the environment to gain maximum yields with minimum **inputs**.

PRESSURE AREAS Areas within a **national park** which are already under threat from large numbers of visitors.

PREVAILING WIND The wind which most frequently blows into an area, from one particular direction.

PRIMARY DATA A first-hand source of information.

PRIMARY EFFECT A direct consequence.

PRIMARY GOODS Natural resources produced by the land or the sea.

PRIMARY RAINFOREST Rainforest that has remained undisturbed for hundreds of years.

PRIMARY SECTOR Industry involving the collection or distribution of **primary goods**.

PROCESSES Continuous actions or activities within a **system** that turn **inputs** into **outputs**.

PUSH–PULL MODEL A representation of the issues that attract and repel people.

PYRAMIDAL PEAK A sharp mountain point caused by two **corries** forming back to back.

PYROCLASTICS Violent volcanic eruptions of **ash** and rock fragments.

QUATERNARY SECTOR Industries or jobs involving research and the supply of information to other sectors, e.g. the ICT industry.

RAINSHADOW An area of low rainfall on the leeward (sheltered) side of high land.

RAPIDS Part of a river where the water is very shallow and the bed is very steep, making the flow fast and turbulent.

RECREATION Leisure activity.

REFUGEES People displaced from their homes by natural disasters such as **earthquakes** or floods, or as a result of war or persecution.

REGELATION Re-freezing of ice that was forced to melt under pressure.

REMITTANCE A sum of money sent to someone.

REMOTE SENSING A way of getting information about something without being close to it.

RENEWABLE ENERGY An energy source that can be used over and over again without it eventually running out.

RESERVOIR Artificial lake sometimes created by building a dam across a river valley. It is used to store water.

RICHTER SCALE A scale used to measure the strength of an **earthquake**. Each point on the scale is ten times greater than the point below.

RIVER SYSTEM An **open system** including a river and its **tributaries** with **inputs** in the form of **precipitation** and outputs in the form of **evaporation** and **discharge** into the sea.

ROCHE MOUTONÉE A rock mass forming a small hill.

RUN-OFF Water that flows straight off the surface of the land rather than soaking in.

SALINISATION **Deposition** of salts into the **soil**.

SALTMARSH A coastal wetland which has been colonised by salt-tolerant plants and is flooded daily by the tides.

SAVANNA Areas of tropical grassland found in the area between the **tropical rainforests** and **deserts**.

SEASONAL MIGRATION Movement of people which is directly linked to the time of year.

SECONDARY DATA Information that other organisations or people have collected.

SECONDARY EFFECT An indirect consequence.

SECONDARY SECTOR Manufacturing employment that turns raw materials into finished products.

SEDIMENT Small grains and particles of rock.

SEMI-ARID ENVIRONMENT An area characterised by very low rainfall and little vegetation.

SERVICE SECTOR Employment that assists the needs of the primary and secondary industries but doesn't provide a finished product.

SET-ASIDE To take out of production.

SETTLEMENT A place where people live.

SETTLEMENT HIERARCHY Classification of places where people live in order of size.

SHANTY TOWN An area of a city where people, usually poor migrants from the countryside, settle and build their own homes.

SHIFTING CULTIVATION A form of agriculture practised in areas of **tropical rainforest**, in which a patch of forest is cleared by burning or chopping down the trees, crops are grown for a few years until the natural soil **nutrients** are depleted and then the cultivators move to another part of the forest, leaving the original area to regenerate.

SILT Fine-grained **soil** and sand often deposited by a river in flood.

SMOG A combination of smoke and fog.

SOFT ENGINEERING Management of natural features to reduce the danger of, e.g., flooding or coastal erosion.

SOIL The upper layer of the ground, which consists of **humus** and the **weathered** remains of the underlying rocks.

SOIL CONSERVATION Measures such as **terracing** and **mulching** to prevent **soil erosion** and **desertification**.

SOIL CREEP The slow movement of **soil** down a slope.

SOIL EROSION The removal of **soil** by the action of water and wind and the movement of soil down a slope.

SOIL PROFILE A vertical section through the **soil**, showing the different layers.

SOLAR RADIATION Heat from the sun.

SOLUTION When something is dissolved and carried in water.

SOURCE The origin of a stream or river.

SQUATTER SETTLEMENT Shanty town where people live illegally.

STAPLE The main component of a diet.

STORES Where something is kept and from where it is collected.

STORM SURGE A rise in sea level caused by the intense **low pressure** systems of **tropical storms**. During high tides or in low-lying areas the effects can be devastating.

STRIATIONS Scratches on the surfaces of ice-worn rocks.

SUBDUCTION ZONE The area where one **plate** moves underneath another and is destroyed as it is carried back into the **mantle**.

SUBSIDY Money given by the state to make an economic activity possible or to keep down the price of goods.

SUBSISTENCE FARMER A farmer who grows only enough food to feed his or her family.

SUBURBANISATION The growth of suburbs at the city edge, spreading into the surrounding rural area.

SURFACE RUN-OFF Water running over the top of the ground.
SUSTAINABLE Something that can continue because it is not wasteful.
SUSTAINABLE DEVELOPMENT Making improvements and advancements that meet present and future needs.
SUSTAINABLE FOREST MANAGEMENT The management of forest areas so that the resources are used in a controlled way and the trees are allowed to regenerate.
SWASH The movement of water up a beach after a wave has broken.
SWELL WAVES Rounded, gentle waves, formed far out at sea, which do not break.
SYNOPTIC CHART A map which shows conditions in the **atmosphere**, such as **precipitation**, temperature and air pressure.
SYSTEM A set of components that, together with **inputs, processes** and **outputs**, make a unified whole. Changes in one part of the system have effects on other parts.

TARN Small mountain lake filled by rainwater.
TEMPERATE CLIMATE A mild climate without great extremes of hot or cold.
TEMPERATE DECIDUOUS FOREST The natural **ecosystem** in Britain and north-west Europe. Deciduous trees grow in the summer and lose their leaves in the winter.
TEMPERATE INVERSION When warm air is compressed and descends, trapping the air below, e.g. so that pollution cannot escape.
TEMPORARY MIGRATION People moving on a short-term basis.
TERMINAL MORAINE A half-moon shaped deposit of loose rocks and boulders.
TERRACES A series of artificial steps on hillsides that allow land to be farmed and help to reduce **soil erosion**.
TERTIARY SECTOR Service industries.
THROUGHFLOW The movement of water through the soil.
TOWNSHIPS Part of a city set aside for non-white groups of people in South Africa during the **apartheid** years.
TRANSFERS Movement from one place to another.
TRANSHUMANCE The system in which farmers migrate with the seasons so that they can farm all year round.
TRANSNATIONAL CORPORATION (TNC) A large company with operations in more than one country.
TRANSPIRATION The process by which plants return moisture to the **atmosphere** through their leaves.
TRIBUTARY A smaller stream or river which joins a larger river.
TROPICAL CLIMATE High average temperatures with a low seasonal range. High rainfall throughout the year.
TROPICAL RAINFOREST The natural **ecosystem** in areas with a tropical climate. They are areas of great **biodiversity** with many species of plants and animals.
TROPICAL STORM An intense **low pressure** system over tropical oceans, associated with strong winds and high rainfall.
TRUNCATED SPUR A **spur** that has been eroded by a valley **glacier**.

U-SHAPED VALLEY A valley with a flat floor and steep sides.
UNDERDEVELOPMENT Where a country is not making the best use of its natural and economic resources
UNSUSTAINABLE An activity that cannot continue because it is wasteful.
URBAN PLANNING Designing of a built-up area.
URBAN REDEVELOPMENT A plan to renew and revive built-up areas.
URBAN REGENERATION A plan to replace or restore poor-quality buildings in a city.
URBAN ZONES Built-up areas in a town or city.
URBANISATION An increase in the proportion of people living in towns and cities.

V-SHAPED VALLEY A valley eroded by a river.
VOLCANO An opening in the Earth's **crust** through which molten rock, **lava** and **ash** erupt. It mainly occurs at **plate** margins.
VOLUNTARY MIGRATION A movement of people that is based on the attraction of another place rather than a problem with the current area.
VULNERABLE AREAS Areas where natural wildlife habitats are at risk from increased visitor numbers.

WARM FRONT A line where warm air rises over cold air, giving prolonged rain.
WATER TABLE The level below which **permeable rock** is saturated.
WATER VAPOUR Water in the form of a gas, produced by **evaporation**.
WATERSHED The imaginary line drawn around a **drainage basin** that separates it from neighbouring **river systems**.
WAVE-CUT NOTCH Wave erosion at the bottom of a cliff.
WAVE-CUT PLATFORM Wave erosion of the rock at the base of a cliff cutting out a flat platform.
WEATHER Short-term changes in atmospheric conditions.
WEATHERING The breaking-down of rocks by chemical, biological or physical processes so as to form smaller particles.
WIND FARM A group of wind turbines used to generate electricity from wind.
WORLD HERITAGE SITE A natural or cultural site that is recognised internationally by the United Nations Educational, Scientific and Cultural Organization (UNESCO) as being of sufficient importance to be protected for the future of the global community.

YIELD MAPPING A way of measuring the amount of crops produced from a particular area of a field.

Index